BULBOUS PLANTS
Biotechnology

BULBOUS PLANTS
Biotechnology

Editors

K.G. Ramawat

Formerly, Professor and Head, Botany Department
M.L. Sukhadia University, Udaipur, India

J.M. Mérillon

Faculty of Pharmacy, University of Bordeaux 2
Institute of Wine and Vine
Villenave d'Ornon, France

CRC Press
Taylor & Francis Group
Boca Raton London New York

CRC Press is an imprint of the
Taylor & Francis Group, an **informa** business
A SCIENCE PUBLISHERS BOOK

CRC Press
Taylor & Francis Group
6000 Broken Sound Parkway NW, Suite 300
Boca Raton, FL 33487-2742

Printed in the United States of America on acid-free paper

International Standard Book Number: 978-1-4665-8967-4 (Hardback)

Library of Congress Cataloging-in-Publication Data

Bulbous plants : biotechnology / editors: K.G. Ramawat, J.M. Mérillon.
p. cm.
Includes bibliographical references and index.
ISBN 978-1-4665-8967-4 (hardcover : alk. paper) 1. Bulbs (Plants)--Biotechnology. I. Ramawat, K. G. II. Mérillon, J. M.

SB425.B853 2014
584'.146--dc23

2013035177

Visit the Taylor & Francis Web site at
http://www.taylorandfrancis.com

CRC Press Web site at
http://www.crcpress.com

Science Publishers Web site at
http://www.scipub.net

Preface

We are pleased to present this unique book on plants with special storage organs such as tubers, corms and bulbs. Tubers, corms and bulbs are various types of modified plant structures that are enlarged to store nutrients. They are used by plants to survive the winter or dry months and provide energy and nutrients for regrowth during the next growing season. Multiple storage organs are produced by such plants, therefore they represent a method of asexual reproduction. Both *stem* and *root* modifications can produce such organs. These are mostly underground. Root crops, tuber crops and corm crops are special terms used to refer to specific types of farm crops collectively called root and tuber crops. A bulbous plant is an ornamental or edible plant, herbaceous or perennial species which produce fleshly storage organs including true bulbs as well as corms, tubers, rhizomes, and tuberous roots. Plants from diverse taxa produce these structures which may provide food, medicine or flowers and saplings for the next generation. Therefore, this generalized term is used for the book.

Starchy root and tuber crops, also referred to as *roots* and *tubers* and *tuberous* crops, are plants that are grown for their modified, thickened root or stem which generally develop underground. These organs are rich in carbohydrate and are commonly used as staple, livestock feed, or as raw materials for the production of industrial products such as starch and alcohol, or processed into various food products. With the ever increasing world population, increasing the potential of these plants and agrotechnology for higher yield is focus of current research.

Root crops and tuber crops have very high yield potential although their protein, mineral and vitamin content are generally low compared to cereals. However, taro and yams contain up to 6% protein of good quality and the potato tuber provides some minerals and vitamin C. The orange-fleshed sweet potato is also well known as a rich source of beta-carotene, a precursor of vitamin A. These crops are known for their high dry matter production, with a potential of 2.1 tons per hectare from underground storage organs. In contrast, cereals have achieved about 1.5 tons.

In any review of tuberous plants, flower and medicinal plants having storage organs needs to be included. Several tuber/bulb bearing plants are excellent ornamentals. A section in the book is devoted to such plants by bringing together in one place the information growers need to propagate and cultivate their eye-catching plants. Dahlia tubers are a tuberous perennial plant much appreciated for the colorful Dahlia flowers. Dahlia flowers make great cut flowers.

A section is devoted to medicinal plants having such storage organs. Medicinal plants are sources of important therapeutic aid for healing human diseases. The depletion of the wild resources has prompted conservation, propagation, and enhancement of resources for medicinal plants. Micropropagation offers an alternate method to propagate and improve medicinal plants through selection of high-yield lines and their efficient cloning.

Biotechnological approaches such as cloning, haploid production, process of tuber formation, vegetative seeds production, use of molecular markers in genetics and breeding, genetic transformation for improvement have been employed for development and production of desired characters. This requires a deep understanding of structure and functioning of tuberous plants then only conservation, characterization and improvement technologies can be developed.

This book aimed at providing comprehensive information about biology, propagation strategies, genetics and genetic resources, plant growth and development, tuber quality, crop management, biological understanding and improvement through various biotechnological and molecular tools to all those who are interested in propagation, cultivation, improvement and protection of such crops. Important new tools have been developed in the area of molecular genetics, and our understanding of tuber plants physiology has been revolutionized due to amenability of the plants to genetic transformation. The chapters are written by recognized international specialists in their respective fields of research.

The book is divided in following sections to encompass entire spectrum of developments in this field-General biological aspects, edible plants, medicinal plants, and flowering plants. This book will be useful to botanists, biologists, foresters, industrialists, environmentalists, policy makers, conservationists, NGOs working in rural areas for village people, and biotechnologist involved in micropropagation.

We would like to acknowledge cooperation, patience and support of our contributors who have put their serious efforts to ensure the high scientific quality of this book with up to date information. We wish to thank our publishers for their support and timely publication.

K.G. Ramawat
J.M. Mérillon

Contents

Section B: Flowering Plants

Section C: Medicinal Plants

Section A

Edible Crops

1

Hormonal Regulation of Tuber Formation in Potato

Nina P. Aksenova,[1] *Lidiya I. Sergeeva,*[2] *Oksana O. Kolachevskaya*[1] and *Georgy A. Romanov*[3,*]

ABSTRACT

Tubers of potato (*Solanum tuberosum* L.) plants serve as organs for vegetative propagation and stock of matter and energy for the next generation. Tuber formation is a complex process involving several successive stages: stolon formation and growth, induction of tuberization, tuber initiation and further growth. Tuber formation is affected by several external and internal factors, first of all photoperiod, temperature, levels of carbohydrates and nitrogen. Adaptive response of tuberization to these cues is a part of the seasonal strategy of potato reproductive development.

Plant hormones control all stages of potato tuber formation. The hormonal regulation of tuberization in this important crop was studied

[1] Timiryazev Institute of Plant Physiology, Russian Academy of Sciences, Botanicheskaya 35, Moscow 127276, Russia.
Emails: ifr@ippras.ru; vatrushbox@mail.ru

[2] Laboratory of Plant Physiology, Wageningen University and Research Centre, PO Box 386, 6700 AJ Wageningen, The Netherlands; Timiryazev Institute of Plant Physiology, Russian Academy of Sciences, Botanicheskaya 35, Moscow 127276, Russia.
Email: lsergeeva@mail.ru

[3] Timiryazev Institute of Plant Physiology, Russian Academy of Sciences, Botanicheskaya 35, Moscow 127276; Belozersky Institute of Physico-Chemical Biology, Moscow State University, Moscow, Russia.
Email: gar@ippras.ru; gromanov@yahoo.com

* Corresponding author

intensively for a long time. Recent findings obtained with the use of large arsenal of modern molecular and genetic methods greatly extended our knowledge on the mechanisms of phytohormone action in the tuberization control.

This review considers successive tuberization stages and their hormonal regulation. Special attention is paid to the effects of gibberellins, cytokinins, auxins, jasmonic and abscisic acids on tuber formation. Prominent progress in identification of a long-distance proteinaceous tuber-inducing stimulus (tuberigen) is highlighted. Physiological, molecular and genetic aspects of hormonal action during photoperiodic induction, tuber initiation and growth are discussed. The analysis of current data led to the conclusion that the hormonal control of tuberization occurs as a coordinated in time and space crosstalk between different phytohormone signals.

Introduction

Potato tubers serve as vegetative propagation organs, as well as energy and essential substance stocks for the progeny. Tuberization has evolved in *Solanum tuberosum* species as a result of its adaptation to severe habitat conditions in its ancestral homeland. At the same time, tuberization is of prime importance not only for the survival of plant organisms themselves, but also for human life sustenance. Potato (like wheat, rice and maize) is now one of the most important phytoproducers of food and industrial resources.

The transition to tuber formation in potato depends on certain environmental and internal cues. External factors, primarily photoperiod, temperature and nutrition supply strongly affect potato tuber formation (Jackson 1999).

Like in most tuberiferous species, potato tuberization requires short days (SD);[1] however, various potato subspecies and cultivars differ in the extent of tuberization dependence on the day length. This difference is due to their origin and breeding strategies. Potato cultivars from the subspecies *tuberosum* manifest weak SD response of the quantitative type and can produce tubers at various photoperiods though with different efficiency. Wild potato forms and plants belonging to the subspecies *andigena* manifest obligate photoperiodic response to tuberization: they transit to tuber formation only upon induction by SD (Chailakhyan 1984, Ewing 1995, Prat 2004).

Another environmental factor favoring production and growth of potato tubers is a moderate decrease of temperature to 14–20°C, especially at night time. An increase in the night temperature during tuberization to

[1] At present long night is considered as an effective period for tuber induction.

25–27°C reduces sharply the number and weight of the developing tubers (Struik 2007a). Therefore conditions most favorable for tuberization are a long dark period of short-day cycle with lower night temperatures in comparison with day temperatures.

Carbohydrates, first of all sucrose, are necessary for tuber growth and storage starch biosynthesis. Sucrose is also an efficient inducer of tuberization. This has been shown in early studies on *in vitro* tuberization (Gregory 1956, Palmer and Smith 1970) and confirmed in ensuing works (Chincinska et al. 2008, Fisher et al. 2008) performed on transgenic potato plants (subspecies *andigena* and *tuberosum*) having intensified sucrose biosynthesis in leaves and its enhanced transport to stolons. Monosaccharides (glucose, fructose) can promote tuberization *in vitro* too, but their effect is less pronounced than that of sucrose (Ewing 1995). It has recently emerged that non-reducing disaccharide trehalose and its phosphates play an essential role as signal molecules in coordinating diverse metabolic and developmental processes (Ponnu et al. 2011, Debast et al. 2011). It is possible that trehalose signaling also play an important role in regulation of potato tuberization.

Plant nitrogen nutrition markedly affects tuberization as well (Prat 2004). Under field conditions excessive nitrogen fertilizers stimulate growth of up-ground shoots (haulm) and retard tuber formation. In hydroponic culture (Krauss 1985), the removal of nitrogen compounds from the nutrient solution resulted in tuber formation while the subsequent addition of nitrogen fertilizers induced tuber sprouting and active growth of secondary stolons.

Tuberization is an integrated response to various simultaneous factors: day length, temperature and levels of carbohydrate and nitrogen nutrition, and has an adaptive character constituting the seasonal strategy for potato development. Tubers are formed by the end of summer when an extended dark period and decrease in night temperatures serve as signals of season change. A dependence of tuberization on carbohydrate and nitrogen nutrition allows plants to "choose" internal conditions most favorable for tuber initiation and growth.

Tuber formation is a complex process consisting of several successive stages. The boundaries between these stages are rather conventional, and researchers define them sometimes differently. Most common subdivisions are as follows (Fig. 1): stolon formation and growth, induction of tuberization, tuber initiation and tuber growth (Ewing and Struik 1992). Thereafter, tubers enter maturation and dormancy, whereupon buds release dormancy and new plants develop.

Numerous studies show that phytohormones affect markedly all stages of tuberization (Ewing 1995, Rodríguez-Falcón et al. 2006). The successive stages of tuberization and their hormonal regulation are given below.

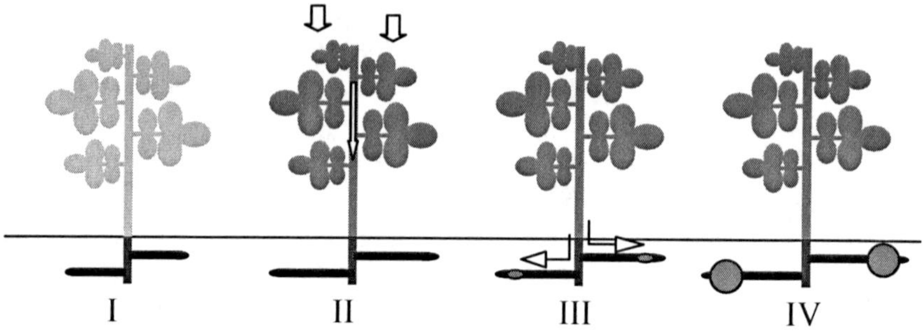

Fig. 1. Stages of tuberization in potato plants. Stage I, stolon growth. Stage II, induction of tuberization: favorable cue(s) affect leaves which produce stimulus for tuberization. Stage III, tuber initiation: stimulus of tuberization arrives to stolon tips and activates the subapical zone. Stage IV, tuber growth: intense cell division and enlargement in the subapical zone of stolons.

Stolon Formation and Growth

Tuber formation in potato is usually preceded by stolon formation. Stolons are underground lateral shoots arising from the basal lateral buds and growing diageotropically. Under common field conditions tubers are formed due to swelling of the subapical part of the stolon. However, not all stolons produce tubers; in some cases (for example, in the *in vitro* culture or in darkness) tubers may arise also from stem buds or on previously developed tubers. Therefore, some researchers do not consider stolon development as an obligatory stage of tuberization (Martínez-García et al. 2002a). During stolon growth, cell divisions are confined mainly to the apical bud and subapical zone; they occur primarily perpendicularly to the axis of stolon growth. Stolon elongation ceases after tuber initiation on it.

Gibberellins (GA) generally enhance stem organ elongation, stimulate potato stolon initiation, growth and branching as well (Vreugdenhil and Struik 1989). Conditions favorable for stolon growth are unfavorable for tuber initiation since the latter requires the ceasing of stolon growth and cell division in its apex. Cessation of stolon elongation is accompanied by a decrease in the intrinsic GA content (Xu et al. 1998a). As a rule, all conditions promoting tuber initiation, e.g., SD, lowering temperature and limited nitrogen supply, reduce sharply the GA content in stolons (Ewing 1995).

Abscisic acid (ABA) is also a factor controlling stolon growth. The cessation of stolon apical growth in potato is accompanied by an increase of the ABA/GA ratio (Mokronosov 1990). The endogenous content of ABA and especially ABA/GA ratio increased in *andigena* potato under SD conditions; hereupon stolon growth was retarded (Macháčková et al. 1998).

Ethylene may be also involved in the regulation of stolon growth. In experiments on stem explants from potato cv. Bintje, the inhibitory action of ethylene and its precursor ethephon on stolon elongation together with swelling effect were observed (Vreugdenhil and van Dijk 1989).

The role of cytokinin (Ck) in stolon formation and growth is poorly studied. Experiments by Kumar and Wareing (1972) showed that treatment of diageotropically growing stolons with Ck resulted in the formation from their tips of leafy shoots with negative geotropism. On the other hand, stolon growth cessation at the onset of tuber formation was accompanied by the accumulation of Ck in the stolon (Hannapel 2007).

Induction of Tuberization

The transit to tuberization in potato is based on the formation of a certain signal which reaches the stolon tip and causes tuber initiation there. Tuber induction stage proceeds beyond stolon tips and occurs, according to known examples, in the leaves. During tuber induction, the plant translates external and internal cues into a readiness to form tubers, yet without any visible signs of tuberization (Ewing and Struik 1992).

The most obvious example of tuber induction is photoperiodic SD induction which leads to formation in potato leaves a movable stimulus, providing the possibility to initiate tubers. However, not only the photoperiod, but other factors can lead to the induction of tuber formation as well. For example, in a recent work (Nookaraiu et al. 2012), a number of external conditions (short day, low temperature, light intensity) and internal conditions (Ca^{2+}-, nitrogen-, hormones-, and sucrose content) are considered as putative inducing factors. However, potato lines with obligate SD-dependent tuberization remain the preferred model for investigation of the process of tuber induction.

Photoperiodic response of tuberization has much in common with the flowering response to the day length (Jackson 1999). Leaves of tuberiferous plants are organs which recognize the day length (Razumov 1931). For photoperiodic induction of tuberization, continuous darkness during SD cycle is required; even short-term light interruption of the long night cancels the favorable effect of SD (Mokronosov and Lundina 1959). Phytochrome is involved in this response because darkness interruption with red light suppressed tuber formation whereas subsequent illumination with far red light completely restored the process of tuber formation (Batutis and Ewing 1982). Involvement of phytochrome in the photoperiodic regulation of tuberization was confirmed in experiments performed on transgenic potato plants of the *andigena* subspecies. These plants were transformed with the Arabidopsis *PHYB* gene, encoding phytochrome B apoprotein in antisense orientation (Jackson et al. 1996). This resulted in reduction in plants of phytochrome B content and acquirement of a capability of tuber formation under both SD and long days (LD) conditions. In other works (Thiele et al.

1999, Aksenova et al. 2002) potato cv. Désirée plants were transformed with normally oriented *PHYB* gene from *Arabidopsis*. This increased the content of *PHYB* in transgenic plants and enhanced inhibition of tuberization under LD conditions. These results showed that phytochrome B is involved in the photoperiodic control of tuberization and mediates the inhibitory effect of unfavorable LD on tuber formation.

The stimulus appearing at photoperiodic induction of tuberization is stored in the leaves and can be transmitted via grafts to non-induced plants. Thus, when leafy cuttings of *andigena* plants preliminarily kept under SD conditions were grafted onto plants preliminary kept under LD conditions, the latter acquired a capacity to form tubers (Gregory 1956). These experiments showed that an unknown stimulus was produced in the leaves of plants on SD. This substance, a putative hormone, was called *tuberigen*. It can be transported across grafts to the stem buds and induce tuber formation, i.e., it can play a role of long-distance stimulus of tuberization.

It was shown in further experiments that tubers could be produced on non-induced potato plants upon their grafting with non-tuberiferous plants such as tomato, sunflower and others. In these cases, the photoperiodic response of the scions was of major importance. The detailed experiments demonstrating the role of phototoperiodic response of non-tuberiferous scion in tuberization of potato stock were carried out by Chailakhyan and co-workers (Chailakhyan et al. 1981). Short-day, long-day and day-neutral tobacco plants served scions while potato species with obligate SD response of tuberization but kept under LD served rootstocks. Tuber formation by the rootstock was shown to occur only when the photoperiodic regime was favorable for scion flowering. Thus, photoperiodic conditions favorable for tobacco cultivar flowering were also suited for the production in their leaves of a tuberization stimulus, tuberigen. These experiments showed a clear relationship between the formation of two hormones—florigen and tuberigen, both hypothetical at that time.

The results of these experiments and the discovery of molecular mechanisms for transition to flowering in *Arabidopsis* allowed a suggestion on the occurrence of common elements in the genetic control of photoperiodic induction of flowering and tuberization (Rodríguez-Falcón et al. 2006). This suggestion relied on the identification in *andigena* plants of the homologs of the main genes involved in the photoperiodic induction of *Arabidopsis* flowering: *CO* (*CONSTANS*), and *FT* (*FLOWERING LOCUS T*).

The photoperiodic regulation of flowering by CO relies on the daily oscillation pattern of its mRNA and the control of CO protein levels by light. When the CO protein coincides with the light of the LD, it is stabilized and it activates transcription of FT in the LD-plant *Arabidopsis*. But there are differences in the mechanisms by which *CO* orthologs operate in LD- and

SD-plants. *CO* ortholog *Hd1* represses flowering in SD-plant rice under LD conditions. *FT* and its orthologs play a key role in flowering time control both in LD- and SD-plants. Once *FT* is transcribed in the leaves, the *FT* protein is transported through the phloem to the shoot apical meristem, where it interacts with other proteins to induce the transition to flowering (Zeevaart 2008).

A *CO*-like gene *StCO* was identified in potato ssp. *andigena* (González-Schain et al. 2012) and found to be highly similar to a potato ssp. *tuberosum* gene *StICOL1* (Drobyazina and Klavkin 2006, 2011). Sequencing of two diploid potato genomes led to identification of the *FT*-like paralogs, *StSP6A*, *StSP5G* and *StSP3D* (Navarro et al. 2011).

It was shown that constitutive overexpression in *andigena* potato plants of the *Arabidopsis AtCO* impairs tuberization under SD inducive conditions. Grafting experiment using these *AtCO* overexpressing lines indicated that *AtCO* exerts its inhibitory effect on tuber formation by acting in leaves (Martínez-García et al. 2002b). *Andigena* plants overexpressing *StCO* tuberized later than wild-type plants under a weakly-inductive photoperiod. *StCO* silencing promoted tuberization under both repressive LD and weakly-inductive photoperiods, but had no effect under strongly inductive SD, demonstrating that *StCO* represses tuberization in a photoperiod-dependent manner. In addition, *StCO* transcript levels followed a daily oscillation, indicating that day length affects pattern of *StCO* expression (González-Schain et al. 2012).

Strict SD *andigena* plants were transformed with *Hd3a* gene, the *FT* ortholog in rice. The expression of *Hd3a* allowed *andigena* potato to tuberize in LD (Navarro et al. 2011). Evidence is provided showing that the potato floral and tuberization transitions are controlled by two different FT-like paralogs (StSP3D and StSP6A) that respond to independent environmental cues. RT-PCR revealed that *StSP6A* gene expression occurred in leaves and stolons and strongly correlated with tuberization: high level expression was observed in SD-induced plants. *StSP6A* overexpression rendered plants able to tuberize under non-inductive LD, while *StSP6A* silencing, in turn, strongly delayed tuber formation in SD, pointing to an essential role for *FT*-like protein in tuberization promotion. *StSP6A* expression analysis in commercial cultivars with early (Jaerla), late (Baraka) and intermediate (Kennebec) tuberization periods show that the accumulation of these transcripts in leaves correlated with the tuberization time of these cultivars, indicating that this *FT*-paralog is involved in tuberization control even in non-photoperiodic cultivars (Navarro et al. 2011).

Thus, recent experimental results indicate that *StCO* and *StSP6A* play an important role in potato plants controlling photoperiodic response which involves long-distance regulatory signal (tuberigen), a putative trigger of the tuber initiation.

Data of recent years point to a possible involvement of homeotic genes from *BEL* and *KNOX* families in both photoperiodic induction of tuber formation and transport of the inducing signal (Banerjee et al. 2006, Jackson 2009). These genes encode transcription factors controlling many morphogenetic processes in plants; proteins BEL and KNOX interact and produce heterodimers involved in selective regulation of other gene expressions. The *KNOX* gene isolated from potato was called *POTH1*, whereas the potato *BEL* gene was called *StBEL5* (Rosin et al. 2003, Chen et al. 2003). It might be that *StBEL5* transcripts participate in transport of the tuberization stimulus from leaves to stolons, as full-length mRNAs (comprising also untranslated regions) play an active role in this process (Hannapel 2010). The *in situ* RNA analysis shows that *StBEL5* transcripts are localized in the leaf phloem and can move along the phloem over long distances. Their transport in the leaf veins, petiols and stems to the stolon tips was markedly enhanced under SD. The enhanced transport of *StBEL5* RNA was accompanied by substantial activation of tuber formation. Study of transgenic plants overexpressing *StBEL5* and grafted to wild-type rootstocks evidenced for a transmission of these transcripts across the graft. Transgenic *andigena* potato lines overexpressing *StBEL5* could overcome the inhibitory effect of LD photoperiod and produced tubers under both SD and LD conditions. Levels of *StBEL5* mRNA were affected by *StCO* expression, as leaves of *StCO*-silencing lines of *andigena* plants displayed higher levels of *StBEL5* transcript abundance (Gonzalez-Schain et al. 2012). These results point to a close crosstalk between *StCO, StSP6A* and *StBEL* genes in the control of tuber induction in potato.

Studies by Martin et al. (2009) indicated an involvement of the micro-RNA172 (miR172) in the photoperiodic induction of tuber formation in *andigena* potato. This miR172 is known to participate in the regulation of *Arabidopsis* development and suppresses the *APETALA2* gene which is involved in the flowering control (Aukerman and Sakai 2003). miR172 was found in all potato organs: leaves, stems and stolons; under inductive SD conditions its content was much higher than under LD. miR172 overexpression in transgenic *35S::MIR172* plants stimulated not only tuber formation under LD but also flowering. This demonstrates ones again the interrelation between molecular mechanisms for photoperiodic regulation of potato tuberization and flowering. An ability for tuber formation under LD in *35S::MIR172* plants could be transmitted through the grafts from transformed onto untransformed plants.

Not only stimuli of tuberization under SD but also inhibitors of tuber initiation under LD take part in the photoperiodic regulation of tuber formation (Sarkar 2010). One well-known inhibitor is GA. Yet in early studies the inhibitory effect of GA on induction of tuberization has been demonstrated. Indeed, under LD GA activity and content in potato leaves was found to be several times higher than under favorable for induction SD (Vreugdenhil and Sergeeva 1999). Under SD tuberization can be retarded by plant treatment with GA (Kumar and Wareing 1974); under LD tuber formation can be stimulated by blocking GA synthesis with corresponding inhibitors (Simko 1994, Jackson and Prat 1996).

Further experiments with transgenic and mutant *andigena* potato plants were aimed at the elucidation of the involvement of various GA forms and enzymes of their metabolism in the photoperiodic control of tuberization. The main biologically active GA version in potato plants is believed to be GA_1 (Van der Berg et al. 1995a). However, plant treatment with other GA forms, for example, GA_3, GA_4, GA_7 and some others also exerted clear phenotypic responses. The biosynthetic pathway from the precursor of diterpene *trans*-geranylgeraniol diphosphate to GA_1 includes functioning of many genes and encoded enzymes. In potato the basic, but not the only pathway of GA biosynthesis in its final stage is 13-hydroxylation pathway with the formation of successive products: $GA_{12} \rightarrow GA_{53} \rightarrow GA_{44} \rightarrow GA_{19} \rightarrow GA_{20} \rightarrow GA_1$.

The key enzymes of biosynthesis (dioxygenases) are GA20-oxidase, involved in the formation of GA_{20} precursors and of GA_{20} itself, and GA3-oxidase catalyzing the reaction $GA_{20} \rightarrow GA_1$. The main enzyme inactivating bioactive GA is GA2-oxidase inactivating GA_{20} and GA_1 in the reactions $GA_{20} \rightarrow GA_{29}$ and $GA_1 \rightarrow GA_8$ (Hedden and Phillips 2000).

Experiments with dwarf potato mutant *ga-1*, deficient in hydroxylation of GA_{12} aldehyde at the formation of GA_{53}, confirmed the general inhibitory action of GA on photoperiodic induction of tuber formation. A total decline in the level of endogenous GA permitted the mutant to produce tubers in LD, although after keeping the plant under these conditions for a long time (three months). At the same time, in spite of the decrease in the total GA content, these mutant plants responded noticeably to the favorable day length and rapidly (sooner than a week) produced tubers under SD conditions (Van den Berg et al. 1995b). This corroborated the suggestion on the complex nature of photoperiodic regulation of tuberization involving both the negative GA signaling under LD and the positive tuberization stimulus under SD.

Three genes encoding GA20-oxidase were isolated from *andigena* potato plants and their transcripts were detected manifesting tissue specificity of the expression. Among them, the *StGA20ox1* gene was actively expressed only in the leaves. The level of this gene transcription declined in darkness and increased in light, especially blue light. The daily rhythm of this *StGA20ox1* gene expression depended on the photoperiodic regime and changed markedly under unfavorable condition for the tuberization regimes: LD or SD with the night interruption. However, the total level of mRNA for this oxidase did not differ drastically in plants growing under each of these three photoperiodic regimes (Carrera et al. 1999, Jackson et al. 2000). The authors ascribe this to the autoregulation of GA synthesis, i.e., feedback inhibition of *StGA20ox1* gene expression by excessive GA synthesized by GA20-oxidase in LD (Carrera et al. 1999).

Meanwhile, enhanced expression of the *StGA20ox1* gene in the transgenic *andigena* plants and antisense suppression of this gene expression markedly affected GA content in shoots and the photoperiodic induction of tuberization (Jackson et al. 2000). In these experiments the level of *StGA20ox1* gene expression was negatively correlated with the time of tuber formation in transgenic plants under SD. Enhanced transcription of this gene retarded tuber formation, and suppressed expression accelerated it. However, a decrease in the GA level due to the suppression of *StGA20ox1* gene transcription did not neutralize the inhibitory effect of LD, i.e., did not result in tuber formation under LD (Carrera et al. 2000). It might be that not only GA signaling but also other mechanisms are involved in the inhibition of tuberization under LD.

One of such mechanisms is the inhibition of photoperiodic induction of tuberization in LD with the involvement of phytochrome B (Rodríguez-Falcón et al. 2006). As was mentioned, transgenic *andigena* plants with blocked synthesis of phytochrome B produced tubers rapidly and extensively under both SD and LD (Jackson and Prat 1996). It was proposed that phytochrome B participates in the regulation of tuber formation by two different ways. One of them is the inhibition of biosynthesis of tuberigen in potato leaves under LD. The second pathway is related with an influence of phytochrome B on GA metabolism. The content of GA_1 in leaves and whole shoots of plants with suppressed *PHYB* increased by 3–5 times as compared with wild-type plants; the levels of GA_{20} and GA_8 increased as well, but that of GA_{29} was reduced. This indicates that phytochrome B, involved in the suppression of tuberization under LD, affects the content, ratios and metabolism of bioactive GA forms (Martínez-García et al. 2002a). One of the possible ways by which phytochrome can influence GA synthesis was demonstrated in the study of dynamics of light-induced transcripts in Solanaceae plants. It was found that in potato *andigena* phytochrome-dependent inhibition of tuberization under LD was accompanied by enhanced expression of the *ENT-KAURENOLIC ACID OXIDASE (KAO)* gene, which is involved in the early stage of GA biosynthesis (kaurene→ GA_{12}) (Rutitzky et al. 2009).

As the dynamic and functions of various GA forms in potato come to be better understood, the complex role of these compounds in the photoperiodic induction of tuber formation becomes more evident. In experiments on transgenic *andigena* potato lines with enhanced activity of GA3-oxidase catalyzing $GA_{20} \rightarrow GA_1$ conversion, the substantial accumulation of GA_1 in shoots and decrease of the content of GA_{20} in shoots and stolons were observed. Such transgenic plants demonstrated accelerated tuber formation under SD and increased tuber yield. In this study, spraying leaves of control plants with the GA_1 solution resulted in increased and not reduced tuber yield under SD conditions (Rodríguez-Falcón et al. 2006).

In further experiments (Bou-Torrent et al. 2011) another GA3-oxidase encoding gene, the *StGA3ox2* was isolated from *andigena* plants. The expression of this gene was increased in the aerial parts and repressed in stolons after transfer of plants to the tuber-inductive SD conditions. The *StGA3ox2* has been used to produce gene constructs for leaf-specific, tuber-specific and constitutive (35S CAMV promoter) transgene expression in potato.

The leaf-specific or the 35S overexpression of *StGA3ox2* resulted in earlier tuberization. In contrast, the tuber-specific expression leads to slightly delayed tuberization. The results were explained by different transport of various GAs. The main transported GA was shown to be GA_{20}, a precursor of GA_1. Therefore, overexpression of *StGA3ox2* in the leaves would vigorously convert GA_{20} to GA_1. This leads to lowering the amount of GA_{20} which is transported to stolon tips. As a consequence, reduced GA_{20} availability in the stolon tips will result in reduced GA_1 content that can lead to earlier tuberization. These experiments show that various GA versions affect tuberization differently; the distribution of these GA versions among plant organs is essential (Bou-Torrent et al. 2011).

The GA role in the photoperiodic induction of tuberization is still ambiguous because of the interference of effects of environmental cues and internal metabolic factors on the synthesis of various GA forms, their signaling pathways and plant sensitivity to these signals (Hedden and Phillips 2000). For example, it was shown that one of the function of *PHOR1 (Photoperiod Responsive 1)* gene, identified on the basis of cDNA from potato leaves induced by SD, is related to gibberellin signaling (Amador et al. 2001, Rodríguez-Falcón et al. 2006). *PHOR1* gene, encoding a transcription factor, responds to the day length and is actively expressed in potato leaves under SD conditions. *PHOR1* gene suppression with an antisense construct resulted in the retardation of potato stem growth, rapid tuberization under SD and plant partial insensitivity to the treatment with GA. The analysis of PHOR1 protein subcellular localization indicated its presence in the cytosol and rapid migration to the nucleus upon treatment with GA. It was also found that PHOR1 favored degradation of DELLA proteins, which are repressors of GA-induced growth, and thus favored potato stem elongation.

In more recent time, data were published about substantial effects of GA on potato gene expression related to carbohydrate (Chincinska et al. 2008, Fischer et al. 2008) and protein (País et al. 2010) metabolism in leaves, including genes somehow affecting the photoperiodic regulation of tuberization (Sarkar 2008). It was shown also that there is a link between GA effect on tuberization and some metabolic processes and multiple signaling systems, such as Ca^{2+} and calmodulin (Nookaraju et al. 2012), activity of protein phosphatases (País et al. 2010), oxidative burst and aquaporins (Sarkar 2008). Further experiments would elucidate the exact mechanisms of these connections and their participation in tuber induction.

In studies of hormonal regulation of tuberization induction, greatest attention was paid to GA action; however the involvement of other plant hormones do not have to be neglected. ABA, which, a distinct from GA, retards plant growth, generally has a positive impact on induction of tuberization. There are data that ABA content increased under inductive conditions (Ewing 1995). In *andigena* potato leaves the content of ABA increased substantially under SD as compared to LD (Macháčková et al. 1998). Treatment of potato leaves with ABA stimulated tuber formation and counteracted GA inhibitory influence (Abdullah and Ahmad 1980, Xu et al. 1998a). However, the process of tuber formation *per se* is obviously not related directly to the ABA accumulation in leaves. This is evident from experiments with *droopy* mutants of *S. phureja* potato deficient in ABA synthesis. In spite of the blockage of ABA synthesis, this mutant transits to tuber formation normally under SD conditions (Quarrie 1982). This suggests that ABA does not play a key role in tuberization induction and its stimulatory effect may be due to its antagonism to GA signaling.

Ck exert in general the positive effect on tuberization (Ewing 1995, Hannapel 2007); however, their involvement in the photoperiodic induction of tuber formation is not proven so far. Data on the effect of the day length on Ck content in leaves are ambiguous. In early studies (Mauk and Langille 1978), it has been reported that in extracts from potato shoots kept under favorable day length and temperature, the content of zeatin riboside was 25–30% higher than in non-induced plants. The highest Ck level in shoots was observed in 4–6 days after setting inductive conditions, and then it declined again. In another study (Macháčková et al. 1998) content of some Ck (zeatin, zeatin riboside, isopentenyladenosine) in *andigena* potato leaves growing under non-inductive LD was elevated. In the leaves of *in vitro* growing potato cv. Désirée, a decrease in content of zeatin and zeatin riboside was observed after plant transfer from LD to SD (Aksenova et al. 2005). Attempts to affect tuberization by potato shoot treatment with Ck gave uncertain and sometimes even negative results. The whole plant treatment with Ck was usually inefficient (Chailakhyan 1984), the immersion of leafy potato cuttings in the N^6-benzyladenine (BA) solution did not accelerate but slightly delayed tuberization (McGrady et al. 1986). The treatment of *andigena* plants with Ck failed to induce tuber formation under the non-inductive day length (Rodríguez-Falcón et al. 2006). Even so, treatment with kinetin of *tuberosum* stem cuttings growing *in vitro* resulted in tuber appearance without SD induction (Forsline and Langille

1976). Other results were obtained in experiments with regulation of cytokinin catabolism in cultured *in vitro* explants of potato cv. Désirée (Raspor et al. 2012). Cytokinin catabolism was performed by cytokinin oxidase/dehydrogenase (CKX), which catalyzes irreversible cytokinin degradation. *AtCKX2* gene driven by 35S promoter was used to construct AtCKX2-overexpressing plants. The CKX activity was remarkably enhanced in all *AtCKX2* transgenic lines. All transgenic lines showed a substantial decrease in amount of bioactive cytokinins (*trans-*, *cis-*, dihydrozeatins, isopentenyladenine and corresponding ribosides). The most interesting result concerning AtCKX2-overexpressing plants was a promotion of tuber formation under LD conditions, indicating that cytokinin deficiency overrules inhibitory consequences of LD treatment for tuber formation. However, as 35S promoter drives gene expression throughout the whole plant, it is unclear what stage of tuberization process, induction or initiation, was promoted by overexpression of *AtCKX2* gene.

In general, there is no sufficient evidence so far for active Ck participation in the formation of tuberization stimulus in leaves and its long-distance transport. The role of Ck and particular Ck versions in tuberization induction is not studied enough; the role of enzymes for Ck synthesis and inactivation in tuberization induction is poorly studied; only few studies on tuber induction were done using transgenic potato plants with changed cytokinin status.

Jasmonic acid and its derivatives, JA methyl ester (tuberonic acid, TA) and glucoside of tuberonic acid (GTA), are compounds stimulating tuber formation and growth of potato explants cultivated *in vitro* (Yoshihara et al. 1989, Koda 1997). It has been supposed earlier that just these compounds are long-distance stimuli of tuberization (tuberigens). However, the involvement of these compounds in the generation of the tuberization stimulus in leaves is not proven so far. JA, TA and GTA are synthesized in plants with the involvement of lipoxygenases (LOX); LOX catalyze oxygen insertion in polyunsaturated fatty acids (FA), mainly into linoleic and linolenic acids, with the formation of corresponding hydroperoxides. In potato leaves, 13-LOX activity was detected; this enzyme attaches oxygen to the 13th carbon atom of FA. In tubers, 9-LOX activity was also detected; it produces 9-hydroperoxides (Hannapel 2007, Sarkar 2008). JA is synthesized in the leaves via the C13-hydroperoxide pathway and then is metabolized into TA and GTA. It was found that in potato leaves, 13-LOX was substantially activated under SD and low temperature conditions favorable for tuberization. This stimulated JA synthesis and accumulation of JA, TA and GTA in potato leaves and shoots under conditions promoting tuber formation (Helder et al. 1993, Nam et al. 2005, Nam et al. 2008). It was suggested that under inductive photoperiods not only LOX but also enzymes hydroxylating JA and glucosylating TA are activated in potato leaves. These products generated in leaves can be transported to stolons and stimulate tuber formation (Sarkar 2008).

However, this hypothesis is not yet experimentally proven. Treatment of non-induced *andigena* plants leaves with JA did not induced tuberization under LD (Jackson and Willmitzer 1994), and *S. demissium* treatment of aboveground shoot with LOX inhibitor did not prevent tuber formation under SD (Rodríguez-Falcón et al. 2006). However, the inhibitor used, salicylhydroxamic acid, is not sufficiently specific and suppresses also other oxidative enzymes in potato (Shugaev and Sokolova 2001). Thus, the possible involvement of JA and its derivatives in the long-distant signaling of tuberization needs further investigation.

On the whole, available data show that the photoperiodic induction of tuberization is based on systemic crosstalk between signals inducing tuberization under SD and inhibiting this process under LD. Hormonal, metabolic and genetic regulatory systems are involved in this interaction (Fig. 2). It was demonstrated that other effective environmental cues (temperature, nitrogen nutrition) also act through changing the balance of phytohormones, mainly GA and ABA (Krauss 1985, Menzel 1985).

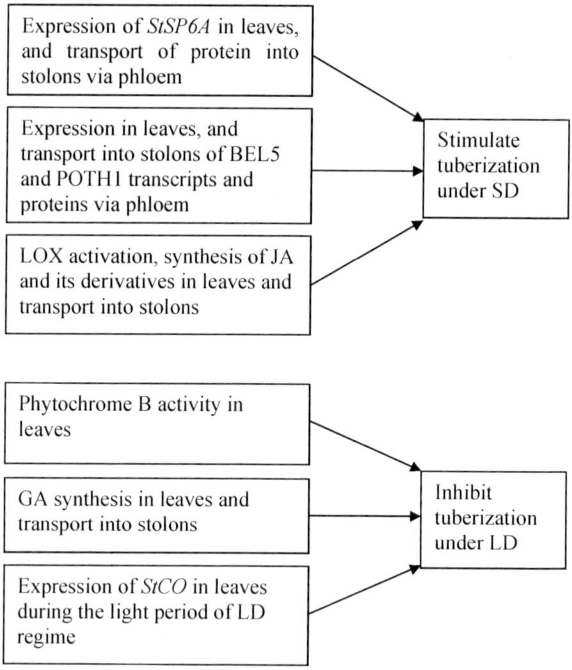

Fig. 2. Factors controlling photoperiodic induction of tuber formation. StSP6A—ortholog of FLOWERING LOCUS T protein; BEL5—potato transcription factor of BEL type; POTH1—potato transcription factor of KNOX type; LOX—lipoxygenases; JA—jasmonic acid; GA—gibberellins; StCO—ortholog of CO-like protein in potato.

Tuber Initiation

Tuber initiation follows tuber induction stage; as a rule, tuber initiation occurs in stolon tips. Stolons are underground organs, and therefore they cannot perceive the photoperiod signal. Thus, tuber initiation in potato plants with an obligate photoperiodic dependence starts only after the inflow of tuberization stimulus synthesized in aerial organs during photoperiodic induction.

Potato tuber is a modified stem with radially extended axis, shortened internodes and strongly thickened as compared to the shoot; its tissues contain much starch (up to 75% of dry weight), tuber-specific proteins patatins, and proteinase inhibitors (altogether up to 40% of total tuber protein). Under natural conditions tubers develop on the stolons, but potentially any potato bud and especially stem apex are capable of tuber formation (Ewing 1995).

Tuber initiation on different stolons of a single potato plant is not a synchronous process: on some stolons tubers appear earlier, on others later, and on some stolons they are not formed at all (Vreugdenhil and Struik 1989, Martínez-García et al. 2002a). This indicates that tuber initiation is a local process evidently related to the metabolic conditions in tissues of particular stolon and their correspondent competence. On the other hand, uneven distribution among stolons of a tuberization stimulus, coming from haulm, is not excluded resulting in different stolon behavior.

Tuber initiation is connected with the transition of stolon meristem from indeterminate to determinate growth. Indeterminate self-renewing meristems have to continue cell divisions and maintain unchanged general cell organization. Cells constituting determinate meristems undergo differentiation and yield a terminal, usually reproductive, structures. In the case of stolons these structures are tubers. Over the past decade, extensive research led to identify the *CETS* family genes responsible for the fate of indeterminate and determinate meristems (Shalit et al. 2009, McGarry and Ayre 2012). The *CETS* members *FT* of Arabidopsis, *SFT* in tomato and potato *SP6A* are important in promoting the transition to determinate growth, while Arabidopsis *TFL1* and tomato *SP* oppose this activity by maintaining meristems in an indeterminate state. A new model emerged for the timing of stolon meristem transition from indeterminate to determinate growth through a balance of *FT*-like and *TEL1*-like gene activities (McGarry and Ayre 2012). This model might explain the non-synchronous tuber initiation on different stolons. Important processes required for tuber initiation are reorientation of cell divisions and changes in their intensity in apical and subapical zones of the stolon (Xu et al. 1998b, Vreugdenhil et al. 1999). At tuber initiation, cell divisions cease in the stolon apex but start in the subapical zone changing their direction from transverse to longitudinal toward the axis of stolon growth. This initiates radial swelling of the subapical part of the stolon leading to its gradual conversion into a tuber. Change in the direction of cell division characteristic for tuber initiation is related to the shift in orientation of cortical microtubules controlling deposition of cellulose microfibrils in the cell walls of dividing cells (Fujino et al. 1995).

Simultaneously with the change in the direction of cell divisions and the increase in cell number and sizes, substantial rearrangement of carbohydrate metabolism occurs in the developing tuber (Appeldorn et al. 2002). The characteristic process is also an enhanced supply of developing tuber with sucrose; besides, the carbohydrate transport to the tuber changes from apoplastic to more efficient symplastic one (Viola et al. 2001). The synthesis of starch is markedly activated and activities and ratio of involved enzymes are changed as well. Sucrose synthase, fructokinase, and amyloplastic starch synthase are activated, whereas functioning of acid invertase is almost completely blocked. An elevated content of sucrose activates expression of a great number of genes related to the starch synthesis (Kloosterman et al. 2005). As a result of all these structural and metabolic changes, potato storage organ with a high sink capacity, tuber, appears.

Numerous changes in gene expression, which correspond to structural and metabolic changes occurring at tuber initiation compared to non-induced stolons, were observed; however, genes determining tuber identity are not found yet (Bachem et al. 2000, Kloosterman et al. 2005). In addition, during early stages of tuber formation, substantial changes were detected in the expression of genes responding to hormonal signals. This indicates the important role of hormonal regulation in the process of tuber initiation. During tuber formation in *andigena* potato, the expression of GA-sensitive and IAA-dependent genes was reduced, whereas Ck-inducible genes demonstrated enhanced expression (Hannapel 2007).

The role of GA in tuber initiation is studied in more detail. Already in early studies, it has been shown that GA inhibits not only tuber induction but also their initiation (Ewing 1995). Treatment of detached potato stolons, cuttings or cultivated plantlets with GA suppressed tuber initiation *in vitro*. Activities of GA-like compounds were markedly lower in tubers than in stolons. A detailed comparison of GA content in the tips of stolons and developing tubers showed that the level of endogenous GA was high in growing stolons and strongly reduced before visible tuber initiation.

In *andigena* potato, substantial changes in activities of enzymes of GA metabolism as well as expression of corresponding genes were observed during tuber initiation (Rodríguez-Falcón et al. 2006, Kloosterman et al. 2007, Bou-Torrent et al. 2011). Under LD unfavorable for tuber formation, mRNA of *StGA3ox2* gene accumulated in stolons; this gene is involved in GA_1 biosynthesis. Plant transition to SD resulted in tuber initiation and also in complete blocking expression of *StGA3ox2* and activation of expression of *StGA2ox1* gene encoding the enzyme of bioactive GA catabolism. Such changes in expression of *StGA3ox2* and *StGA2ox1* in stolons occurred at the earliest steps of tuber initiation, yet before the appearance of the first visible signs of tuberization. In *andigena* plants cultivated *in vitro*, overexpression of *StGA2ox1* gene directed by 35S promoter accelerated the process of tuber formation, whereas transgenic plants with suppressed expression of this gene contained more GA_{20} and were characterized by delayed tuberization. In general, these results indicate a crucial role of GA (especially GA_1) content reduction in the stolon tips during transition to tuber initiation.

One more gene, whose expression changes the level of GA activity during tuber initiation, was isolated from potato cv. Karnika and named *StGAN* (Bachem et al. 2001). This gene encodes a protein homologous to steroid dehydrogenases and it actively expressed in stolon tips during the early period of tuber initiation. Suppression of this gene expression by antisense constructs resulted in marked stolon elongation and production of strongly elongated tubers in potato explants cultivated *in vitro*. Plants with suppressed *StGAN* gene expression had elevated (by 2–7 times) levels of GA_1, GA_4, GA_9 and GA_{20}. These results indicate the possibility for one more pathway regulating bioactive GA levels during tuber initiation with the involvement of protein encoded by *StGAN*.

Inducing signals coming under favorable photoperiod from the leaves also markedly affect GA content in stolons. As was noted above, phloem transport of *StBEL5* mRNA from leaves to stolons is stimulated under SD conditions (Rosin et al. 2003, Chen et al. 2003). Further studies (Chen et al. 2004, Hannapel 2010) showed that, in the stolon tips, transcription factors StBEL5 and POTH1 interact and their tandem is involved in suppression of the *GA20ox1* gene. Transcription factor duet BEL/KNOX (BEL/POTH1) repressed *GA20ox1* gene by blocking this gene promoter. As a result, level of bioactive GA in stolon tips declined substantially at tuber initiation.

The mechanism for the regulation by GA of the tuber initiation process is not fully deciphered, but a number of suggestions were put forward. It might be that a decrease of the GA content in stolons before the start of tuber initiation induces reorientation of cell divisions in stolon tips. This assumption corresponds to GA capacity of cortical microtubules reorientation in elongating cells into the transverse position relatively to the axes of growing organs. A decrease in the GA content after plant tissue treatment with an inhibitor of GA synthesis unicazol led to reorientation of cortical microtubules in elongating cells from transverse to longitudinal (Fujino et al. 1995). In experiments with single node potato explants cultivated *in vitro* (Xu et al. 1998b), direction of cell divisions in the subapical stolon zone depended on the presence of GA in a cultural medium. GA addition resulted in the transverse divisions and further stolon growth, while in the absence of GA longitudinal divisions started, which was positively correlated with tuber initiation.

It is also possible that the inhibitory GA effect on tuber initiation is partially related to negative influence of these phytohormones on the rearrangement of carbohydrate metabolism in developing tubers. Treatment of potato plant with GA reduced starch content in tubers and suppressed activities of enzymes of starch biosynthesis (Obata-Sasamoto and Suzuki 1979). The activity of invertase related to stolon elongation positively correlated with GA content in culture medium (Palmer and Barker 1972). GA addition to the medium suppressed markedly sucrose synthase in developing tubers of potato cv. Désirée (Sergeeva et al., unpublished data). The interrelationship between GA-induced suppression of tuber initiation and carbohydrate metabolism was shown in the experiments with single node potato cuttings cultivated *in vitro*. An increase in the content of sucrose in medium markedly weakened GA_1 inhibitory effect on tuber initiation (Xu et al. 1998a). The same was observed on sucrose- and GA-treated potato shoot segments (Simko 1994). The data showed the systemic counteraction between GA inhibitory and sucrose inducing effects at their combined action on tuber initiation.

As distinct from GA, Ck stimulates tuber formation in many tuberiferous plant species (Ewing 1995). As mentioned earlier, treatment of the aboveground organ with Ck exerts no clear positive effect on tuber formation but Ck directly promotes tuber initiation in stolons. The addition of kinetin or BA to the medium for cultivation of detached stolons markedly accelerated tuber formation (Palmer and Smith 1970). It was found that for the acceleration of tuber initiation, the presence of kinetin in a culture medium was required only during the first 3–4 days of cultivation although visible tubers appeared not earlier than 10–12 days after onset of kinetin treatment. This indicates that Ck stimulates tuber initiation at its earliest steps. For the manifesting of kinetin stimulatory action, sufficient amount of sucrose must be present in the culture medium. Thus, as distinct from GA, Ck together with sucrose stimulates tuber initiation. The interaction between kinetin and sucrose in their influence on tuber initiation was also found in the *in vitro* experiments performed on stem explants of six potato cultivars and transgenic lines of potato cv. Désirée (Aksenova et al. 2000, Romanov et al. 2000). In these experiments kinetin stimulated tuber initiation accelerating tuber formation and increasing the number of tubers. The stimulatory effect of kinetin depended not only on potato genotype but also on the content of sucrose in the nutrient medium. The highest kinetin effect on tuber initiation was manifested at the suboptimal sucrose concentration. Under these conditions, stimulatory effect of Ck on tuber initiation complemented suboptimal effect of sucrose when its content was not sufficient for maximal action.

When potato plants transited to tuber formation, activity of endogenous Ck increased substantially in the stolons (Obata-Sasamoto and Suzuki 1979) and the developing tubers (Mauk and Langille 1978). In transgenic potato cv. Miranda plants expressing *ipt* gene for Ck biosynthesis tuber initiation was enhanced, but only in lines with a moderate increase in Ck content. These lines could develop tubers *in vitro* even at much lower content of sucrose in culture medium than wild-type plants (Macháčková et al. 1977).

Favorable Ck effect on tuber initiation can evidently be exerted in several ways. Ck are known to stimulate plant cell division (Romanov 2009). In the cell division cycle, they regulate transition from G1 to S phase implying the onset of DNA biosynthesis and partially act via induction of D-type cyclins (Francis 2007). Since tuber initiation is tightly associated with cell division, Ck are evidently necessary partners in this process.

In addition, it was shown that during potato tuber initiation, Ck activated enzymes of starch biosynthesis, which resulted in starch accumulation and enhanced sink capacity of the developing tubers. In the *in vitro* culture of stolon segments transiting to tuberization, kinetin treatment stimulated starch synthesis and reduced invertase activity (Palmer and Barker 1973); cytokinin also activated phosphorylase and ADP-glucose pyrophosphorylase (Mingo-Castel et al. 1976). Ck-induced activation of starch biosynthesis and related increase in the sink capacity were clearly demonstrated on transgenic tobacco plants expressing *ipt* gene locally in stem lateral buds (Guivarach et al. 2002). Enrichment in Ck induced morphological and biochemical changes in transgenic tobacco buds resembling those occurring during potato tuber initiation. In these buds, reorientation of cell divisions, tuber-like swelling and intense starch deposition were observed. This effect can be associated with Ck antagonism to GA signaling (Brenner et al. 2005, Romanov 2009, Fleishon 2011).

Some researchers showed the involvement of Ck in the stolon response to signals, coming from leaves and controlling tuber initiation. In fact, transcription factors StBEL5 and POTH1 whose content increases in stolons due to inflow of stimulatory signals from leaves under SD conditions, elevate Ck content in stolons. In transgenic *andigena* potato plants overexpression of *StBEL5* and *POTH1* accelerated tuber initiation and also increased the content of zeatin-type Ck by 3–5 times and isopentenyl-type Ck by 2.5–3.0 times (Hannapel 2007). Inhibition of tuber initiation by LD in transgenic potato plants cv. Désirée *in vitro* was reinforced by expressing *35S::PHYB* construct. Addition of kinetin to the culture medium almost completely neutralized this inhibition and counteracted the unfavorable LD action on tuber initiation (Aksenova et al. 2005).

IAA is one of the most important phytohormones directly involved in the control of most processes of plant growth and morphogenesis. However, the role of auxins in potato tuber initiation remains unclear.

Auxin treatments of intact potato plants and their detached shoots and cuttings gave contradictory results; in general, auxins exerted a rather negative effect on tuber initiation (Ewing 1995). High content of auxins in the culture medium inhibited tuber initiation on stem cuttings from induced *andigena* potato plants (Kumar and Wareing 1974). IAA addition to the culture medium differently affected tuber initiation by explants of potato cv. Désirée and its transgenic lines with changed hormonal balance (*B33::rolB*- and *B33::rolC* transformants). The IAA effect on tuber initiation depended on both plant genotype and sucrose content in the culture medium. When sucrose content was low—close to the threshold for tuberization value—IAA addition to the medium increased the number of produced tubers, whereas when the medium contained 5–8% of sucrose, IAA reduced the number of tubers in most cases (Aksenova et al. 2000).

Further molecular and transgenic studies corroborated the role of IAA in tuber initiation. It was shown (Faivre-Rampant et al. 2004) that the transcription factor ARF6, one of the key regulators of potato auxin-responsive gene transcription, took part in the early stages of tuber initiation. It was found that the transition in the stolon tips to tuber initiation was accompanied by a stable reduction in the ARF6 level. Accordingly, the expression of IAA-dependent genes was reduced during early tuber formation in *andigena* potato and coincided with a visible onset of swelling in the stolon tips (Hannapel 2007). In contrast, a detailed study using a microarray approach (Kloosterman et al. 2005, Roumeliotis et al. 2012a, b) showed that several auxin-related genes exhibited a peak in expression after the onset of tuber initiation. Examples of such genes are two *StPIN*-like genes that are involved in auxin transport, auxin biosynthesis gene *StYUC*-like, *arcA*-like (auxin regulated) genes and some others. Based on these results and on scoring of auxin content in stolons and tubers, the authors concluded that auxin has an important role to play in tuber development by regulating a plan of cell division and cell growth during the early stages of tuber initiation (Roumeliotis et al. 2012a). A physiological model for the combined action of GAs and auxin during the early period of tuberization was suggested.

The role of auxins in tuber formation was studied also in potato Bintje and in *andigena* plants (Roumeliotis et al. 2012b). It was shown that auxin content increased in stolons prior to tuberization and remained relatively high during subsequent tuber growth. The expression of auxin biosynthesis gene (*StYUK-like 1*) increased 200-fold in *andigena* stolons after transition of transgenic plants to SD conditions. Synthetic strigolactone (GR24), which has an inhibitory role on shoot branching, was also tested in *in vitro* experiments. GR24 application led to a marked inhibition of stolon axillary bud overgrowth and subsequent potato tuber formation (Roumeliotis et al. 2012b). This indicates that strigolactone along with auxins takes part in the regulation of stolon bud formation and growth.

It is possible that other phytohormones interfere with auxin effects on tuber initiation. In fact, in experiments with potato cv. Miranda and its transgenic lines bearing *ipt* gene from *Agrobacterium tumefaciens,* it was found that the ratio of IAA to Ck affected tuber initiation substantially (Macháčková et al. 1997). The highest intensity of tuber initiation was detected in transgenic lines with moderately increased content of Ck and IAA with their ratio close to that in wild-type plants. In another work (Kolachevskaya et al. 2012), the agrobacterial gene *tms1* under B33 promoter was introduced into Désirée potato plants. This gene encodes the enzyme converting tryptophan to indolyl-3-acetaldehyde in the pathway for auxin biosynthesis in agrobacteria. In cultured *in vitro* transgenic plantlets the content of IAA increased (by 1.5–2.9 times) and the acceleration of tuber initiation was observed. It is worth noting the changes in the hormonal balance caused by *tms1* expression: contents of jasmonic acid and salicylic acid increased, the content of ABA reduced and the levels of several cytokinins rose. So the modification of IAA content led to changes in the whole hormonal balance in plants. To sum up, the mode of auxin involvement in the regulation of tuber initiation remains unclear and needs further investigation, on the molecular level in particular.

Already in early studies it has been shown that when jasmonates (both JA itself and products of its metabolism, TA and GTA) come to stolon, they can induce tuber initiation therein (Pelacho and Mingo-Castel 1991). During early steps of tuber initiation, the level of JA in potato stolons increased (Abdala et al. 2002). JA and its derivates were supposed to change the direction of cell division and radial extension in stolon tips during the transition to tuber initiation. It was established that JA and TA induced similar changes in the cortical microtubule orientation in stolons as inhibitors of GA synthesis unicazol and affected cell division in a way that is opposite to the action of GA (Matsuki et al. 1992, Abdala et al. 2002). Treatment with jasmonates completely reversed the GA_3 inhibitory action on tuber formation in *in vitro* cultivated potato explants (Castro et al. 2000). GA inhibited the activity of lipoxygenase (9-LOX) (Lemeza et al. 2010) normally involved in jasmonate biosynthesis in potato tubers and stolons. During transition to tuber formation, the expression of gene encoding tuber-specific 9-LOX is activated. By *in situ* hybridization, mRNA of tuber-specific LOX was found in the tips of stolons producing tubers; it was located mainly in the tuber perimedullary zone where most active cell division and cell growth are concentrated (Kolomiets et al. 2001).

The above data indicate the involvement of jasmonates and enzymes of their biosynthesis in processes of cell division and growth at the tuber initiation. The effects of JA and its derivatives are opposite to the inhibitory influence of GA.

The effect of ABA on tuberization was studied by Wareing and Jennings (Wareing and Jennings 1980). They showed that treatment of intact *andigena* and *tuberosum* potato plants with ABA solutions accelerated tuber formation. As distinct from GA, ABA suppressed stolon growth and stimulated tuber initiation. It was also shown that the stimulatory effect of ABA was dose-dependent with the optimum ABA concentrations depending on potato cultivar (Hussey and Stacey 1984). In the *in vitro* experiments performed on single-node potato cuttings (Xu et al. 1998a) addition of ABA to the culture medium slightly (by one day) accelerated tuber initiation. In these experiments, ABA stimulated tuberization at both high and low sucrose content in the medium. When ABA and GA were added to the culture medium simultaneously, ABA weakened the inhibitory GA influence on tuber initiation. However, no substantial changes in the content of endogenous ABA related to tuber initiation were observed. Only small increase in ABA activity was observed in potato stolons during the early period of tuber initiation (Ewing 1995). In the *in vitro* experiments, the level of endogenous ABA in stolons even declined during tuber initiation (Xu et al. 1998a).

In view of all these data, it was supposed (Ewing 1995, Rodríguez-Falcón et al. 2006) that ABA is likely not among key endogenous regulators of tuber initiation; however, it can affect tuberization as a negative regulator of GA activity.

Thus, the whole complex of plant hormones participates in the hormonal regulation of tuber initiation, and each of them has its specific role. The hormonal regulation of tuber initiation is provided by a coordinated in time and space crosstalk between various groups of phytohormones (Fig. 3).

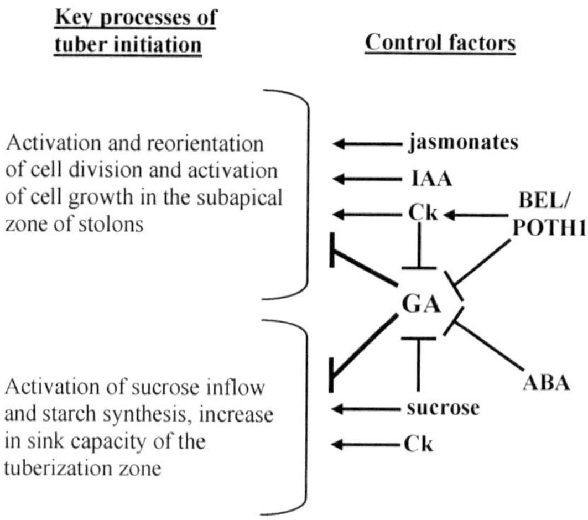

Fig. 3. The key processes of tuber initiation and scheme of its regulation.
← Stimulation; ⊢ inhibition. Other designations are as in Fig. 2.

Tuber Growth

The growth of tubers follows their initiation: the number and size of tuber cells increase, and the tuber grows isodiametrically in all directions. During tuber growth, starch is actively synthesized and deposits as starch grain in amyloplasts of tuber cells. The intense assimilate inflow plays an important role in tuber growth. In this period, general changes in the source-sink relations occur in the plant organism. A strong competition for assimilates between individual tubers and especially between growing tubers and growing shoots occurs (Struik 2007b). Therefore, factors stimulating aboveground shoot growth (enhanced nitrogen nutrition, LD, treatment with GA and others) retard tuber growth and, *vice versa*, factors stimulating tuber growth retard growth of aboveground plant parts (haulm) (Mokronosov 1990, Ewing and Struik 1992).

It is difficult to adjust exactly time limits for the tuber growth period. Already during tuber initiation, cell divisions and growth in the subapical zone of the stolon are activated and coordinated switches in carbohydrate metabolism occurs, leading to the enhancement of starch biosynthesis. All these processes persist during tuber growth. At the end of the growth period, the tuber enters in the dormant state also gradually. The end of tuber growth coincides with the end of potato plant growth and haulm senescence (Struik 2007b).

GA retard growth of the developing tubers. Treatment with GA of intact plants *in vivo* or tubers *in vitro* suppressed tuber growth and activated buds on tubers, i.e., secondary growth or regrowth resulting in tuber involution (Vreugdenhil and Sergeeva 1999). Spraying the *andigena* plants with GA_3 resulted in the formation of elongated tubers similar to overgrown stolon regions (Vreugdenhil and Sergeeva 1999). Addition of GA_3 to culture medium reduced tuber sink capacity in potato cv. Désirée and decreased substantially the yield index (the ratio of tuber weight to the weight of shoots and roots) (Aksenova et al. 2009).

Treatment of growing tubers with GA reduced the content of sucrose, starch and patatin (Hannapel et al. 1985, Vreugdenhil and Sergeeva 1999). Under natural conditions, content of free GA, which was declined in stolons before tuber formation, remained on very low level in tubers during the entire growth period (Claassens and Vreugdenhil 2000).

The elevated content of Ck was detected in tubers only during the early stages of growth (4–6th days after the start of initiation). Thereafter, the content of Ck decreased markedly and remained low during the entire further period of tuber growth (Ewing 1995, Claassens and Vreugdenhil 2000). In potato cv. Malakhit the content of zeatin and zeatin riboside declined during tuber growth in both tuber cortex and pith (Borzenkova and Borovkova 2003). Kinetin addition to the culture medium did not affect substantially *in vitro* tuber growth on explants of various potato cultivars (Romanov et al. 2000); kinetin influenced tuber size in transgenic lines of cv. Désirée differently, depending on plant genotype and sucrose content in the culture medium (Aksenova et al. 2000). BA addition to the culture medium reduced the size of tubers produced on stem explants of three Indian potato cultivars (Sarkar et al. 2006). Treatment of disks excised from tubers with kinetin activated ^{14}C-glucose incorporation into starch during the initial period of tuber growth but did not affect this process in grown tubers (Borzenkova et al. 1998). The presence of kinetin in medium for *in vitro* cultivation of cv.Désirée stem explants did not affect the starch content in microtubers on these explants (Gukasyan et al. 2005). Generally, it was accepted that Ck favor tuber growth during its early stage by stimulation of cell division in young tubers, but they are not the main regulators of the later period of tuber growth (Ewing 1995, Rodríguez-Falcón et al. 2006).

At the same time, it is well known that Ck enhance the sink capacity of plant organs and tissues, including potato (Ron'zhina and Mokronosov 1994). The sink capacity plays an important role in tuber growth. In the period of tuber growth, assimilate flows are clearly shifted toward tubers, whereas growth of aboveground organs slows down (Ewing and Struik 1992, Struik 2007b). Kinetin addition to the medium for the cultivation of potato cuttings cv. Désirée resulted in a redistribution of the organic mass between aboveground shoots and tubers. Kinetin markedly increased the proportion of tubers at the expense of shoots in the total plant biomass (Aksenova et al. 2009).

Growth of individual tubers on the plant depends on their particular sink capacity. During tuber formation, the potato plant initiates excessive number of tubers; some of them do not develop because of competition for metabolite inflow with other growing tubers (Struik 2007b). It was shown that the sink capacity of particular tubers depends strongly on the Ck inflow from the roots positioned near the corresponding stolon (Mokronosov 1990). One can conclude that Ck is an essential participant in processes leading to the enhancement of the tuber sink capacity.

Early studies of auxin activity during tuber development gave contradictory results. It was reported that the level of endogenous IAA increased before tuber initiation, then declined, and thereafter remained low during the entire period of further tuber growth (Obata-Sasamoto and Suzuki 1979, Claassens and Vreugdenhil 2000). By contrast, a positive correlation was observed between the rate of tuber growth and the content of IAA in tubers (Marschner et al. 1984). The results of more detailed analysis of correlation between IAA content and tuber growth were published mainly by Russian researchers. In tubers of potato cv. Skoroplodnyi grown in soil, the IAA concentration was higher in larger tubers compared to smaller ones. In this case, treatment of plant stolons with auxin resulted in the increased rate of tuber growth (Puzina et al. 2000). In the *in vitro* experiments, IAA addition to the culture medium induced substantial (sometimes more than twofold) increase in the tuber size in the subspecies *andigena*, several potato cultivars from the subspecies *tuberosum*, as well as in some lines of transgenic potato (Romanov et al. 2000, Aksenova et al. 2000). IAA distribution between various tuber zones in relation to growth and starch deposition in these zones was studied in small (young) and large (more mature) potato tubers grown in soil (Borzenkova and Borovkova 2003). During tuber growth, its pith predominantly enlarges. Growth of this zone occurred mainly due to cell expansion and was accompanied by acceleration of starch synthesis, increase in the size of starch grains, and a substantial IAA accumulation. The higher starch content and enlarged starch grains in tubers of cv. Désirée grown *in vitro* were observed at IAA addition to the medium (Gukasyan et al. 2005). Thus, all above mentioned data show that IAA favors tuber growth. Its stimulatory effect on growth is correlated with the increase in the cell size in tubers, an enhanced starch biosynthesis, and its deposition as larger grains. It might be that IAA also increases tuber sink capacity because the conversion of sucrose into storage starch facilitates the inflow of new assimilate portions in growing tubers (Mokronosov 1990).

The involvement of the other phytohormones in the regulation of tuber growth was also reported. It was shown (Kolomiets et al. 2001) that the regulatory pathway starting from the expression of *LOX1* class lipoxygenase genes and resulting in the accumulation of corresponding enzymes is possibly involved in tuber growth stimulation. Though addition of JA and methyl jasmonate to the culture medium did not affect the number of tubers developing on stem explants of three potato cultivars, the tuber growth was enhanced in the jasmonate presence (Sarkar et al. 2006).

The content of ABA during tuber growth was shown to be low. In this period, a negative correlation was observed between the growth rate and the content of ABA in tubers (Claassens and Vreugdenhil 2000). Therefore, it seems likely that ABA retards tuber growth. On the other hand, there are some data on the positive correlation between ABA content and the intensity of starch biosynthesis in the cortex and pith parenchyma of growing tubers (Borzenkova and Borovkova 2003). In addition, a negative effect of ABA on the activity of GA inhibiting tuberization is known (Mokronosov 1990). Therefore, a stimulatory effect of ABA via its antagonism towards GA is not excluded.

Overall, the data show that plant hormones are closely involved in the control of tuber growth. However, the hormonal regulation of tuber growth is studied much less than the regulation of tuber induction and initiation. This is particularly evident from the small number of studies performed at molecular and genetic levels.

Conclusion

Since potato is one of the major crops in the world, the regulation of tuber formation and the role of phytohormones in this process were studied extensively for a long time. Already in the early works and by using classic physiological methods, it was shown that plant hormones participate in the control of potato tuberization. The changes in phytohormone content during tuber development were studied, and the key role of GA as a main inhibitory factor in the hormonal control of tuber formation was revealed. The effects of various phytohormones on successive stages of tuber formation are presented in outline in Table 1.

At present, a tendency is apparent for the application of the wide arsenal of modern molecular and genetic methods for elucidation of specific role of phytohormones in tuberization, especially in the processes of tuber induction and initiation. Recent findings substantially improved our knowledge on the mechanisms of phytohormone participation in the control of tuberization.

The prominent progress was associated with the identification of StSP6A protein as tuberigen or its important part (Navarro et al. 2011). It was shown that the StSP6A acts as a mobile long-distance proteinaceous hormone. The StSP6A protein is formed in the potato leaves in response to inductive conditions, then it is transported through the phloem to the stolon tips where it induces the transition to tuber formation. Further research is required to identify the stolon-specific StSP6A target genes and StSP6A-interacting partners necessary for tuber initiation.

More and more evidence appears concerning the complexity of phytohormone biosynthesis and catabolism, the perception of hormonal signals and the involvement of numerous proteins (enzymes) and encoding genes in these processes. Also, mechanisms of interaction between phytohormones at the level of their signaling pathways started to emerge (Brenner et al. 2005, Fleishon et al. 2011, Vanstraelen and Benková 2012). The examples for participation of different hormones in regulating the tuberization process are: antagonism between GA and ABA (Xu et al. 1998a) and GA and Ck (Hannapel 2010) in tuber initiation; a crosstalk of auxin and GA and of auxin and strigolactone during tuber initiation and growth (Roumeliotis et al. 2012a, b); the interrelation of IAA and Ck in regulating tuber formation (Macháčková et al. 1997); the antagonizing action of Ck and jasmonates in tuber growth (Sarkar et al. 2006). Also, it becomes evident that experimental change in the content of a single hormone can lead to the change in the content of other hormones affecting the whole net of the hormonal control of tuberization. To investigate such a problem, modern molecular and profiling methods are to be used.

Table 1. Effect of phytohormones at various stages of tuber formation.

Tuberization stage	Plant hormone	Physiological effect	Source
Stolon formation and growth	GA	stimulate	Ewing 1995, Vreugdenhil and Sergeeva 1999
	ethylene	retards elongation, stimulates stolon thickening	Vreugdenhil and Struik 1989, Vreugdenhil and van Dijk 1989
	ABA	retards elongation by functioning as anti-GA	Mokronosov 1990, Ewing 1995
Induction of tuberization	StSP6A protein	tuber-inducing proteinaceous hormone	Navarro et al. 2011, González-Schain et al. 2012
	GA	inhibit	Martínez-García et al. 2002, Bou-Torrent et al. 2011
	jasmonates	possible involvement in positive signaling	Yoshihara et al. 1989, Sarkar 2008
	ABA	accelerates by functioning as anti-GA	Abdullah and Ahmad 1980, Ewing 1995
Tuber initiation	GA	inhibit	Rodríguez-Falcón et al. 2006, Kloosterman et al. 2007
	Ck	stimulate	Palmer and Smith 1970, Hannapel 2007
	jasmonates	stimulate	Koda 1997, Abdala et al. 2002
	IAA	stimulates cell division and enlargement	Faivre-Rampant et al. 2004, Roumeliotis et al. 2012b
	ABA	slightly stimulates	Xu et al. 1998a, Rodríguez-Falcón et al. 2006
Tuber growth	GA	inhibit	Ewing and Struik 1992, Vreugdenhil and Sergeeva 1999
	IAA	stimulates	Marschner et al. 1984, Romanov et al. 2000
	jasmonates	stimulate	Kolomiets et al. 2001, Sarkar et al. 2006
	Ck	enhance sink capacity of growing tubers	Obata-Sasamoto and Suzuki 1979, Macháčková et al. 1997

In the last few years, changes in the expression of hundreds of genes and proteins occurring during tuber appearance and development were demonstrated (Bachem et al. 2000, Kloosterman et al. 2005, Agraval et al. 2008). The relationship between hormone action and gene expression is extensively analyzed. However, despite the large amount of new data, these studies have not yet led to unambiguous conclusions about cause-and-effect relationships between phytohormone effects on tuberization and differential gene expression.

In general, the investigation of the regulatory hormonal network in the course of tuber formation provides perspectives to influence this process by affecting potato hormonal status via various approaches including the directed modification of the genome.

Acknowledgement

This work was supported by the Presidium of Russian Academy of Science (program *Molecular and Cellular Biology*) by the Centre for Biosystems Genomics (CBSG) The Netherlands, and by the Russian Foundation for Basic Research (project no. 10.04.00638).

References

Abdala, G., G. Castro, O. Miersch and D. Pearce. 2002. Changes in jasmonate and gibberellin levels during development of potato plants (*Solanum tuberosum*). Plant Growth Regul. 36: 121–126.

Abdullah, Z.U.N. and R. Ahmad. 1980. Effect of abscisic acid and gibberellic acid on tuberization and some chemical constituents of potato *Solanum tuberosum* L. Plant Cell Physiol. 21: 1343–1346.

Agrawal, L., S. Chakraborty, D.K. Jaiswal, S. Gupta, A. Datta and N. Chakraborty. 2008. Comparative proteomics of tuber induction, development and maturation reveal the complexity of tuberization process in potato (*Solanum tuberosum* L.). J. Proteome Res. 7: 3803–3817.

Aksenova, N.P., T.N. Konstantinova, S.A. Golyanovskaya, I.A. Gukasyan, C. Gatz and G.A. Romanov. 2002. Tuber formation and growth of *in vitro* cultivated transgenic potato plants overproducing phytochrome B. Russ. J. Plant Physiol. 49: 478–483.

Aksenova, N.P., T.N. Konstantinova, S.A. Golyanovskaya, J. Kossmann, L. Willmitzer and G.A. Romanov. 2000. Transformed potato plants as a model for studying the hormonal and carbohydrate regulation of tuberization. Russ. J. Plant Physiol. 47: 370–379.

Aksenova, N.P., T.N. Konstantinova, V.N. Lozhnikova, S.A. Golyanovskaya, I.A. Gukasyan, C. Gatz and G.A. Romanov. 2005. Photoperiodic and hormonal control of tuberization in potato plants transformed with the *PHYB* gene from *Arabidopsis.* Russ. J. Plant Physiol. 52: 623–628.

Aksenova, N.P., T.N. Konstantinova, V.N. Lozhnikova, S.A. Golyanovskaya and L.I. Sergeeva. 2009. Interaction between day length and phytohormones in the control of potato tuberization in the *in vitro* culture. Russ. J. Plant Physiol. 56: 454–461.

Amador, V., E. Monte, J.L. García-Martinez and S. Prat. 2001. Gibberellins signal nuclear import of PHOR1, a photoperiod responsive protein with homology to *Drosophila armadillo*. Cell 106: 343–354.

Appeldoorn, N.J.G., L.I. Sergeeva, D. Vreugdenhil, L.H.W. van der Plas and R.G.F. Visser. 2002. *In situ* analysis of enzymes involved in sucrose to hexose-phosphate conversion during stolon-to-tuber transition of potato. Physiol. Plant. 115: 303–310.

Aukerman, M.J. and H. Sakai. 2003. Regulation of flowering time and floral organ identity by a microRNA and its APETALA-like target genes. Plant Cell 15: 2730–2741.

Bachem, C.W.B., R. van der Hoeven, J. Lucker, R. Oomen, E. Casarini, E. Jacobsen and R.G.F. Visser. 2000. Functional genomic analysis of potato tuber life-cycle. Potato Res. 43: 297–312.

Bachem, C.W.B., B. Horvath, L. Trindade, M. Claassens, E. Davelaar, W. Jordi and R.G.F. Visser. 2001. A potato tuber-expressed mRNA with homology to steroid dehydrogenases affects gibberellin levels and plant development. Plant J. 25: 595–604.

Banerjee, A.K., M. Chatterjee, Y. Yu, S.G. Suh, W.A. Miller and D.J. Hannapel. 2006. Dynamics of a mobile RNA of potato involved in a long distance signaling pathway. Plant Cell 18: 3443–3457.

Batutis, E.J. and E.E. Ewing. 1982. Far-red reversal of red light effect during long night induction of potato (*Solanum tuberosum* L.) tuberization. Plant Physiol. 69: 672–674.

Borzenkova, R.A. and M.P. Borovkova. 2003. Developmental patterns of phytohormone content in the cortex and pith of potato tubers as related to their growth and starch content. Russ. J. Plant Physiol. 50: 119–124.

Borzenkova, R.A., E.A. Sobyanina, A.A. Pozdeeva and M.Yu.Yashkov. 1998. Effect of phytohormones on starch-synthesizing capacity in growing potato tubers. Russ. J. Plant Physiol. 45: 472–480.

Bou-Torrent, J., J.F. Martínez-García, J.L. García-Martínez and S. Prat. 2011. Gibberellin A_1 metabolism contributes to the control of photoperiod-mediated tuberization in potato. PLoS ONE 6: 9: e24458.

Brenner, W.G., G.A. Romanov, J. Kollmer, L. Burkle and T. Schmülling. 2005. Immediate-early and delayed cytokinin response genes of *Arabidopsis thaliana* identified by genome-wide expression profiling reveal novel cytokinin-sensitive processes and suggest cytokinin action through transcriptional cascades. Plant J. 44: 314–333.

Carrera, E., S.D. Jackson and S. Prat. 1999. Feedback control and diurnal regulation of gibberellin 20-oxidase transcript levels in potato. Plant Physiol. 119: 765–773.

Carrera, E., J. Bou, J.L. García-Martínez and S. Prat. 2000. Changes in GA 20-oxidase gene expression strongly affect stem length, tuber induction and tuber yield of potato plants. Plant J. 22: 247–256.

Castro, G., G. Abdala, C. Aguero and R. Tizio. 2000. Interaction between jasmonic and gibberellic acids on *in vitro* tuberization of potato plants. Potato Res. 43: 83–88.

Chailakhyan, M.Kh. 1984. Photoperiodic and hormonal control of tuber formation in plants (in Russian). Nauka, Moscow.

Chailakhyan, M.Kh., L.I. Yanina, A.G. Devedjan and G.N. Lotova. 1981. Photoperiodism and tuberization in grafts of tobacco on potato plants. Dokl. Akad. Nauk SSSR 257: 1276–1280.

Chen, H., A.K. Banerjee and D.J. Hannapel. 2004. The tandem complex of BEL and KNOX partners is required for transcriptional repression of *ga20ox1*. Plant J. 38: 276–284.

Chen, H., F.M. Rosin, S. Prat and D.J. Hannapel. 2003. Interacting transcription factors from the three-amino acid loop extension superclass regulate tuber formation. Plant Physiol. 132: 1391–1404.

Chincinska, I.A., J. Liesche, U. Krugel, J. Michalska, P. Geigenberger, B. Grimm and C. Kühn. 2008. Sucrose transporter StSUT4 from potato affects flowering, tuberization, and shade avoidance response. Plant Physiol. 146: 515–528.

Claassens, M.M.J. and D. Vreugdenhil. 2000. Is dormancy breaking of potato tubers the reverse of tuber initiation? Potato Res. 43: 347–369.

Debast, S., A. Nunes-Nesi, M.R. Hajirezaei, J. Hofmann, U. Sonnewald, A.R. Fernie and F. Börnke. 2011. Altering trehalose-6-phosphate content in transgenic potato tubers affects tuber growth and alters responsiveness to hormones during sprouting. Plant Physiol. 156: 1754–1771.

Drobyazina, P.E. and E.E. Khavkin. 2006. A structural homolog of CONSTANS in potato. Russ. J. Plant Physiol. 53: 698–701.

Drobyazina, P.E. and E.E. Khavkin. 2011. The structure of two CONSTANS-LIKE genes in potato and its wild relatives. Gene. 471: 37–44.

Ewing, E.E. 1995. The role of hormones in potato (*Solanum tuberosum* L.) tuberization. *In:* P.J. Davies (ed.). Plant Hormones. Physiology, Biochemistry and Molecular Biology. Kluwer, Dordrecht, pp. 698–724.

Ewing, E.E. and P.C. Struik. 1992. Tuber formation in potato: induction, initiation and growth. Horticult. Reviews 14: 89–198.

Faivre-Rampant, O., L. Cardle, D. Marshall, R. Viola and M.A. Taylor. 2004. Changes in gene expression during meristem activation processes in *Solanum tuberosum* with a focus on the regulation of an auxin response factor gene. J. Exp. Bot. 55: 613–622.

Fischer, L., H. Lipavska, J.F. Hausman and Z. Opatrny. 2008. Morphological and molecular characterization of a spontaneously tuberizing potato mutant: an insight into the regulatory mechanisms of tuber induction. BMC Plant Biol. 8: 117.

Fleishon, S., E. Shani, N. Ori and D. Weiss. 2011. Negative reciprocal interactions between gibberellin and cytokinin in tomato. New Phytol. 190: 609–617.

Forsline, P.L. and A.R. Langille. 1976. An assessment of the modifying effect of kinetin on *in vitro* tuberization of induced and non-induced tissues of *Solanum tuberosum* L. Can. J. Bot. 54: 2513–2516.

Francis, D. 2007. The plant cell cycle—15 years on. New Phytol. 174: 261–278.

Fujino, K., Y. Koda and Y. Kikuta. 1995. Reorientation of cortical microtubules in the sub-apical region during tuberization in single-node stem segments of potato in culture. Plant Cell Physiol. 36: 891–895.

González-Schain, N.D., M. Díaz-Mendoza, M. Zurczak and P. Suárez-López. 2012. Potato CONSTANS is involved in photoperiodic tuberization in a graft-transmissible manner. The Plant J. 70: 678–690.

Gregory, L.E. 1956. Some factors for tuberization in potato. Am. J. Bot. 41: 281–288.

Guivarch, A., J. Rembur, M. Goetz, T. Roitsch, M. Noin, T. Schmülling and D. Chriqui. 2002. Local expression of the *ipt* gene in transgenic tobacco (*Nicotiana tabacum* L., cv. SR1) axillary buds establishes a role for cytokinins in tuberization and sink formation. J. Exp. Bot. 53: 621–629.

Gukasyan, I.A., S.A. Golyanovskaya, E.V. Grishunina, T.N. Konstantinova, N.P. Aksenova and G.A. Romanov. 2005. Effect of *rol* transgenes, IAA, and kinetin on starch content and the size of starch granules in tubers of *in vitro* potato plants. Russ. J. Plant Physiol. 52: 809–813.

Hannapel, D.J. 2007. Signalling in induction of tuber formation. *In:* D. Vreugdenhil (ed.). Potato Biology and Biotechnology Elsevier, Amsterdam, pp. 237–256.

Hannapel, D.J. 2010. A model system of development regulated by the long-distance transport of mRNA. J. Integr. Plant Biol. 52: 40–52.

Hannapel, D.J., J.C. Miller, Jr. and W.D. Park. 1985. Regulation of potato tuber protein accumulation by gibberellic acid. Plant Physiol. 78: 700–703.

Hedden, P. and A.L. Phillips. 2000. Gibberellin metabolism: new insight revealed by genes. Trends Plant Sci. 5: 523–530.

Helder, H., O. Miersch, D. Vreugdenhil and G. Sembdner. 1993. Occurrence of hydroxylated jasmonic acid in leaflets of *Solanum demissum* plants grown under long- and short-day conditions. Physiol. Plant 88: 647–653.

Hussey, G. and N.J. Stacey. 1984. Factors affecting the formation of *in vitro* tubers of potato (*Solanum tuberosum* L.). Ann. Bot. 53: 565–578.

Jackson, S.D. 1999. Multiple signaling pathways control tuber induction in potato. Plant Physiol. 119: 1–8.

Jackson, S.D. 2009. Plant responses to photoperiod. New Phytol. 181: 517–531.

Jackson, S.D., A. Heyer, J. Dietze and S. Prat. 1996. Phytochrome B mediates the photoperiodic control of tuber formation in potato. Plant J. 9: 159–168.

Jackson, S.D., P.E. James, E. Carrera, S. Prat and B. Thomas. 2000. Regulation of transcript levels of a potato gibberellin 20-oxidase gene by light and phytochrome B. Plant Physiol. 124: 423–430.

Jackson, S. and S. Prat. 1996. Control of tuberization in potato by gibberellins and phytochrome B. Physiol. Plant 98: 407–412.

Jackson, S.D. and L. Willmitzer. 1994. Jasmonic acid spraying does not induce tuberization in short-day-requiring potato species kept in non-inducing conditions. Planta 194: 155–159.

Kloosterman, B., C. Navarro, G. Bijsterbosch, T. Lange, S. Prat, R.G.F. Visser and C.W.B. Bachem. 2007. StGA2ox1 is induced prior to stolon swelling and controls GA levels during potato tuber development. Plant J. 52: 362–373.

Kloosterman, B., O. Vorst, R.D. Hall, R.G.F. Visser and C.W. Bachem. 2005. Tuber on a chip: differential gene expression during potato tuber development. Plant Biotechnol. J. 3: 505–519.

Koda, Y. 1997. Possible involvement of jasmonates in various morphogenic events. Physiol. Plant. 100: 639–646.

Kolachevskaya, O.O., V.V. Alekseeva, S.N. Lomin, L.I. Sergeeva, Y.I. Buryanov and G.A. Romanov. 2012. Manipulation of endogenous hormone status and tuber formation in potato plants. *In*: 3rd International Symposium "Intracellular Signaling and Bioactive Molecules Design" (Abstract book), Lviv, Ukraine, p. 25.

Kolomiets, M.V., D.J. Hannapel, H. Chen, M. Tymeson and R.J. Gladon. 2001. Lipoxygenase is involved in the control of potato tuber development. Plant Cell 13: 613–626.

Krauss, A. 1985. Interaction of nitrogen nutrition, phytohormones and tuberization. *In:* P.H. Li (ed.). Potato Physiology. Acad. Press INC, Orlando, Florida, pp. 209–230.

Kumar, D. and P.F. Wareing. 1972. Factors controlling stolon development in the potato plant. New Phytol. 71: 639–648.

Kumar, D. and P.F. Wareing. 1974. Studies on tuberization of *Solanum andigena*. II. Growth, hormones and tuber induction. New Phytol. 73: 833–840.

Lemeza, O.V., Ya.O. Zubo and V.V. Kusnetsov. 2010. Regulation of lipoxygenase gene expression in potato mini-tubers by phytohormones. Russ. J. Plant Physiol. 57: 715–719.

Macháčková, I., T.N. Konstantinova, L.I. Sergeeva, V.N. Lozhnikova, S.A. Golyanovskaya, N.D. Dudko, J. Eder and N.P. Aksenova. 1998. Photoperiodic control of growth, development and phytohormone balance in *Solanum tuberosum*. Physiol. Plant. 102: 272–278.

Macháčková, I., L. Sergeeva, M. Ondrej, O. Zaltsman, T. Konstantinova, J. Eder, J. Ovesna, S. Golyanovskaya, V. Rakitin and N. Aksenova. 1997. Growth pattern, tuber formation and hormonal balance in *in vitro* potato plants carrying *ipt* gene. Plant Growth Regul. 21: 27–36.

Marschner, H., B. Sattelmacher and F. Bangerth. 1984. Growth rate of potato tubers and endogenous contents of indolylacetic acid and abscisic acid. Physiol. Plant. 60: 16–20.

Martin, A., H. Adam, M. Díaz-Mendoza, M. Zurczak, N.D. González-Schain and P. Suárez-Lopez. 2009. Graff-transmissible induction of potato tuberization by the microRNA miR172. Development 136: 2873–2881.

Martínez-García, J.F., J.L. García-Martínez, J. Bou and S. Prat. 2002a. The interaction of gibberellins and photoperiod in the control of potato tuberization. Plant Growth Regul. 20: 377–386.

Martínez-García, J.F., A. Virgós-Soler and S. Prat. 2002b. Control of photoperiod regulated tuberization in potato by the *Arabidopsis* flowering time gene CONSTANS. Proc. Natl. Acad. Sci. USA 99: 15211–15216.

Matsuki, T., H. Tazaki, T. Fujimori and T. Hogetsu. 1992. The influences of jasmonic acid methyl ester on microtubules in potato cells and formation of potato tubers. Biosci. Biotechnol. Biochem. 56: 1329–1333.

Mauk, C.S. and A.R. Langille. 1978. Physiology of tuberization in *Solanum tuberosum* L. *cis*-Zeatin riboside in potato plant: its identification and changes in endogenous levels as influenced by temperature and photoperiod. Plant Physiol. 62: 438–442.

McGarry, R.C. and B.G. Ayre. 2012. Manipulating plant architecture with member of the CETS gene family. Plant Sci. 188–189: 71–81.

McGrady, J.J., P.C. Struik and E.E Ewing. 1986. Effects of exogenous application of cytokinin on the development of potato (*Solanum tuberosum* L.) cuttings. Potato Res. 29: 191–205.

Menzel, C.M. 1985. Tuberization in potato at high temperatures: interaction between temperature and irradiance. Ann. Bot. 55: 35–39.

Mingo-Castel, A.M., R.E. Young and O.E. Smith. 1976. Kinetin-induced tuberization of potato *in vitro*: on the mode of action of kinetin. Plant Cell Physiol. 17: 557–570.

Mokronosov, A.T. 1990. Tuberization and sink–source relations in potato. *In:* M.Kh. Chailakhyan and A.T. Mokronosov (eds). Regulation of Potato Growth and Development (in Russian). Nauka, Moscow, pp. 6–12.

Mokronosov, A.T. and T.N. Lundina. 1959. The role of dark and light daytime in potato photoperiodic response. Dokl. Akad. Nauk SSSR 127: 924–928.

Nam, K.-H., F. Kong, H. Matsuura, K. Takahashi, K. Nabeta and T. Yoshihara. 2008. Temperature regulates tuber-inducing lipoxygenase-derived metabolites in potato (*Solanum tuberosum*). J. Plant Physiol. 165: 233–238.

Nam, K.-H., C. Minami, F. Kong, H. Matsuura, K. Takahashi and T. Yoshihara. 2005. Relation between environmental factors and the LOX activities upon potato tuber formation and flower bud formation in morning glory. Plant Growth Regul. 46: 253–260.

Navarro, C., J.A. Abelenda, E. Cruz-Oró, C.A. Cuéllar, S. Tamaki, J. Silva, K. Shimamoto and S. Prat. 2011. Control of flowering and storage organ formation in potato by FLOWERING LOCUS T. Nature 478: 119–122.

Nookaraju, A., S.K. Pandey, C.P. Upadhyaya, J.J. Heung, H.S. Kim, S.C. Chun, D.H. Kim and S.W. Park. 2012. Role of Ca^{2+}-mediated signaling in potato tuberization: an overview. Bot. Studies. 53: 177–189.

Obata-Sasamoto, H. and H. Suzuki. 1979. Activities of enzymes relating to starch synthesis and endogenous levels of growth regulators in potato stolon tips during tuberization. Physiol. Plant. 45: 320–324.

País, S.M., M.N. García, M.T. Téllez-Iñón and D.A. Capiati. 2010. Protein phosphatases type 2A mediate tuberization signaling in *Solanum tuberosum* L. leaves. Planta 232: 37–49.

Palmer, C.E. and W.G. Barker. 1972. Changes in enzyme activity during elongation and tuberization of stolons of *Solanum tuberosum* L. cultured *in vitro*. Plant Cell Physiol. 13: 681–688.

Palmer, C.E. and W.G. Barker. 1973. Influence of ethylene and kinetin on tuberization and enzyme activity in *Solanum tuberosum* stolons cultured *in vitro*. Ann. Bot. 37: 85–95.

Palmer, C.E. and O.E. Smith. 1970. Effect of kinetin on tuber formation on isolated stolons of *Solanum tuberosum* L. cultured *in vitro*. Plant Cell Physiol. 11: 303–314.

Pelacho, A.M. and A.M. Mingo-Castel. 1991. Jasmonic acid induces tuberization of potato stolons cultured *in vitro*. Plant Physiol. 97: 1253–1255.

Ponnu, J., V. Wahl and M. Schmid. 2011. Trehalose-6-phosphate: connecting plant metabolism and development. Frontiers in Plant Sci. 2: 70.

Prat, S. 2004. Hormonal and daylength control of potato tuberization. *In:* P.J. Davies (ed.). Plant Hormones. Biosynthesis, Signal Transduction, Action! Kluwer, Dordrecht. pp. 538–560.

Puzina, T.I., I.G. Kirillova and N.I. Yakushkina. 2000. Indoleacetic acid dynamics in potato organs during ontogeny and its role in the regulation of tuber development. Izv. Akad. Nauk, Ser. Biol. 2: 170–177.

Quarrie, S.A. 1982. *Droopy*: a wilty mutant of potato deficient in abscisic acid. Plant Cell Environ. 5: 23–26.

Raspor, M., V. Motyka, E. Žižková, P.I. Dobrev, A. Trávníčková, S. Zdravković-Korać, A. Simonović, S. Ninković and I.C. Dragićević. 2012. Cytokinin profiles of *AtCKX2*-overexpressing potato plants and the impact of altered cytokinin homeostasis on tuberization *in vitro*. J. Plant Growth Regul. 31: 460–470.

Razumov, V.I. 1931. Localization of photoperiodic stimulation. Tr. Prikl. Botanik., Genetik. i Selekt. 27: 249–280.

Rodríguez-Falcón, M., J. Bou and S. Prat. 2006. Seasonal control of tuberization in potato: conserved elements with the flowering response. Annu. Rev. Plant Biol. 57: 151–180.

Romanov, G.A. 2009. How do cytokinins affect the cell? Russ. J. Plant Physiol. 56: 269–290.

Romanov, G.A., N.P. Aksenova, T.N. Konstantinova, S.A. Golyanovskaya, J. Kossmann and L. Willmitzer. 2000. Effect of indole-3-acetic acid and kinetin on tuberization parameters of different cultivars and transgenic lines of potato *in vitro*. Plant Growth Regul. 32: 245–251.

Ron'zhina, E.S. and A.T. Mokronosov. 1994. Source-sink relations and the role of cytokinins in the regulation of transport and partitioning of organic substances in plants. Russ. J. Plant Physiol. 41: 396–406.

Rosin, F.M., J.K. Hart, H.T. Horner, P.J. Davies and D.J. Hannapel. 2003. Overexpression of a *Knotted*-like homeobox gene of potato alters vegetative development by decreasing gibberellin accumulation. Plant Physiol. 132: 106–117.

Roumeliotis, E., R.G. Visser and C.W. Bachem. 2012a. A crosstalk of auxin and GA during tuber development. Plant Signaling and Behavior 7: 1360–1363.

Roumeliotis, E., B. Kloosterman, M. Oortwijn, W. Kohlen , H.J. Bouwmeester, R.G. Visser and C.W. Bachem. 2012b. The effects of auxin and strigolactones on tuber initiation and stolon architecture in potato. J. Exp. Bot. 63: 4539–4548.

Rutitzky, M., H.O. Ghiglione, J.A. Curá, J.J. Casal and M.J. Yanovsky. 2009. Comparative genomic analysis of light regulated transcripts in the Solanaceae. BMC Genomics 10: 60.

Sarkar, D. 2008. The signal transduction pathways controlling in planta tuberization in potato: an emerging synthesis. Plant Cell Rep. 27: 1–8.

Sarkar, D. 2010. Photoperiodic inhibition of potato tuberization: an update. Plant Growth Regul. 62: 117–125.

Sarkar, D., S.K. Pandey and S. Sharma. 2006. Cytokinins antagonize the jasmonates action on the regulation of potato (*Solanum tuberosum* L.) tuber formation *in vitro*. Plant Cell Tiss. Organ Cult. 87: 285–295.

Shalit, A., A. Rozman, A. Goldshmidt, J.P. Alvarez, J.L. Bowman, Y. Eshed and E. Lifschitz. 2009. The flowering hormone florigen functions as a general systemic regulator of growth and termination. Proc. Natl. Acad. Sci. USA 106: 8392–8397.

Shugaev, A.G. and S.V. Sokolova. 2001. The changes in the pathways of mitochondrial oxidation at the initial period of tuber development. Russ. J. Plant Physiol. 48: 45–50.

Simko, I. 1994. Sucrose application causes hormonal changes associated with potato tuberization. J. Plant Growth Regul. 13: 73–77.

Struik, P.C. 2007a. Responces of the potato plant to temperature. *In:* D. Vreugdenhil (ed.). Potato Biology and Biotechnology. Advances and perspectives. Elsevier, Amsterdam, pp. 367–394.

Struik, P.C. 2007b. Above-ground and below-ground plant development. *In:* D. Vreugdenhil (ed.). Potato Biology and Biotechnology. Advances and Perspectives. Elsevier. Amsterdam, pp. 367–394.

Thiele, A., M. Herold, J. Lenk, P.H. Quail and C. Gatz. 1999. Heterologous expression of Arabidopsis phytochrome B in transgenic potato influences photosynthetic performance and tuber development. Plant Physiol. 120: 73–81.

Viola, R., A.G. Roberts, S. Haupt, S. Gazzani, R.D. Hancock, N. Martiroli, G.C. Machray and K.J. Oparka. 2001. Tuberization in potato involves a switch from apoplastic to symplastic phloem unloading. Plant Cell 13: 385–398.

Van den Berg, J.H., P.J. Davies, E.E. Ewing and A. Halinska. 1995a. Metabolism of gibberellin A_{12} and A_{12}-aldehyde and the identification of endogenous gibberellins in potato (*Solanum tuberosum* ssp. *andigena*) shoots. J. Plant Physiol. 146: 459–466.

Van den Berg, J.H., I. Simko, P.J. Davies, E.E. Ewing and A. Halinska. 1995b. Morphology and [14C]gibberellin A_{12} metabolism in wild type and dwarf *Solanum tuberosum* ssp. *andigena* grown under long and short photoperiods. J. Plant Physiol. 146: 467–473.

Vanstraelen, M. and E. Benková. 2012. Hormonal interactions in the regulation of plant development. Annu. Rev. Cell Dev. Biol. 28: 463–87.

Vreugdenhil, D. and W. van Dijk. 1989. Effects of ethylene on the tuberization of potato (*Solanum tuberosum*) cuttings. Plant Growth Regul. 8: 31–39.

Vreugdenhil, D. and L.I. Sergeeva. 1999. Gibberellins and tuberization in potato. Potato Res. 42: 471–481.

Vreugdenhil, D. and P.C. Struik. 1989. An integrated view of the hormonal regulation of tuber formation in potato (*Solanum tuberosum*). Physiol. Plant. 75: 525–531.

Vreugdenhil, D., X. Xu, C.S. Jung, A.A.M. van Lammeren and E.E. Ewing. 1999. Initial anatomical changes associated with tuber formation in single node potato (*Solanum tuberosum* L.) cuttings: a re-evaluation. Ann. Bot. 84: 675–680.

Wareing, P.F. and A.M.V. Jennings. 1980. The hormonal control of tuberization in potato. *In:* F. Skoog (ed.). Plant Growth Substances. Springer-Verlag, New York, pp. 293–300.

Xu, X., A.A.M. van Lammeren, E. Vermeer and D. Vreugdenhil. 1998a. The role of gibberellin, abscisic acid and sucrose in the regulation of potato tuber formation *in vitro*. Plant Physiol. 117: 575–584.

Xu, X., D. Vreugdenhil and A.A.M. van Lammeren. 1998b. Division and cell enlargement during potato tuber formation. J. Exp. Bot. 49: 573–582.

Yoshihara, T., E.-L.A. Omer, H. Koshino, S. Sakamura, Y. Kikuta and Y. Koda. 1989. Structure of a tuber-inducing stimulus from potato leaves (*Solanum tuberosum* L.). Agric. Biol. Chem. 53: 2835–2837.

Zeevaart, J.A.D. 2008. Leaf-produced floral signals. Curr. Opin. Plant Biol. 11: 541–547.

2

Bioreactor Technologies for Mass Propagation of Potato: Future Prospects

Rokka Veli-Matti,[1] *Kämäräinen-Karppinen Terttu,*[2] *Virtanen Elina*[3] and *Pirttilä Anna Maria*[4,*]

ABSTRACT

Vegetatively propagated crop plants, such as potato, are ideal material to develop efficient commercial applications for cloning by *in vitro* techniques. To enable profitable production of plants true to type, extensive proliferation rates along with large-scale culture systems are needed. The traditional *in vitro* propagation techniques are however laborious and time consuming, and therefore the cloning process should be as automated as possible, keeping in mind the starting material, proliferation method, and further handling. Because cultivated potato (*Solanum tuberosum* L.) has the capacity to produce microtubers *in vitro*, it is an exploitable trait for application in the seed potato production industry. Microtubers have already been applied for mass production of planting materials in various types of semisolid media and in

[1] MTT Agrifood Research Finland, Biotechnology and Food Research, Myllytie 1, FI-31600 Jokioinen, Finland.
[2] Pohjoisen Kantayhtiö, Leppiojantie 11, FI-91800 Tyrnävä, Finland.
[3] MTT Agrifood Research Finland, Biotechnology and Food Research, P.O. Box 413, FI-90014 University of Oulu, Finland.
[4] Department of Biology, University of Oulu, P.O. Box 3000, FI-90014 Oulu, Finland.
* Corresponding author: Am.pirttila@oulu.fi

liquid cultures including different temporary immersion techniques, such as ebb-and-flow technique in glass fermentors, Rita® and twin-flask systems, nutrient mist bioreactors, and also in plastic vessel cultivations, such as the Liquid Lab Rocker™ technique. Currently, the liquid cultures are considered superior compared to applications using semi-solid medium, because the liquid culture medium can be replaced simultaneously for a number of explants instead of transferring single plantlets one by one to a fresh medium. Furthermore, bioreactors can easily be applied on liquid culture systems for large-scale propagation of potato microtubers. Bioreactors enable further process automation, quality control systems and synchronous material development, which pave the way for efficient, profitable production processes for cultivated potato. Bioreactors combined with techniques for inducing somatic embryogenesis may offer a promising tool for mass propagation systems in the future. Regardless of the high potential of somatic embryogenesis, only a few systems have so far been used for commercial propagation of plants with economical value. In the future, improved protocols and maintenance of embryogenic status in cultures could be achieved by combining information from gene expression studies with selection of highly embryogenic tissues and medium supplementations.

Keywords: seed potato, bioreactor, microtuber, tissue culture

Introduction

The cultivated potato (*Solanum tuberosum* ssp. *tuberosum* L.) is a tuber-forming crop plant that ranks as the world's fourth most important food crop after maize, wheat and rice. Potato, together with other tuberous crop plants, plays an important role in developing countries, because their carbohydrate yield per unit of land is higher than in other major crops across a wide range of ecological zones. Potato belongs to the genus *Solanum* with at least 1,000 other species, including tomato and eggplant. As a crop plant, potato is vegetatively multiplied through tubers, because the botanical seeds (true potato seeds, TPS) cannot be routinely used by commercial seed potato industry. The tubers grown from TPS are less uniform, and for production of high-quality tubers, the commercial multiplication must be done through seed tubers, which are genetically true to type. However, vegetatively propagated plant materials are vulnerable to plant diseases, such as ring rot, blackleg, viruses and other pathogens. Therefore, the initiative materials for cloning need to be completely disease-free, confirmed by appropriate analyses, and thereby providing suitable materials for certified seed potato production (Dean 1994).

For production of seed potato, nuclear stock materials typically originating from *in vitro*-grown plantlets (pre-basic tissue culture materials)

are used. The *in vitro* production of pre-basic material is convenient, as the plantlets can be manipulated easily to eradicate any latent pathogens present in the tissues. The most serious pathogens in potato are viruses, as some of them can be latent and even undetectable in the seed tubers. In potato, infection with *Potato virus Y* (PVY) may ultimately result even in 10–100% loss of yield.

Thermotherapy, where, for example, meristem cultures are kept in high temperatures for long periods, is one of the most used methods to obtain virus-free plant material for further multiplication (Espinoza et al. 1986). Thermotherapy is based on the inhibition of the virus replication at conditions where plant cell division and meristem growth occurs. For potato, the maximum of three weeks at high temperatures can be applied, but longer periods of months can also be used at slightly elevated growth temperatures (Mellor and Stace-Smith 1987). Alternatively, chemotherapy, i.e., the use of antivirals is also suitable for eradication of the viruses from potato tissue. The most typically used antiviral is ribavirin (virazole, l-β-D-ribofuranosyl-1H-l, 2,4-triazole-3-carboxamide), which is effective in concentrations of 5–20 mg/l on cultured axillary bud tips (Wambugu et al. 1986, Conrad 1991). A third method for obtaining virus-free plant material is cryotherapy of shoot tips (Brison et al. 1997). Viruses such as *Potato leafroll virus* (PLRV) and *Potato virus Y* (PVY) have been eliminated from potato by cryotherapy, and this method combined with thermotherapy has proved an efficient tool for virus eradication in potato (Wang et al. 2006, 2008). The treated plant material, tested negative for pathogens, can further be used as nuclear stocks for certified seed potato production.

The pre-basic seed material typically originates from *in vitro*-grown nuclear stocks, started from disease-free single-node cuttings. The *in vitro*-grown stock materials can be used directly for minituber production, or multiplied by microtuber formation to produce basic seeds. The propagation through single-node cuttings is still most popular among seed potato producers, even though the method is laborious, time consuming and more expensive compared to microtuber production. This method is also dependent on the season, because a required number of *in vitro* plants need to be propagated in a given time for minituber production. Compared to *in vitro*-grown explants, microtubers have many advantages. For example, the small size significantly facilitates the handling, storage and transport compared to *in vitro* plants. The production of microtubers is not season-dependent, because they can be produced throughout the year, and the greatest advantage of microtubers is the short production period for obtaining the first generation of pre-basic seeds (Ranalli 2007).

Furthermore, microtubers are a convenient choice for long-term preservation of germplasm and for distribution and international exchange of disease-free material (Thieme 1992, Mix-Wagner 1999), and for selection of material in plant breeding programmes (Ranalli 2007). The first protocol for *in vitro* production of microtubers was described by Barker (1953). Since then, these methods have been developed further, and probably the most common techniques for microtuber production are based on the culture of single-node explants (e.g., Lê 1999) on solidified culture media with high sucrose content (Otroshy et al. 2009).

Bioreactors for Production of Seed Potato Material

Whereas traditional agar-based tissue culture techniques via single-node cuttings are typically used for microtuber production, *in vitro* hydroponic cultivations present an inexpensive alternative. The highest costs in tissue culture media result from the use of agar, which makes up to two thirds of the price of a liter of cultivation medium. Liquid medium is used in hydroponic cultures in a form of a very thin layer, instead of submerging the plant material. This technique utilizes a large number of relatively small growth vessels that are manually assembled from different inexpensive parts. For example, hydroponic culture in simple plastic vessels with several cultivation layers was tested for potato microtuber production by Nhut et al. (2006). In that experiment, cheap paper disks or cotton plugs were connected in each layer by a paper bridge submerged in nutrient solution in the plastic vessel, and stacked to separate layers with the aid of plastic support. The method was fast, as in 30 days one microtuber per explant was obtained (Nhut et al. 2006).

Mass propagation of microtubers in bioreactors is typically done in liquid cultures using various systems (Piao et al. 2003, Nhut et al. 2006). Several types of semisolid media in different cultivation vessels have been tested (Donnelly et al. 2003), but liquid cultures have several advantages over solid or semisolid media. The use of liquid cultures results in increased shoot length, higher number of internodes, and the capacity to form microtubers from all plant nodes (Jiménez et al. 1999, Jova et al. 2005). Growth in liquid culture results in better growth because a larger area of the explant is in contact with the medium. However, there are disadvantages in using liquid cultures, such as the high cost of the traditional bioreactor equipment, and hyperhydricity resulting from poor gas exchange, as the explants are continuously immersed in medium (Eibl and Eibl 2008, Etienne and Berthouly 2002).

A number of techniques have been tested for producing microtubers in bioreactors: plain liquid cultures (Estrada et al. 1986), liquid cultures together with temporary immersion techniques (Piao et al. 2003), including ebb-and-flow in glass fermentors (Akita and Takayama 1994), Rita® system (Teisson and Alvard 1999) and twin flask system (Jiménez et al. 1999). Also nutrient mist bioreactor (Hao et al. 1998) and plastic bag cultivations

(Grigoriadou and Leventakis 2003) have been used for testing production of microtubers. These techniques have proved useful, but there are shortcomings. For example, their application can require machinery of relatively high cost, or the use of air pressed through filters, which can cause contamination of cultures.

Potato microtubers have been produced in bioreactors with large vessels connected to the temporary immersion system (T.I.S.), which is often performed in a twin-flask system (Jiménez et al. 1999, Etienne and Berthouly 2002). In T.I.S., the idea is that the liquid medium is in contact with the plant material only for a short period of time to avoid hyperhydricity, lack of oxygen, and other problems that are typically associated with liquid cultures. When undifferentiated single cells of microbes, animal or plant cells are cultured, the gas exchange can be maintained by mechanical stirring, which does not work on whole plant cultivations. Temporary immersion systems consist of two vessels, one containing the plant material, connected to another, which contains the nutrition medium. Usually a pump is used to transfer the liquid to the vessel containing the plant material. The timing and duration of the immersion can be controlled. Also the twin-flask system is operated in this way, and then the flasks can be large in size. However, with increased size of the vessels, the material becomes vulnerable to contamination. Several prototypes of simple T.I.S. have been reported (Jiménez et al. 1999, Piao et al. 2003, Ebadi et al. 2007, Perez et al. 2008). Jiménez et al. (1999) obtained 2.6 microtubers per explant, from a total of 390, in a 10-liter bioreactor. In a study by Peréz et al. (2008), 229 tubers from 80 explants were harvested in 600 ml of medium in a 5-liter bioreactor, whereas Piao et al. (2003) reported 80 microtubers from 50 explants in 1.5 l MS medium in a 10-liter T.I.S. bioreactor. Another factor to consider in a bioreactor microtuber cultivation study is the size of the tubers. Piao et al. (2003) considered this, and reported that most of the microtubers obtained in the experiment were larger than 1.1 g by size, which they considered sufficient for direct planting in the field and to sustain storage (Piao et al. 2003) but there are different opinions about the adequate size (Alix et al. 2001, Kämäräinen-Karppinen et al. 2010).

Besides the large glass vessels, T.I.S. can be done also in small inexpensive plastic containers. The commercial T.I.S. product (Rita®) utilizes transparent polysulfone vessels. In the Rita® system, multiple bioreactors can be connected together and operated on the same line where pressured air from one pump can maintain several, even tens of reactors. The bioreactors of the Rita® system are small, a volume of 1 liter, and therefore inexpensive. However, the cleaning, autoclaving and starting the cultures (cutting and placing the plant material into the bioreactor) are laborious and time consuming. Teisson and Alvard (1999) tested the Rita® system and obtained 48–90 microtubers per vessel within a period of ten weeks. They gained a maximum of three microtubers per single-node explant, depending on the potato variety. The majority of the microtubers were >0.5 g of size, but they were not tested on the field (Teisson and Alvard 1999). Another T.I.S. system utilizing small plastic vessels called Plantima® is commercially available, but although tested for tuber formation in *Dioscorea* sp. (Yan et al. 2011), it has to our knowledge not yet been utilized for production of potato microtubers.

Jar fermentor cultures resemble the T.I.S. techniques, because they utilize two vessels, where the plant material is kept in one and the nutrient solution remains mainly in another separate vessel. Akita and Ohta (1994) have earlier used the jar fermentors with an airlift. In such fermentors, the aeration of the cultures was obtained by air spray from the bottom of the jar. In these bioreactors utilizing such semi-continuous surface control technique, 500 to 960 tubers were obtained from 100 explants in 10-liter fermentors containing 6 liters of nutrient solution (Akita and Ohta 1994). Another type of bioreactor used for production of potato microtubers is the nutrient mist bioreactors. In the vessels of these bioreactors the nutrient solution is sprayed onto the plants in a form of mist. However, these bioreactors were producing only one microtuber per exlant (Hao et al. 1998).

Various kinds of other bioreactors have been developed to lower the production costs per produced microtuber. These low-cost bioreactors typically have plastic vessels that lack ventilation generated by pumping. Yu et al. (2000) developed an inexpensive bioreactor that utilized rotating plastic vessels with filtered lids. In these bioreactors, microtubers of >1 g in 40% of the 50 explants were obtained, whereas the total of 100 microtubers was obtained in 10 weeks of culture in the tuberization media (Yu et al. 2000). Akita and Ohta (1998) had earlier reported a similar rotating system, where 100 microtubers per vessel were obtained with 200 ml of tuber-induction medium.

Another, new system using the thin-layer cultivation method is LiquidLab™ (L-L; Adelberg 2005). In LiquidLab™, a tilting movement is produced by the machinery, resulting in the liquid inside the cultivation vessels to move from side to side. The illumination and the timing of the tilting can easily be controlled. In the LiquidLab™ system, the vessels have porous patches attached to each side, for aeration of the cultures. Up to 200 vessels can be operated simultaneously by the machinery. The system has been tested for production of microtubers, where 75 microtubers from 50 explants with 200 ml cultivation medium per vessel were produced (Fig. 1, Kämäräinen-Karppinen et al. 2010).

A

B

Fig. 1. Microtubers grown in the Liquid Lab™ (L-L) system. (A) Seedlings with developing microtubers in the L-L™ vessel after 14 weeks. (B) Harvested microtubers on a Petri dish. Scale bar = 1 cm. Photo credit Terttu Kämäräinen-Karppinen.

Color image of this figure appears in the color plate section at the end of the book.

Low-cost Bioreactors for Developing Countries

The choice of technology

There are two main choices for technology for production of microtubers. One approach is to use high technology with relatively large apparatus, and another way is to use multiple relative low cost vessels each capable of producing only small numbers of microtubers. In countries where plant liquid culture methods have developed to a high level (for example in Japan), the techniques have been used for potato microtuber production

(Akita and Ohta 1994). For example, the mist bioreactor and ebb and flow bioreactor use expensive stainless steel tanks equipped with sensors to measure oxygen and pumps to control the amount of nutrients available for the plants. These systems require highly educated personnel to run efficiently, and the initial costs to build such a system are high. However, the results, i.e., quantity of products, obtained with these systems are promising. Therefore, when the initial costs as well as well-paid staff to run the system can be afforded, they are an option to consider.

Less expensive, so-called 'self-made' systems, such as *in vitro* hydroponics or the use of rotating vessels have also been developed for the production of potato microtubers. The advantage of such systems is low cost of material and personnel, because the operation is easier and straightforward, without the need for high education to run or repair. The less sophisticated systems are the choice for entities with lack of money and infrastructure, such as developing countries. For example, companies or enterprises, which are not capable or willing to invest a large amount of capital for the development and introduction of high technology systems for the production of microtubers, the inexpensive 'self-made' systems offer an affordable option.

There are few examples of commercially available systems for potato microtuber production that operate a large number of relatively small vessels independently from each other. The advantages of these systems are the relatively simple use and the possibility to grow several potato varieties simultaneously. This is an advantage compared to high-technology based large units, where only one variety of potato is grown at a time. The risk of contamination associated with liquid cultures is also lower in a system like L-L™ or Rita®, where the production of microtubers is restricted to small units of several hundreds, compared to high-cost, large single units. However, the operation of a system using compressed air to aerate cultures sterilized with filters, like in the Rita® or Plantima® systems, enhances the risk of airborne contaminants entering the cultures. Cultivation in systems such as L-L™ that operate without compressed air in several small-sized cultivation units significantly lowers the risk of losing the entire microtuber yields due to contaminations. As a result, there are no strict requirements for the L-L™ system to be operated in sterile laboratory premises. Furthermore, the compact size of the Liquid-Lab Rocker machinery and the fact that it contains a lighting system in itself makes the system a profitable choice for production of microtubers of potato in developing, as well as developed, countries.

The Choice of Material for Propagation

Seed potato production systems are fairly expensive, and therefore the cost of potato seed material is high. Generally, costs of propagation of potato tissues, tuber storage and transportation can be too high for many countries or seed potato-producing companies to be profitable. The fact that microtubers can be produced in potato has several benefits for application in seed potato production. Microtubers can offer a practical and inexpensive pre-basic seed material, and disease-free starting material for production of seed tubers in countries with a high burden of pathogens. For example, basic seed potatoes have been imported with high cost from The Netherlands to Pakistan, because the production of high-quality seed tubers was problematic (Rafique et al. 2004). Also for countries where climatic conditions allow only one outdoor growing season of *in vitro* plantlets and all labour is required for the multiplication, microtubers can be produced and stored throughout the year, saving costs in labor and machinery.

However, comparison of the cost-effectiveness between using *in vitro* cuttings or microtubers as the starting material for seed potato production via various cultivation systems is difficult. For example, there is no need for storing tubers (or microtubers) in areas where crops can be produced throughout the year. Furthermore, the concept of applicable microtuber, i.e., usable on the field for potato production, is not stable. The applicability of a microtuber varies between potato cultivars as well as growing site and depends on the growth conditions. The adequate size of a microtuber considered sufficient for further application varies (Alix et al. 2001, Kämäräinen-Karppinen et al. 2010). Larger tubers can survive the storage better and produce a higher number of minitubers. Most of the techniques and methods tested for production of microtubers result in tubers with median size less than 0.5 g (Yu et al. 2000).

Future Prospects

Microtubers are at the moment the most promising material for the initiative production stages in seed potato industry. However, in the future other applications can become more profitable, as the tools in biotechnology develop further. For example, somatic embryogenesis offers promising material for mass propagation systems utilizing bioreactors. Somatic embryos originate from a single plant cell or a cell cluster that can be also capsulated and used as artificial seeds (Ducos et al. 2008). Somatic embryos have a bipolar structure, and they generate both a shoot apex and a discrete radical end, which is the most distinctive anatomic characteristic of an embryo (Haensch 2004). Somatic embryos can further be regenerated into plants, which are genetically true-to-type with their nuclear plants (Tomar and Gupta 1988). Direct embryogenesis is common in species with sporophytic polyembryony, such as nucellar cells in *Citrus*, and in protocorm-producing plants, such as protocorm-like bodies of orchids (George and Debergh 2008).

Somatic embryogenesis is generally induced by high auxin level on the medium in explants. Exogenous auxin is often needed for embryogenic cultures since it improves morphological and physiological development of embryos (Carman 1990). Auxin plays an important role both in the induction of embryos and in the subsequent morphogenesis of embryo development (Zimmerman 1993). However, the further development and germination of somatic embryos has often been induced by transferring the cultures onto medium with high cytokinin levels, i.e., thidiazuron or BA. Similar combinations of growth regulators have previously been used for induction of proembryogenic callus and indirect somatic embryogenesis in potato (Seabrook and Douglass 2001, Sharma and Millam 2004) and lettuce (*Lactuca sativa* L.) (Seabrook and Douglass 2003). Secondary somatic embryogenesis has also great potential for large-scale propagation, being important for woody plants with long generation cycles, and because of their low capacity for somatic embryogenesis (Li et al. 2002).

Regardless of the high potential of somatic embryogenesis, only a few systems have been used in commercial propagation. The reason could be that non-embryogenic callus is frequently formed during the culture, especially when indirect embryogenesis systems are used (Hoenemann et al. 2010). The most suitable and economic method for propagating plants of particular species could well change with time (George and Debergh 2008). Improvements will come from a better understanding of factors controlling morphogenesis and genetic stability *in vitro* (George and Debergh 2008). Further knowledge of fundamental regulatory mechanisms of morphogenesis can be transferred to development of regeneration protocols (Nuti Ronchi 1990). For example, improved protocols and maintenance of embryogenic status in cultures could be achieved by combining information from gene expression studies with selection of highly embryogenic tissues and medium supplementations (de Oliveira Santos et al. 2005, Hoenemann et al. 2010). In *Coffea* spp., somatic embryogenesis induced in bioreactors is even now a powerful method compared to other vegetative mass propagation techniques, as reported from a pilot scale experiment (Ducos et al. 2007). Significant progress has been made towards commercial implementations via fully automatized systems (Ducos et al. 2007) with a production capacity of 2.5–3 million embryos from *Coffea canephora* per year (Ducos et al. 2008).

Formation and utilization of accessory embryos can be a method in micropropagation, which will be effectively developed in the future (George and Debergh 2008). Somatic embryos probably are the best material in tissue culture, for plant propagation to become economically deployed on extensively planted field crops. However until now, only the first attempts for successful formation of somatic embryos in potato have been made (Seabrook and Douglass 2001, Sharma and Millam 2004).

References

Adelberg, J. 2005. Efficiency in thin-film liquid system for *Hosta* micropropagation. Plant Cell, Tissue and Organ Culture 81: 359–368.

Akita, M. and Y. Ohta. 1998. A simple method for mass propagation of potato (*Solanum tuberosum* L.) using a bioreactor without forced aeration. Plant Cell Reports 18: 284–287.

Akita, M. and S. Takayama. 1994. Stimulation of potato (*Solanum tuberosum* L.) tuberization by semicontinous liquid medium surface level control. Plant Cell Reports 13: 184–187.

Alix, M.J., S. Savvides, J. Blake, R. Herrmann and R. Hornung. 2001. Effects of illumination source, culture, ventilation, and sucrose on potato (*Solanum tuberosum*) microtuber production under short days. Annals of Applied Biology 139: 175–187.

Barker, W.G. 1953. A method for *in vitro* culturing of potato tubers. Science 118: 384–385.

Brison, M., M.T. de Boucaud, A. Pierronnet and F. Dosba. 1997. Effect of cryopreservation on the sanitary state of a cv. *Prunus* rootstock experimentally contaminated with plum pox potyvirus. Plant Science 123: 189–196.

Carman, J.G. 1990. Embryonic cells in plant tissue cultures: occurrence and behavior. *In vitro* Cellular and Developmental Biology 26: 746–753.

Conrad, P.L. 1991. Potato virus S-free plants obtained using antiviral compounds and nodal segment culture of potato. American Potato Journal 68: 507–513.

de Oliveira Santos, M., E. Romano, K.S.C. Yotoko, M.L.P. Tinoco, B.B.A. Dias and F.J.L. Aragão. 2005. Characterisation of the cacao *somatic embryogenesis receptor-like kinase* (*SERK*) gene expressed during somatic embryogenesis. Plant Science 168: 723–729.

Dean, B.B. 1994. *Managing the potato production system* (edition). Binghamton, NY: Food Products Press.

Donnelly, D.J., W.K. Coleman and S.E. Coleman. 2003. Potato microtuber production and performance: a review. American Journal of Potato Research 80: 103–115.

Ducos, J.-P., C. Lambot and V. Pétiard. 2007. Bioreactors for coffee mass propagation by somatic embryogenesis. International Journal of Plant Development Biology 1: 1–12.

Ducos, J.P., B. Terrier, D. Courtois and V. Pétiard. 2008. Improvement of plastic disposable bioreactors for plant science needs. Phytochemistry Reviews 7: 607–613.

Ebadi, M., A. Iranbakhsh and G.B. Khaniki. 2007. Shoot micropropagation and microtuberization in potato (*Solanum tuberosum* L.) by the semi-continuous bioreactor. Pakistan Journal of Biological Sciences 10: 861–867.

Eibl, R. and D. Eibl. 2008. Design of bioreactors suitable for plant cell and tissue cultures. Phytochemistry Reviews 7: 593–598.

El-Amin, S.M., J.P.T. Valkonen, K. Bremer and E. Pehu. 1994. Elimination of viruses and hypersensitivity to potato virus Y (PVY^{o}) in an important sudanese potato stock (Zalinge). American Potato Journal 71: 267–272.

Espinoza, N.O., R. Estrada, D. Silva-Rodriguez, P. Tovar, R. Lizarraga and J.H. Dodds. 1986. The Potato: A model crop plant for tissue culture. Outlook on Agriculture 15: 21–26.

Estrada, R., P. Tovar and J.H. Dodds. 1986. Induction of *in vitro* tubers in a broad range of potato genotypes. Plant Cell, Tissue and Organ Culture 7: 3–10.

Etienne, H. and M. Berthouly. 2002. Temporary immersion systems in plant micropropagation. Plant Cell, Tissue and Organ Culture 69: 215–231.

Etienne-Barry, D., B. Bertrand, N. Vasquez and H. Etienne. 1999. Direct sowing of *Coffea arabica* somatic embryos mass-produced in a bioreactor and regeneration of plants. Plant Cell Reports 19: 111–117.

George, E.F. and P.C. Debergh. 2008. Micropropagation: Uses and methods. *In:* E.F. George, M.A. Hall and G.-J. De Klerk (eds.). Plant Propagation by Tissue Culture, 3rd edn., Vol. 1, The Background, Springer, Dordrecht, pp. 29–64.

Golmirzaie, A.M., P. Malagamba and N. Pallais. 1994. Breeding potatoes based on true potato seed propagation. *In:* J.E. Bradshaw and G.R. Mackay (eds.). *Potato Genetics*. Cambridge: University Press, pp. 499–513.

Grigoriadou, K. and N. Leventakis. 2003. Comparative use of commercial bioreactor system and conventional micropropagation for the production of potato microtubers and grape myrtle (*Lagerstroemia indica*) microshoots. *In:* A.S. Economou and P.E. Read (eds.). Proceedings of International Symposium on Acclimatization and Establishment of Micropropagated Plants. Acta Horticulturae 616: 369–371.

Hao, Z., F. Ouyang, Y. Geng, X. Deng, Z. Hu and Z. Chen. 1998. Propagation of potato tubers in a nutrient mist bioreactor. Biotechnology Techniques 12: 641–644.

Haensch, K.-T. 2004. Thidiazuron-induced morphogenetic response in petiole cultures of *Pelargonium x hortorum* and *Pelargonium x domesticum* and its histological analysis. Plant Cell Reports 23: 211–217.

Hoenemann, C., S. Richardt, K. Krüger, A.D. Zimmer, A. Hohe and S.A. Rensing. 2010. Large impact of the apoplast on somatic embryogenesis in *Cyclamen persicum* offers possibilities for improved development control *in vitro*. BMC Plant Biology 10: 77.

Jiménez, E., N. Pérez, M. de Feria, R. Barbón, A. Capote, M. Chávez, E. Quiala and J.C. Pérez. 1999. Improved production of potato microtubers using a temporary immersion system. Plant Cell, Tissue and Organ Culture 59: 19–23.

Jova, M.C., R.G. Kosky, M.B. Pérez, A.S. Pino, V.M. Vega, J.L. Torres, A.R. Cabrera, M.G. García and J. De La Caridad Ventura. 2005. Production of yam microtubers using a temporary immersion system. Plant Cell, Tissue and Organ Culture 83: 103–107.

Kämäräinen-Karppinen, T., E. Virtanen, V.-M. Rokka and A.M. Pirttilä. 2010. Novel bioreactor technology for mass propagation of potato microtubers. Plant Cell, Tissue and Organ Culture 101: 245–249.

Khuri, S. and J. Moorby. 1996. Nodal segments or microtubers as explants for *in vitro* microtuber production of potato. Plant Cell, Tissue and Organ Culture 45: 215–222.

Lê, C.L. 1999. *In vitro* microtuberization: an evaluation of culture conditions for the production of virus-free seed potatoes. Potato Research 42: 489–498.

Li, X., S.F. Krasnyanski and S.S. Korban. 2002. Somatic embryogenesis, secondary somatic embryogenesis, and shoot organogenesis in *Rosa*. Journal of Plant Physiology 159: 313–319.

Lommen, W.J.M. and P.C. Struik. 1994. Field performance of potato minitubers with different fresh weights and conventional seed tubers: Multiplication factors and progeny yield variation. Potato Research 37: 301–313.

Mellor, F.C. and R. Stace-Smith. 1987. Virus-free potatoes through meristem culture. *In:* Y.P.S. Bajaj (ed.). Biotechnology in Agriculture and Forestry 3. Potato. Springer-Verlag, Berlin, pp. 30–39.

Mix-Wagner, G. 1999. The conservation of potato cultivars. Potato Research 42: 427–436.

Murashige, T. and F. Skoog. 1962. A revised medium for rapid growth and bioassays with tobacco tissue cultures. Physiologia Plantarum 15: 473–497.

Nhut, D.T., N.H. Nguyen and D.T.T. Thuy. 2006. A novel *in vitro* hydroponic culture system for potato (*Solanum tuberosum* L.) microtuber production. Scientia Horticulturae 110: 230–234.

Norris, D.O. 1954. Development of virus-free stock of Green Mountain potato by treatment with malachite green. Australian Journal of Agricultural Research 5: 658–663.

Nuti Ronchi, V. 1990. Organogenesis and embryogenesis: Genetic and physiological approaches. Acta Horticulturae 280: 1–10.

Otroshy, M., F. Nazarian and P.C. Struik. 2009. Effects of temperature fluctuation during *in vitro* phase of *in vitro* microtuber production in different cultivars of potato (*Solanum tuberosum* L.). Plant Cell, Tissue and Organ Culture 98: 213–218.

Pennazio, S. and P. Redolfi. 1973. Factors affecting the culture of potato meristem tip. Potato Research 16: 20–29.

Peréz, N.M., D.C. Restrepo, J.D. García and D.R. Giraldo. 2008. *In vitro* tuberization of potato (*Solanum tuberosum* L.) variety Diacol Capiro in temporary immersion bioreactors and grow evaluation in field. Ciencia 16: 288–295.

Piao, X.C., D. Chakrabarty, E.J. Hahn and K.Y. Paek. 2003. A simple method for mass production of potato microtubers using a bioreactor system. Current Science 84: 1129–1132.

Rafique, T., M.F. Jaskani, H. Raza and M. Abbas. 2004. Microtuber induction in potato. International Journal of Agriculture and Biology 6: 375–377.

Ranalli, P. 2007. The canon of potato science: 24. Microtubers. Potato Research 50: 301–304.

Seabrook, J.E.A. and L.K. Douglass. 2001. Somatic embryogenesis on various potato tissues from a range of genotypes and ploidy levels. Plant Cell Reports 20: 175–182.

Seabrook, J.E.A. and L.K. Douglass. 2003. Somatic embryogenesis of lettuce from mature tissues. *In:* F.A. Hammerschlag and P. Saxena (eds.). XXVI International Horticultural Congress: Biotechnology in Horticultural Crop Improvement, Achievements, Opportunities and Limitations, Acta Horticulturae Vol. 625, International Society for Horticultural Sciences, Leuven, pp. 217–221.

Sharma, S.M. and S. Millam. 2004. Somatic embryogenesis in *Solanum tuberosum* L.: a histological examination of key developmental stages. Plant Cell Reports 23: 115–119.

Sharma, S.K., G.J. Bryan, M.O. Winfield and S. Millam. 2007. Stability of potato (*Solanum tuberosum* L.) plants regenerated via somatic embryos, axillary bud proliferated shoots, microtubers and true potato seeds: a comparative phenotypic, cytogenetic and molecular assessment. Planta 226: 1449–1458.

Stace-Smith, R. and F.C. Mellor. 1968. Eradication of potato virus X and S by thermotherapy and axillary bud culture. Phytopathology 58: 199–203.

Teisson, C. and D. Alvard. 1999. *In vitro* production of potato microtubers in liquid medium using temporary immersion. Potato Research 42: 499–504.

Thieme, R. 1992. An *in vitro* potato cultivar collection: microtuberization and storage of microtubers. FAO/IBPGR Plant Genetic Resources Newsletter 88/89: 17–19.

Tomar, U.K. and S.C. Gupta. 1988. Somatic embryogenesis and organogenesis in callus cultures of a tree legume—*Albizia richardiana* King. Plant Cell Reports 7: 70–73.

Wambugu, F.M., G.A. Secor and N.C. Gudmestad. 1986. Eradication of potato virus Y and S from potato by chemotherapy of cultured axillary bud tips. American Potato Journal 62: 667–672.

Wang, Q., Y. Liu, Y. Xie and M. You. 2006. Cryotherapy of potato shoot tips for efficient elimination of *Potato leafroll virus* (PLRV) and *Potato virus Y*. Potato Research 49: 119–129.

Wang, Q., W.J. Cuellar, M.-L. Rajamäki, Y. Hirata and J.P.T. Valkonen. 2008. Combined thermotherapy and cryotherapy for efficient virus eradication: relation of virus distribution, subcellular changes, cell survival and viral RNA degradation in shoot tips. Molecular Plant Pathology 9: 237–250.

Yan, H., L. Yang and Y. Li. 2011. Improved growth and quality of *Dioscorea fordii* Prain et Burk and *Dioscorea alata* plantlets using a temporary immersion system. African Journal of Biotechnology 10: 19444–19448.

Yu, W.-C., P.J. Joyce, D.C. Cameron and B.H. McCown. 2000. Sucrose utilization during potato microtuber growth in bioreactors. Plant Cell Reports 19: 407–413.

Zimmerman, J.L. 1993. Somatic embryogenesis: a model for early development in higher plants. Plant Cell 5: 1411–1423.

3

Genetic Diversity and Population Structure in Five Cultivated *Dioscorea* spp. (Dioscoreaceae) using Molecular Markers

Elizabeth Ann Veasey,[1,*] *Eduardo de Andrade Bressan,*[2] *Marcos Vinícius Bohrer Monteiro Siqueira,*[3] *Aline Borges,*[1] *Wellington Ferreira do Nascimento,*[1] *Thiago Fonseca Mezette,*[1] *Lidinalva de Resende Gomes da Silva,*[1] *Danielle Muniz da Silva*[1] and *Gabriel Dequigiovanni*[1]

ABSTRACT

Yams are considered an important staple food and are found throughout the tropics. In this chapter we provide the reader with an up-to-date scenery on genetic diversity studies held on five cultivated yam species (*D. alata, D. cayenensis, D. rotundata, D. trifida* and *D. bulbifera*) using molecular markers. Isozymes, RAPD (Random Amplified Polymorphic DNA), AFLP (Amplified Fragment Length Polymorphism), microsatellites, ISSR (Inter Simple Sequence Repeats)

Authors' affiliations given at the end of the chapter.

and some modifications of these markers have been applied to look at the genetic diversity of local varieties, cultivars and plant breeding materials, but also to build genetic maps, QTL studies (quantitative trait loci) linked to resistant genes and other subjects in *Dioscorea* spp. But one of the main focuses here was to highlight, through the use of molecular markers, the importance of traditional farming systems in the management and maintenance of genetic variability of yam species, in several countries, or in the domestication process of these species, contributing, therefore, for *in situ/on farm* conservation practices.

Introduction

Dioscorea genus belongs to the Dioscoreaceae family, presenting more than 600 species throughout the tropics. This genus had world wide dispersion at the end of the Cretaceous period, occurring soon after an evolution with different courses in the Old and New World, originating distinct species. The separation of the Asian and African species occurred later in the Miocene (Coursey 1967).

Ten species of *Dioscorea* are considered staple yams, belonging to five sections: Enantiophyllum (*D. alata* L., *D. cayenensis* Lam., *D. rotundata* Poir., *D. nummularia* Lam., *D. opposita* Thunb. and *D. transversa* R.Br.), Combilium (*D. esculenta* Burk.), Opsophyton (*D. bulbifera* L.), Macrogynodium (*D. trifida* L.) and Lasiophyton (*D. pentaphylla* L.). The different species of edible yams have been domesticated independently in America, Africa, Madagascar, South and Southeast Asia. The African originated species are *D. cayenensis*, *D. rotundata* and *D. bulbifera*. The Asian and Melanesian originated species are *D. alata, D. esculenta, D. bulbifera, D. nummularia* and *D. pentaphylla*. *D. opposita* was originated in Japan and China, *D. transversa* in Australia and Melanesia, while *D. trifida* was originated in America. An observation should be made for *D. bulbifera,* for being the only species of this genus distributed in the wild state in both Asia and Africa. *D. alata, D. cayenensis* and *D. rotundata* are by far the major cultivated species (Lebot 2009).

Yams are herbaceous plants, creepers (in general) when they find support, with subterranean tubers (air tubers in some species in the leaves axils), with the air stem presenting thorns or not; the leaves show heterophily (they are narrow and have a shape of a knife tip), with white, yellow or light green flowers (Montaldo 1991). Yam species present both sexual, through an outcrossing mating system, and vegetative propagation, with propagation carried out by seeds or/and tubers. They are mainly dioic species, although monoic plants have been observed in some cultivated species (Abraham and Nair 1990, Zoundjihekpon 1997). The inflorescences are in the form of axillary spikes, generally solitary, rarely occurring in pairs, with flowers of 4 to 6 mm of diameter. The yam fruit is a capsule with three loci, each one with two small and winged seeds. Rhizomes can be solitary or in groups. The tubers weight varies from 0.1 kg and in some *D. alata* varieties they can reach up to 5 kg (Hasan et al. 2008).

Polyploidy is a strong characteristic among *Dioscorea* species, which present different chromosome numbers even within species, varying from 2n = 18 to 2n = 140 chromosomes for the 10 most important cultivated species (Lebot 2009). The basic chromosome numbers are x = 9 and x = 10, the two numbers are present in America, Europe and Africa, while x = 10 is present in Asia and Oceania (Essad 1984). However, new findings suggest a chromosome base number of x = 20 for some species, such as *D. rotundata* (Scarcelli et al. 2005), *D. trifida* (Bousalem et al. 2006, 2010), and *D. alata* (Arnau et al. 2009). These findings are important to be considered in plant breeding programmes, genetic diversity studies and to elucidate the phylogenetic origins of these species (Arnau et al. 2009).

Yams are now cultivated on 5 million hectares in about 47 tropical and subtropical countries, but the statistics are still dispersed and irregular (IITA 2009; FAOSTAT 2012). About 94% of the world annual production, estimated at 48.3 million ton, is cultivated in Western and Central Africa, with Nigeria being the world largest producer (FAOSTAT 2012). Some yam species are cultivated not only for their nutritional value, but also for their medicinal properties such as the steroidal sapogenins, used in the production of cortisone and synthetic steroid hormones (Pedralli 2002a, Narula et al. 2007). A study was conducted to investigate the effects of yam (*D. alata*) ingestion on lipids, antioxidant status, and sex hormones in postmenopausal women, reaching the conclusion that, although the exact mechanism is not clear, replacing two thirds of the staple food with yam for 30 days improves the status of sex hormones, lipids and antioxidants, reducing the risk of breast cancer and cardiovascular diseases in postmenopausal women (Wu et al. 2005). Ethnobotanical studies held in local traditional communities of Enugu, Nigeria, revealed the great applicability of *D. alata* leaves in reducing fever and of *D. cayenensis* tubers to combat diarrhea (Aiyeloja and Bello 2006). As for the nutritional quality, the yam tubers are rich in carbohydrates, protein, vitamins (A, B1, B2, C e D) and minerals (Ca, Cu, Fe, K, Mg, Na, P, and Zn) (Baah et al. 2009). The starch is similar to that of corn, in taste, texture and color. The flour can be added to the wheat flour for making bread or can be used in different sweet or salty dishes (Abramo 1990).

Yams are an important subsistence crop associated with traditional agriculture systems worldwide (Bressan et al. 2005, Scarcelli et al. 2006a, Bustamante and Buitrago 2006, Tamiru et al. 2008, Martins and Oliveira 2009), known to maintain a wide genetic reservoir represented by several varieties bearing different folk names (Bressan et al. 2005, Tamiru et al. 2008, Veasey et al. 2010). Traditional farming systems include shifting cultivation, also known as slash-and-burn agriculture, where each field acts as their basic evolutionary unit. The species composition in this system is not totally random, but represents an association of species to minimize competition and maximize the use of resources, which are limited (Martins 1994, Martins and Oliveira 2009). Traditional farming systems, therefore, represent an important source for agrobiodiversity conservation (Martins 1994, Amorozo 2008), also known as *in situ/on farm* conservation (Jarvis et al. 2000). Substantial differences in the composition and diversity of plants as well as in the origin of materials are found between the different fields (Coomes and Ban 2004).

In Brazil, a survey made by Peroni and Martins (2000) on the Vale do Ribeira, on São Paulo State, in the domain of the Atlantic Forest, showed that the number of edible species utilized by traditional farmers is high when compared with studies carried out in other world regions. Peroni and Hanazaki (2002) found 29 yam varieties cultivated among traditional farmers in six communities that practice traditional agriculture in the Vale do Ribeira. A survey carried out by Bressan et al. (2005), in 24 local communities, showed that four *Dioscorea* species (*D. trifida*, *D. cayenensis*, *D. alata* and *D. bulbifera*) are grown in the same area, mainly for the farmers' own consumption. Further collection expeditions made in four regions in Brazil (south, southeast, northeast and central-west), in 36 municipalities and 66 local communities, found these four species, plus *D. rotundata* (which may be part of a *D. cayenensis/D. rotundata* species complex), being cultivated in areas of traditional agriculture, although a few were represented by commercial plantings as well (Veasey et al. 2010). Although the importance of yams may be registered in these studies, yams and other crops are still neglected and underutilized by researchers and inhabitants of large cities in Brazil (Siqueira 2011a).

Molecular markers are important tools in ecological, evolutionary, taxonomical, phylogenic and genetic diversity studies of plant sciences (Agarwal et al. 2008). Among the different molecular markers, the first to be established were the isozymes, which exhibit Mendelian inheritance, codominant expression, complete penetrance, and absence of pleiotropic and epistatic interactions (Weeden and Wendel 1989). Following the isozymes, the most used markers are the DNA based marker techniques such as RAPD (Random Amplified Polymorphic DNA), microsatellites or SSR (Simple Sequence Repeats), ISSR (Inter-Simple Sequence Repeats) and AFLP (Amplified Fragment Length Polymorphism), and even more advanced techniques, primarily derived from a combination of earlier basic techniques, such as retrotransposons, mitochondrial and chloroplast based microsatellites, SNPs, RAD sequencing, whole-genome resequencing, are revealing genetic variation through increased genome coverage and have been routinely used in genetic studies, for many different purposes (Agarwal et al. 2008, Allendorf et al. 2010). Microsatellites are among the most important molecular markers, and even with the advent of new technologies its use has not been affected due to their cost effectiveness and use in large-scale genotyping. Microsatellites are involved not only in genetic diversity, population genetics and evolutionary studies, but also in fundamental research like genome analysis, gene mapping and marker-assisted selection (Kalia et al. 2011). Microsatellites and chloroplast DNA have recently been used in a phylogeographic study for the Chilean endemic *D. biloba* (Phil.) Caddick & Wilkin which occurs in the Atacama Desert (Viruel et al. 2012). Chloroplast DNA as well as AFLP have also been used to study the genetic diversity in 53 accessions of *D. dumetorum* (Kunth) Pax from six countries in West and Central Africa (Sonibare et al. 2010). Scarcelli et al. (2011) recently developed a new set of 100 primer pairs of chroloplast DNA optimized for amplification in monocotyledons, and used to study population genetics and phylogeny of 13 species of monocotyledons, among which the *Dioscorea* species such as *D. abyssinica* Hochst. ex Kunth, *D. praehensilis* Benth., *D. rotundata, D. dumetorum* and *D. bulbifera.*

The following topics of this review will focus on genetic diversity studies using molecular markers on the most important staple yams (*D. alata, D. cayenensis/D. rotundata, D. trifida* and *D. bulbifera*) (Fig. 1). One of the main focuses here was to highlight the importance of traditional farming systems in the management and maintenance of genetic variability of yam species in several countries, or in the domestication process of these species, contributing, therefore, for *in situ/on farm* conservation practices.

D. alata

Greater yam (*Dioscorea alata* L.), also known as water yam, is the most widely distributed species of the genus in the humid and semi-humid tropics. It is an important food crop in Africa, Caribbean, and especially Melanesia where it has considerable social and cultural importance (Coursey 1976). The crop plays a vital role in local subsistence in the so-called 'yam belt' of West Africa, in Ethiopia (Tamiru et al. 2008), as well as in Nigeria, along with *D. cayenensis* and *D. rotundata* (Amusa et al. 2003). Nevertheless, its diversity has not been studied in detail in many areas of Africa. Tropical America, which is also considered one of the main centers of yam domestication, in the case of *D. trifida*, and diversity, deserves more studies. Although it is a commercialized crop in Brazil and was considered to be in the 40th place among 57 horticultural crops in order of commercial value (Monteiro 2002), its importance is also connected with traditional agriculture systems (Peroni and Hanazaki 2002, Bressan et al. 2005, Veasey et al. 2010, Siqueira and Veasey 2011, Siqueira et al. 2013), mostly as a subsistence crop.

Fig. 1. Adult plants of *Dioscorea cayenensis* (left and above), *D. trifida* (right and above), *D. alata* (left and below) and *D. bulbifera* (right and below) showing a few aerial tubers.

Color image of this figure appears in the color plate section at the end of the book.

The first studies on *D. alata* molecular characterization were based on isozymes (Twyford et al. 1990, Lebot et al. 1998, Bressan et al. 2011). Twyford et al. (1990) used isozymes for the identification of shoot culture materials of four yam species, detecting differences between four clones of *D. alata*, otherwise difficult to distinguish based on morphological traits alone. The genetic diversity of 269 cultivars of *D. alata* originated from the South Pacific, Asia, Africa, Caribbean and South America, studied by Lebot et al. (1998), revealed 66 isozyme phenotypes, or zymotypes, each uniquely characterized by the presence or absence of 27 electromorphs. The authors could not find any correlation between patterns of morphological variation, isozyme polymorphism and geographic origins.

Using six isozyme systems and scoring a total of 71 polymorphic bands, Bressan et al. (2011) assessed 16 *D. alata* local varieties from the Vale do Ribeira, comparing them with 19 commercial varieties collected in markets of six municipalities in São Paulo State, Brazil. A cluster analysis classified the accessions in six groups, showing a clear separation of the commercial varieties, all classified in group IV, from the local varieties, which were classified into five groups showing higher genetic diversity. Also, a molecular variance analysis (AMOVA) showed that most of the total isozymatic variation (54%) was distributed within households fields, compared to the variation between households within communities (20%) and between communities (26%). This data shows clearly the importance of traditional farmers in maintaining the genetic diversity of this species, which can be noticed also by a total of nine folk names given to *D. alata* varieties in the Vale do Ribeira (Bressan et al. 2011). Siqueira et al. (2013) identified 21 folk names in a survey made in four regions in Brazil: South, Southeast, Northeast and Central-West, registering different ways in which this crop is popularly called, such as *cará* (in the Southern and Central-West regions) and *inhame* (in the northeast region), which shows the higher diversity in this crop's nomenclature. The highest percentage of citation was for the name *inhame de São Tomé* (31%), followed by *cará roxo* (16%), *cará branco* (14%), and simply *cará* (13%). This general denomination, whether *cará* or *inhame*, is also valid for the other yam species in Brazil (Veasey et al. 2010).

Molecular characterization with RAPD markers were conducted by Asemota et al. (1996), Cheng and Liu (1996), Mignouna et al. (2002a) and Zannou et al. (2009). Asemota et al. (1996) used nine primers to characterize eleven cultivars of *D. alata*, *D. cayenensis*, *D. rotundata* and *D. trifida* from Jamaica, concluding that RAPD was a reliable marker to discriminate between all Jamaican cultivars investigated. Considerable variation among lines of *D. alata*, even between lines of the same region, was found by Cheng and Liu (1996) using 14 RAPD primers. Mignouna et al. (2002a) found two RAPD markers linked to anthracnose (*Colletotrichum gloeosporioides*) resistant *D. alata* genotypes among 34 breeding lines, indicating the potential of this marker for use in marker-assisted selection. Twenty cultivars of *D. alata*, as well as 70 cultivars of *D. cayenensis*/*D. rotundata*, collected from farmers throughout the transitional Guinea-Sudan zone of Benin, were assessed using 67 amplified bands from six RAPD primer combinations by Zannou et al. (2009). The AMOVA analysis detected that most of the variation (52.7%) was found among individuals within varietal groups. This study suggested that the Guinea Sudan zone of Benin is a very large gene-pool of yam varieties and also shows the significant contribution of farmers in Benin in the enrichment and the maintenance of the genetic diversity of these species.

AFLP markers were used to investigate the genetic diversity of 53 accessions of *D. alata* of diverse geographic regions in West and Central Africa (Benin, Ivory Coast, Ghana, Nigeria and Chad) and Puerto Rico (Egesi et al. 2006), where no correlation was found between genetic diversity and geographic origins, indicating that geography has not played a significant role in the differentiation of this species. A considerable intraspecific genetic variation was found in this study, with the genetic similarity coefficients ranging from 0.29 to 0.95.

The genetic relationship between *D. alata* and nine other *Dioscorea* species, including *D. abyssinica*, *D. bulbifera*, *D. cayenensis/D. rotundata*, *D. esculenta*, *D. nummularia*, *D. pentaphylla*, *D. persimilis* Prain. et Burk., *D. transversa* and *D. trifida* was assessed with AFLP markers (Malapa et al. 2005). This study revealed that Enantiophyllum section's species (*D. alata*, *D. abyssinica*, *D. cayenensis*, *D. nummularia*, *D. persimilis*, and *D. transversa*) are clearly distinguished from each other and are genetically distant from species from other sections. It also showed that, within the Enantiophyllum section, African species (*D. abyssinica*, *D. cayenensis/D. rotundata*) are distant from the southeast Asian-Oceanian species, *D. alata*, *D. nummularia*, *D. persimilis* and *D. transversa*. Another interesting observation found by Malapa et al. (2005) is that the non-distinction between Asian, African and Melanesian cultivars of *D. alata* strengthens the inferences based on other studies (Lebot et al. 1998, Egesi et al. 2006) indicating that clones of this species have been widely distributed.

AFLP markers were also employed to construct a genetic linkage map from an F1 cross from presumed heterozygous parents (Mignouna et al. 2002b). The 469 markers were mapped on 20 linkage groups with a total map length of 1,233 cM. QTL mapping revealed one AFLP marker associated with anthracnose resistance, explaining 10% of the total phenotypic variance. This map was the first linkage map reported for *D. alata*, providing a tool for further genetic analysis of traits of agronomic importance and for using marker-assisted selection in *D. alata* breeding programs. A more recent linkage map was developed for *D. alata* (Petro et al. 2011) generating 523 polymorphic markers from 26 AFLP primer combinations, resulting in a map with total length of 1,538 cM, including 20 linkage groups. QTL analysis for anthracnose resistance was also performed and the total amount of phenotypic variation for anthracnose resistance explained by all significant QTLs varied from 26.4 to 73.7%.

A total of 16 microsatellite primers were developed for *D. alata*, *D. abyssinica* and *D. praehensilis* (Tostain et al. 2006), using a microsatellite-enriched genomic library methodology, with 12 primers showing successful amplification for *D. alata*. A microsatellite-enriched genomic library methodology (Billotte et al. 1999) was also used to develop nine polymorphic primers for *D. alata*, from which six markers showed transferability for the species *D. bulbifera*, *D. cayenensis-D. rotundata* and *D. trifida* (Siqueira et al. 2011). The new SSR markers together with three primers developed by Tostain et al. (2006) were used to conduct a molecular analysis of 72 *D. alata* local accessions from four regions in Brazil (South, Southeast, Northeast and Central-West) and 17 commercial varieties (Siqueira 2011b). The study showed a high intraspecific diversity among local varieties from different regions in Brazil. However, population structuring between sampling regions was rather low, confirmed by an UPGMA cluster analysis, similar to the findings of Egesi et al. (2006) with AFLP. Only the accessions from the central-western region showed an apparent regional clustering.

Another study in Brazil aimed at fingerprinting 36 water yam accessions collected in markets from several municipalities in Brazil, including accessions from the Agronomic Institute (IAC) germplasm bank, introduced from Puerto Rico, Singapore and the Democratic Republic of Congo, using nine microsatellite loci (Siqueira et al. 2012). High polymorphism was found among loci verified by the high PIC values (0.57–0.77), and moderate genetic variability was shown among accessions, with a magnitude of the Jaccard's similarity coefficient ranging from 0.63 to 1.0. Results suggested the hypothesis of different origins for the varieties currently under cultivation in Brazil. Also, similar accessions acquired in different municipalities revealed the commercialization of the same accessions on different locations.

Six SSR markers were employed by Arnau et al. (2009) to study the segregation patterns from four different progenies produced from 2n = 40 types of *D. alata*, providing support to the hypothesis that *D. alata* accessions with 2n = 40 chromosomes would be diploid and not tetraploid as usually assumed. Obidiegwu et al. (2009a) used 13 microsatellite loci to evaluate 89 accessions from nine countries in Africa. Results showed highest gene diversity (0.678) for accessions from Nigeria, whereas those from Côte d'Ivoire showed the lowest diversity, with 0.596. The authors found no relationship between relatedness of the accessions and their geographical area of collection, similar to the studies cited above conducted in West and Central Africa (Egesi et al. 2006) with AFLP and in Brazil (Siqueira 2011b) using microsatellites.

Nine microsatellite loci were also used to map *D. alata* populations for anthracnose disease (Sartie and Asiedu 2011), cooking quality and oxidation. The authors underlined that SSR marker analysis of DNA from parents and six randomly selected progenies of each population indicated that parental crosses were successful and the progenies were true hybrids. Narina et al. (2011) developed a comprehensive annotated transcriptome data set in *D. alata* to enrich the EST information in public databases. The results aimed to help the identification of unique transcripts in each library for disease resistance.

D. cayenensis and *D. rotundata*

Among the cultivated yam species, *D. cayenensis* and *D. rotundata*, also known as Guinea yams, are the most important yams in West and Central Africa where they are indigenous (Mignouna et al. 2003), especially in the region known as 'yam belt', which extends from Cameroon up to the Ivory Coast, where this vegetatively propagated yam crop provides the main food source (Coursey 1967, 1976). It is the world's second most important yam species after *D. alata*, including some countries in tropical America (Montaldo 1991). African yam cultivars with yellow tuber color are known as *D. cayenensis*, while white-fleshed cultivars are distinguished as *D. rotundata*. But, in addition, a large number of intermediate cultivars cannot be clearly considered either one or the other (Martin and Rhodes 1978). There are 10 other morphological characteristics that can be used to differentiate both species (Martin and Rhodes 1978), but several authors still consider these species as part of a polyploid species complex (Ayensu and Coursey 1972, Hamon and Bakary 1990, Chaïr et al. 2005), taking into account the existence of thousands of morphotypes and, so far, no satisfying classification tool, either morphological or molecular (Lebot 2009).

Isozyme markers have been used to study the genetic diversity of *D. cayenensis/D. rotundata* species complex (Hamon and Toure 1990, Zoundjihekpon et al. 1994, Dansi et al. 2000a, Mignouna et al. 2002c, Bressan et al. 2013), as well as segregation patterns for isozymic loci in *D. rotundata* (Scarcelli et al. 2005). Results obtained by Hamon and Toure (1990) when evaluating 453 accessions of the *D. cayenensis/D. rotundata* complex collected in West Africa, using five isozyme systems, showed an adequate agreement between the isozymic polymorphism and morphological variation. Zoundjihekpon et al. (1994) assessed progenies obtained from crosses of tetraploid varieties of the complex *D. cayenensis/D. rotundata* in southern Ivory Coast, observing different segregation patterns for six enzymatic systems, giving important information on the genetics of the systems analyzed.

The band patterns obtained from seven isozyme systems led to the identification of 227 different cultivars, from a total of 467 accessions of the complex *D. cayenensis/D. rotundata* of Benin Republic (Dansi et al. 2000a). Another study allowed the identification of 28 genotypes within 45 accessions collected in 1996 in different localities of Cameroon with six enzymatic systems (Mignouna et al. 2002c). The clustering of accessions based on isozyme data was very close to the one obtained using morphological characters in this study.

Isozymes were also used to study the process of domestication of wild yams of the section Enantiophyllum by African farmers in Benin (Mignouna and Dansi 2003). The domestication process consists of bringing into cultivation selected individuals which go through intense vegetative multiplication and selection procedures (over a lengthy but variable period of time) that induce morphological and biochemical changes in the plant mainly at the tuber level. Individuals resulting from these manipulations were found to be either similar or identical to known local varieties or completely new based on both morphological and isozyme analysis. The authors discussed the potential of this domestication process in yam breeding, which appears to be a strategy that could be useful to breeders, while developing a methodology for yam participatory breeding.

In Brazil, isozyme markers were employed for the study of the genetic diversity of 21 local varieties of *D. cayenensis* collected in the Vale do Ribeira, São Paulo State, from traditional farming households and two accessions identified as *D. rotundata* originating from northeast Brazil, using seven isozyme systems (Bressan et al. 2013). A total of 65 polymorphic bands were found allowing the discrimination of 17 from the 21 local varieties analyzed, besides the two *D. rotundata* accessions. High genetic variability was detected, considering the wide amplitude of the Jaccard's similarity coefficient, varying from 0.2 to 1.0. A considerable exchange of accessions between the three municipalities (Iguape, Cananéia and Eldorado), where the plant collections were made, was observed, as the varieties were not classified according to their original municipality. The two *D. rotundata* accessions were classified in a separate cluster, showing some similarities with one *D. cayenensis* variety from Cananéia municipality (Bressan 2005). Three folk names were attributed to the set of 21 varieties in this area: '*cará de espinho*' (thorn yam), '*cará guaçu*' (Guaçu yam) and '*cará D'Angola*' (Angola's yam) (Bressan et al. 2005).

RAPD markers were used to assess 23 late-maturing cultivars of the species complex *D. cayenensis/D. rotundata* from the Benin Republic that could not be separated using isozyme markers (Dansi et al. 2000b). Using 12 RAPD primers, which yielded 63 amplified DNA bands, the great majority of the cultivars was separated with the combinations of polymorphic bands (75%) and duplicates and cultivar misclassifications were identified. In another study, two RAPD markers successfully identified genotypes resistant to *Yam Mosaic Virus* (YMV) among 12 *D. rotundata* varieties, showing the potential of this marker in marker assisted selection (Mignouna et al. 2002d). RAPD was also applied by Zannou et al. (2009) to access 70 cultivars of the *D. cayenensis/D. rotundata* complex collected from farmers throughout the transitional Guinea-Sudan zone of Benin. Results indicated a higher genetic diversity found for these species complex than for *D. alata* cultivars (20 cultivars). The study also showed the significant role of the Benin farmers in the enrichment and the maintenance of the genetic diversity of yam.

The debated issue of whether *D. cayenensis* and *D. rotundata* should be considered as two separate taxon or as a species complex was examined with both RAPD and microsatellite-primed PCR (MP-PCR) markers, as well as RAMPO (Random Amplified Microsatellite Polymorphism) technique (Ramser et al. 1997), and the results supported the concept that both are distinct species. This conclusion was confirmed by another study with RAPD and double stringency PCR (DS-PCR) (Mignouna et al. 2005) and in a study with AFLP, where *D. cayenensis* accessions where clearly separated from *D. rotundata* accessions in a cluster analysis, although the separation of the two species was relatively weak in the coordinate analysis (Tamiru et al. 2008). Obidiegwu et al. (2009b), using 15 microsatellite primer pairs, also concluded that *D. cayenensis* and *D. rotundata* are distinct species with intermediate hybrid forms. However, the same result was not found by Asemota et al. (1996) using RAPD, considering that *D. cayenensis* and *D. rotundata* cultivars, which shared 10.3% of all bands, were not grouped according to their species designation.

In a comparison of three molecular markers (RAPD, AFLP and SSR) for germplasm characterization of *D. rotundata*, AFLPs showed the highest efficiency in detecting polymorphism and revealed genetic relationships that most closely reflected morphological classification (Mignouna et al. 2003). AFLP markers were used to obtain a genetic linkage map of *D. rotundata*, constructed based on 341 co-dominantly scored AFLP markers segregating in a F1 intraspecific cross (Mignouna et al. 2002e). Three and one quantitative trait loci with effects of resistance to *Yam Mosaic Virus* (YMV) were identified on the maternal and paternal linkage maps, respectively.

AFLP markers were also used to study the domestication process of yams, as discussed above, as a 'traditional farmers' practice in Benin, West Africa (Scarcelli et al. 2006a). A total of 213 accessions, consisting of 32 predomesticated yams, 70 *D. cayenensis/D. rotundata*, 86 *D. abyssinica* and 25 *D. praehensilis* yams were analyzed, using 91 AFLP markers, resulting in the distinction of three groups of accessions, broadly corresponding to the above botanical species. Of the 32 predomesticated accessions, 16% were clustered with *D. praehensilis*, 37% with *D. abyssinica* and the remaining 47% with *D. cayenensis/D. rotundata*, which shows the use of wild plants by farmers in their domestication process. This process was further investigated using 11 SSR markers, demonstrating on yam tubers from a small village in northern Benin, that wild and cultivated hybrids are spontaneously formed (Scarcelli et al. 2006b). This study confirmed that farmers create new varieties with new genetic combinations via sexual reproduction of wild and cultivated yams.

Ten polymorphic SSR markers were used to assess 146 accessions of *D. rotundata* from Benin, showing an average of 8.4 alleles per locus and 0.57 mean heterozygosity (Tostain et al. 2007). An excess of heterozygotes found in this study may be the result of farmer's selection during the domestication process and also an accumulation of somatic mutations can be expected. According to the authors, in the study area, in Benin, yams have to survive very harsh conditions, with poor soils, floods and long periods of drought, and the strong selection carried out by farmers, considering that yams are vegetatively propagated plants, could favor heterozygote genotypes. A low, but significant, correlation between geographical and genetic distances showed that genetic diversity in this region was structured according to physiological characteristics, such as tuberization, reflecting an unbalanced distribution of early and late cultivars across regions, resulting directly from the cultivation practices and specific farming techniques applied (Tostain et al. 2007). A more recent study analyzing 219 accessions of Guinea yam germplasm from Benin, Congo, Côte d' Ivoire, Equatorial Guinea, Gabon, Ghana, Nigeria, Sierra Leone and Togo using 15 SSR loci, showed no relationship between relatedness of the accessions and their geographical area of origin (Obidiegwu et al. 2009b). Similar genetic parameters to the study of Tostain et al. (2007) was found, that is, an average of 8.1 alleles per locus and an observed mean heterozygosity value of 0.563, suggesting, according to Obidiegwu et al. (2009b), that spontaneous hybridization must have contributed to the ancestry of some of the accessions and improvement by farmers must have been far more often by selection of somatic mutants.

Using a microsatellite-enriched genomic library methodology (Billotte et al. 1999), eleven specific SSR primers were developed for *D. cayenensis*, showing high transferability to *D. rotundata* (Silva 2012). Nine of these primers were polymorphic, which together with an heterologous polymorphic SSR primer developed by Tostain et al. (2006), were used to quantify the genetic variability of 48 accessions (22 *D. cayenensis* and 26 *D. rotundata*) collected in three regions (South, Southeast and Northeast) in Brazil. The species identification of the accessions was carried out in accordance with a morphological analysis performed, considering mainly the tuber color (yellow for *D. cayenensis* and white for *D. rotundata*) and leaf shape (orbicular for *D. cayenensis* and ovate for *D. rotundata*) (Martin and Rhodes 1978). Accessions identified as *D. cayenensis* were mostly originated from the Southeast region, while those identified as *D. rotundata* were originated from the Northeast, which was an interesting finding in this study (Silva 2012). A separation of accessions originated in the Northeast from those originated in the Southeast region, with few exceptions, was observed in a cluster analysis with 10 microsatellite primers, while the accessions from the South were allocated in either group. Most of the genetic variability was found between regions and between species, and this variability was spatially structured, with high and significant correlation ($r = 0.64$; $P<0.01$) between genetic and geographical distances (Silva 2012).

D. trifida

Among the cultivated edible yams, *D. trifida* is the only American species domesticated by the Amerindians (Lebot 2009). Known as American yam and 'cush-cush' in some places, this species originated from the border areas between Brazil and Guyana, Suriname and French Guyana. It has been cultivated since the last century in the Antilles, in coastal areas on Brazil and by Indians in the Central-West area of the country (Pedralli 1991). Showing a Neotropical distribution, in Central and South America, *D. trifida* can be found in tropical rainforests, Brazilian savannas in the central plateau of Brazil, in the 'campos rupestres' (literally, rocky fields) of Serra do Espinhaço (Espinhaço Mountains) and seasonal forests adjacent to these fields (Pedralli 2002b).

Although tubers are of good quality, *D. trifida* has not been used for commercial purposes in Brazil, except in a few regions such as the municipality market of Manaus, in the Northern State of Amazonas, and in local markets in the Southeastern municipalities of Ubatuba, São Paulo State, and Joinville, Santa Catarina State (Veasey et al. 2010). In a survey made in the Vale do Ribeira, São Paulo State, *D. trifida* was the most cultivated species, along with *D. alata, D. cayenensis* and *D. bulbifera,* among the local traditional agriculturists (Bressan et al. 2005).

Despite the economic and geo-cultural importance of *D. trifida*, very little is known about its origin, phylogeny, diversity and genetics. However, a more recent study, representing the first direct proof of the Amazonian origin of this species (Bousalem et al. 2010), presented the putative progenitor of *D. trifida*, wild diploid ($2n = 40$) *D. trifida* accessions, while cultivated forms are autotetraploids ($2n = 80$).

The first study with isozyme markers in *D. trifida* was conducted in Brazil, with 25 local varieties from the municipalities of Iguape and Cananéia, in the Vale do Ribeira (Bressan 2005, Veasey et al. 2012). Seven enzymatic systems were used, revealing a total of 64 polymorphic bands, allowing the discrimination of 22 local varieties. Therefore, high genetic diversity was found among *D. trifida* varieties in this particular region. A wide range of folk names were given for the local varieties in this area (Bressan et al. 2005). A cluster analysis using the Jaccard's similarity coefficient and UPGMA method, classified the accessions in two great groups: the first group contained the local varieties from Iguape municipality and the second group those from Cananéia municipality. Therefore, a clear separation was observed between municipalities, which are about 80 km apart. Also, the

AMOVA analysis showed that most of the isoenzymatic variability occurred between households within communities (48%) (Bressan 2005, Veasey et al. 2012). The main conclusion from this study is that traditional agriculturists of the Vale do Ribeira maintain and manage high genetic variability in their agricultural system for the cush-cush crop.

Eight SSR primers have been developed for *D. trifida* (Hochu et al. 2006). The primers developed for *D. alata*, *D. abyssinica* and *D. praehensilis* (Tostain et al. 2006) were also tested for *D. trifida*, with seven primers presenting successful amplification for this species. Six from the nine polymorphic SSR primers developed for *D. alata* by Siqueira et al. (2011) also showed transferability for *D. trifida*. Bousalem et al. (2006) used a combination of the primers developed by Tostain et al. (2006) and Hochu et al. (2006), in a total of eight primers, to study the inheritance patterns in *D. trifida* by examining parental and progeny genotypes from controlled crosses. Their results provided genetic evidence supporting a tetrasomic behaviour of *D. trifida* genome.

From a set of 21 microsatellite primers tested, 12 primers (seven developed by Hochu et al. 2006 and five developed by Tostain et al. 2006) were selected to assess 21 local varieties of *D. trifida* originated from three Brazilian States: Iguape and Cananéia municipalities, in the Vale do Ribeira, Southern coast of São Paulo State; Ubatuba municipality on the Northern coast of São Paulo State; Brasnorte municipality in Mato Grosso State; and Manacapuru, Itapiranga and Manaus municipalities in Amazonia State (A.B., unpublished data). A total of 57 bands with 86% polymorphism were found, showing considerable genetic diversity that was also detected by the relatively wide amplitude of the Jaccard's similarity coefficient, varying from 0.46 to 1.0. Data from this study were analyzed using both a cluster analysis, with Jaccard's similarity coefficient, as binary data was used to discriminate the accessions, and UPGMA method, and a principal coordinate analysis (PCoA), both conducted with NTSYS-PC software (Rohlf 1992). In the cluster analysis two groups were well defined, confirmed by the bootstrap analysis (100%), classifying in group I the varieties from São Paulo State, together with the two varieties from Mato Grosso State, and in group II the varieties from Amazonas State (Fig. 2). This result showed that there seems to be little or no exchange of *D. trifida* from the Amazon and those from traditional farmers in the coastal areas in the State of São Paulo. Also, both the cluster analysis and the principal coordinate analysis (Fig. 3) showed that within the State of São Paulo, a separation was observed among varieties from the South (Vale do Ribeira) and Northern coastal areas (Ubatuba municipality), as well as a separation of the two accessions from Mato Grosso. The AMOVA analysis, comparing the distribution of genetic variability between the two groups and within groups, showed that most of the genetic diversity (65%) was distributed within groups (Table 1).

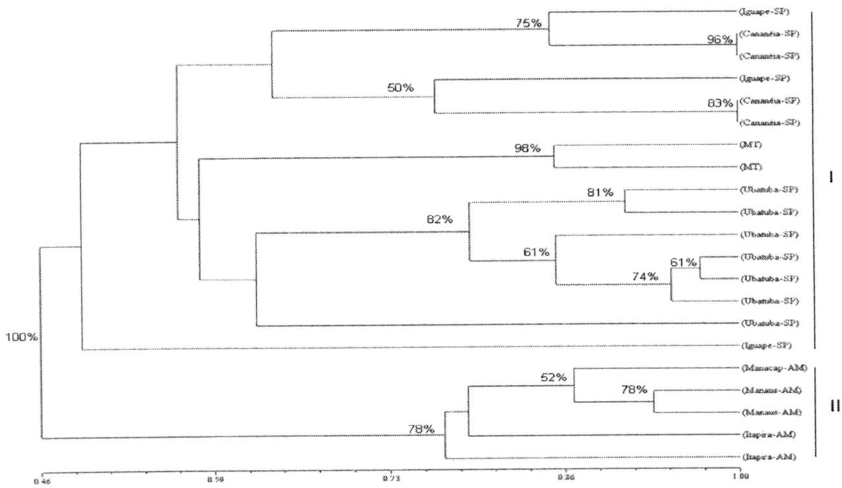

Fig. 2. Dendrogram using UPGMA method and Jaccard´s similarity coefficient, including 21 *D. trifida* local varieties, determining two well defined groups: group I—accessions from São Paulo State (Iguape and Cananéia municipalities, and Ubatuba municipality) and two accessions from Mato Grosso State (Brasnorte municipality); and group II—accessions from Amazonas State (municipalities of Manacapuru, Manaus and Itapiranga). Percentage numbers refer to a bootstrap analysis.

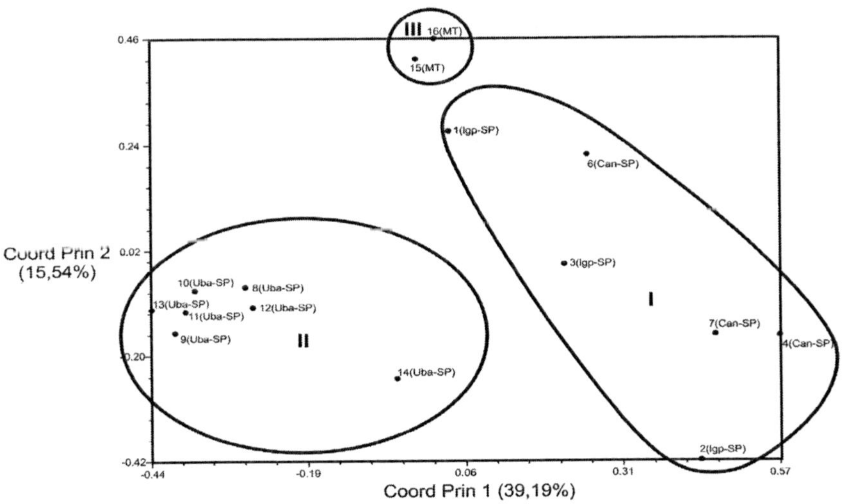

Fig. 3. Scatter graph for the principal coordinate analysis, corresponding to 54.7% of total variation for the 16 *D. trifida* accessions classified in group I in the cluster analysis, showing the clustering of accessions in three sub-groups (I—accessions from Iguape and Cananéia municipalities, in the Vale do Ribeira; II—accessions from Ubatuba municipality; and III—accessions from the State of Mato Grosso.

Table 1. Molecular variance analysis (AMOVA) for 21 *D. trifida* accessions classified into two groups defined in the cluster analysis (Fig. 2). Group I originated from the States of São Paulo and Mato Grosso and group II originated from Amazonas State.

Source of variation	DF	SS	Percentage of total variation	P^1 value
Among groups	1	9.605	35.18	0.0019
Within groups	19	35.538	64.82	0.0000
Total	20	45.143	-	-

[1] Permutation number = 1023

Similar results were obtained in a study conducted with ISSR primers used to evaluate accessions of *D. trifida* originating in traditional communities in the States of Santa Catarina, São Paulo, Mato Grosso and Amazonas (Nascimento et al. 2013). A cluster analysis classified the accessions into four well-defined groups: group I (accessions from Ubatuba-SP, Northern coast of São Paulo), group II (accessions from Iguape-SP, Southern coast of São Paulo, and Santa Catarina-SC), group III (accessions from Mato Grosso-MT) and group IV (accessions from the Amazon). In this study, as well as in the previous one cited above, a separation of accessions from the Southern and Northern coasts of São Paulo State was observed. Also, a positive correlation identified between genetic and geographical distances showed that *D. trifida* accessions are spatially structured in the geographic area sampled (Nascimento et al. 2013).

D. bulbifera

D. bulbifera, also known as aerial yam, is found in the wild state in Africa, Asia and Oceania, and is now cultivated throughout the tropics (Lebot 2009). Differing from other cultivated *Dioscorea* species, the organ of *D. bulbifera* used as food is in most cases the bulbil or air tuber (Terauchi et al. 1991). It represents a potential substitute to *D. deltoidea* Wall as a source of diosgenin, a steroidal sapogenin required for the synthesis of cortisones, corticosteroids, sex hormones (progesterone) and oral contraceptives (Narula et al. 2007). Distributed in the wild state in both Asia and Africa, as stated above, it shows an apparent difference in the morphology between these two continents, with globular bulbils in Asia and angular bulbils in Africa. A RFLP analysis of chloroplast DNA from 15 accessions of *D. bulbifera* from Africa and Asia, showed that the African and Asian chloroplast genomes have diverged from each other since very ancient times, indicating independent domestication (Terauchi et al. 1991, Lebot 2009).

Seven isozyme systems were employed in genetic diversity studies of 17 local varieties and six commercial varieties of *D. bulbifera* in Brazil (Bressan 2005). The local varieties were originated from the Vale do Ribeira,

São Paulo south coast area, and five commercial varieties were obtained from a research institution in São Paulo (Agronomic Institute - IAC) and one from a municipal market of São João dos Reis, State of Minas Gerais. The wide range for the Jaccard's similarity coefficient in the cluster analysis (Fig. 4) shows high genetic variability among *D. bulbifera* accessions, scored in a total of 67 polymorphic bands. Five groups can be depicted from the cluster analysis, showing the separation of the local varieties from the municipalities of Iguape (groups I and II) and Cananéia (groups IV and V), from the commercial varieties (group III) (Bressan 2005). Although Peropava is a community of Iguape municipality, according to the farmer, at the moment of the sampling collection, the tubers had been brought from CEAGESP (General Warehouses Company of São Paulo) in São Paulo, which is the main distribution center of horticultural crops in Brazil. Therefore, he probably planted in his home garden a commercial material, similar to those from IAC. The seven accessions from Peropava were all collected from the same farmer, and were proved to be clones. The materials obtained from IAC were also clones, except for one accession differing from the others. Also, the accession obtained in the market of São João del Rey, in Minas Gerais, showed high similarity to the local varieties collected in Cananéia municipality.

The intraspecific variability of *D. bulbifera* accessions originated from Africa, Asia and Polynesia were assessed with 10 RAPD primers, resulting in a total of 375 informative fragments (Ramser et al. 1996). Similar groupings

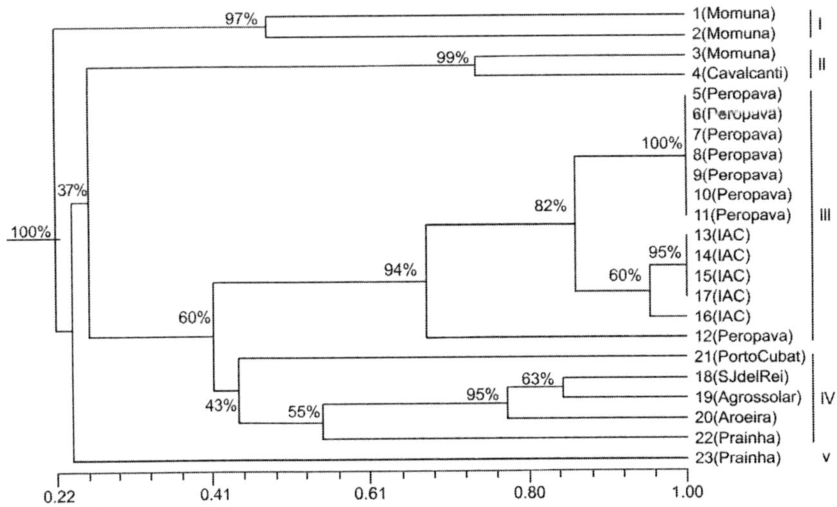

Fig. 4. Cluster analysis using the UPGMA method and Jaccard's similarity coefficient, to study 17 local varieties from the Vale do Ribeira and six commercial varieties of *D. bulbifera* in Brazil, based on 67 isozymic bands, using the NTSYS-pc software (Rohlf 1992).

reflected the geographical origin of the samples. The African accessions formed a distinct isolated group, whereas Asian and Polynesian accessions proved to be more heterogeneous. RAPD markers were also used to assess the genetic stability of cryopreserved-derived and *in vitro* grown plantlets, with great success (Dixit et al. 2003, Narula et al. 2007).

AFLP was also used to assess five African *D. bulbifera* local varieties, together with 48 yam accessions from Ethiopia, and the other cultivated species *D. alata* (one accession), *D. cayenensis* (three accessions) and *D. rotundata* (five accessions) (Tamiru et al. 2007). In this study, ten AFLP primer combinations generated 900 fragments, of which 97% were polymorphic. On average, the AFLP markers revealed a relatively higher proportion of polymorphic bands within the unidentified accessions from Ethiopia (44%) than within the accessions of *D. rotundata* (40%) or *D. cayenensis* (38%), while a very low level of polymorphism was detected within the accessions of *D. bulbifera* (6%). However, the authors pointed out that this comparison is not exhaustive due to unequal group sizes. Still in this study, five groups were obtained in a cluster analysis, and the accessions of *D. bulbifera* were grouped into a separate cluster.

Seven primers were recently developed for *D. bulbifera* using a microsatellite-enriched genomic library methodology, with six polymorphic primers (Silva 2013). Using the seven primers as well as 10 heterologous primers developed by Tostain et al. (2006) and Hochu et al. (2006), which showed successful amplification for this species, in a total of 17 primers, considerable genetic diversity was found among 42 accessions obtained in several municipalities in Brazil, but mainly from the states of São Paulo and Minas Gerais (Silva 2013). A cluster analysis, using the Jaccard's index and UPGMA method, indicated no separation among the accessions according to their origin, showing no genetic spatial structure.

Concluding Remarks

This brief review presents data on population genetics, germplasm genetic diversity and structure, the construction of genetic linkage maps and QTL analysis for disease resistance with several molecular markers used so far for the main cultivated species of *Dioscorea* (*D. alata*, *D. cayenensis*, *D. rotundata*, *D. trifida* and *D. bulbifera*), including some other *Dioscorea* species as well. The studies referred mostly to molecular markers such as isozymes, RAPD, AFLP, microsatellites, ISSR, including chloroplast DNA markers, aiming at phylogeny or even phylogeographic studies. Most of the studies emphasized the special role of traditional farmers in maintaining and amplifying high diversity among local varieties of these species, through their management practices and cultural and social habits. One of the main conclusions that can be made is that some of these species do not show any spatial structure among its local populations, such as *D. alata* and *D. bulbifera*, probably due to exchange of plant material among farmers and among regions, while others show a strong spatial structure, such as *D. trifida*.

However, much information still needs to be generated to better exploit the full potential of these crops. Furthermore, the cultivated species of *Dioscorea* add many beneficial nutritional factors, showing great potential as a food security crop. Incentives to study yam species can contribute to better understanding of the ecological and evolutionary processes linked to these crops and thus make possible more efficient cultivation practices while maintaining the existing variability.

Acknowledgements

The authors wish to thank the Brazilian institutions FAPESP, CAPES and CNPq for the financial support and grants given for some of the research reported above. We wish to thank the agriculturists from several municipalities in Brazil who contributed with plant material and provided valuable information acquired over the years. Also, we wish to thank many researchers which have collaborated in these studies, such as Nivaldo Peroni, Edson Ferreira da Silva, Lin Chau Ming, José Carlos Feltran, Maria Imaculada Zucchi, José Baldin Pinheiro, Eliane Gomes Fabri, Roland Vencovsky, Antônio Henrique dos Santos, the technicians Ronaldo José Rabello, Domingos de Salvio Amaral and Marcos Cella, during plant collections, SSR development, field practices and further discussions.

References

Abraham, K. and P.G. Nair. 1990. Floral biology and artificial pollination in *Dioscorea alata* L. Euphytica 48: 45–51.

Abramo, M.A. 1990. Taioba, cará e inhame. Ícone, São Paulo.

Agarwal, M., N. Shrivastava and H. Padh. 2008. Advances in molecular marker techniques and their applications in plant sciences. Plant Cell Rep. 27: 617–631.

Aiyeloja, A.A. and O.A. Bello. 2006. Ethnobotanical potentials of common herbs in Nigeria: A case study of Enubu state. Educ. Res. Rev. 1: 16–22.

Allendorf, F.W., P.A. Hohenlohe and G. Luikart. 2010. Genomics and the future of conservation genetics. Nature Rev. 11: 697–710.

Amorozo, M.C.M. 2008. Maintenance and management of agrobiodiversity in small-scale. Funct. Ecosys. and Communities 2: 11–20.

Amusa, N.A., A.A. Adegbite, S. Muhammed and R.A. Baiyewu. 2003. Yam diseases and its management in Nigeria. Afr. J. Biotechnol. 2: 497–502.

Arnau, G., A. Nemorin, E. Maledon and K. Abraham. 2009. Revision of ploidy status of *Dioscorea alata* L. (Dioscoreaceae) by cytogenetic and microsatellite segregation analysis. Theor. Appl. Genet. 118: 1239–1249.

Asemota, H.N., J. Ramser, C. Lopéz-Peralta, K. Weising and G. Kahl. 1996. Genetic variation and cultivar identification of Jamaican yam germplasm by random amplified polymorphic DNA analysis. Euphytica 92: 341–351.

Ayensu, E.S. and D.G. Coursey. 1972. Guinea yams. The botany, ethnobotany, use and possible future of yams in West Africa. Econ. Bot. 26: 301–318.

Baah, F.D., B. Maziya-Dixon, R. Asiedu, I. Oduro and W.O. Ellis. 2009. Nutritional and biochemical composition of *D. alata* (*Dioscorea* spp.) tubers. J. Food Agric. Environ. 7: 373–378.

Billotte, N., P.J.L. Lagoda, A.M. Risterucci and F.C. Baurens. 1999. Microsatellite-enriched libraries: applied methodology for the development of SSR markers in tropical crops. Fruits 54: 277–288.

Bousalem, M., V. Viader, C. Mariac, R. Gomez, I. Hochu, S. Santoni and J. David. 2010. Evidence of diploidy in the wild Amerindian yam, a putative progenitor of the endangered species *Dioscorea trifida* (Dioscoreaceae). Genome 53: 371–383.

Bousalem, M., G. Arnau, I. Hochu, R. Arnolin, V. Viader, S. Santoni and J. David. 2006. Microsatellite segregation analysis and cytogenetic evidence for tetrasomic inheritance in the American yam *Dioscorea trifida* and a new basic chromosome number in the Dioscoreae. Theor. Appl. Genet. 113: 439–451.

Bressan, E.A. 2005. Diversidade isoenzimática e morfológica de inhame (*Dioscorea* spp.) coletados em roças de agricultura tradicional do Vale do Ribeira – SP. M.S. Thesis, Luiz de Queiroz College of Agriculture, University of São Paulo, Piracicaba, São Paulo.

Bressan, E.A., E.A. Veasey, N. Peroni, A.P. Felipim and K.M.P. Santos. 2005. Collecting yam (*Dioscorea* spp.) and sweet potato (*Ipomoea batatas*) germplasm in traditional agriculture small-holdings in the Vale do Ribeira, São Paulo, Brazil. Plant Genet. Resour. Newsl. 144: 8–13.

Bressan, E.A., T. Briner Neto, M.I. Zucchi, R.J. Rabello and E.A. Veasey. 2011. Morphological variation and isozyme diversity in *Dioscorea alata* L. local varieties from Vale do Ribeira, Brazil. Sci. Agric. 68: 494–502.

Bressan, E.A., T. Briner Neto, M.I. Zucchi, R.J. Rabello and E.A. Veasey. 2013. Genetic structure and diversity of the *Dioscorea cayenensis/D. rotundata* complex revealed by morphological and isozyme markers. Genet. Mol. Res., in press.

Bustamante, S. and H. Buitrago. 2006. Molecular characterization of Colombian yam germoplasm by selective amplification of microsatellite polymorphic loci (SAMPL). Rev. Col. Bio. 8: 60–66.

Chaïr, H., X. Perrier, C. Agbangla, J.L. Marchand, O. Dainou and J.L. Noyer. 2005. Use of cpSSRs for the characterization of yam phylogeny in Benin. Genome 48: 674–684.

Cheng, K.T. and S.Y. Liu. 1996. Genetic relationships in *Dioscorea alata* L. revealed by RAPD analysis. Chin. Pharm. J. 48: 197–205.

Coomes, O.T. and N. Ban. 2004. Cultivated plant species diversity in home gardens of an Amazonian peasant village in Northeastern Peru. Econ. Bot. 58: 420–434.

Coursey, D.G. 1967. Yams. Longmans, London.

Coursey, D.G. 1976. Yams. *In*: N.W. Simmonds (ed.). Evolution of crop plants. Longman Group Limited, London, pp. 70–74.

Dansi, A., H.D. Mignouna, J. Zoundjihékpon, A. Sangare, R. Asiedu and N. Ahoussou. 2000a. Using isozyme polymorphism to assess genetic variation within cultivated yams (*Dioscorea cayenensis/Dioscorea rotundata* complex) of the Republic of Benin. Genet. Resour. Crop Evol. 47: 371–383.

Dansi, A., H.D. Mignouna, J. Zoundjihékpon, A. Sangare, N. Ahoussou and R. Asiedu. 2000b. Identification of some Benin Republic's Guinea yam (*Dioscorea cayenensis/rotundata* complex) cultivars using Randomly Amplified Polymorphic DNA. Genet. Resour. Crop Evol. 47: 619–625.

Dixit, S., B.B. Mandal, S. Ahuja and P.S. Srivastava. 2003. Genetic stability assessment of plants regenerated from cryopreserved embryogenic tissues of *Dioscorea bulbifera* L. using RAPD, biochemical and morphological analysis. Cryoletters 24: 77–84.

Egesi, C.N., R. Asiedu, G. Ude, S. Ogunyemi and J.K. Egunjobi. 2006. AFLP marker diversity in water yam (*Dioscorea alata* L.). Plant. Genet. Resour. 4: 181–187.

Essad, S. 1984. Variation géographique des nombres chromosomiques de base et polyploidie dans le genre *Dioscorea* à propos du dénombrement des espéces *transversa* Brown, *pilosiuscula* Bert. et *trifida* L. Agronomie 4: 611–617.

[FAOSTAT] Food and Agriculture Organization of the United Nations. 2012. Production: crops. Available at: http://faostat.fao.org/site/567/default.aspx#ancor.

Hamon, P. and T. Bakary. 1990. The classification of the cultivated yams (*Dioscorea cayenensis-rotundata*) of West Africa. Euphytica 47: 179–187.

Hamon, J.R. and B. Toure. 1990. Characterization of traditional yam varieties belonging to the *Dioscorea cayensis-rotundata* complex by their isozymic patterns. Euphytica 46: 101–107.

Hasan, S.M.Z., A.A. Ngadin, R.M. Shah and N. Mohamad. 2008. Morphological variability of greater yam (*Dioscorea alata* L.) in Malaysia. Plant Genet. Resour. 6: 52–61.

Hochu, I., S. Santoni and M. Bousalem. 2006. Isolation, characterization and cross-species amplification of microsatellite DNA loci in the tropical American yam *Dioscorea trifida*. Mol. Ecol. Notes 6: 137–140.

[IITA] International Institute of Tropical Agriculture. 2009. Yam. Available at: http://www.iita.org/cms/details/research_summary.aspx?articleid=268&zoneid=63.

Jarvis, D.I., L. Myer, H. Klemick, L. Guarino, M. Smale, A.H.D. Brown, M. Sadiki, B. Sthapit and T. Hodgkin. 2000. A training guide for *in situ* conservation on-farm. Version 1. IPGRI, Rome.

Kalia, R.K., M.K. Rai, S. KALIA, R. Singh and A.K. Dhawan. 2011. Microsatellite markers: an overview of the recent progress in plants. Euphytica 177: 309–334.

Lebot, V. 2009. Tropical root and tuber crops: cassava, sweet potato, yams and aroids. Crop production science in horticulture series: 17. CABI, Wallingford.

Lebot, V., B. Trilles, J.L. Noyer and J. Modesto. 1998. Genetic relationships between *Dioscorea alata* L. cultivars. Genet. Resour. Crop Evol. 45: 499–509.

Malapa, R., G. Arnau, J.L. Noyer and V. Lebot. 2005. Genetic diversity of the greater yam (*Dioscorea alata* L.) and relatedness to *D. nummularia* Lam. and *D. transversa* Br. as revealed with AFLP markers. Genet. Resour. Crop Evol. 52: 919–929.

Martin, F.W. and A.M. Rhodes. 1978. The relationship of *Dioscorea cayenensis* and *D. rotundata*. Trop. Agr. 55: 193–206.

Martins, P.S. 1994. Biodiversity and agriculture: patterns of domestication of Brazilian native plants species. An. Acad. Bras. Cienc. 66: 219–224.

Martins, P.S. and G.C.X. Oliveira. 2009. Dinâmica evolutiva em roças de caboclos amazônicos. *In*: I.G.C. Vieira, J.M.C. Silva, D.C. Oren and M.A. D'incao (eds). Diversidade biológica e cultural da Amazônia. Museu Paraense Emílio Goeldi, Belém, pp. 373–391.

Mignouna, H.D. and A. Dansi. 2003. Yam (*Dioscorea* ssp.) domestication by the Nago and Fon ethnic groups in Benin. Genet. Resour. Crop Evol. 50: 519–528.

Mignouna, H.D., M.M. Abang, A. Onasanya and R. Asiedu. 2002a. Identification and application of RAPD markers for anthracnose resistance in water yam (*Dioscorea alata*). Ann. Appl. Biol. 141: 61–66.

Mignouna, H.D., R.A. Mank, T.H.N. Ellis, N. Bosch, R. Asiedu, M.M. Abang and J.A. Peleman. 2002b. Genetic linkage map of water yam (*Dioscorea alata* L.) based on AFLP markers and QTL analysis for anthracnose resistance. Theor. Appl. Genet. 105: 726–735.

Mignouna, H.D., A. Dansi and S. Zok. 2002c. Morphological and isozymic diversity of the cultivated yams (*Dioscorea cayenensis/Dioscorea rotundata* complex) of Cameroon. Genet. Resour. Crop Evol. 49: 21–29.

Mignouna, H.D., M.M. Abang, A. Onasanya and R. Asiedu. 2002d. Identification and potential use of RAPD markers linked to Yam mosaic virus resistance in white yam (*D. rotundata*). Ann. Appl. Biol. 141: 61–66.

Mignouna, H.D., R.A. Mank, T.H.N. Ellis, N. van den Bosch, R. Asiedu, S.Y.C. Ng and J. Peleman. 2002e. A genetic linkage map of Guinea yam (*Dioscorea rotundata* Poir.) based on AFLP markers. Theor. Appl. Genet. 105: 716–725.

Mignouna, H.D., M.M. Abang and S.A. Fagbemi. 2003. A comparative assessment of molecular marker assays (AFLP, RAPD and SSR) for white yam (*Dioscorea rotundata*) germplasm characterization. Ann. Appl. Biol. 142: 269–276.

Mignouna, H.D., M.M. Abang, N.W. Wanyera, V.A. Chikaleke, R. Asiedu and G. Thottappilly. 2005. PCR marker-based analysis of wild and cultivated yams (*Dioscorea* spp.) in Nigeria: genetic relationships and implications for *ex situ* conservation. Genet. Resour. Crop Evol. 52: 755–763.

Montaldo, A. 1991. Cultivo de raíces y tubérculos tropicales. Instituto Interamericano de Ciências Agrícolas de la OEA, Lima.

Monteiro, D.A. 2002. Situação atual e perspectivas do inhame no Estado de São Paulo. *In*: C.A.S. Carmo (ed.). Inhame e taro: sistemas de produção familiar. Incaper, Vitória, pp. 85–92.

Narina, S.S., R. Buyyarapu, K.R. Kottapalli, A.M. Sartie, I.A. Mohamed, R. Asiedu, J.D. Mignouna, B.L. Sayre and B.E. Scheffler. 2011. Generation and analysis of Expressed Sequence Tags (ESTs) for marker development in yam (*Dioscorea alata* L.). BMC Genomics 12: 100.

Narula, A., S. Kumar and P.S. Srivastava. 2007. Genetic fidelity of *in vitro* regenerants, encapsulation of shoot tips and high diosgenin content in *Dioscorea bulbifera* L., a potential alternative source of diosgenin. Biotechnol. Lett. 29: 623–629.

Nascimento, W.F., J.F. Rodrigues, S. Koehler, P. Gepts and E.A.Veasey. 2013. Spatially structured genetic diversity of the Amerindian yam (*Dioscorea trifida* L.) assessed by SSR and ISSR markers in Southern Brazil. Genet. Resour. Crop Evol., on line publication. DOI 10.1007/s10722-013-0008-y.

Obidiegwu, J.E., R. Asiedu, E.E. Ene-Obong, C.O. Muoneke and M. Kolesnikova-Allen. 2009a. Genetic characterization of some water yam (*Dioscorea alata* L.) in West Africa with simple sequence repeats. J. Food Agric. Environ. 7: 132–136.

Obidiegwu, J.E., M. Kolesnikova-Allen, E.E. Ene-Obong, C.O. Muoneke and R. Asiedu. 2009b. SSR markers reveal diversity in Guinea yam (*Dioscorea cayenensis/D. rotundata*) core set. Afr. J. Biotechnol. 8: 2730–2739.

Pedralli, G. 2002a. Distribuição geográfica e taxonomia das famílias Araceae e Dioscoreaceae no Brasil. *In*: C.A.S. Carmo (ed.). Inhame e taro: sistemas de produção familiar. Incaper, Vitória, pp. 15–26.

Pedralli, G. 2002b. Levantamento florístico das Dioscoreaceae (R. Br.) Lindley da Cadeia do Espinhaço, Minas Gerais e Bahia, Brasil. Bol. Bot. Univ. São Paulo 20: 63–119.

Pedralli, G. 1991. Collecting yam germplasm in Brazil. Plant Genet. Resour. Newsl. 83/84: 22.

Peroni, N. and P.S. Martins. 2000. Influência da dinâmica agrícola itinerante na geração de diversidade de etnovariedades cultivadas vegetativamente. Interciência 25: 22–29.

Peroni, N. and N. Hanazaki. 2002. Current and lost diversity of cultivated varieties, especially cassava, under swidden cultivation systems in the Brazilian Atlantic Forest. Agricult. Ecosys. Environ. 92: 171–183.

Petro, D., T.J. Onyeka, S. Etienne and S. Rubens. 2011. An intraspecific genetic map of water yam (*Dioscorea alata* L.) based on AFLP markers and QTL analysis for anthracnose resistance. Euphytica 179: 405–416.

Ramser, J., K. Weising, C. Lopez-Peralta, W. Terhalle, R. Terauchi and G. Kahl. 1997. Molecular marker based taxonomy and phylogeny of Guinea yam (*Dioscorea rotundata–D. cayenensis*). Genome 40: 903–915.

Ramser, J., C. López-Peralta, R. Wetzel, K. Weising and G. Kahl. 1996. Genomic variation and relationships in aerial yam (*Dioscorea bulbifera* L.) detected by random amplified polymorphic DNA. Genome 39: 17–25.

Rohlf, J. 1992. NTSYS-pc: Numerical Taxonomy and Multivariate Analysis System, Version 1.70 (Software). Stony Brook, New York.

Sartie, A. and R. Asiedu. 2011. Development of mapping populations for genetic analysis in yams (*Dioscorea rotundata* Poir. and *Dioscorea alata* L.). Afr. J. Biotechnology 10: 3040–3050.

Scarcelli, N., A. Barnaud, W. Eiserhardt, U.A. Treier, M. Seveno, A. d'Anfray, Y. Vigouroux and J.C. Pintaud. 2011. A set of 100 chloroplast DNA primer pairs to study' population genetics and phylogeny in monocotyledons. PLoS ONE 6: e19954.

Scarcelli, N., S. Tostain, C. Mariac, C. Agbangla, O. Da, J. Berthaud and J.L. Pham. 2006a. Genetic nature of yams (*Dioscorea* sp.) domesticated by farmers in Benin (West Africa). Genet. Resour. Crop Evol. 53: 121–130.

Scarcelli, N., S. Tostain, Y. Vigouroux, C.O. Agbangla, O. Dainou and J.L. Pham. 2006b. Farmers' use of wild relative and sexual reproduction in a vegetatively propagated crop. The case of yam in Benin. Mol. Ecol. 15: 2421–2431.

Scarcelli, N., O. Daïnou, C. Agbangla, S. Tostain and J.L. Pham. 2005. Segregation patterns of isozyme loci and microsatellite markers show the diploidy of African yam *Dioscorea rotundata* (2n=40). Theor. Appl. Genet. 111: 226–232.

Silva, L.R.G. 2012. Desenvolvimento de marcadores microssatélites e caracterização genética de etnovariedades de inhame do complexo *Dioscorea cayenensis/D. rotundata*. Ph.D. Thesis, Luiz de Queiroz College of Agriculture, University of São Paulo, Piracicaba, São Paulo.

Silva, D.M. 2013. Diversidade genética de cará-do-ar (*Dioscorea bulbifera* L.) originados de roças de agricultura tradicional por meio de marcadores microssatélites. M.S. Thesis, Luiz de Queiroz College of Agriculture, University of São Paulo, Piracicaba, São Paulo.

Siqueira, M.V.B.M. 2011a. Yam: A neglected and underutilized crop in Brazil. Hort. Brasil 29: 6–20.

Siqueira, M.V.B.M. 2011b. Caracterização da diversidade genética de inhame (*Dioscorea alata* L.) utilizando marcadores microssatélites. Ph.D. Thesis, Luiz de Queiroz College of Agriculture, University of São Paulo, Piracicaba, São Paulo.

Siqueira, M.V.B.M. and E.A. Veasey. 2009. Raíces y tubérculos tropicales olvidados o subutilizados en Brasil. Ver. Colomb. Cienc. Hortíc. 3: 110–125.

Siqueira, M.V.B.M., T.G. Marconi, M.L. Bonatelli, M.I. Zucchi and E.A. Veasey. 2011. Development and characterization of microsatellite markers in water yam (*Dioscorea alata*). Am. J. Bot. 98: 142–146.

Siqueira, M.V.B.M., G. Dequigiovanni, M. Corazon-Guivin, J.C. Feltran and E.A. Veasey. 2012. DNA fingerprinting of water yam (*Dioscorea alata* L.) cultivars in Brazil based on microsatellite markers. Hort. Brasil. 30: 653–659.

Siqueira, M.V.B.M., W.F. Nascimento, L.R. Gomes, A.B. Ferreira, E.F. Silva, L.C. Ming and E.A. Veasey. 2013. Distribution, management and diversity of yam local varieties in Brazil: a study on Dioscorea alata L. Braz. J. Biol., in press.

Sonibare, M.A., R. Asiedu and D.C. Albach. 2010. Genetic diversity of *Dioscorea dumetorum* (Kunth) Pax using Amplified Fragment Length Polymorphisms (AFLP) and cpDNA. Biochem. Sys. Ecol. 38: 320–334.

Tamiru, M., H.C. Becker and B.L. Maass. 2007. Genetic diversity in yam germplasm from Ethiopia and their relatedness to the main cultivated *Dioscorea* species assessed by AFLP markers. Crop Sci. 47: 1744–1753.

Tamiru, M., H.C. Becker and B.L. Maass. 2008. Diversity, distribution and management of yam local varieties (*Dioscorea* spp.) in Southern Ethiopia. Genet. Resour. Crop Evol. 55: 115–131.

Terauchi, R., T. Terachi and K. Tsunewaki. 1991. Intraspecific variation of chloroplast DNA in *Dioscorea bulbifera* L. Theor. Appl. Genet. 81: 461–470.

Tostain, S., N. Scarcelli, P. Brottier, J.L. Marchand, J.L. Pham and J.L. Noyer. 2006. Development of DNA microsatellite markers in tropical yam (*Dioscorea* sp.). Mol. Ecol. Notes 6: 173–175.

Tostain, S., C. Agbangla, N. Scarcelli, C. Mariac, O. Daïnou, J. Berthaud and J.L. Pham. 2007. Genetic diversity analysis of yam cultivars (*Dioscorea rotundata* Poir.) in Benin using simple sequence repeat (SSR) markers. Plant Genet. Resour. 5: 71–81.

Twyford, C.T., A.M. Viana, A.C. James and S.H. Mantell. 1990. Characterization of species and vegetative clones of *Dioscorea* food yams using isoelectric focusing of peroxidase and acid phosphatade isoenzymes. Trop. Agr. 67: 337–341.

Veasey, E.A., M.V.B.M. Siqueira, L.R.G. Silva, W.F. Nascimento, A.B. Ferreira, D.M., Silva, E.F. Silva, L.C. Ming, N. Peroni and A.H. Santos. 2010. Ocorrência e diversidade de espécies cultivadas do gênero *Dioscorea* em diversos agroecossistemas brasileiros. *In*: L.C. Ming, M.C.M. Amorozo and C.W. Kffuri (eds.). Agrobiodiversidade no Brasil. NUPPEA, Recife, pp. 45–74.

Veasey, E.A., E.A. Bressan, M.V.B.M. Siqueira, A. Borges, J.R. Queiroz-Silva, K.J.C. Pereira, G.H. Recchia and L.C. Ming. 2012. Genetic characterization of cassava (*Manihot esculenta* Crantz) and yam (*Dioscorea trifida* L.) local varieties in swidden agriculture systems in Brazil. *In*: P. Gepts, T. Famula, R. Bettinger, C. Qualset, P. McGuire, S. Brush and A. Damania (eds.). Biodiversity in agriculture: domestication, evolution and sustainability. Cambridge University Press, Cambridge, pp. 344–360.

Viruel, J., P. Catalán and J.G. Segarra-Moragues. 2012. Disrupted phylogeographical microsatellite and chloroplast DNA patterns indicate a vicariance rather than long-distance dispersal origin for the disjunct distribution of the Chilean endemic *Dioscorea biloba* (Dioscoreaceae) around the Atacama Desert. J. Biogeography 39: 1073–1085.

Weeden, N.F. and J.F. Wendel. 1989. Genetics of plant isozymes. *In*: D.E. Soltis and P.S. Soltis (eds.). Isozymes in plant biology. Chapman and Hall Ltd, London, pp. 46–72.

Wu, W.H., L.Y. Liu, C.J. Chung, H.J. Jou and T.A. Wang. 2005. Estrogenic effect of yam ingestion in healthy postmenopausal women. J. Am. Coll. Nutr. 24: 235–243.

Zannou, A., E. Agbicodo, J. Zoundjihékpon, P.C. Struik, A. Ahanchédé, D.K. Kossou and A. Sanni. 2009. Genetic variability in yam cultivars from the Guinea-Sudan zone of Benin assessed by random amplified polymorphic DNA. Afr. J. Biotechnol. 8: 26–36.

Zoundjihekpon, J., P. Hamon, M. Noirot, B. Tio-Touré and S. Hamon. 1997. Flowering synchronization between male and female West African cultivated yams (*Dioscorea cayenensis-rotundata* complex). Euphytica 95: 371–375.

Zoundjihekpon, J., S. Hamon, B. Tio-Toure and P. Hamon. 1994. First controlled progenies checked by isozymic markers in cultivated yams *Dioscorea cayenensis-rotundata*. Theor. Appl. Genet. 88: 1011–1016.

[1] Genetics Department, Luiz de Queiroz College of Agriculture, University of São Paulo, Caixa Postal 83, 13400-970, Piracicaba, SP, Brazil.
Emails: eaveasey@usp.br; alinebor@hotmail.com; nascimentowf@yahoo.com.br; thiagomezette@gmail.com; lidinalvar@yahoo.com.br; danielle83rj@gmail.com; gabriel.dequi@gmail.com

[2] Agriculture Nuclear Energy Center (CENA/USP), Piracicaba, São Paulo, Brazil.
Email: ebress@hotmail.com

[3] São Paulo's Agency for Agribusiness Technology (APTA), Polo Regional Centro Sul, Piracicaba, São Paulo, Brazil.
E-mail: mvbsiqueira@gmail.com

* Corresponding author: eaveasey@usp.br

4

Improvement Strategies for Cocoyam (*Xanthosoma sagittifolium* (L.) Schott)

Conventional Breeding to Tissue Culture Technology

Tsafack Takadong Julie Judith[1,2,3,*] and *Charles Gilbert*[3,*]

ABSTRACT

Cocoyam (*Xanthosoma sagittifolium* (L.) Schott) is a tuberous plant, widely consumed all over the tropical regions of the world but is a more-or-less neglected crop and mainly grown by small-scale farmers. Botany and biology of the species is reviewed in the light of the main limitations encountered for the improvement of this vegetatively propagated crop. The major problem encountered with conventional breeding methods includes very limited sexual reproduction or incompatibility in different cultivars. After the development of mutation breeding methods, the

[1] Institute of Medical Research and Medicinal Plant Studies P.O. Box 6163 Yaounde, Cameroon.

[2] Laboratory of Plant Physiology, University of Yaounde I, Higher Teacher's Training College P.O. Box 47 Yaounde, Cameroon.

[3] Géoarchitecture EA2219, Département de Biologie, Faculté des Sciences et Techniques–Université de Brest, 6 av. Victor le Gorgeu - CS93837 29238 Brest Cedex 3, France.

* Corresponding authors: juliejudith@yahoo.fr and gilbert.charles@univ-brest.fr

tissue culture techniques open up many possibilities for sustainable production and improvement of this crop. Research programmes conducted with meristem culture, direct or indirect multiplication, somatic embryogenesis, medium and long-term conservation and microtuberisation are discussed with their respective advantages and perspectives of use.

Introduction

Cocoyam (*Xanthosoma* spp.) is one of the six most important root and tuber crops widely consumed all over the tropical and subtropical regions of the world (Jennings 1987, Onwueme and Charles 1994, Bown 2000). Edible varieties of this herbaceous monocotyledonous crop are grouped into few species, including the most widely grown *X. sagittifolium* (L.) Schott. They are mainly cultivated by small-scale farmers (Onwueme and Charles 1994) in Asia, Africa and Latin America (Wilson 1984). The corm and cormels of cocoyam are an important source of carbohydrates for human nutrition or animal feed, but the leaves are also consumed (Nyochembeng and Garton 1998). In spite of the importance of this staple food in many countries, cocoyam has received very little research attention (Goenaga and Heperly 1990), and is regarded as an under-exploited and insufficiently studied crop (Nguyen and Nguyen 1987, Giacometti and León 1994, Watanabe 2002).

Commercial varieties of cocoyam are generally highly susceptible to diseases such as cocoyam root rot caused by *Pythium myriotylum* or the Dasheen mosaic virus. The vegetative propagation of planting material from tuber fragments increases the distribution of pathogens. The lack of resistant genotypes and the poor efficiency of the conventional breeding methods make the production of clean planting material the only immediate possible solution to increase the yield and maintain the production areas (Castro 2006).

In vitro technologies are already applied for the production and multiplication of healthy clones but need to be developed for conservation and gene banking, to mass propagate high-quality materials and to increase the breeding possibilities within the germplasm resources (Manner 2011). Research works carried out with meristem culture, direct or indirect multiplication, somatic embryogenesis, medium and long-term conservation and microtuberisation are reviewed in this chapter with their advantages and perspectives of use.

Origin and Geographic Distribution

Cocoyam (*Xanthosoma sagittifolium*) is native to tropical South America where it has been cultivated since pre-Columbian times (Wilson 1984).

According to Coursey (1968), Portuguese missionaries introduced it in Africa only in 1840 in the Gold Coast (now Ghana) but, for Bown (2000), cocoyam reached West Africa between the 16th and 17th centuries and was spread further by traders, missionaries and other travellers. It soon became more popular in African cultivation and diet than the "old" cocoyam (taro, *Colocasia esculenta*) that had been introduced from Southeast Asia. In the 19th century, it was widely spread from the West Indies throughout the tropical world (Catherinet 1965). The distribution of the genus *Xanthosoma* spp. is pantropical with a vast growing area at the equator between the 15th degree of south latitude and the 30th degree of north latitude and over a fairly wide range of temperature, from 13° to 29°C (Lopez et al. 1995). This area covers Central and South America, the West and Central Africa, South-East Asia, Oceania and New Caledonia (Vickery 1984, Bown 2000).

The local name varies from region to region. Thus macabo is called in English: tan(n)ia, yautia, new cocoyam, tanier, cocoyam (Cameroon); in Spanish: yautia, malanga (Antilles), macal (Mexico), quiscamote (Honduras), tiquisque (Costa Rica), oto (Panama), okumo (Venezuela), uncucha (Peru), gualuza (Bolivia), malagay (Colombia); in Portuguese: taioba, mangareto, mangarito, mangaras (Brazil); in French: chou Caraïbe (West Indies) or macabo (Cameroon) (Giacometti and Leon 1994, Lopez et al. 1995, Reyes 2006). The name 'new cocoyam' reflects this late introduction into areas where *Colocasia* (cocoyam) was previously established (Kay 1987).

Botany

Xanthosoma sagittifolium is an herbaceous monocotyledonous crop. The stem is a starch-rich underground structure, the corm, from which offshoots called cormels develop. Two other genera cultivated for their edible corms and cormels belong to the same family: *Alocasia* and *Colocasia*. A detailed description of tannia (*Xanthosoma* spp.) plants is given by Kay (1987) that helps to distinguish the genus *Xanthosoma* with that of *Colocasia* (tannias look like coarse taros and create confusion with cocoyam). Tannia plants can reach a height of about 2 m and have a short erect stem and large, sagittate or hastate leaves, which are 50–75 cm long with a prominent marginal vein and pointed lobes at the tip. The leaf stalks are about 1 m long and join the blade at the margin between the lobes, whereas in *Colocasia*, the stalk is connected into the surface of the blade and the tips of the lobes are rounded. In *Xanthosoma*, mature leaves are angled about 30° off the petiole but for *Alocasia*, the leaf blade and the petiole are in the same plane. Inflorescences sprout between the leaves in a spadix, with a pale green spathe which is about 20 cm long. The flowers are unisexual with the female flowers located at the base of the spadix, the male flowers at the top and the sterile ones between them. Flowering is more prone to occur in wet regions (Purseglove 1972) but some cultivars never flower. The inflorescence of *Xanthosoma* is protogynous, and the pistillate flowers are normally receptive 2 to 4 days before pollen is shed (Wilson 1984). The spadices are seldom fertile and produce few viable seeds (Giacometti & León 1994). There is no evidence of seed dormancy in tannia. The seeds can be stored at low moisture contents and temperature, but longevity is reported to be comparatively short (Wilson 1980). A corm is produced at the base of the plant and this bears usually 10 or more lateral corms (cormels), each 10–25 cm long, which are harvested. The growth cycle lasts from 9 to 11 months before harvest, although cocoyam is a perennial crop. The etymology of the genus *Xanthosoma* comes from the Greek words *xanthos* (yellow) and *soma* (body), due to the yellowish colour of the corm pulp in several species (Mayo et al. 1997).

Genetic Diversity and Taxonomy

Xanthosoma sagittifolium belongs to the tribe Caladieae of the family Araceae. The systematics of this monophyletic family was recently revised: it comprises about 118 genera and 3500 reported species, number may reach up to approximately 6000 species (Boyce and Croat 2011). This classification was comprehensively reviewed by Mayo et al. (1997), who carried out a cladistic analysis from morphological and anatomical characters. French et al. (1995) published the first cladistic analysis of the family Araceae based on chloroplast DNA restriction-site (RFLP). Cabrera et al. (2008) used a matrix of chloroplast DNA sequences from 102 aroid genera and more recently, Cusimano et al. (2011) published a synthetic work combining the study of morpho-anatomical characters and molecular phylogenies as a contribution to a new formal classification of the Araceae. The possible base chromosome number and evolution of the family were also analyzed (Cusimano et al. 2012).

The genus *Xanthosoma* includes nearly 60 species of which 57 are found to be cultivated (Thompson and Dewet 1983, Bown 2000) but the taxonomic situation of different species is confusing. According to IBPGR descriptors (1989), edible and cultivated varieties are grouped into few species, including *X. sagittifolium* (L.) Schott, by far the most widely grown, *X. atrovirens* Koch and Bouché, with yellow tubers and favored in Puerto Rico and Dominica, *X. violaceum* Schott, a large plant grown in the Pacific Islands but of little value for food, *X. caracu* Koch and Bouché, seldom-mentioned but with varieties of high quality, and *X. brasiliense* Engl., cultivated for its edible leaves (Morton 1972, Kay 1987, Milian et al. 2001). According to the World Checklist of Selected Plant Families (WCSP 2013), *X. violaceum* (syn. *X. nigrum*) are only synonyms for *X. sagittifolium* and a list of accepted names in the genus is proposed, but according to Bown (2000), these are only two main species in the genus.

Krishnan (1977) reports various chromosomal counts for edible cultivars (clones) of *Xanthosoma* ($2n$ = 24, 26 or 39). In *X. sagittifolium*, all the 13 bivalents ($n = 13$) could be assigned to 12 chromosomal types, but 8 of them could be regarded as homeologous, reducing the basic number to $n = 8$.

Improvement Strategies of Cocoyam

Cocoyam is considered as a neglected crop mainly grown by small-scale farmers, and attempts to improvement have therefore been limited.

Diseases

Cocoyam root rot disease (CRRD) caused by the oomycete *Pythium myriotylum* Dreschler is the most important production constraint on cocoyam yield in Africa and particularly in Cameroon (Pacumbaba et al. 1992, Tambong and Höfte 2001, Perneel et al. 2006) reducing the crop yield up to 90% (Nzietchueng 1983). More recently, the disease was also observed in Sri Lanka (Tojo et al. 2005) and indicates a progression of the pathogen.

Dasheen mosaic virus (DsMV) is the most important viral pathogen of cultivated aroids world-wide (Chen et al. 2001), and was first reported by Zettler et al. in 1970. In Central America DsMV was first identified in Costa Rica by Ramírez (1985). The recurrent infection of the clones of cocoyam leads to gradual decrease in the yield (around 25% in Nicaragua due to the DsMV present in all plantations). The main objective of the cocoyam programmes is to improve agronomic qualities and resistance to those diseases that cause significant economic losses. Cultural practices, chemical, biological, genetic and biotechnology control methods have been developed to improve cocoyam production.

Cultural practices, chemical and biocontrol of diseases

The conventional planting material includes portions of the central corm (100–150 g) or whole small cormels. Corms and cormels display a dormancy-like period of approximately five weeks, during which sprouting does not occur (Wilson 1984) but the removal of the apical bud accelerates the sprouting of the lateral buds. Ramesh et al. (2007) presented various aspects of crop production, with special emphasis on soil fertility and plant nutrition. Cocoyam requires well-drained soil for optimum growth and cannot withstand water logging. The crop is suited to high rainfall areas receiving an annual precipitation of 140–200 cm and the mean soil temperature for their optimum growth must exceed 20°C. Among the aroids, cocoyam is relatively more tolerant to dry conditions but performs better under irrigation (Onwueme and Charles 1994).

The disease is spread through soil and planting material (Nzietchueng 1983) and can persist in the soil for many years. In order to reduce the root rot incidence, cocoyam farmers recommend several cultural practices such as soil drainage, hill planting, crop rotation, multiple cropping, early planting and early harvesting (Onwueme and Charles 1994).

Chemical control is most effective, especially when multiple treatments are applied. Yet this method presents a range of negative side effects. The use of fungicides such as Ridomil, Aliette and methyl bromide is costly and induces environmental pollution and detrimental health effects for farmers and consumers (Martin and Lopper 1999). Besides, some pesticides such as methyl bromide induce emergence of resistant pathogen strains and have been banned since 1999 by the Montreal protocol on protecting the ozone layer (Duffy and Defago 1999).

An alternative approach in the fight against pathogenic agents was initiated with the biological control of the root rot disease caused by *P. myriotylum* by fast-growing bacteria (*Pseudomonas fluorescens* and *P. aeruginosa*) which are the most important rhizobacteria, ubiquitous in soil, involved in the biocontrol of plant diseases (Dowling and O'Gara 1994). Their mechanism of action is the production of antibiotics (2,4-diacetylphloroglucinol, Pyoluteorin, Pyrrolnitrin and Phenazine) that inhibit mycelial growth (Turner and Messenger 1986, Tambong and Höfte 2001).

The antagonistic and mycoparasitic potential of isolates of *Trichoderma* strains were also evaluated. *Trichoderma* spp. are fungi present in nearly all soils, also known to control damping-off and other root diseases in seedlings. Some species are producing relatively high concentrations of enzymes that degrade the cell wall such as β-1,3-glucanases and chitinases (Xu et al. 1995). Recently, the ability of *T. asperellum* to protect cocoyam plants from *P. myriotylum* infection was tested *in vitro* and *in vivo*: two strains could reduce *P. myriotylum* infection by 50% after soaking the plants in conidial suspensions (Mbarga et al. 2012). These methods are promising but still experimental, and they could also present some limitations such as the production of isolates, and their dissemination could be risky to the ecosystem (Xu et al. 1995, Tambong and Höfte 2001).

The use of chemicals (BTH) as defence elicitor to trigger resistance in cocoyam against CRRD was also recently reported by Mbouobda et al. (2010). A combination of cultural management, chemical and biological application could enhance local production until cocoyam breeders succeed in developing cultivars productive and tolerant to the disease by conventional plant breeding and biotechnology procedures (Onokpise et al. 1999).

Genetic improvement through conventional breeding and biotechnology

Sexual propagation is essential to obtain the variability required for genetic improvement but cocoyam rarely flowers naturally. The flowers are protogynous, which makes the use of classical breeding methods difficult, but treatments were successfully applied to induce flowering with artificial pollination techniques. Day length seems to have no discernible effect on flower induction (McDavid and Alamu 1979), hence it was fruitfully induced by gibberellic acid (Alamu et al. 1982, Goenaga and Hepperly 1990, Tambong and Meboka 1994), by removal of leaves, by stress induced by drought and high temperature, and by removal of cormels, suckers and stolons (Lebot 2009). Techniques for pollen storage and germination were also developed for sexual crossings and viable pollen could be obtained after 28 days in storage at 5°C and 30% relative humidity (Agueguia and Fatokun 1987).

In Cameroon, there are mainly three cocoyam cultivars distinguished by the colour of the tuber flesh, the level of ploidy, and the productivity as well as their resistance to the root rot disease due to *Pythium myriotylum*. The highly productive and widely preferred White and the less productive Red cocoyam types are diploid ($2n=2x=26$) but susceptible to the root rot disease, whereas the much less productive Yellow type is tetraploid ($2n=4x=52$) but shows some level of resistance (Ngouo 1988, Boudjeko et al. 2005, Mbouobda et al. 2007, Oumar et al. 2011).

The production of pollen grains was observed in the White and Red cultivars, but rarely in the Yellow cultivars. Hybridization resulted in the production of seeds from White x White, and White x Red crosses. Virtually, no viable seeds were produced from the resistant Yellow cocoyam with the highly preferred diploid White or Red cultivars using conventional techniques, probably due to ploidy differences (Onokpise et al. 1999).

It has been undertaken to explore resistance markers with the aim of cloning their genes and transferring them in the Red and White cultivars (Boudjeko 2003, Boudjeko et al. 2005). The resistance to root rot disease was determined mainly as a result of additive gene effects with a predominance of general combining ability over specific one (Agueguia et al. 1994). Breeding for resistance to *P. myriotylum* requires a test to assess genotypes. After inoculation of sensitive and tolerant accessions, an increase in peroxidase activity was observed in roots of tolerant genotypes and might help the breeders to screen large progenies and identify genes of interest (Nyochembeng et al. 2007).

The characterization of the genetic diversity disposable among the accessions of germplasms collected in various regions is also of great value for breeding programmes and conservation purposes. Offei et al. (2004) analyzed the genetic structure of 70 cocoyam accessions from two regions of Ghana with RAPD markers. They identified a most polymorphic accession suitable for crossing and suggested from the results that there might have germplasm movements between the two regions.

Polyploidy induction has often been used to develop new cultivars (Peng et al. 2008). In order to increase the genetic variability in cocoyam, some treatments inducing a polyploidization of the genome were implemented. Thus, colchicine was applied to the diploid cultivars (White and Red) for inducing tetraploidy (Esnard et al. 1993, Tambong et al. 1998). More recently, the efficiency of colchicine or oryzalin treatments for doubling the chromosome number of White cocoyam was evaluated, and the ploidy level determined using flow cytometry (Oumar et al. 2011). Tetraploid, octoploid and myxoploid plants were obtained in these experiments, but no evaluation of the resulting plants was reported.

Since no gene or group of genes for resistance have been identified to date and no valuable tetraploid or hybrid cocoyam plant have so far been reported, the challenge for breeders turned to tissue culture techniques.

Till date, no results have been reported with genetic transformation of cocoyam but results with other aroid tuber crops are already published, particularly with taro plants after direct (Fukino et al. 2000) or indirect gene transfer (He et al. 2008). Many species have already been transformed with the coat protein-coding region of viruses, such as cassava, another major root crop severely affected by two main viral diseases (Vanderschuren et al. 2012). The coat protein-coding region from DsMV was isolated, and this information could be of further use in the production of transgenic plants resistant to the virus (Castro 2006).

In addition to genetic transformation, the technique of somatic hybridization via protoplast fusion is also worthwhile to explore but till date, no results have been reported on cocoyam. However, this strategy was developed successfully with taro plants (Murakami et al. 1998) and should encourage similar study with cocoyam.

Enhancing genetic variability through alternative methods such as somaclonal variation or induced mutations also increases the scope of cocoyam improvement. Gupta (1985) reported various morphological changes in callus-derived cocoyam plants. *In vitro* mutation breeding may contribute to genetic improvement as cocoyam plants with different levels of tolerance towards CRRD have been obtained after irradiation of apices with gamma rays (Saborio et al. 2004, Danquah et al. 2006, Ndzana et al. 2008). A field evaluation of agronomic traits is required after irradiation to assess the nature and the stability of the variation.

In vitro Approaches

Cocoyam is usually propagated vegetatively from tuber fragments, which are the principal source of pathogen dissemination. The tissue culture is considered as a practical means for the production of pathogen-free plants, the development of new varieties and the mass production of a planting stock (Zettler and Hartmann 1987). Therefore, development of an efficient *in vitro* tissue culture system for sustainable mass propagation of disease-free planting material should be a main objective.

Micropropagation via caulogenesis or embryogenesis

Many results have been obtained from shoot tip or axillary buds on the growth, proliferation and plant regeneration of various *Xanthosoma* species (Asokan et al. 1984, Nguyen and Nguyen 1987, Omokolo et al. 1995, Zok et al. 1998, Kanmegne and Omokolo 2003). Sama et al. (2012) obtained the greatest proliferation from 3–5mm shoot tips with 0.5 mm basal corm tissue of white plants in stationary liquid (for economic reasons) modified Gamborg's B5 mineral salts supplemented with 0.05 µM 1-naphthaleneacetic acid (NAA) and 20 µM Benzylaminopurine (BAP): an average rooted shoots number of 9.7 was achieved after six weeks of culture.

Improvement of the technology is still required to develop the scale-up production at a commercial level using a temporary immersion bioreactor. Using the method described by Santos Pino et al. (2011), Niemenak et al. (2013) obtained a higher proliferation rate (68) in MS-based liquid medium (Murashige and Skoog 1962) with 20 g L^{-1} sucrose (considered a determinant factor) and 10 µM BAP (shoots were immersed for 1 mn every 6 h).

Indirect neoformation of cocoyam plants was achieved by Nyochembeng and Garton (1998) who initiated a protocol of plant regeneration from petiole explants with dicamba and thidiazuron. Similar work was reported with genotypes of Bangladesh using BAP and NAA for callus induction (Paul and Bari 2007). Such results could be of help for further genetic engineering research or for the induction of somaclonal changes as source of variability. Selective pressures exerted to the cultured cells and tissues may also increase the tolerance of the regenerated plants for biotic or abiotic stress and fit a demand for such characteristics (Watanabe 2002).

Recently, Verma and Cho (2010) reported somatic embryogenesis in taro plants. The neoformation of cocoyam through this process was also reported earlier (Gomez et al. 1992) but would require new studies to completely develop the technology. Somatic embryos present several advantages for the production of non-chimerical plants, after mutagenic or transgenic treatments, in relation to the putative unicellular origin of the embryos. They are also able to ensure considerable multiplication rates in model systems and commonly used for conservation after alginate encapsulation.

Shoot tip and meristem culture

The eradication of virus from regenerated clones by meristem-tip culture is one of the most interesting applications of tissue culture. Since the first success on virus-infected dahlias (Morel and Martin 1952), a lot of vegetatively propagated species have been cleaned of their viruses. For genotypes that cannot be propagated from true seeds, it may be the sole means to eliminate viruses from planting stock and make possible safe exchanges of germplasm between countries after virus indexing, according to FAO/IBPGR guidelines for safe movement of aroid germplasm (Zettler et al. 1989).

Hartman (1974) eliminated the Dasheen mosaic virus and other phytopathogens from various aroids (*Caladium*, taro and cocoyam) by culture of shoot tips, starting from only the apical dome with one or two leaf primordia remaining.

An additional advantage of the *in vitro* plants is the re-invigoration frequently experienced in plants passed through a tissue culture phase (Pierik 1990). Meristem-derived plants of cocoyam performed better in terms of yield than virus-infected *in vitro* plants (Reyes et al. 2006). According to Zok et al. (1998) cormels from meristem-derived plants grow faster and the tuberization starts earlier in comparison with conventionally propagated plants. Of course this technology does not prevent the reinfestation, mainly via aphids for the DsMV, and the healthy plants for distribution should be multiplied in safe regions or directly in large quanitities from *in vitro* plants to meet the demand of farmers.

Germplasm conservation

Many horticultural plants, being vegetatively propagated or having recalcitrant seeds, require field genebank for their conservation but on long term, they are expensive, difficult to manage and are vulnerable to natural disasters and pathogen damages. The alternative approaches are *in vitro* conservation and cryopreservation.

Medium-term in vitro conservation

Tissue culture techniques are routinely used for conservation of vegetatively propagated crops (Tyagi and Yusuf 2003) with no risks of losses of germplasm during the maintenance in the field. Various slow growth techniques are used for medium-term conservation, mainly through combinations of growth retardants such as low temperature, osmoticums (sorbitol, mannitol) or lowering the gaseous exchanges with a film of mineral oil. *Xanthosoma sagittifolium, X. robustum* and *X. brasiliense* could be stored in the dark for at least two years at 13°C and by adding mannitol (3%) to the medium (Zandvoort et al. 1994).

Long-term in vitro conservation

The long-term conservation occurs ideally in liquid nitrogen at ultra-low temperature (–196°C), as longevity of shoot tips held at temperatures between –4 and –80°C is short. Cryopreservation requires small space and very limited maintenance. Several treatises on cryopreservation of germplasm were already published with details for recipes (Engelmann and Takagi 2000, Razdan and Cocking 2000, Towill and Bajaj 2002). Takagi et al. (1997) presented a protocol for cryopreservation of shoot tips of taros by vitrification with 80% recovery.

Genetic stability has to be assessed after meristem culture, long-term storage and cryopreservation that can induce variations, due to the possible callusing of the meristem tissues and the stress induced by the preservative treatment with cryoprotectants.

In vitro tuber formation and applications

The production of healthy material through plant micropropagation techniques has been established, but the handling and transportation of *in vitro* plantlets remain very difficult. Also, maintenance of cocoyam shoot stock requires generally frequent subcultures and they remain fragile under slow growth protocols.

Technologies for microtuber production show a great potential as alternatives for micropropagation in tuber crop (Coleman et al. 2001, Jova et al. 2012). Moreover, they can be used as experimental research tools for biochemical, physiological and genetic studies of plants (Charles et al. 1993, Coleman et al. 2001). Increasing attention has been recently paid to *in vitro* production of cocoyam microtubers. Shoots obtained *in vitro* under controlled conditions were used as explants for seed tuber formation and the effects of various factors were described. In their study, Omokolo et al. (2003) obtained the higher tuberization rate with 30 µM BAP and 8% sucrose.

Tsafack et al. (2009) showed the influence of the photo- and thermoperiodic regimes on the tuberization of cocoyam on MS-based medium (Murashige and Skoog 1962), free of plant growth regulators and supplemented with 8% sucrose. Shoot tips were cultured under a short photoperiod (8 hr lighting), at 25/20°C day/night temperature for 10 days, followed by a continuous darkness at 20°C for 50 days as the most effective treatment. The application of low night temperature during this process produced a more synchronous development and a higher tuberization rate. The best response was observed with White cultivar (70% tuberized explants with 15 microtubers on each explant) followed by Red and Yellow cv. (29% response with 5 tubers for Yellow) (Tsafack et al. 2009). The small tubers harvested sprouted well (up to 98% after 30 d) and resulted in vigorous plants.

The production of microtubers could facilitate the transfer of germplasm as seed tubers to local stations for multiplication and further distribution of healthy planting material for agriculture. However, in order to ensure their role as practical means for germplasm conservation and distribution, it is necessary to improve their conditions of storage for a longer conservation period. The application of a combination of sucrose (2%) and osmoticums (sorbitol and mannitol, 2% each) increased the storage period up to 8 months at 25°C and delayed the sprouting of the microtubers for 9 days (Tsafack 2010).

Conclusion and Perspectives

The improvement of cocoyam production through distributing disease-free planting material would help to ensure immediate better agronomic performances of the few genotypes commonly propagated by farmers. The establishment of germplasm collections, with the help of conservation techniques, are also necessary for further exchanges (after sanitization of the material) and improvement. The evaluation of the disposable diversity collected and studies at molecular level would be valuable for a better understanding of their relationship and taxonomy, but also to identify markers of tolerance or resistance to main diseases. Manipulating the ploidy level is an approach for creating variability and overcome the problems encountered with conventional breeding, but the poor fertility frequently observed in the genus is a limiting factor. As for somaclonal variation, mutagenic treatments used for improving cultivars remain hazardous, even associated with selective pressure. Alternative methods to sexual breeding such as somatic hybridization or genetic transformation would require more investigations to develop the high potential of this under-exploited crop.

Abbreviations

CRRD: Cocoyam root rot disease
DsMV: Dasheen Mosaic Virus

References

Alamu, S., C.R. McDavid and E.J. Duncan. 1982. Production of viable seed in gibberellic acid-treated tannia (*Xanthosoma sagittifolium* (L.) Schott) plants. Tropical Agriculture 59(4): 333–334.

Agueguia, A. and C.A. Fatokun. 1987. Pollen storage in cocoyam (*Xanthosoma sagittifolium* [L.] Schott). Euphytica 39: 195–198.

Agueguia, A., C.A. Fatokun and S.K. Hahn. 1994. The genetics of résistance to cocoyam root rot blight complex disease in *Xanthosoma sagittifolium* (L.) Schott. Acta Hort. (ISHS) 380: 438–442.

Asokan, M.P., S.K. O'Hair and R.E. Litz. 1984. Rapid multiplication of *Xanthosoma caracu* by *in vitro* shoot tip culture. Hort. Science 19: 885–886.

Boudjeko, T. 2003. Quelques indicateurs de résistance dans la relation parasitaire *Xanthosoma sagittifolium* L. Schott (macabo)/*Pythium myriotylum*. Thèse de Doctorat/Ph.D. Universités de Yaoundé I et Rouen. 104 p.

Boudjeko, T., N.D. Omokolo, A. Driouch and A.P. Balange. 2005. Peroxydases and pectin methyl esterase activities in cocoyam (*Xanthosoma sagittifolium* L. Schott) roots upon *Pythium myriotylum* inoculation. Journal of Phytopathology 153: 409–416.

Bown, D. 2000. Aroids: Plants of the Arum Family. 2nd edition. Timber Press. Portland, Oregon, USA, pp. 392.

Boyce, P.C. and T.B. Croat. 2011. The Überlist of Araceae, Totals for Published and Estimated Number of Species in Aroid Genera http://www.aroid.org/genera/111109uberlist.pdf.

Cabrera, L.I., G.A. Salazar, M.W. Chase, S.J. Mayo, J. Bogner and P. Davila. 2008. Phylogenetic relationships of aroids and duckweeds (Araceae) inferred from coding and noncoding plastid DNA. Am J. Bot. 95: 1153–1165.

Castro, G.R. 2006. Studies on cocoyam in Nicaragua, with emphasis on Dasheen mosaic virus. Doctoral thesis. Swedish University of Agricultural Sciences, Uppsala. ISBN: 91-576-7056-0.

Catherinet, M. 1965. Note sur la culture du macabo et du taro au Cameroun. Agronomie Tropicale 20: 717–724.

Charles, G., L. Rossignol and M. Rossignol. 1993. A synchronous model perfecting for fundamental studies on the tuberization process. J. Plant Physiol. 142: 474–479.

Chen, J., J. Chen, J. Chen and M.J. Adams. 2001. Molecular characterization of an isolate of Dasheen mosaic virus from *Zantedeschia aethiopica* in China and comparisons in the genus *Potyvirus*. Archives of Virology 146: 1821–1829.

Coleman, W.K., D.J. Donnelly and S.E. Coleman. 2001. Potato tuber as research tools. American Journal of Potato Research 78: 47–55.

Coursey, D.G. 1968. The edible aroids. World Crops 20: 25–30.

Cusimano, N., J. Bogner, S.J. Mayo, P.C. Boyce, S.Y. Wong, M. Hesse, W.L.A. Hetterscheid, R.C. Keating and J.C. French. 2011. Relationships within the Araceae: comparison of morphological patterns with molecular phylogenies. Am. J. Bot. 98: 654–668.

Cusimano, N., A. Sousa and S.S. Renner. 2012. Maximum likelihood inference implies a high, not a low, ancestral haploid chromosome number in Araceae, with a critique of the bias introduced by 'x'. Annals of Botany 109: 681–692.

Danquah, A., S.K. Offei, E.T. Blay, E. Asare and E.Y. Danquah. 2006. Characterization of a Mutant Population of Cocoyam (*Xanthosoma sagittifolium* L. Schott). International Journal of Botany 2: 128–132.

Dowling, D.N. and F. O'gara. 1994. Metabolites of Pseudomonas involved in the biocontrol of plant disease. Trends in Biotechnology 12: 133–141.

Duffy, B.K. and G. Defago. 1999. Environmental factors modulating antibiotic and siderophore biosynthesis by *Pseudomonas fluorescens* biocontrol strains. Applied and Environmental Microbiology 65: 2429–2438.

Engelmann, F. and H. Takagi (eds.). 2000. Cryopreservation of tropical plant germplasm: current research progress and application. Japan International Research Center for Agricultural Sciences, Tsukuba, Japan 496pp.

Esnard, J., F. Ferwerda, E. Rivera-Amador and P.R. Hepperly. 1993. Induction of tetraploidy in the tanier cultivar "Inglesa" (*Xanthosoma sagittifolium* (L.) Schott). Plant Breeding 111: 335–338.

French, J.C., M.G. Chung and Y.K. Hur. 1995. Chloroplast DNA phylogeny of the Ariflorae. *In*: P.J. Rudall, P. Cribb, D.F. Cutler and C.J. Humphries (eds.). Monocotyledons: Systematics and evolution, Royal Botanic Gardens, Kew, UK 1: 255–275.

Fukino, N., K. Hanada and H. Ajisaka. 2000. Transformation of taro (*Colocasia esculenta* Schott) using particle bombardment. JARQ 34(3): 159–165.

Giacometti, D.C. and J. León. 1994. Tannia, Yautia (*Xanthosoma sagittifolium*). *In:* Neglected Crops: 1492 from a different perspective. Plant Production and Protection Series No. 26. J.E. Hernando-Bermejo and J. León (eds.). FAO, Rome, Italy, pp. 253–258.

Goenaga, R. and P. Hepperly. 1990. Flowering induction, pollen and seed viability and artificial hybridization of taniers (*Xanthosoma* spp.). J. Agric. Univ. Puerto. Rico. 74(3): 253–260.

Gomez, L., R. Valverde, O. Arias and T. Thorpe. 1992. Regeneration of *Xanthosoma sagittifolium* by somatic embryogenesis. Agron. Costarric 16: 219–223.

Gupta, P.P. 1985. Plant regeneration and variabilities from tissue cultures of cocoyams (*Xanthosoma sagittifolium* and *X. violaceum*). Plant Cell Reports 4: 88–91.

Hartman, R.D. 1974. Dasheen Mosaic Virus and Other Phytopathogens Eliminated from Caladium, Taro and Cocoyam by Culture of Shoot Tips. Phytopath. 64: 237–240.

He, X., S.C. Miyasaka, M.M.M. Fitch, P.H. Moore and Y.J. Zhu. 2008. *Agrobacterium tumefaciens*-mediated transformation of taro (*Colocasia esculenta* (L.) Schott) with a rice chitinase gene for improved tolerance to a fungal pathogen *Sclerotium rolfsii*. Plant Cell Rep. 27(5): 903–909.

IBPGR. 1989. Descriptors for *Xanthosoma*. International Board for Plant Genetic Resources, Rome.

Jennings, D.L. 1987. Starch crops. *In:* B.R. Christie (ed.). CRC Handbook of Plant Science in Agriculture. Volume II. CRC Press, Inc. Boca Raton, Florida, USA, pp. 137–143.

Jova, M.C., R.G. Kosky, E.E. Cuellar and A.E. Cuellar. 2012. Efficiency of semi-automated culture systems on microtuber formation of yam. Biotechnol. Agron. Soc. Environ. 16(1): 45–47.

Kanmegne, G. and N.D. Omokolo. 2003. Changes in phenol content and peroxidase activity during *in vitro* organogenesis in *Xanthosoma sagittifolium* L. Plant Growth Regulation 40(1): 53–57.

Kay, D.E. (revised by Gooding, E.G.B.) 1987. Crop and Product Digest No. 2 - Root Crops, 2nd edition. Tropical Development and Research Institute, London. xv & 380 pp.

Krishnan, R. and M.L. Magoon. 1977. Edible aroids—new insights into phylogeny. *In:* Proceedings of the 3rd Symposium of the International Society for Tropical Root Crops (with IITA, Ibadan, Nigeria 1973). C.L.A. Leakey (ed.). pp. 58–60.

Lebot, V. 2009. Tropical Root and Tuber Crops: Cassava, Sweet Potato, Yams and Aroids. Wallingford, UK: CABI. 413 pp.

López, M., E. Vásquez and F. López. 1995. Raíces y tubérculos. Pueblo Educacion, Universidad Central de Las Villas, Cuba 312 pp.

Manner, H.I. 2011. Farm and Forestry Production and Marketing Profile for Tannia (*Xanthosoma* spp.). *In:* C.R. Elevitch (ed.). Specialty Crops for Pacific Island Agroforestry. Permanent Agriculture Resources (PAR), Holualoa, Hawai'i. http://agroforestry.net/scps.

Martin, F.N. and J.E. Lopper. 1999. Soil-borne plant disease caused by Pythium spp: Ecology, epidemiology and prospects for biological control. Critical Reviews in Plant Science 18: 111–181.

Mayo, S.J., J. Bogner and P.C. Boyce. 1997. The Genera of Araceae. The Trustees, Royal Botanic Gardens, Kew, U.K. xi + 370 pp.

Mbarga, J.B., G. Martijn Ten Hoopen, J. Kuaté, A. Adiobo, M.E.L. Ngonkeu, Z. Ambang, A. Akoa, P.R. Tondje and B.A.D. Begoude. 2012. *Trichoderma asperellum*: A potential biocontrol agent for *Pythium myriotylum*, causal agent of cocoyam (*Xanthosoma sagittifolium*) root rot disease in Cameroon. Crop Protection 36: 18–22.

Mbouobda, H.D., T. Boudjeko, P.F. Djocgoue, T.J.J. Tsafack and N.D. Omokolo. 2007. Morphological characterization and agronomic evaluation of cocoyam *Xanthosoma sagittifolium* L. Schott germplasm in Cameroon. J. Biol. Sci. 7(1): 27–33.

Mbouobda, H.D., P.F. Djocgoue, N.D. Omokolo, I. El Hadrami and T. Boudjeko. 2010. Benzo-(1,2,3)-thiadiazole-7-carbothioic S-methyl ester (BTH) stimulates defense reactions in *Xanthosoma sagittifolium*. Phytoparasitica 38(1): 71–79.

McDavid, C.R. and S. Alamu. 1979. Effect of daylength and gibberellic acid on the growth and promotion of flowering in tannia (*Xanthosoma Sagittifolium*). Tropical Agri. 56(1): 17–23.

Milian, M., I. Sanchez, M. Garcia, D. Guerra and A. Corrales. 2001. Variabilidad genética en el género *Xanthosoma* en Cuba. Plant Genetic Resources Newsletter 127: 11–14.

Morel, G.T. and C. Martin. 1952. Guérison de dahlias atteints d'une maladie à virus. CR Hebd. Seances Acad. Sc. 235: 1324–1325.

Morton, J.F. 1972. Cocoyams (*Xanthosoma caracu, X. atrovirens* and *X. nigrum*), ancient root- and leaf- vegetables, gaining in economic importance. Proceedings of the Florida State Horticultural Society 85: 85–94.

Murakami, K., J. Nishioka and S. Matsubara. 1998. Production of somatic hybrids in taro (*Colocasia esculenta* Schott). Journal of the Japanese Society for Horticultural Science 67(4): 607–612.

Murashige, T. and F. Skoog. 1962. A revised medium for rapid growth and bioassays with tobacco tissue cultures. Physiol. Plant 15: 473–497.

Ndzana, X., S. Zok and A.E. Sama. 2008. Preliminary Study on radiation sensitivity of *in vitro* cultures of *Xanthosoma* (macabo) in Cameroon. Plant Mutation Reports. 2(1): 10–12.

Ngouo, L.V. 1988. Contribution à l'étude des macabo (*Xanthosoma sagittifolium*) au Cameroun: identification et analyse de quelques obstacles à la production d'hybrides. Thèse 3ème cycle, Université de Yaoundé, Cameroun 128 p.

Nguyen, T.Q. and V.U. Nguyen. 1987. Aroid propagation by tissue culture: shoot tip culture and propagation of *Xanthosoma violaceum*. Hort. Science 22: 671–672.

Niemenak, N., A.M. Noah and D.N. Omokolo. 2013. Micopropagation of cocoyam (*Xanthosoma sagittifolium* L Schott) in temporary immersion bioreactor. Plant Biotechnol. Rep. 1–8.

Nyochembeng, L.M. and S. Garton. 1998. Plant regeneration from cocoyam callus derived from shoot tips and petioles. Plant Cell, Tissue and Organ Culture 53(2): 127–134.

Nyochembeng, L.M., C.A. Beyl and R.P. Pacumbaba. 2007. Peroxidase Activity, Isozyme Patterns and Electrolyte Leakage in Roots of Cocoyam Infected with *Pythium myriotylum*. Journal of Phytopathology 155: 454–461.

Nzietchueng, S. 1983. La pourriture racinaire du macabo (*Xanthosoma sagittifolium*) au Cameroun: symptomatologie de la maladie. Agronomie Tropicale 38: 321–325.

Offei, S.K., I.K. Asante and E.Y. Danquah. 2004. Genetic structure of seventy cocoyam (*Xanthosoma sagittifolium* (L.) Schott) accessions in Ghana based on RAPD. Hereditas 140(2): 123–8.

Omokolo, N.D., N.G. Tsala, G. Kanmegne and A.P. Balange. 1995. Production of multiple shoots, callus, plant regeneration and tuberization in *Xanthosoma sagittifolium* cultured *in vitro*. C. R. Acad. Sci. 318: 773–778.

Omokolo, N.D., T. Boudjeko and T.J.J. Tsafack. 2003. *In vitro* tuberization of *Xanthosoma sagittifolium* (L.) Schott: effects of phytohormones, sucrose, nitrogen and photoperiod. Sci. Horti. 98: 337–345.

Onokpise, O.U., J.G. Wutoh, X. Ndzana, J.T. Tambong, M.M. Mebeka, A.E. Sama, L. Nyochembeng, A. Agueguia, S. Nzietchueng, J.G. Wilson and M. Borns. 1999. Evaluation of macabo cocoyam germplasm in Cameroon. *In*: J. Janick (ed.). Perspectives on New Crops and New Uses. V.A. Alexandria ASHS Press, USA, pp. 394–396.

Onwueme, I.C. and W.B. Charles. 1994. Cultivation of cocoyam. *In:* FAO (ed.). Tropical Root and Tuber Crops: Production, Perspectives and Future Prospects. Plant Production and Protection Paper, Rome 126. pp. 139–161.

Oumar, D., E.A. Sama, A. Adiobo and S. Zok. 2011. Determination of ploidy level by flow cytometry and autopolyploid induction in cocoyam (*Xanthosoma sagittifolium*). African Journal of Biotechnology 10(73): 16491–16494.

Pacumbaba, R.P., J.G. Wutoh, A.E. Sama, J.T. Tambong and L.M. Nyochembeng. 1992. Isolation and pathogenicity of rhizosphere fungi of cocoyam in relation to cocoyam root rot disease. Journal of Phytopathology 135: 265–273.

Paul, K.K. and M.A. Bari. 2007. Protocol establishment for micro propagation and *in vitro* callus regeneration of Maulavi Kachu (*Xanthosoma sagittifolium* (L.) Schott) from cormel axillary bud meristem. J. Plant Sci. 2(4): 398–406.

Peng, H., J. Zhang and X. Wu. 2008. The ploidy effects in plant gene expression: progress, problems and prospects. Sci. China Ser. C-Life Sci. 51(4): 295–301.

Perneel, M., J.T. Tambong, A. Abiobo, C. Floren, F. Sabario, A. Léveque and M. Hofte. 2006. Intraspecific variability of *Pythium myriotylum* isolated from cocoyam and other host crops. Mycological Research 110: 583–593.

Purseglove, J.W. 1972. Tropical crops. *In: Monocotyledons* 1. Longman, London. pp. 58–75.

Ramesh, V., K. Susan John, C.S. Ravindran and S. Edison. 2007. Agrotechniques and Plant Nutrition of Tannia (*Xanthosoma* sp.): An Overview. Journal of Root Crops 33(1): 1–11.

Ramírez, P. 1985. Aislamiento y caracterización del virus del mosaico del dasheen (DMV) en Costa Rica. Turrialba 35(3): 279–283.

Razdan, M.K. and E.C. Coking (eds.). 2000. Conservation of plant genetic resources *in vitro*, Vol 2. Applications and limitations. Science Publisher, Enfield, NH.

Reyes, G. 2006. Studies on cocoyam (*Xanthosoma* spp.) in Nicaragua, with emphasis on Dasheen Mosaic Virus. Doctoral Thesis No. 2006:7, ISSN 1652-6880, ISBN 91-576-7056-0.

Reyes, G., A.C. Rönnberg-Wästljung and M. Nyman. 2006. Comparison of field performance between Dasheen mosaic virus-free and virus-infected *in vitro* plants of cocoyam (*Xanthosoma* spp.) in Nicaragua. Experimental Agriculture 42(3): 301–310.

Saborio, F., G. Umana, W. Solano, P. Amador, G. Munoz, A. Valerin, A. torres and R. Valverde. 2004. Induction of genetic variation in *Xanthosoma* spp. *In:* Genetic improvement of under-utilized and neglected crops in low-income food deficit countries through irradiation and related techniques. International Atomic Energy Agency (eds.). Vienna, Austria, pp. 143–154.

Sama, A.E., H.G. Hughes, M.S. Abbas and M.A. Shahba. 2012. An efficient *in vitro* propagation protocol of cocoyam (*Xanthosoma sagittifolium* (L.) Schott). Sci. World J. ID 346595. 10p. doi:10.1100/2012/346595.

Santos Pino, A., J.M. Cabrera, K.F. Gómez, T.J. López, C.M. Rayas, P.M. Basail, M.M. Pérez et al. 2011. Multiplicación en sistema de inmersión temporal del clon de malanga "Viequera" (*Xanthosoma* spp.). Rev. colomb. Biotecnol. 13(2): 97–106.

Takagi, H., N. Tien Thinh, O.M. Islam, T. Senboku, A. Sakai. 1997. Cryopreservation of *in vitro*-grown shoot tips of taro (*Colocasia esculenta* (L.) Schott) by vitrification. 1. Investigation of basic conditions of the vitrification procedure. Plant Cell Reports 16(9): 594–599.

Tambong, J.T. and M. Meboka. 1994. Cocoyam (*Xanthosoma sagittifolium* (L.)) hybridization studies: pollen viability and seed germination. Acta Horticulturae 380: 448–459.

Tambong, J.T., V.T. Sapra and S. Garton. 1998. *In vitro* induction of tetraploid in colchicine-treated cocoyam plantlets. Euphytica 104: 191–197.

Tambong, J.T. and M. Hofte. 2001. Phenazines are involved in biocontrol of *Pythium myriotylum* on cocoyam by *Pseudomonas aeruginosa* PNA1. European Journal of Plant Pathology 107: 511–521.

Thompson, S.A. and J.M.J. Dewet. 1983. *Xanthosoma* (*Araceae*): A taxonomic and ethnobotanic conspectus. 6th Symp. ISTRC Lima (peru) ; 20-25 Feb. (Roneotype) 17 p.

Tojo, M., H. Ono, C. Nakashima, S. Yoneyama and J.A.S. Jayakody. 2005. First report of Root Rot of Cocoyam Caused by *Pythium myriotylum* in Sri Lanka. Plant disease 89(10): 1132.

Tsafack, T.J.J., G. Charles, A. Hourmant, N.D. Omokolo and M. Branchard. 2009. Effect of photoperiod and thermoperiod on microtuberization and carbohydrate levels in Cocoyam (*Xanthosoma sagittifolium* L. Schott). Plant Cell Tiss. and Organ Cult. 96: 151–159.

Tsafack, T.J.J. 2010. Microtuberisation chez *Xanthosoma sagittifolium* L. Schott (Macabo) et analyse de quelques aspects physiologiques et biochimiques. M.S. Thesis, University of Yaoundé I, Cameroun, and University of Western Brittany- Brest, France.

Turner, J.M. and A.J. Messenger. 1986. Occurrence, biochemistry and physiology of phenazine pigment production. Advances in Microbial Physiology 27: 211–275.

Tyagi, R.K. and A. Yusuf. 2003. *In vitro* medium term storage of germplasm. *In:* Conservation Biotechnology for Plant Germplasm. B.B. Mandal, R. Chaudury, F. Engelmann, B. Mal, K.L. Tao and B.S. Dhillon (eds.). NBPGR, New Delhi/FAO, Rome, pp. 115–121, 157–161.

Vanderschuren, H., I. Moreno, R.B. Anjanappa, I.M. Zainuddin and W. Gruissem. 2012. Exploiting the Combination of Natural and Genetically Engineered Resistance to Cassava Mosaic and Cassava Brown Streak Viruses Impacting Cassava Production in Africa. PLoS ONE 7(9): e45277. doi:10.1371/journal.pone. 0045277.

Vickery, M.C. 1984. Ecology of tropical plants. Ed. John Wiley & Sons. Chichester—New York 170 p.

Watanabe, K.Z. 2002. Challenges in biotechnology for abiotic stress tolerance on root and tubers. JIRCAS Working Reports 75–83.

WCSP. 2013. World Checklist of Selected Plant Families. Facilitated by the Royal Botanic Gardens, Kew. Published on the Internet; http://apps.kew.org/wcsp/ Retrieved 2013-04-02.

Wilson, J.E. 1980. Cocoyam breeding by flower induction, pollination and seed germination. Manual series No. 4, International Institute for Tropical Agriculture, Ibadan, Nigeria 15 pp.

Wilson, J.E. 1984. Cocoyam. *In:* The Physiology of Tropical Field Crop. P.R. Goldsworthy and N.M. Fisher (eds.), John Wiley and Sons Ltd. New York, London, pp. 589–605.

Xu, T., N.D. Omokolo, N.G. Tsala and M.E. Ngonkeu. 1995. Identification of the causal agent of Cocoyam root rot disease in Cameroon. Acta Mycologica Sinica 14(1): 37–45.

Zandvoort, E.A., M.J.H. Hulshof and G. Staritsky. 1994. *In vitro* storage of *Xanthosoma* spp. under minimal growth conditions. Plant Cell, Tiss. Organ. Cult. 36(3): 309–316.

Zettler, F.W., M.J. Foxe, R.D. Hartman, J.R. Edwardson and R.G. Christie. 1970. Filamentous virus infecting dasheen and other araceous plants. Phytopathology 60: 983–987.

Zettler, W.F. and R.D. Hartman. 1987. Dasheen mosaic virus as a pathogen of cultivated aroids and control of the virus by tissue culture. Plant Dis. 71: 956–963.

Zettler, F.W., G.V.H. Jackson and E.A. Frison (eds.). 1989. FAO/IBPGR technical guidelines for the safe movement of edible aroid germplasm. International Board for Plant Genetic Resources, Rome (Italy). Food and Agriculture Organization of the United Nations, Rome (Italy). FAO/IBPGR. 24 p.

Zok, S., A.E. Sama, L. Nyochembeng, J.T. Tambong, X. Ndzana and J.G. Wutoh. 1998. Culture *in vitro* et multiplication rapide des plantes à tubercules et racines au Cameroun. Cahiers Agriculture 7. 63–66.

5

Status and Prospects of Biotechnology Applications to Conservation, Propagation and Genetic Improvement of Yam

Morufat Balogun[1,*] and *Badara Gueye*[2]

ABSTRACT

Yam plays a critical role in providing good quantity and quality yield and source of income to combat the challenge of food insecurity in the tropics. Although significant advances in science and technology have improved efficient management and improvement of their genetic resources, the mission of fully exploiting their potentials is yet to be maximally achieved. Lack of formal seed systems increases germplasm losses and reduces yields. Research is needed in the control of tuber dormancy in yam. Biotechnological approaches, especially *in vitro* culture of meristem/shoot tip combined with thermo/cryotherapy are promising methods to clean yam seeds of pathogens like viruses. The multiplication of clean yam seeds so obtained can be propagated

[1] Department of Crop Protection and Environmental Biology, University of Ibadan.
[2] International Institute of Tropical Agriculture, P.M.B. 5320, Ibadan, Nigeria.
* Corresponding author: kemtoy2003@yahoo.com

using improved systems such as temporary immersion bioreactors, aeroponics and photoautotrophic systems but knowledge gaps still exist on their use. Although the relative importance of clean materials, survival on farmers' fields, production cost and multiplication ratio should be determined and considered in deciding the propagation technique to adopt, conventional and improved tissue culture techniques will be indispensable in the production of clean seed yams. Marker-assisted selection, embryo culture, genetic transformation and genome sequencing have been initiated to support conventional genetic improvement, but investigations into the presence of transposable elements, site targeted mutagenesis, somatic embryogenesis and haploid plant production will be necessary to fast track the genetic improvement. These biotechnological approaches will not only enhance the use of disease-free, quality-declared planting materials but also facilitate germplasm exchange and speed up genetic improvement while providing excellent means for conservation.

Introduction

Yam (*Dioscorea* spp.) is a tuber-producing plant and is popular in the humid and sub-humid tropics, particularly in Africa, West Indies, parts of Asia and South and Central America. Knuth (1924) estimated that there are about 600 species in the genus *Dioscorea*. Yam species are annual or perennial vines and climbers with annual or perennial underground tubers. The Guinea yam (*Dioscorea rotundata* Poir. and *D. cayenensis* Lam.), are most preferred in Africa but in the Caribbean and Pacific, *D. alata* L. and *D. esculenta* (Lour.) Burk. are preferred. *D. bulbifera* L. also forms small aerial tubers or bulbils in the leaf axils. *D. hispida* Dennst. (Asia), *D. Dumetorum* (Knuth.) Pax. (Africa) and *D. trifida* L. (Central America and the Caribbean), *D. opposite* Thunb. and *D. japonica* Thunb. (China and Japan) are other edible species. Yam is a primary source of income in West Africa, from where 94% of global yam production emanates. The edible portion is the underground tuber which is a major source of carbohydrate. Nigeria produces 71% (FAO 2006) of global yam production, equivalent to 37 million tonne. The consumption of yam on a per capita basis on the West African coast is highest in Togo, followed by Cote d'Ivoire, Ghana, Benin Republic and Nigeria (Onwueme 1978, Kalu and Erhabor 1992). The most preferred form of eating yam is the boiled and pounded form. It is also fried in palm oil or roasted and eaten with oil, while yam flakes and chips are taken as snack (Orkwor 1998).

Yam production is constrained by abiotic factors, influence of pests and diseases and scarcity of planting material. In addition, tuber dormancy, which prevents year-round production and uncontrolled sprouting after dormancy break, which causes storage losses, also hampers productivity and reduces incomes and profits (Craufurd et al. 2001, Lang 1996). Consequently, its genetic resources should be used to support crop improvement programmes followed by production, propagation and distribution of quality planting materials to farmers for optimal yields. Biotechnology is the more reliable tool to achieve this. This chapter presents the state-of-the-art, constraints and way forward in yam genetic resources (GR) conservation, propagation and improvement.

Losses in Yam GR

Yam is susceptible to genetic erosion and variety loss. The losses are caused by multiple factors and are often related to their vegetative mode of propagation. The combination of population increase, negative consequences of climate change, and increased incidence of pest and diseases and natural disasters is a serious threat. *D. alata* anthracnose disease caused by *Colletotrichum gloeosporioides* (Penz.) Penz and Sac. (Mignouna et al. 2001, Pérez et al. 2003) caused losses in excess of 90% (Winch et al. 1984). *D. alata* and *D. rotundata* vine and foliage are also attacked by the beetle *Crioceri slivida,* the cricket *Gymnogryllus lucens* eats the tubers and cut the vines while the beetle *Heterolygusmeles* damages the tubers (Emehute et al. 1998). The nematode *Scutellonema bradys* (Steiner and Lehew) damages the tubers by forming superficial lesions, and this facilitates entrance of fungi. Chlorosis, vein-banding, flecking and leaf pluckering in *D. alata* indicate infection by the water yam virus (Thottaphilly 1992) while the yam mosaic virus was isolated from *D. rotundata* (Thottaphilly 1992) and *D. alata* (IITA 1993). In cases of severe infection, the whole plant appears stunted. Storage rots are caused principally by *Botryodiplodia theobromae* (Emehute et al. 1998), *Penicillium oxalicum, P. sclerotigenum, Aspergillus niger, Rhizoctonia solani* and *Fusarium oxysporum* (Ikotun 1983).

Climate and soil also affect yam production (Orkwor and Asadu 1998). Idah, an ancient town in Igala Kingdom in Nigeria, bordered by the River Niger produces about 20% of yams consumed in the country. However, in 2012, homes and farmlands were washed away according to the All Farmers Association of Nigeria (AFAN) and crops worth more than N400 million were lost in nine local government areas of Kogi state (Patrick 2012). At the other extreme, desertification is fast encroaching on Lake Chad basin of Northern Nigeria where farming activities have virtually ceased. The recent flood has been attributed to changes in rainfall pattern due to climate change, which is a serious challenge to agriculture and livelihood in sub-Saharan Africa. This is because agriculture is mostly rainfed in contrast to advanced countries where irrigation and out-of-season farming is common place and coping with climate change more practical. Consequently, food security of millions of people, whose lives rely on these staple, is also endangered.

Conservation of Yam GR

Safeguarding genetic resources of yam is critical for tackling plant biodiversity losses and ensuring their sustainable conservation and use as raw materials for crop improvement for global food security.

The conservation and use of yam GR follows the main policy development that impact the conservation of plant genetic resources in genebanks and lies within the context of availability and distribution of germplasm. The policies arose from the adoption of various international agreements including the Convention on Biological Diversity (CBD), the International Treaty on Plant Genetic Resources for Food and Agriculture (ITPGRFA), the International Plant Protection Convention (IPPC) and the World Trade Organization/Sanitary and Phytosanitary Agreement (WTO/SPS). In 2010, the CBD adopted the Nagoya Protocol on Access to Genetic Resources and Equitable Sharing of Benefits Arising from their Utilization, which has potential for impact upon germplasm exchange. Thus, the key principles genebank operations are the preservation of germplasm identity, maintenance of viability and genetic integrity, and the promotion of access. This includes associated information to facilitate use of the stored plant material in accordance with relevant national and international regulatory instruments. In general, it's about developing an efficient, effective, rational and transparent global system of *ex situ* conservation that provides optimal maintenance of germplasm and genetic integrity in gene banks, thereby ensuring access to, and use of, high quality germplasm of conserved plant genetic resources.

Untill now, the trend of biotechnology was to increase crop yield of clonal crops by propagation of elite/improved genotypes. In the context of climate change however, focus should be put on the maintenance of the adaptive potential of clonal crops by the understanding and the improvement of their mixed clonal/sexual reproductive system. This, of course, needs to take advantage of standardized farmer knowledge on sexual reproductive biology of clonal crops.

Vegetatively propagated crops like yam produce recalcitrant or intermediate heterogeneous seeds in few numbers and require a long life-cycle to generate planting materials. Consequently, they cannot be conserved as orthodox seeds and require other methods of conservation. Yam is conserved as live plants in field gene banks, as potted plants in enclosed structures or different plant parts are conserved into *in vitro* either in slow growth (medium-term) or cryopreservation (Long-term). DNA or pollen bank are also other conservation methods to maintain genes. The collection, conservation and duplication of yam has to be prioritized according to the repository size, resources and the material type (local landraces, introduced accessions, improved and breeding material). In case of large collection, a core collection can be defined with high priority in terms of availability for evaluation and breeding.

In situ, on Farm/Field Conservation of Yam GR

The first conservation system for yam is the *in situ* conservation, especially for local and crop wild relatives. However, this was highly under-documented and inadequately managed. A small portion of it was maintained in gene banks although *in situ* conservation is an important potential contributor to the overall conservation effort (Brush 1991). The yam collections are badly represented in protected areas and the global priority genetic reserve locations for wild relatives of yam are located in West Africa. The relevance of *in situ* conservation is about the direct involvement of farmers/smallholders, combining socio-cultural and ecological factors needed to understand characters and processes for the management of genetic diversity. For example, West African yams' genetic diversity management is a recombination between wild and cultivated forms which is an ongoing process that farmers manage (Hamon et al. 1995).

The procedures of yam field bank must be based on the maintenance of general principles like germplasm identity, purity and health, its reproductive capacity, its documentation availability and security. Yam is principally conserved vegetatively in field gene banks.

Agronomic parameters are important for yam field banking. Weed control is one of the most important factors. The effect of weed competition was observed to be responsible for the high yield loss, both the number and size (tuber weight), especially in drought environment (Wall and Friesen 1990). On white guinea yam (*Dioscorea rotundata*) field, chemical control gave better results in term of yield than hoeing on Congo grass (*Imperata cylindrica*) weed (Chikoye et al. 2006).

The major challenges of field conservation of yam are pests and diseases, mislabelling and duplication leading to germplasm losses. These obstacles have been major problems for international exchanges for diversification and global agriculture, especially for such root and tuber crops which are vegetatively propagated. This is due to the risk of spreading plant pathogens like virus, fungi, bacteria, phytoplasma and other diseases caused by microbes as well as insects and nematodes. However, many studies were carried out to explore the control of these clonal crop aggressors. The development of new and improved disease control method for viral diseases of vegetatively propagated staple food crops have to take into account the evolutionary response of the virus itself, through its biology and transmission (Van Den Bosch et al. 2007).

In vitro (tissue culture) Slow Growth Conservation and Cryopreservation

To support safer and longer conservation of yam genetic diversity for sustainable utilization, *in vitro* slow growth conservation and cryopreservation are the best methods. Biotechnological approaches will help to address future economical and environmental demands on yam (Pilatti et al. 2011). Conserving true seed in seed banks and cryobanks provides an additional safeguard against the risks (e.g., loss due to disease, climate change) of field conservation and permit the long-term conservation of a wider genetic base which offsets the labour and space intensive costs of conserving in the active growing state. Yam pollen storage at 0% relative humidity and –5°C for up to one year (Akoroda 1983) and at –80°C for more than two years (Daniel et al. 2002) were reported. Pollen storage using the wet–freeze procedure was recommended for the conservation of haploid gene pool of yams in base collections (Daniel et al. 1999). However, yam seed conservation can only be applied to female plants while pollen conservation is only applicable to male plants and non-flowering genotypes can only be conserved vegetatively (Balogun 2009).

Slow growth conservation leads to the reduction of loss risks associated with the field banks, and constitute a viable alternative to complement and reduce the large size required for field banks. Yam gene banks around the world have *in vitro* tissue culture facilities as a complementary conservation system, giving the possibility to clean the germplasm from diseases and pest via meristem culture and/or thermo-treatment. This conservation method requires technical expertise, facilities and operating budget. They are generally more economical and less risky in a long-term perspective; as compared to field collections. Plant tissue culture is a powerful tool for safer and faster way to multiply large quantity of material for distribution, duplication in other gene banks and international exchange (easier plant material transport); and also for breeding purposes. Slow growth storage is however for short- to medium-term conservation, after which the plantlets are subcultured when signs of deterioration/necroses are visible (Balogun 2009). The *in vitro* slow growth conservation of yam needs small quantities of material and allows longer duration between two regenerations or subcultures. The principle is to place the *in vitro* plantlets under slow growth conditions, through growth adaptation to cultural growth factors (light, temperature, culture medium, growth retardants). The *in vitro* conservation procedures details are available in IITA Manual (Dumet et al. 2007), also accessible at www.iita.org (IITA website).

Yam conservation is also done *in vitro* for safety duplication in some gene banks, using slow growth conditions. Apical and axillary buds or nodal cuttings are mainly used as initial explants for *in vitro* culture, which increases the possibility to initiate pathogen-free yam plantlets. Yam conservation duration under *in vitro* slow growth conservation varies according to species/genotype. Most of *D. alata* and *D. rotundata* can be stored up to 2 years under 16–18°C, 12 hr light a day, as described in IITA yam *in vitro* processing manual (Dumet et al. 2007). Taylor (1996b) reported yam conservation duration of 9–12 months at 20°C and Zamora and Paet (1996) only one year at 25–28°C for different yam species.

Cryopreservation, almost systematically associated with *in vitro* conservation, is another conservation method for yam germplasm. It is about the plant material maintenance at ultra-low temperature (in liquid nitrogen at –196°C) using cryogenic techniques. At such low temperature, plant cell biological activities and metabolism are stopped, eliminating the need to regularly rejuvenate or regenerate the plant. It is currently a supplementary tool to improve conservation of germplasm in a longer term perspective. Cryopreservation is the most reliable technique for long-term storage of plant genetic resources (Popov et al. 2005). It avoids the disadvantages of irreversible loss of totipotent competencies caused by *in vitro* ageing process (Benson 2008), time and labour consumption. Many studies confirmed that it is economically more competitive compared to other conservation systems (Harvengt et al. 2004, Reed et al. 2004, Keller et al. 2008). Cryopreservation helps to overcome many of the *in vitro* maintenance disadvantages such as labour-intensive subculturing, potential elimination of fungal and bacterial contaminants and somaclonal variation. It also ensures the safe long-term conservation of genetic resources. Thus, cryoconservation techniques have been increasingly used for Long Term Storage (LTS). In the last 25 years, several cryogenic techniques have been developed, especially those based on vitrification method (the transition of water directly from the liquid phase into an amorphous or "glassy" phase, whilst avoiding the formation of crystalline ice) such as encapsulation-dehydration, preculture-dehydration, and encapsulation/vitrification. Therefore, the main requirement for using cryopreservation method is that it should be simple, economical, reproducible and should allow relatively high regrowth rate (Leunufna and Keller 2003).

Various cryopreservation processes were described in the past for yam (Mandal et al. 1996, 2000, Malaurie et al. 1998, Leunufna and Keller 2005) with various levels of success depending on accessions and approach. To date, there is no generic process for cryopreservation of yam genetic resources.

Yam cryopreservation trials were already carried out using separate vitrification methods (Leunufna and Keller 2003, 2005, Gonzalez-Arnao et al. 2007) or encapsulation-dehydration (Malaurie et al. 1998). A comparative study between vitrification, encapsulation-dehydration and encapsulation-vitrification methods was done on yam apices by Mandal (2000).

Yam cryopreservation protocol setting is ongoing at IITA. High variability was observed in the ability of meristems to recover after exposure to liquid nitrogen, from one accession to another, within and across species (Dumet et al. 2012). However, the essence of the yam collection lies in its diversity. Somehow, the diversity of accession performance during the cryopreservation process echoes the diversity of the entire collection. The more diverse a collection, the more valuable it is, and the more likely adjustments will be needed to cryopreserve successfully the entire collection.

Propagation of Yam

In Africa, yam farmers still depend on the informal seed sources, including farm-savings from previous year (self supply), local markets and neighbours leading to use of poor quality seeds that accelerates build up of seed-borne diseases. This is because certified seeds are scarce. An efficient, cost-effective propagation system must therefore be developed (Asiedu et al. 1998, Quin 1998).

Traditional yam propagation has been done by field planting of whole tubers or large pieces ("sett") of between 200 to 500 grams. Setts may be pre-germinated in moist sawdust or coconut coir and more recently, carbonized rice husk. Thus, a large quantity of otherwise consumable tubers is reserved for planting yearly and this causes scarcity of propagules (Okoli et al. 1982), which alone, accounts for 50% of production costs (Nweke et al. 1991, Akoroda and Hahn 1995).

The "milking" technique (Okigbo and Ibe 1973) is also used, where tubers are harvested two-thirds into the growing season without destroying the root system. This provides early yam for home consumption and market. The parent plant then regenerates fresh small tubers from the corm at the base of the vine which are used as planting materials for the following season. This system therefore logically doubles the propagation ratio at the least relative to traditional method. These multiplication ratios are still very low (traditional: 1:6; Milking: 1:12), relative to other staple crops like cassava (1: 10) and some cereals (1:200) (Mbanaso et al. 2011). This relatively low multiplication rate reduces production and also delays the release of improved varieties. True seeds are also used in some cases when found on vines but not all genotypes flower. In addition, plants can be obtained from sexual seeds but the produced tubers are small compared to those from plants raised from tubers, due probably to small amount of stored food reserves in the seed (Okonkwo 1985).

The modified minisett technique (Ikeorgu et al. 2000, 2001, 2003, 2007, Kalu and Erhabor 1992), has reduced the production cost of yam (Okoli et al. 1982, Otoo et al. 1987) but the rate of adoption is still low (Kalu and Erhabor

1992). Increasing contact with extension workers, education, income and farm sizes of yam farmers were however suggested for a higher adoption (Nlerum 2009). In the vine rooting technique (Acha et al. 2004, Kikuno et al. 2007, Agele et al. 2010), about 20 cm long 3-node vines are rooted in carbonized rice husk, transplanted to the field to produce minitubers after 8 months. However, response varied among genotypes and percentage rooting and survival should be increased. To determine the propagation ratio of this technique, the number of nodes available per plant three months after planting should be determined. The differential response of genotypes may be a consequence of their being at different physiological stages at the time of vine collection since this was not pre-determined. Growth stages of yam, at which plant growth regulators were applied, were reported to affect rate of microtuber production *in vitro* (Balogun 2005). Use of histological techniques in determining growth phases of specific genotypes should be explored rather than the time after planting or sprouting since crop growth durations differ among genotypes. Multiplication rates are doubled using the partial sectioning technique (Nwosu 1975), but enormous manpower is needed for frequent examination and digging out to excise sprouted tuber sections for field planting. Also, the layering technique is unsuitable for farm use although applicable to some genotypes (Acha et al. 2004). Irrespective of the propagation technique, tuber dormancy, the programmed inability for growth in various types of plant meristematic apices, often in spite of suitable environmental conditions (Lang 1996) occurs in yam for about 5 months after harvesting and this limits out-of-season production (Craufurd et al. 2001).

Tissue culture techniques have been used for yam multiplication. Organogenesis from pre-formed meristems (Malaurie et al. 1995a, 1995b) have been reported in *D. zingiberensis* (Chen et al. 2003), shoot organogenesis from immature leaves (Kohmura et al. 1995) and roots (Twyford and Mantell 1996) of *D. opposita*, shoot (nodes) culture and microtuber formation in *D. composite, D. rotundata* and *D. alata* (Alizadeh et al. 1998, Balogun et al. 2006, John et al. 1993, Salazar and Hoyos 2007, Ovono 2007). Tuber pieces were reported not to produce *in vitro* plantlets in *D. rotundata, D. trifida* and *D. cayenensis* (Mitchell et al. 1995) while it did in *D. alata* (Fosto et al. 2013). Meristem culture combined with heat therapy has been successfully used to produce virus-tested plantlets, which are used in rapid multiplication of superior clones (Mantell et al. 1980, Ng 1984, 1992). This technique ensures that the viral inoculum is not passed on to subsequent generations. Gibberellin inhibitors were reported to induce multiple shoots from nodal explants in *D. oppositifolia* and *D. pentaphylla* (Poornima et al. 2007) and numerous axillary nodes in *D. alata* and *D. rotundata* (Bimbaun et al. 2002, Balogun 2005). Younger and more vigorous mother plants had higher rate of micropropagation *in vitro* (Mitchell et al. 1995a).

The use of synthetic seeds (Standardi and Piccioni 1998) for yam propagation is a possibility, although it will require conversion. This is because shoots and buds do not have root meristems and they must regenerate roots in order to be able to convert (Piccioni 1997). Synthetic seeds can be used to time production cycles in micropropagation laboratories if the development of the plant could be properly directed towards proliferation and rooting. In contrast, somatic embryos are bipolar structures and have been reported in *D. floribunda, D. composita, D. alata* and *D. bulbifera* (Ammirato 1978, 1982). Recently, (Padron et al. 2011) somatic embryos were induced from leaf tissues of *D. rotundata* in medium containing 2.4-D and incubated in darkness. However, low induction frequencies (<30%) were recorded and protocols will have to be optimized. In *D. alata* and *D. opposite*, embryogenic cell masses were induced from root explants in liquid MS supplemented with 2.4-D and cultured in light (Twyford and Mantell 1996, Nagasawa and Finer 1989). Germination of somatic embryos of *D. alata* increased in the presence of GA3 (Deng and Cornu 1992, Twyford and Mantell 1996). Plantlet recovery from somatic embryos of *D. rotundata* was enhanced at 4.5% sucrose but not affected by benzylaminopurine (Okezie et al. 1994, Pandro et al. 2011). These reports point to probable genotype-dependent protocol for yam embryogenesis.

Meristem and/or shoot tip culture combined with thermotherapy (Kartha and Gamborg 1975) cleaned cassava plantlets of leaf distortion symptoms. Cryotherapy (Wang et al. 2009) has also been shown to be capable of eliminating some viral diseases. However, this is yet to be integrated into the operational seed distribution system. It is also possible to increase multiplication rates using advanced tissue culture technology like temporary immersion systems.

For yam, low multiplication rates limit the use of *in vitro*-produced, virus-tested plantlets in conventional tissue cultures (about 1:4; Malaurie et al. 1995a,b, Medero et al. 1999, Chu and Ribeiro 2002, Borges et al. 2004, Ondo et al. 2007) in addition to losses during transplanting. In germplasm transfer where conditions of transit are unpredictable (Ng 1988), losses are encountered. Microtubers produced from *in vitro* plantlets were therefore suggested to be adequate for conservation (Balogun 2009) due to their hardiness, small size and easy establishment in the soil. In addition, they can be kept for a longer period due to microtuber dormancy (Ng 1988). Plant raised from microtubers produced more tubers on the field than plants from tuber crowns (Jova et al. 2011). This will not only facilitate international germplasm exchange, but also provide excellent means of storage. However, research is needed in the area of further increasing the size of microtubers, control of microtuber dormancy and applicability to more of the economically important genotypes.

The use of temporary immersion bioreactor systems (TIBs) will reduce culture losses by contamination from handling due to automation of the system while optimizing the rate of plantlet and microtuber production and sprouting (Jova et al. 2011, Watt 2012). In TIBs, growth is enhanced (Escalona 2006) since there is lack of continuous immersion in liquid medium, contamination is controlled while aeration is timed. In addition, microtubers from TIBs can be grown on the field and used in original seed production programmes. Due to enhancement of growth in TIBs, it can be explored for the production of medicinal secondary metabolites like yam steroidal diosgenins (Raju and Rao 2012, Tripathi and Tripathi 2003) as done for fenugreek (Sharareh 2011). Shoots grown in TIBs had enhanced growth and the leaves had higher photosynthetic pigment content than other techniques (Jova et al. 2011, 2012, Cabrera et al. 2011).

Photoautotrophic propagation (PAP) is another system that should be tested for yam. In PAP, explants are directed towards autotrophy while in culture by reducing or completely substituting sucrose (which directs towards heterotrophy) with carbon dioxide. Plantlet growth is enhanced in photoautotrophic more than heterotrophic conditions if environmental control is adequate (Hazarika et al. 2003, Xiao 2004, Afreen 2005). However, after about 4 weeks, plants may stop responding to carbon dioxide and growth is reduced due to slight closing of the stomata on the bottom of the leaves as plants sense high CO_2. Use of CO_2 enrichment for 2 weeks, then a week off intermittently will ensure higher yields as the plant continuously seeks as much CO_2 as possible (Andrew 2002). It may be worthwhile to investigate the performance of *in vitro* yam plantlets in a combination of TIBs and PAP. In potatoes, immense work has been done on PAP and photomixotrophic propagation (Mohamed and Alsadon 2010).

Aeroponics, being used in potato propagation (CIP 2008), is yet to be reported for other root and tuber crops like yam. In the case of potato, yield in aeroponics was 14–18 times greater than from the conventional methods (Badoni and Chauhan 2010, Muthoni et al. 2011, Chiipanthenga et al. 2012). Pumps are used to spray or produce a mist of nutrient solution around the root zone using micro-nozzles. Interplant root contact is minimal and this aids sanitation while the soilless technique avoids soil-associated plant pests. The system is a much higher aerated environment than traditional hydroponics and this reduces the risk of microbial contamination which limits productivity in tissue cultures. In addition, the technique has proven to be an efficient system for use of biofertilizer, e.g., growing vesicular arbuscular micorrhizal fungi inoculum without a physical substrate (Hung and Sylvia 1988) which has the potential of reducing the use of inorganic fertilizers. Irrespective to the propagation technique used, it is necessary to determine the relative importance of clean materials, percentage survival after transplanting, cost, and multiplication ratio in deciding the technology to adopt.

Biotechnology for Yam Improvement

Smallhold farmers need access to innovative ways of coping with production constraints of yam and cassava. Selection and breeding for genotypes with high nutrient use efficiency, nutrient profile and adaptation to abiotic factors (drought, flood, etc.) in addition to adequate control of duration of yam tuber dormancy to achieve year-round production is indispensable for smallhold farmers to produce optimally.

The creation of genetic variations, followed by selection and identification of those that are desirable remain the basis of viable improvement programmes. Yam genotypes exhibit wide variations (Ng and Ng 1997) which provides a broad germplasm base for selection and breeding when conserved. Njualem et al. (2007) reported farmer/researcher participatory evaluation of 53 local landraces of yam in the Cameroon (Otoo 2003) while Oke et al. (2013) reported variations in pasting of starch from *D. alata* genotypes. In terms of duration of dormancy, some genotypes of *D. rotundata* from the drier area of the savannah keep for longer time (can be stored for 3–4 months and sometimes for longer periods (Coursey 1967).

Conventional yam hybridization (Lopez et al. 2012) breeding involves production of botanic seeds from bi-parental crosses or in poly-cross fields among genotypes. This is followed by seedling evaluation and selection in nurseries, and identification of superior genotypes in clonal trials. Consequently, populations developed for specific traits are improved over years, principally through recurrent selection. In collaboration with the National Agricultural Research Systems, this method has led to the release of 16 high-yielding, disease-resistant cultivars and good organoleptic attributes in Nigeria and Ghana in the last decade (Lopez et al. 2012). However, this method takes 6–9 years and is from intraspecific crosses. Production of haploid plants through anther culture of yams will shorten the breeding time by doubling haploids and early selection. There is a great challenge in inter-specific hybridization due to lack of synchronization of flowering and cross-compatibility, especially in crossing either of *D. rotundata* or *D. cayenensis* to *D. alata*. Flowering also varied with season and location (Hamadina et al. 2009). Flowering was induced in tissue culture of inflorescences of *Dioscorea zingberensis* C.H. Wright (Yang et al. 2009), which if developed for economically important genotypes will allow for location-independent, *in vitro* screening of flowering responses among genotypes. Biotechnology can overcome these limitations in the area of development of protocols for embryo rescue for interspecific crosses (Amazue 2009) genetic transformation and protoplast fusion. Genetic transformation (Tor et al. 1993) using a biolistic particle gun to deliver DNA into intact yam cells of *D. alata* resulted in the recovery of the transformed cell. It is however necessary that protocols for successful regeneration of cultivated yams through somatic embryogenesis is developed for field evaluation of transformed plants.

In creating new genetic variations, mutation breeding was explored at the National Root Crops Research Institute and putative mutants were selected (Nwachukwu and Obi 2001). Exploiting the use of site-targeted mutagenesis (Kunkel 1985, Suprasanma et al. 2012) will speed up the mutation breeding process. Transposable elements (TEs) (McClintock 1950, Kidwell 1992, Schnable and Peterson 1988, McCarty et al. 2005) if found in yam may facilitate genetic modification for desirable traits. The TEs can be used to generate insertion mutations, map them and facilitate gene cloning and generation of transgenic plants. Although TEs (P elements) were used to modify genes in Drosophila by inserting into developing embryo *in vivo* (Rubin et al. 1982, Spradling et al. 1982), immature embryo/endosperm culture may also have a role to play in the genetic modification of yam using transposable elements. Somatic embryogenesis can be explored in generating somaclonal variants with desirable traits.

Identification of molecular markers that are linked to desirable traits will facilitate selection in breeding programmes. However, these require mapping populations. In yam, only two mapping populations were reported for each of *D. alata* and *D. rotundata* until 2003 (Mignouna et al. 2002a, 2002b, 2003). A suitable marker system is also being developed at the IITA for *D. alata* and *D. rotundata*. In 2011, Alieu and Asiedu reported 3 F1 full-sibs mapping populations in *D. rotundata* for virus disease, multiple tuber production, cooking quality or tuber texture and 4 in *D. alata* for anthracnose disease, cooking quality and oxidation. However, sizes of the mapping populations reduced due to plant death and tuber rotting. It was reported from the sequence repeat marker analysis that the parental crosses were successful and the progeny were true hybrids. Sequencing of the Dioscorea genome has been initiated (Bhattacharjee et al. 2012), conclusion of which should fast-track genetic improvement of yams by application of molecular techniques in marker-assisted selection and site-specific mutagenesis, among others.

In vitro screening will enhance selection for desirable traits such as tolerance to abiotic factors, disease resistance and production of secondary metabolites. In fenugreek, Rezaeian (2011) found higher diosgenin content in calli produced from leaf than root or shoot explants, such that improving callus production efficiency in medicinal yam (Gomez et al. 2004) for use by pharmaceutical companies may be worthwhile. Yam grown in TIBs can be used to investigate cellular pathways and processes (Ivanov et al. 2012, Tripathy and Tripathy 2003) as in exo/endogenous control of tuber dormancy by stage-wise application of growth factor.

Conclusion

Yams are in urgent need of research for the optimization of protocols for sustainable conservation, formal seed systems, efficient micro/macro-propagation and genetic improvement. Biotechnological approaches of *in vitro* medium-term storage and cryopreservation (long-term conservation) have increased efficiency of conservation of yam genetic resources. However, it is necessary to increase the genotype inspecificity, duration of storage and percentage survival after recovery, especially for cryopreservation. In improving production and propagation rates of disease-free plantlets, many aspects of conventional tissue culture like meristem culture, shoot tip culture, direct organogenesis combined with thermo or cryotherapy has been reported for yam. It is necessary to develop protocols toward more automation and macropropagation systems, ranging from temporary immersion bioreactors through aeroponics to photoautotrophic propagation so as to reduce transplanting losses and boost yam seed system for economically important genotypes. Marker-assisted selection, embryo culture/rescue, genetic transformation, genome sequencing have been reported but need to be optimized and concluded for yam. Investigations into the presence of transposable elements, site targeted mutagenesis, somatic embryogenesis and haploid plant production will be necessary to fast track genetic improvement. More importantly, research should focus on protocol setting and adaptation for the conservation, propagation and improvement techniques with respect to recalcitrant genotypes of yam.

Acknowledgement

The authors are grateful to Dr. Norbert Maroya and Prof. Michael Abberton of the International Institute of Tropical Agriculture, Nigeria for reviewing the manuscript.

References

Acha, I.A., H. Shiwachi, R. Asiedu and M.O. Akoroda. 2004. Effect of auxins on root development in yam (*Dioscorea rotundata*) vine. Tropical Science 44: 80–84.

Afreen, F. 2005. Physiological and anatomical characteristics of *in vitro* photoautotrophic plants. *In:* T. Kozai, F. Afreen and S.M.A. Zobayed (eds.). Photoautotrophic (sugar-free medium) micropropagation as a new propagation and transplant production system. 1st edn. Dordrecht, Springer, pp. 61–90.

Agele, S.O., T.G. Ayankanmi and H. Kikuno. 2010. Effects of synthetic hormone substitutes and genotypes on rooting and mini tuber production of vines cuttings obtained from white yam (*Dioscorea rotundata*, Poir). African Journal of Biotechnology Vol. 9(30), pp. 4714–4724.

Akoroda, M.O. 1983. Long-term storage of yam pollen. Sci. Hortic. 20: 225–230.

Akoroda, M.O. and S.K. Hahn. 1995. Yams in Nigeria: status and trends. A Workshop Report in African J. of Root and Tuber Crops (AJRTC), Vol. 1, No. 1, 38–41.

Alieu, S. and R. Asiedu. 2011. Development of mapping populations for genetic analysis in yams (*Dioscorea rotundata*, Poir. and *Dioscorea alata* L.). African Journal of Biotechnology Vol. 10(16), pp. 3040–3050.

Alizadeh, S., S. Mantell and A. Viana. 1998. *In vitro* culture and microtuber induction in the steroid yam *Dioscorea composita* Hemsl. Plant Cell, Tissue and Organ Culture 53(2): 107–112.

Amazue. 2009. Developing an effective protocol for embryorescue in *Dioscorea* spp. M. Sc Thesis, University of Nigeria, Nsukka.

Ammirato, P. 1978. Somatic embryogenesis and plantlet development in suspension cultures of the medicinal yam, *Dioscorea floribunda*. American Journal of Botany 65(27): 89–95.

Ammirato, P. 1982. Growth and morphogenesis in cultures of the monocot yam *Dioscorea*. *In:* A. Fujiwara (ed.). Plant Tissue Culture. Proceedings 5th International Congress of Plant Tissue and Cell Culture. Japanese Association for Plant Tissue Culture Maruzen, Tokyo, Japan, pp. 167–170.

Andrew, S. 2002. Carbon dioxide enrichment.www.hydrocentre.com.au. On the internet.

Asiedu, R., S.Y.C. Ng, K.V. Bai, I.J. Ekanayake and N.M.W. Wanyera. 1998. Genetic Improvement. *In:* Food yams. Advances in research. G.C. Orkwor, R. Asiedu and I.J. Ekanayake (eds.). IITA/ NRCRI. pp. 63–104.

Badoni A. and J.S. Chauhan. 2010. Conventional vis-a-vis Biotechnological Methods of Propagation in Potato: A Review. Stem Cell. 1: 1–6.

Benson, E.E. 2008. Cryopreservation of phytodiversity: A critical Appraisal of Theory & practice. Critical Reviews in Plant Sciences 27: 3, 141–219.

Bhattacharjee, R., M. Gedil and A. Lopez-Montes. 2012. R4D Review. International Institute of Tropical Agriculture, Ibadan Issue 8.http://r4dreview.org/2012/05/genomics-for-transforming-yam-breeding/.

Balogun, M.O. 2005. Development of microtuber production and dormancy control protocols for yam (*Dioscorea* sp.) germplasm conservation. Ph.D. Thesis, University of Ibadan.

Balogun, M.O., I. Fawole, S.Y.C. Ng, Q. Ng, H. Shiwachi and H. Kikuno. 2006. Interaction among cultural factors in microtuberization of white yam (*Dioscorea rotundata* Poir). Trop. Sci. 46(1): 55–59.

Balogun, M.O. 2009. Microtubers in yam germplasm conservation and propagation: The status, the prospects and the constraints. Biotechnol. Mol. Biol. Rev. 4(1): 1–10.

Bimbaun, E.H., S.H. Lips and I.E. Nwana. 2002. An integrated approach to improve production of starch tubers. USDA Project Number: C13-153. Final report submitted to the U.S. Agency for International Development; Bureau for Global Problems. Online: http://pdf.usaid.gov/pdf_docs/PNACP601.pdf.

Borges, M., S. Meneses, N. Aguilera and J. Vazquez. 2004. Regeneration and multiplication of *Dioscorea alata* germplasm maintained *in vitro*. Plant Cell Tissue Organ Cult. 76: 87–89.

Brush, S.B. 1991. A Farmer-Based Approach to Conserve Crop Germplasm. Economic Botany, Vol. 45, No. 2 (Apr. - Jun., 1991), pp. 153–165.

Cabrera, M., R. Gómez, E. Espinosa, J. López, V. Medero, M. Basail and A. Santos. 2011. Yam (*Dioscorea alata* L.) microtuber formation in Temporary Immersion System as planting material. Biotecnol. Apl. Vol. 28, 4.

Chen, Y., J. Fan, F. Yi, Z. Lou and Y. Fu. 2003. Rapid clonal propagation of *Dioscorea zingiberensis*. Plant Cell, Tissue and Organ Culture 73(1): 75–80.

Chen, F.Y., D. Wang, X. Gao and L. Wang. 2007. The effect of plant growth regulators and sucrose on the micropropagation of *Dioscorea nipponica* makino. Plant Growth Regul. 26: 38–45.

Chiipanthenga, M., M. Malirol, P. Demo and J. Njoloma. 2012. Potential of aeroponic system in the production of quality potato (*Solanum tuberosum* L.) seed in developing countries. African Journal of Biotechnology. Vol. 11(17), pp. 3993–3999.

Chikoye, D., J. Ellis-Jones, P. Kormawa, E. Udensi, S.E. Ibana and T. Avav. 2006. Options for Cogongrass (*Imperata cylindrica*) Control in White Guinea Yam (*Dioscorea rotundata*) and Cassava (*Manihot esculenta*). Weed Technology, Vol. 20, No. 3, pp. 784–792.

Chu, P.E. and R. Ribeiro. 2002. Growth and cabohydratehanges in shoot cultures of *Dioscorea* species as influenced by photoperiod, exogenous sucrose and cytokinin concentractions. Plant Cell Tissue Organ Cult. 70: 241–249.

CIP. 2008. International Potato Centre. Quality Seed Potato Production Using Aeroponics. Lima. Peru.

Coursey, D.G. 1967. Yams. An account of the nature, origins, cultivation and utilization of the useful members of Discoreaceae. Longmans. London.

Craufurd, P.O., R.J. Summerfield, R. Asiedu and V. Prasad. 2001. Dormancy in Yams. Experimental Agriculture 37: 75–109.

Daniel, I.O., N.Q. Ng ,T.O. Tayo and A.O. Togun. 1999. West African yam seeds stored under dessicated and cold storage conditions are orthodox. Seed Sci. Technol. 27: 969–975.

Daniel, I.O., N.Q. Ng, T.O. Tayo and A.O. Togun. 2002. Wet-cold preservation of West African yam (*Dioscorea* spp.) pollen. J. Agric. Sci. 138: 57–62.

Deng, M. and D. Cornu. 1992. Maturation and germination of walnut somatic embryos. Plant Cell, Tissue and Organ Culture 28(2): 195–202.

Dumet, D., O. Ojuderie and A. Adeyemi. 2007. IITA Genebank Manual Series, Yam *in vitro* genebanking International Institute of Tropical Agriculture, Ibadan, Nigeria.

Dumet, D., E. Diebiru, A. Adeyemi, O. Akinyemi, B. Gueye and J. Franco. 2012. Cryopreservation for the in perpetuity conservation of yam and cassava genetic resources. Cryoletters. In press.

Emehute, J.K.U., T. Ikotun, E.C. Nwauzor and H.N. Nwokocha. 1998. Crop Protection. *In:* Food yams. Advances in research. G.C. Orkwor, R. Asiedu and I.J. Ekanayake (eds.). IITA / NRCRI, pp. 143–186.

Escalona, M. 2006. Temporary inmersion beats traditional techniques on all fronts. Prophyta annual. pp. 48–50.

FAO. 2006. Statistical data on crop, area, yield for 2006. FAOSTAT.www.fao.org Food and Agriculture of the United Nations. Rome.

FAO. 2008. The International Year of Potato. The Global Crop Diversity Trust and FAO's Plant Production and Protection Division. Rome, Italy. www.potato2008.org.

Fotso, Ngo Ngwe Marie Florence Sandrine, Mbouobda Hermann Désiré, Djocgoue Pierre François and OmokoloNdoumou Denis. 2013. Micropropagation of *Dioscorea alata* L. from microtubers induced *in vitro*. African Journal of Biotechnology 12(10): 1057–1067.

Gomez, P., A. Ortuo and J.A. Del Rio. 2004. Ultastructural changes and diosgenin content in cell suspensions of *Trigonella foenum-graecum* L. by ethylene treatment. Plant Growth Regulation 44: 93–99.

Gonzalez-Arnao, M.T., A. Panta, W.M. Roca, R.H. Escobar and F. Engelmann. 2007. Development and large scale application of cryopreservation techniques for shoot and somatic embryo cultures of tropical crops. (http://www.springerlink.com/content/y42117245j364361/fulltext.pdf).

Hamadina, E.I., P.Q. Craufurd and R. Asiedu. 2009. Flowering intensity in White yam (*Dioscorea rotundata*). J. Agric. Sci. pp. 469–477.

Hamon, P., R. Dumont, J. Zoundjihekpon, B. Tio-Toure and S. Hamon. 1995. Wild yams of West Afria. Morphological characteristics. *Ed. De l'Orsom,* Paris, pp: 385–400.

Harvengt, L., A. Meir-Dinkel, E. Dumas and E. Collin. 2004. Establishment of a cryopreserved genebank of European elms. Can. J. For. Res. 34: 43–55.

Hazarika, V., A. Parthasarathy and V. Nagaraju. 2003. Photoautotrophic micropropagation—a review. Agric. Rev., 24(2): 152–156, 2003

Hung, L.L. and D.M. Sylvia. 1988. Production of vesicular-arbuscularmycorrizal fungus inoculum in aeroponic culture. Appl. Environ. Microbiol. 54: 353–357.

Ikeorgu, J.E.G., H.N. Nwokocha and M.C. Ikwelle. 2000. Seed yam production through the minisett technique: recent modifications to enhance farmer adoption. Proc. 12th International Symposium of ISTRC held in Tsukuba Japan. 10–16 Sept 2000, pp. 372–375.

Ikeorgu, J.E.G. and H.N. Nwokocha. 2001. Development of the yam minituber technique for seed yam production. Niger. Agric. J. 32: 97–108.

Ikeorgu, J.E.G. and M.C. Igbokwe. 2003. Seed yam production with minitubers. Niger Agric. J. 34: 63–67.

Ikeorgu, J.G., H. Oselebe, J. Oluwatayo, K. Ugwuoke, U. Ukpabi and R. Asiedu. 2007. Farmer Participatory Evaluation of Four Hybrid Water Yam Clones in the Yam Belt of Nigeria. *In:* B. Nkamleu, D. Annang and N.M. Bacco (eds.). Securing Livelihoods through Yams. Proceedings of a technical workshop on progress in yam research for development in West and Central Africa held in Accra, Ghana, 11–13 September 2007, pp. 226–230.

IITA (International Institute of Tropical Agriculture). 1993. Genetic Resources Unit. Annual Report, 1992. Crop Improvemet Division, IITA Ibadan, Nigeria, 38p.

Ikotun, T. 1983. Post harvest microbial rot of yams in Nigeria. Fitopathologia Brasileira 8: 1–7.

Ivanov, I., V. Georgiev, S. Berkov and A. Pavlov. 2012. Alkaloid patterns in Leucojumaestivum shoot culture cultivated at temporary immersion conditions. J. Plant. Physiol. 169: 206–211.

John, J.L., W.H. Courtney and D.R. Decoteau. 1993. The influence of Plant Growth Regulators and light on microtuber induction and formation in *Dioscorea alata* L. cultures. Plant Cell Tissue and Organ Culture 34: 245–252.

Jova, M.C., R.G. Kosky, M.B. Perez, A.S. Pino, V.M. Vega, J.L. Torres, A.R. Cabrera, M.G. Garcia and J.C. de Venture. 2005. Production of yam microtubers using a temporary immersion system. Plant Cell. Tiss. Organ Cult. 83: 103–107.

Jova, M.C., R. Gómez Kosky and E. Espinosa Cuellar. 2011. Effect of liquid media culture systems on yam plant growth (*Dioscorea alata* L. 'PacalaDuclos') Biotechnol. Agron. Soc. Environ. 15(4): 515–521.

Jova, M.C., R. Gómez Kosky, A.R. Cabrera, M. De Feria, M.B. Perez, V.M. Vega and J. López Torres. 2012. Performance of yam microtubers from temporary immersion system in field conditions. African Journal of Biotechnology 10(46): 9509–9517.

Kalu, B.A. and P.O. Erhabor. 1992. Production and economic evaluation of white Guinea yam (*Dioscorea rotundata*) minisetts under ridge and bed production systems in a tropical Guinea savanna location, Nigeria. Tropical Agriculture 69: 78–82.

Keller, E.R.J., A. Kaczmarczyk and A. Senula. 2008. Cryopreservation for plant Genebank: a matter between high expectations and cautions reservations. Cryolett. 29: 53–62.

Kidwell, M.G. 1992. Horizontal transfer of P elements and other short inverted repeat transposons. Genetica 86(1): 275–286.

Kikuno, H., R. Matsumoto, H. Shiwachi, H. Youohara and R. Asiedu. 2007. Comparative effects of explants sources and age of plant on rooting, shooting and tuber formation of Vine cutting of yams. Japanese J. of Trop. Agric. Vol. 51 Extra issue 2: 71–72.

Knuth, R. 1924. Dioscoreaceae. *In*: A. Engler (ed.). Das Pflanzenreich 87(IV-43): 1–387.

Kohmura, H., H. Araki and M. Imoto. 1995. Micropropagation of 'yamatoimo' Chinese yam (Dioscorea opposite Thum.) from immature leaves. Plant Cell, Tissue and Organ Culture 40: 271–276.

Kunkel, T.A. 1985. HYPERLINK "http://www.ncbi.nlm.nih.gov/pmc/articles/PMC397064/pdf/pnas00342-0237.pdf" Rapid and efficient site-specific mutagenesis without phenotypic selection. Proceedings of the National Academy of Sciences USA 82(2): 488–92.

Lang, G.A. 1996. Plant Dormancy, Physiology, Biochemisry and Molecular Biology. Wallingford: CABI Publishing.

Leunufna, S. and E.R.J. Keller. 2003. Investigating a new cryopreservation protocol for yam yams (*Dioscorea* spp.). Plant Cell Rep. 21: 1159–1166.

Leunufna, S. and E.R.J. Keller. 2005. Cryopreservation of yam using vitrification modified by including droplet method: effects of cold acclimation and sucrose. Cryolett. 26: 93–102.

Lopez-Montes, R. Bhattacharjee and G. Tessema. 2012. Yam breeding at IITA: achievements, challenges, and prospects. R4D Review. International Institute of Tropical Agriculture, Ibadan Issue 8. http://r4dreview.org/2012/05/yam-breeding-at-iita-achievements-challenges-and-prospects/.

Malaurie, B., O. Pungu and M. Trouslot. 1995a. Effect of growth regulators concentrations on morphological development of meristem tips in *Dioscorea cayenensis-D.rotundata* complex and *D. praehensilis*. Plant Cell Tissue Organ Cult. 41: 229–235.

Malaurie, B., O. Pungu and M.F. Trouslot. 1995b. Influence of meristem-tip size and location on morphological development in *Dioscorea cayenensis-D. Rotundata* complex and *D. praehensilis*. Plant Cell Tissue Organ Cult. 42: 215–218.

Malaurie, B., M.F. Trouslot, F. Engelmann and N. Chabrillange. 1998. Effect of pretreatment conditions on the cryopreservation of *in vitro*-cultured yam (*Dioscorea alata* "BrazoFuerte" and *Dioscorea Bulbifera* "NoumeaImboro") shoot apices by Encapsulation-dehydration. Cryolett. 19: 15–26.

Mandal, B.B., K.P.S. Chandel and S. Dwivedi. 1996. Cryopreservation of yam (*Dioscorea* spp.) shoot apices by encapsulation-dehydration CryoLetters 17: 165–174.

Mandal, B.B. 2000. Cryopreservation of yam apices: comparative study with 3 different techniques. *In*: F. Engelmann and Takagi Hiroko (eds.). Cryopreservation of tropical plant germplasm: Current research progress and application. pp. 233–244.

Mantell, S.H., S.Q. Haque and A.P. Whitehall. 1980. Apical meristem tip culture for virus eradication of flexuous viruses in yams (*Dioscorea alata*). Tropical Pest Management. 26(2): 170–179.

Mbanaso, E.N.A., C.N. Egesi, E. Okogbenin, A.O. Ubalua and C.K. Nkere. 2011. Plant Biotechnology for genetic improvement of root and tuber crops. *In:* C.O. Amadi, K.C. Ekwe, G.O. Chukwu, A.O. Olojede and C.N. Egesi (eds.). Root and Tuber Crops Research for food security and empowerment, National Root Crops Research Institute, Umudike, Nigeria pp. 45–64.

McCarty, D.R., A.M. Settles, M. Suzuki, B.C. Tan, S. Latshaw, T. Porch, K.J. Robin, J. Baier W. Avigne, J. Lai, J. Messing, K.E. Koch and L.C. Hannah. 2005. Steady-state transposon mutagenesis in inbred maize. Plant J. 44(1): 52–61.

McClintock, B. 1950. The origin and behavior of mutable loci in maize. Proc. Natl. Acad. Sci. USA 36(6): 344–355.

McKey, D., M. Elias, B. Pujol and A. Duputié. 2010. The evolutionary ecology of clonally propagated domesticated plants. New Phytologist 186(2): 318–332.

Medero, V., L. Del Sol and M. García. 1999. Metodología para la propagación del clon de ñame Blanco o Pelú. Resúmenesdel Bio Cat 99, Granma Cuba 5–7 de Octubre, Proc. Abstract pp. 12.

Mignouna, H., M. Abang, K. Green and R. Asiedu. 2001. Inheritance of resistance in water yam (*Dioscorea alata*) to anthracnose (*Colletotrichum gloeosporioides*). Theoretical and Applied Genetics 103(1): 52–55.

Mignouna, H.D., R.A. Mank, T.H.N. Ellis, N. Van den Bosch, R. Asiedu, S.Y.C. Ng and J. Peleman. 2002a. A genetic linkage map of Guinea yam (*Dioscorea rotundata* L.) based on AFLP markers. Theor. Appl. Genet. 105: 716–725.

Mignouna, H.D., R.A. Mank, T.H.N. Ellis, N. Van den Bosch, R. Asiedu, M.M. Abang and J. Peleman. 2002b. A genetic linkage map of water yam (*Dioscorea alata* L.) based on AFLP markers and QTL analysis for anthracnose resistance. Theor. Appl. Genet. 105: 726–735.

Mignouna, H.D., M.M. Abang and S.A. Fagbemi. 2003. A comparative assessment of molecular marker assays (AFLP, RAPD and SSR) for white yam (*Dioscorea rotundata*) germplasm characterisation. Ann. Appl. Biol. 142: 269–276.

Mitchell, S.A., H.N. Asemota and M.H. Ahmad. 1995a. Factors affecting the *in vitro* establishment of Jamaican Yams (*Dioscorea* spp.) from nodal pieces. J. Sci. Food Agric. 67541–550.

Mitchell, S.A., H.N. Asernota and M.H. Ahmad. 1995b. Effects of Explant source. Culture medium strength and Growth regulators on the *in vitro* propagation of three Jamaican yams (*Dioscoea cavenensis, D. trifida* and *D. rotundata*). J. Sci. Food Agric. 67: 173–180.

Mohamed, M.A.H. and A.A. Alsadon. 2010. Influence of ventilation and sucrose on growth and leaf anatomy of micropropagated potato plantlets. Scientia Horticulturae, Amsterdam, V.123, pp. 295–300.

Muthoni Jane, Miriam Mbiyu and Jackson N. Kabira. 2011. Up-scaling production of certified potato seed tubers in Kenya: Potential of aeroponics technology. Journal of Horticulture and Forestry 3(8): 238–243.

Nagasawa, A. and J. Finer. 1989. Plant regeneration from embryogenic suspension cultures of Chinese yam (*Dioscorea opposita* Thunb.). Plant Science 60: 263–271.

Ng, S.Y.C. 1984. Meristem culture and multiplication. *In:* International Institute of Tropical Agriculture Annual Report. 1983. Ibadan, Nigeria, pp. 133–134.

Ng, S.Y.C. 1988. *In vitro* tuberization in white yam (*Dioscorea rotudatan* Poir). Plant Cell Tissue and Organ Culture 14: 121–128.

Ng, S.Y.C. 1992. Micropropagation of white yam (*D. rotundata*. poir), *In:* Biotechnology in Agriculture Forestry, High-tech and micropropagation III Y.P.S. Bajaj (ed.). Berlin Heidelberg, Springer-Verlag, Vol 19 pp. 135–159.

Ng, S.Y.C. and N.Q. Ng. 1997. Germplasm conservation in food yams (*Dioscorea* spp.): Constraints, Application and Future prospects. *In:* M.K. Razdan and E.C. Cocking (eds.). Conservation of Plant Genetic resources *in vitro*. Volume 1: General Aspects. Science Publishers Inc. USA, pp. 257–286.

Njualem, D.K., F. Ntam, A. Mbairanodji, W.N. Leke, V. Nchinda and D.L. Mapiemfu. 2007. Farmer Participatory Evaluation of Local Landraces of Yams (*Dioscorea* spp.) in Cameroon: A Year's Experience. *In:* B. Nkamleu, D. Annang and N.M. Bacco (eds.). Securing Livelihoods through Yams. Proceedings of a technical workshop on progress in yam research for development in West and Central Africa held in Accra, Ghana, 11–13 September 2007, pp. 244–249.

Nlerum, F.E. 2009. Prediction of adoption of yam minisett technology among yam farmers in Rivers state, Nigeria. Indian J. Agric. Res. 43(4): 294–298.

Nwachukwu, E.C. and I.U. Ubi. 2001. Prelimnary investigation of the sensitivity of white yam (*Dioscorea rotundata*) to gamma radiation. Nigerian Journal of Genetics. Vol. 16.pp.

Nweke, F.I., B.O. Ugwu, C.L.A. Asadu and P. Ay. 1991. Production costs in the yam-based cropping systems of Southwestern Nigeria. Resource and Crop Management Division Research Monograph No. 6. IITA, Ibadan, Nigeria, 29 pp.

Nwosu, N.A. 1975. Recent developments in vegetative propagation of edible yam (*Dioscorea* species). Proceedings, Agricultural society of Nigeria 12: 15.

Okezie, C., S. Okonkwo and F. Nwoke. 1994. Carbon source requiremen for the culture of white yam (*Dioscorea rotundata*) embryos *in vitro*. Acta Horticulturae 380: 329–334.

Okigbo, B.N. and D.G.E. Ibe. 1973. New method of yam propagation. Paper prepared for the Third Symposium of the International Society for Tropical Root Crops, IITA, Ibadan, Nigeria. 2–9 December.

Okoli, O.O., J.U. Nwokoye and C.O. Udugwu. 1984. Economic indices for clonal selection and breeding of yams. *In:* Proceedings of 2nd Triennial Symposium ISTRC-AB. Douala. Cameroon. Aug. 1983.

Okoli, O.O., M.C. Igbokwe, L.S.O. Ene and J.U. Nwokoye. 1982. Rapid multiplication of yam by the minisett technique. Research Bulletin 2. National Root crops research Institute (NRCRI), Unudike, Nigeria, 12 pp.

Okezie, C.E.A. 1984. Indole butyric acid as a pre-sprouting agent for growth and tuberization in *Dioscorea rotundata*. Tropical Agriculture (Trinidad) 62(4): 297–301.

Okonkwo, S.N.C. 1985. The botany of the yam plant and its exploitation in enhanced productivity of the crop. *In:* The Biochemistry and Technology of the Yam Tuber. Osuji (eds.). Biochemical Society of Nigeria and Anambra State University of Technology, pp. 3–29.

Orkwor, G.C. and C.L.A. Asadu. 1998. Agronomy. *In:* Food yams. Advances in research. G.C. Orkwor, R. Asiedu and I.J. Ekanayake (eds.). IITA / NRCRI, pp. 105–141.

Ondo, P., C. Kevers and J. Dommes. 2007. Axillary proliferation and tuberization of *Dioscorea cayenensis-D.rotundata* complex. Plant Cell Tissue Organ Cult. 91: 107–109.

Onwueme, I.C. 1978. The tropical tuber crops. Yam. Cassava. Sweet potato and Cocoyams. John Wiley and Sons, New York. 234pp.

Orkwor, G.C. 1998. The importance of Yams. *In:* Food yams: Advances in Research. G.C. Orkwor, R. Asiedu and I.J. Ekanayake (eds.). IITA/NRCRI, pp. 1–12.

Otoo, E. 2003. Farmer Participatory Breeding—A Waste or a Must: The13th Symposium of International Society of Tropical Root Crops (ISTRC), Arusha, Tanzania 9–15th November 2003.

Ovono, P.O., C. Kevers and J. Dommes. 2007. Axillary proliferation and tuberisation of *Dioscorea cayenensis–D. rotundata* complex. Plant Cell Tissue and Organ Culture. Volume 91, Issue 2, pp 107–114.

Padron, I.E.S., L.A. Torres-Arizal and R. Litz. 2011. Somatic Embryogenesis in Yam (*Dioscorea rotundata*). Rev. Fac. Nal. Agr. Medellín Vol. 64 No.2.

Patrick, I. 2012. Farmers count losses after flood. News Article | Daily times Ng. October 6, 2012. http://www.dailytimes.com.ng/article/farmers-count-losses-after-flood-0

Pérez, L., M. Baquero and J. Beltrán. 2003. Caracterizaciónmorfológica y patogénica de *Colletotrichum* spp. como agente causal de la antracnosis en ñame *Dioscorea* sp. RevistaColombiana de Biotecnología 5(1): 24–35.

Piccioni, E. (1997). Plantlets from encapsulated micropropagated buds of M.26 apple rootstock. Plant Cell, Tissue and Organ Culture 47: 255–260.

Pilatti, F.K., T. Aguiar, T. Simões, E.E. Benson and A.M. Viana. 2011. *In vitro* and cryogenic preservation of plant biodiversity in Brazil. *In vitro* Cell Dev. Biol. - Plant 47(1): 82–98.

Popov, A.S., E.V. Popova, O.N. Nikishina and O.N. Vysotskaya. 2005. Cryobank of plant genetic resources in Russian Academy of Sciences. International Journal of refrigeration. Vol. 28, Issue 6, pp. 1–8.

Poornima, G.N. and R.V. Ravishankar. 2007. *In vitro* propagation of white yams, *Dioscorea oppositifolian* (Linn) and *Dioscorea pentphylla* (Linn.). African Journal of Biotechnology. Vol. 6(20) pp. 2348–2352.

Quin, F.M. 1998. An overview of Yam Research. *In:* Food Yams; Advances in Research. G.C. Orkwor, R. Asiedu and I.J. Ekanayake (eds.). IITA/NRCRI, pp. 215–229.

RajuJayadev and Chinthalapally V. Rao. 2012. Diosgenin, a Steroid Saponin Constituent of Yams and Fenugreek: Emerging Evidence for Applications in Medicine, Bioactive Compounds, *in* Phytomedicine, Iraj Rasooli (ed.). In Tech, Available from: http://www.intechopen.com/books/bioactivecompounds-in-phytomedicine/diosgenin-a-steroid-saponin-constituent-of-yams-and-fenugreek-emergingevidence-for-applications-in.

Reed, B.M., F. Engelmann, M.E. Dulloo and J.M.M. Engels. 2004. Technical guidelines for the Management of Field and *in vitro* Germplasm Collections. IPGRI Handbooks for Genebank. No 7. Rome, IT.

Rubin, G.M. and A.C. Spradling. 1982. Genetic transformation of Drosophila with transposable element vectors. Science 218(4570): 348–353.

Salazar, R. and R. Hoyos. 2007. Multiplicación y tuberización de ñame (*Dioscorea alata* L.) en sistema de inmersión. Revista Facultad Nacional de Agronomía, Medellín 60(2): 3907–3921.

Schnable, P.S. and P.A. Peterson. 1988. The Mutator-Related Cy Transposable Element of Zea mays L. Behaves as a Near-Mendelian Factor. Genetics 120: 587–596.

Seal, S. and E. Muller. 2007. Molecular analysis of a full-length sequence of a new yam badnavirus from *Dioscorea sansibarensis*. Arch. Virol. 152(4): 819–25.

Sharareh, R. 2011. Assessment of diosgenin production by *Trigonella foenum-graecum* L. *In vitro* condition. American Journal of Plant Physiology 6(5): 261–168.

Spradling, A.C. and G.M. Rubin. 1982. Transposition of cloned P elements into Drosophila germ line chromosomes. Science 218(4570): 341–347.

Standardi, A. and E. Piccioni. 1998. Recent perspective on synthetic seed technology using nonembryogenic *in vitro*-derived explants. Int. J. Plant Sci. 159: 968–978.

Suprasanma P. and H. Nakagawa. 2012. Mutation breeding in vegetatively propagated crops. *In:* Q.Y. Shu, Brian P. Forster and Hitoshi Nakagawa. 2012. Plant Mutation Breeding and Biotechnology. International Atomic Energy Agency, Vienna, Austria, pp. 347–358.

Taylor, M. 1996. *In vitro* conservation of root and tuber crops in the South Pacific. Paper presented at the consultation meeting on the management of field and *in vitro* Genebank, 15–20 January, 1996 GAT, Cali.

Thottaphilly, G. 1992. Plant Virus Diseases of importance to African Agriculture. Journal of Cytopathology 134: 265–288.

Tor, M., C.C. Ainsworth and S.H. Mantell. 1993. Stable transformation of the food yam *Dioscorea alata* L. by particle gun bombardment. Plant Cell Rep. 12: 468–473.

Tripathi, L. and J.N. Tripathi. 2003. Role of Biotechnology in Medicinal Plants. Tropical Journal of Pharmaceutical Research 2(2): 243–253.

Twyford, C. and S. Mantell. 1996. Production of somatic embryos and plantlets from root cells of the Greater Yam. Plant Cell, Tissue and Organ Culture 46(1): 17–26.

Van den Bosch, F., G. Akudibilah, S. Seal and M. Jeger. 2006. Host resistance and the evolutionary response of plant viruses. J. Appl. Ecol. 43: 506–516.

Wall, D.A. and G.H. Friesen. 1990. Effect of duration of green foxtail (*Setariaviridis*) competition on potato (*Solanumtuberosum*) yield. Weed Technol. 4: 539–542.

Watt, M.P. 2012. The status of temporary immersion system (TIS) technology for plant micropropagation. African Journal of Biotechnology 11(76): 14025–14035.

Winch, J.E., F.J. Newhook, G.V.H. Jackson and J.S. Cole. 1984. Studies of *Colletotrichumgloeosporioides* disease on yam, *Dioscorea alata*, in Solomon Islands. Plant Pathology 33: 467–477.

Xiao, Y. and T. Kozai. 2004. Commercial Application of a Photoautotrophic Micropropagation system using large vessels with forced ventilation: plantlet growth and production cost. Hort. Science 39(6): 1387–1391.

Yang, B.O., XiaoLong Hu ChunGenYang Bo, JiaLing. 2009. Factors of flower inducing and histological observation in *Dioscorea zingberensis* C.H. Wright *in vitro*. Journal of Wuhan Botanical Research 22 Vol. 27 No. 3 pp. 318–322.

Zamura, A.B. and C.N. Paet. 1996. *In vitro* genebanking activities, Institute of Plant Breeding, College of Agriculture, University of the Philippines at Los Banos. Paper presented at the consultation meeting on the management of field and *in vitro* genebank, 15–20 January 1996. CIAT, Cali, Colombia.

6

In vitro Regeneration of Sugar Beet (*Beta vulgaris* L.)

Songul Gurel[1,*] and *Ekrem Gurel*[2]

ABSTRACT

Sugar beet (*Beta vulgaris* L.) is the most important industrial crop used for sugar production in Europe and one of only two plant sources from which sucrose (i.e., sugar) can be economically produced in the world. Sugarcane and sugar beet contribute to the total sugar production by 77.6% and 22.4%, respectively. Despite its relatively short period of cultivation (ca. 200 years), yield and quality parameters of sugar beet have been significantly improved by conventional breeding methods. However, during the last three decades or so, advanced biotechnological methods including *in vitro* culture and genetic transformation technologies have been combined with classical breeding programmes, which has then lead to the creation of previously non-existing genotypes. Widely applied techniques such as micropropagation and induction of somaclonal variability have already added new germplasm of improved quality and diversity to the sugar beet breeding pool. Organelle exchange, through symetric or asymetric protoplast fusion, and conversion of gametic cells to embryos (i.e., haploid plant production) are other techniques that can be expected to play an increasing role in the improvement of sugar beet. In addition,

[1] Sugar Institute, Department of Plant Breeding, 06930 Etimesgut, Ankara, Turkey. Email: songul_gurel@yahoo.com

[2] Abant Izzet Baysal University, Department of Biology, 14280 Bolu, Turkey. Email: gurel_e@ibu.edu.tr

* Corresponding author

conservation of germplasm materials, production of disease-free plants, development of homozygous diploid lines (i.e., double haploids/ diplohaploids) from haploid plants obtained from pollen or ovary cultures and isolation and culture of highly totipotent protoplasts from guard cells have been achieved. However, regeneration protocols for both shoot organogenesis and somatic embryogenesis of sugar beet are not reproducible, usually resulting in low regeneration frequencies. Also, a high degree of genotypic variation, primarily because of its highly heterozygous nature due to outcrossing, is a serious problem for the regeneration success of several sugar beet materials. Thus, with all these respects sugar beet is known to be a recalcitrant species. In this chapter, the research to date on sugar beet tissue cultures will be presented from a historical point of view, covering the following topics; (i) micropropagation via proliferation of pre-existing (axillary or shoot-tip) meristems, (ii) plant regeneration through development of direct or indirect *de novo* shoots and somatic embryos from different types of tissues/explants, (iii) haploid plant production via mostly gynogenesis but also androgenesis, (iv) protoplast isolation and culture from leaf mesophyll, guard or cell suspension cells, (v) somaclonal variation incurred during indirect shoot organogenesis and somatic embryogenesis from callus and *in vitro* cell selection to be specifically expolited for abiotic stress tolerance, and (vi) long-term storage of sugar beet genetic resources through low-temperature storage and cryo-preservation techniques.

Keywords: Sugar beet, *Beta vulgaris* L., *in vitro* regeneration, clonal propagation, organogenesis, somatic embryogenesis, haploid plant production, protoplast culture and isolation, germpasm storage, genotypic variation

Introduction

Sugar beet (*Beta vulgaris* L.) is a herbaceous dicotyledon belonging to the family *Chenopodiaceae* and it normally completes its life-cycle in two years. There are several different types of beet (*Beta vulgaris* L.), namely; sugar beet (*B. vulgaris* L. var. Saccharifera) which is the commercial beet of commerce for sugar production, garden beet (*B. vulgaris* L. var Esculenta), foliage beet or mangel (*B. vulgaris* L. var. Cicla) and forage beet (*B. vulgaris* L. var. Crassa). During the first year it develops a swollen tap root in which much of the sucrose is stored and during the second year a flowering stem is formed. The mature vegetative beet is an elongated pear-shaped storage organ consisting of a crown, neck and a tap root. The crown is an unelongated stem from which the rosette of leaves emerge. The large leaves are arranged on the crown in a close spiral and are variable in size, shape and colour.

The heterogenous character of sugar beet is a result of many complex crosses between root- and leaf-type genotypes. In addition sugar beet is naturally cross-pollinating, thereby confounding conventional breeding methods. Hybrid seed production is desirable in sugar beet, and a great deal of interest has been directed to the use of male sterility to aid in hybrid seed production. A number of techniques have already moved into the breeding and variety production. The objectives of sugar beet breeding are to create stable, dependable varieties, which give highest possible yield of white sugar per unit area, and which also meet the demands from the growers as well as the sugar producing factories. A number of very important characteristics such as monogermity, polyploidy (induced by the chemical colchicine) and male sterility which permits the production of hybrid varieties have been added to the sugar beet, in order to reshape it into varieties adapted to the agricultural needs. Bolting-resistance, disease-resistance and many other aspects of the beet have been considerably advanced for the benefit of both growers and the sugar processors by combining the genetic factors within beet.

There are many wild relatives of beet which have desirable characters such as disease- and pest-resistance. Because of limited or poor crossability between *Beta vulgaris* and other species, the use of interspecific hybridization in beet breeding has been limited. It has taken nearly two decades to transfer nematode-resistance gene(s) to sugar beet plants. A strong source of virus yellows is known to exist in the species *B. lomatogona*, but the genes can only be transffered to sugar beet by using the 'bridging species' such as *B. macrorhiza*.

Sugar beet is the most important industrial plant used for sugar production in Europe and one of only two plant sources from which sucrose (i.e., sugar) can be economically produced in the world. Sugar cane and sugar beet contribute to the total sugar production by 77.6% and 22.4%, respectively (Turkish Sugar Co. 2011). The ratio has continously changed in favour of sugar cane during the last a couple of decades; for example, it was 67% vs. 37% for sugar cane and sugar beet, respectively, in 1980s (Atanassov 1986). However, sugar beet is now becoming also an important biofuel alternative to fossil fuel energy (Zhang et al. 2008).

Although sugar beet is a relatively new crop relative to other widely cultivated crops, it has undergone significant improvements, given that it only began to be cultivated about 200 years ago (Draycott 2006). Sucrose content of wild beets is 4 to 6%, but sugar beet was selected from fodder beets, which have 12% sucrose content. This level has increased to around 20% in recently developed commercial cultivars. Conventional breeding methods played the major role in these improvements. However, during the last two decades or so, advanced biotechnological methods including *in vitro* culture and genetic transformation technologies have been combined with classical breeding programmes, which has lead to the creation of previously non-existing genotypes. Widely applied techniques such as micropropagation, haploidization and induction of somaclonal variability have already added new germplasm of improved quality and diversity to the sugar beet breeding pool. Organelle exchange, through symetric or asymetric protoplast fusion, and conversion of gametic cells to embryos (i.e., haploid plant production) are other biotechnological techniques that can be expected to play an increasing role in the improvement of sugar beet (for a detailed review, see Gurel et al. 2008).

Micropropagation and conservation of germplasm materials, production of disease-free plants, development of homozygous diploid lines (i.e., double haploids) from haploid plants obtained from pollen or ovary cultures and isolation and culture of highly totipotent protoplasts from guard cells have been achieved (for a detailed review, see Gurel et al. 2008). However, regeneration protocols for both shoot organogenesis and somatic embryogenesis are not reproducible, usually resulting in low regeneration frequencies. Also, a high degree of genotypic variation, primarily because of its highly heterozygous nature due to outcrossing, is a serious problem for the regeneration success of several sugar beet materials. Thus, with all these respects sugar beet is known to be a recalcitrant species.

In this chapter, research on sugar beet tissue cultures till date is presented from a historical point of view, covering the following topics; (i) micropropagation via proliferation of pre-existing (axillary or shoot-tip) meristems, (ii) plant regeneration through development of direct or indirect *de novo* shoots and somatic embryos from different types tissues/explants, (iii) haploid plant production via mostly gynogenesis but also through androgenesis, (iv) protoplast isolation and culture from leaf mesophyll, guard or cell suspension cells, (v) somaclonal variation incurred during indirect shoot organogenesis and somatic embryogenesis from callus and *in vitro* cell selection to be especially expolited for abiotic stress tolerance, (vi) long-term storage of sugar beet genetic resources through low-temperature storage and cryo-preservation techniques.

Micropropagation

Micropropagation from shoot-tips, axillary buds and flower buds

Sugar beet is a cross-fertilizing species and thus each individual plant has a unique genotype which often causes problems in genetic studies and breeding research of this economically important crop. Therefore, developing a protocol for the micropropagation of valuable breeding materials would be of considerable interest to the beet breeding industry since micropropagation would result in the production of identical clones that can be employed to maintain a given genotype constant over several generations, especially useful for combining ability tests (Saunders 1982).

Micropropagation refers to mass propagation of plants through any vegetative means. In this report, however, we will discuss propagation of sugar beet plants using tissues with pre-existing meristems only, such as shoot-tips, axillary buds and flower buds of young inflorescences. *De novo* plant regeneration via direct or indirect adventitious shoot organogenesis and somatic embryogenesis will be described under a separate section below.

Micropropagation is one of the oldest applications of *in vitro* plant culture that is used to mass-propagate genetically identical somatic clones. Since Margara (1970) first obtained adventitious buds from the floral peduncle of sugar beet cultured *in vitro*, much effort has been made to develop micropropagation techniques with the aim of rapid multiplication of improved breeding lines (Atanassov 1986). Axillary and adventitious shoot formation has been obtained from cultured shoot-tips of young seedlings, flower buds and segments of subapical inflorences (Margara 1977, Miedema 1982a,b, Sabir and Ford-Llyd 1991, Zhong et al. 1993a). Direct regeneration of adventitious buds from leaf and petiole explants of young seedlings, cultured shoots or occasionally mature sugar beet plants has been also reported although this appears to be much less common than axillary shoot formation from cultured shoot apices (Mikami et al. 1989a,b, Ritchie et al. 1989, Freytag et al. 1990). Shoots and roots also developed on whole detached leaves when these were inserted into compost (Rogozinska and Goska 1978) and Saunders and Mahoney (1982) induced adventitious shoots on leaf blades and petioles *in vivo* by spraying seedlings at the cotyledonary stage with a high concentration of BAP.

Micropropagation of sugar beet been readily achieved on several defined culture media, however, Murashige and Skoog (1962) medium has been used the most widely one for *in vitro* culture of sugar beet. Adventitious bud formation or axillary bud proliferation mostly occured on media containing a cytokinin, generally BAP, at concentrations of 0.05–3.0 mg/l. The inclusion of additional vitamins and amino acids (Freytag et al. 1988) or an auxin-inhibitor, e.g., TIBA (Detrez et al. 1988, 1989), has been also reported to enhance direct organogenesis from petiole explants. Since micropropagation is very fast and less laborious and the regererants are more genetically stable, shoot-tip explants excised from *in vitro* germinated seedlings have been widely used for transformation by both *Agrobacterium*-mediated methods (Konwar 1994, Zhang et al. 2001a, Hisano et al. 2004, Bekheet and Solliman 2007) and particle bombardment (Mahn et al. 1995). Rooting of multiplied shoots from shoot meristems can be readily achieved, even on a hormone-free medium (Coumans-Gilles et al. 1981). Further, Miedema et al. (1980) achieved rooting of detached sugar beet leaves from

3-week to 4-month-old greenhouse grown plants with no auxin treatment when cultured in pots containing a mixture of sand and compost.

Advantages of micropropagation for sugar beet breeding programmes can be listed as follows (Atanassov 1986, Skaracis and McGrath 2005, Gurel et al. 2008): (i) selected breeding materials can be multiplied quickly at a relatively low cost and in a short time. For example, trisomic breeding material obtained from crossing a diploid and tetraploid sugar beet line were multiplied by micropropagation (Goska and Szota 1992); (ii) self-sterile genotypes can be efficiently preserved for progeny testing; (iii) a selected single plant can be propagated clonally without changing its specific characters; (iv) cytoplasmic male sterile (CMS) genotypes can be maintained before being integrated into a breeding program; (v) wild germplasm, difficult-to-germinate and reproduce, can be propagated, stored and exchanged as needed (Hussey and Hepher 1978, Miedema 1982a).

To restore fertility in CMS sugar beet plants shoot-tip culture was used in combination with heat treatment (Lichter 1978). Plants derived from meristems can be used for *in vitro* cell selection strategies. Multiple bud clumps from cultured immature inflorescence tips were used for *in vitro* production of salt-tolerant (NaCl-tolerant) plantlets by treating buds with γ-radiation (Yang et al. 2004). Also, it is possible to produce polyploid sugar beet plants by treating shoot-tips with an antimitotic agent (Hussey and Hepher 1978).

Although genotypic variation is also evident in micropropagated sugar beet plants (Miedema 1982a, Mezei et al. 1990, Hayakawa et al. 1994), it appears far less frequently than when regeneration involves direct or indirect organogenesis (Gurel et al. 2003a, Mishutkina and Gaponenko 2006) or somatic embryogenesis (Saunders and Tsai 1999). Genetic stability is important when valuable, elite breeding materials are being multiplied or genetically transformed plants are being maintained. The only reliable method of *in vitro* multiplication which can be applied over a wide range of genotypes involves culturing either vegetative shoot-tips or flower buds taken from mature plants (Hussey and Hepher 1978, Atanassov 1980, Miedema 1983).

Micropropagation through shoot organogenesis and somatic embryogenesis

Types of in vitro development

There are three main types of *in vitro* development in plants: *callogenesis* (callus formation), *caulogenesis* (adventitious shoot or root organogenesis) and *somatic embryogenesis* (embryo formation) (Gurel et al. 2008). Formation of adventitous shoots and roots or somatic embryos can be initiated from callus, from cell suspensions derived from callus or from cell colonies of cultured protoplasts (i.e., *indirect regeneration*) or directly from cultured explant tissues without a prior callus formation (i.e., *direct regeneration*) (Fig. 1).

In sugar beet, callus can be obtained from many parts of the sugar beet plant including seedling tissues (hypocotyl and cotyledons), leaves, petioles, roots, inflorescence flower stalks, anthers, embryos and seeds when cultured on a medium supplemented with or without plant growth regulators. Following Margara (1970)'s first report on production of adventitious shoots from callus derived from floral stalks of sugar beet,

Fig. 1. Direct adventitious roots from a leaf lamina disc (top) and direct adventitious shoots from an intact leaf (bottom).

Color image of this figure appears in the color plate section at the end of the book.

efforts accelerated on shoot organogenesis and somatic embryogenesis, mostly by indirect regeneration from callus (Hooker and Nabors 1977, De Greef and Jacobs 1979). Hooker and Nabors (1977) obtained buds on the surface of callus derived from different types of seedling explants cultured on a medium containing an anti-auxin (2,3,5-triiodobenzoic acid, TIBA). De Greef and Jacobs (1979) achieved a single plant only from callus subjected to cold pretreatment for several weeks.

Although early efforts were met certain dificulties, regeneration of adventitious shoots and development of somatic embryos either indirectly from induced callus or directly from a wide range of explant types (including leaf lamina, petiole, hypocotyl, cotyledon, roots, thin cell layers, cell suspensions, protoplasts, immature zygotic embryos and reproductive organs, i.e., ovary or anther) have been achieved by many workers around the globe in the following chronological order; (i) indirect shoot regeneration (ii) indirect somatic embryo regeneration (iii) direct shoot regeneration (iv) direct somatic embryo regeneration (see Gurel et al. 2008 for a full list of references).

It appears to be relatively easy to achieve the development of adventitious shoots and somatic embryos when callus (dedifferentiated cells developed from fully differentiated cells of the cultured explant tissue) is subcultured, i.e., indirect regenertation. This may be due to a higher degree of morphological plasticity in dedifferentiated callus cells as compared to the fully differentiated, and therefore very much determined state of cells in tissues. On the other hand, achieving somatic embryogenesis (especially direct somatic embryogenesis) in sugar beet has been much more difficult than shoot organogenesis; for example, only four groups have achieved direct somatic embryo production in sugar beet so far (Tenning et al. 1992, Kulshreshtha and Coutts 1997). Tenning et al. (1992) was the first to report a high frequency (≥33%) of direct somatic embryogenesis from cultured immature zygotic embryos. They even achieved a much higher frequency (61%) of secondary somatic embryogenesis when primary somatic embryos were subcultured. Kulshreshtha and Coutts (1997) obtained over 90% regeneration frequency when they used zygotic cotyledons from mature embryos excised at the base above the embryonic axis, and followed a sequential culture regime. Zhang et al. (2001b) were able to obtain direct somatic embryogenesis from seedling cotyledons cultured on a medium with either benzyl amino purine (BAP) or thidiazuron (TDZ). Although this approach may have the advantage that readily available seedling explants can be used to produce somatic embryos from which genetically stable plants develop, frequencies of regeneration are considerably lower (7 to 23.6%) than those from immature zygotic embryos (Tenning et al. 1992) or zygotic cotyledons (Kulshreshtha and Coutts 1997), and are found heavily genotype-dependent. None of these direct somatic embryonesis systems has yet to be exploited for genetic transformation studies.

It has been shown that somatic embryos can readily develop into normal plants under almost any culture conditions (Freytag et al. 1988, Tenning et al. 1992, Tsai and Saunders 1995, Kulshreshtha and Coutts 1997), even if the embryos were encapsulated with alginate (i.e., artificial seeds), stored at 4°C for over two months and germinated on either agar or in sterile soil (Tsai and Saunders 1999b). The ease of somatic embryo germination could be attributed to the bipolar structure of somatic embryos with highly determined shoot and root primordia. Therefore, somatic embryogenesis is much preferred over shoot organogenesis for both plant propagation and transformation since regenerated shoots require an extra step of rooting, a time-consuming process and also sometimes limiting the succes of the whole regeneration protocol.

Factors affecting organogenesis and somatic embryogenesis

There are several factors that determine the *in vitro* regenerative success of sugar beet. It has been often reported that the type of initial explant is important for both adventitious root and shoot development; usually petiole explants excised from sterile seedlings are more responsive than leaf lamina or other seedling parts in producing adventitious roots (Gurel 1997) and shoots (Grieve et al. 1997, Gurel et al. 2011). The richness of the vascular tissue of the petiole might be responsible for such a response, at least for direct root organogenesis in sugar beet (Gurel and Wren 1995a), in which it was clearly shown that primordial cells mostly develop from parenchyma cells in the vascular cambium along the vascular bundles of the petiole (Fig. 2). It was also shown that direct adventitious shoots originated from subepidermal parenchyma cells of petiole segments (Detrez et al. 1988), perhaps explaining why thin cell layer explants were successfully used for direct (Detrez et al. 1989) and indirect (Toldi et al. 1996) shoot regeneration. This phenomenon is consistent with the *in vivo* mode of differentiation as the lateral roots are of endogenous origin (i.e., pericycle cells of the root tissue) while leaves (apical meristems) and branches (axillary meristems) are exogenous.

Explant position is also important for the regenerative capacity of sugar beet tissues. Explants from the blade-petiole transition zone, also described as the midrib-petiole junction region separated by the insertion point of the lowest primary vein (Gurel and Wren 1995a), were more productive than either leaf lamina or petiole tissue for direct adventitous shoot (Mikami et al. 1989b) or root organogenesis (Gurel and Wren 1995a). This effect was clearer when midrib-petiole junction segments were cut paradermally into upper, middle and lower sections and cultured separately on a root-

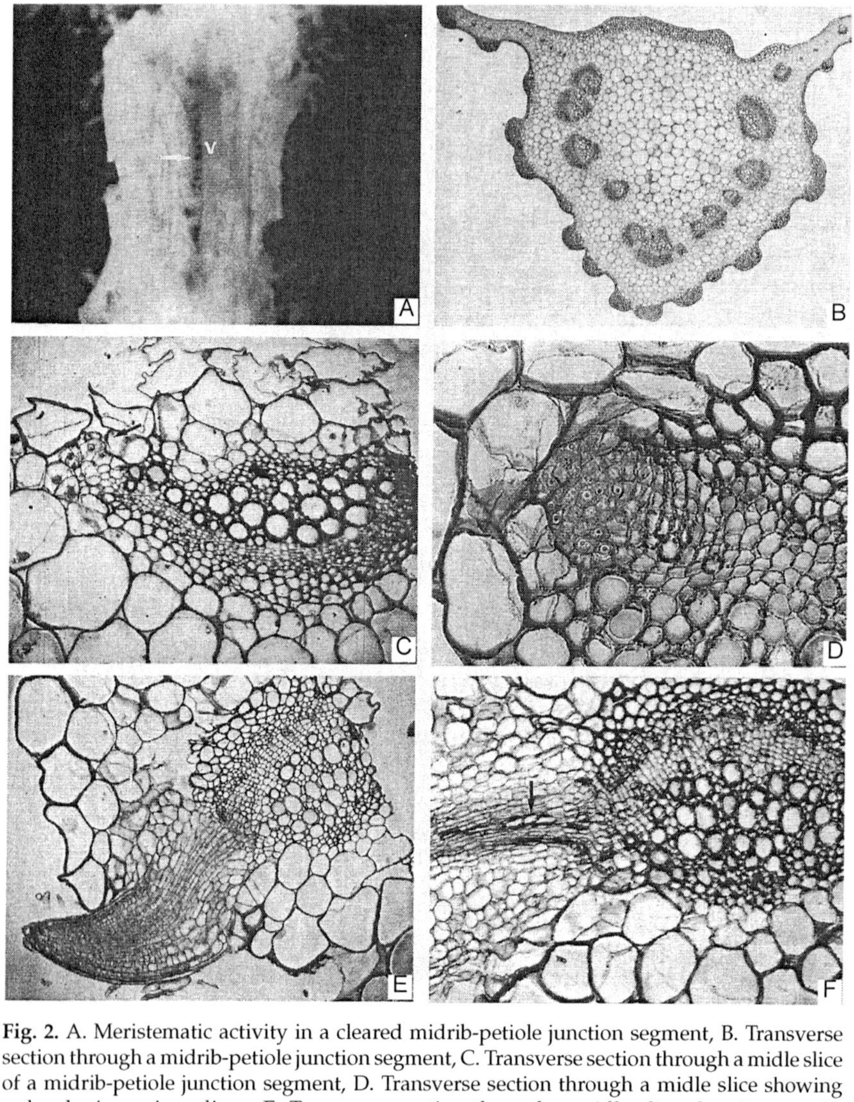

Fig. 2. A. Meristematic activity in a cleared midrib-petiole junction segment, B. Transverse section through a midrib-petiole junction segment, C. Transverse section through a midle slice of a midrib-petiole junction segment, D. Transverse section through a midle slice showing a developing primordium, E. Transverse section through a midle slice showing a well-developed root, F. Longitudinal section through the base of an adventitious root showing the differentiation of vascular tissue (arrow) and connection to the vascular bundle of the original explant tissue (Gurel and Wren 1995a).

induction medium; the middle region forming many more roots in a much faster manner than upper or lower regions. Using a sequential treatment of leaf explants cultured *in vitro* with auxin and cytokinin, Gurel and Wren (1995a) also demonstrated that the length of time during which leaf parenchyma cells are competent to resume meristematic activity, initiate root and shoot primordia and undergo root morphological development were critical (Fig. 3).

Another study in determining the sugar beet leaf material with the greatest potential for growth and organogenesis, discs isolated from different positions on the leaf lamina were cultured on a callus-inducing medium (Gurel 1997). Veinal and interveinal discs from central and marginal positions appeared similar in their capacity for callus production, although discs from the distal tip produced slightly less than those from middle and lower regions. This may be due to the fact that in a growing leaf, cell division and cell expansion cease first at the leaf tip and margins but continue for longer periods at the base; therefore, distal part of the tissues being physiologically older than the remaining lamina. Similar observations were reported by others. Zhang et al. (2004) divided cotyledons of young seedlings into two pieces, i.e., hypo-cotyledon and epi-cotyledon, and found

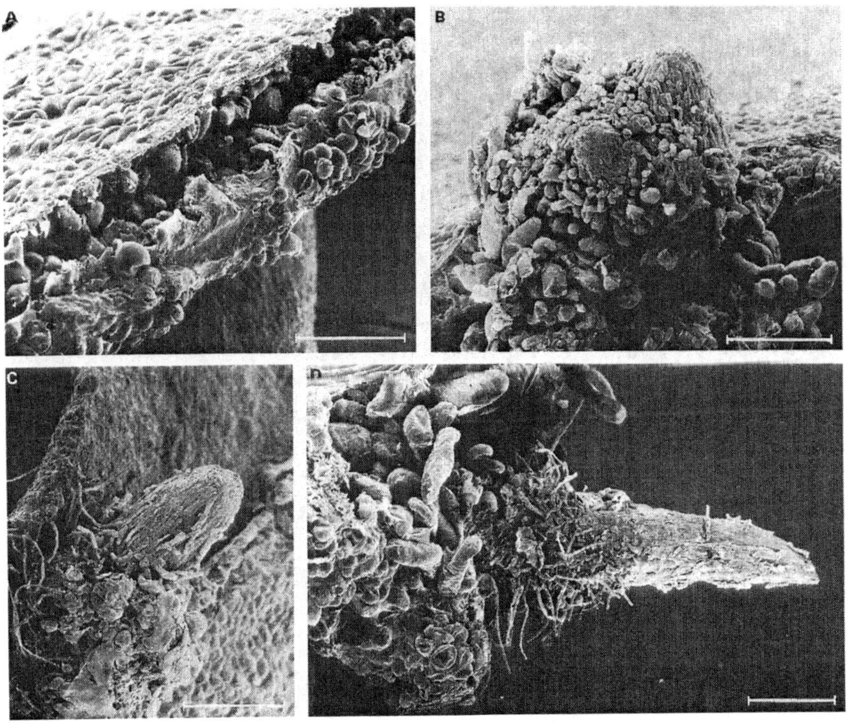

Fig. 3. A. The edge of a freshly cut (0 d) lamina disc showing the intact mesophyll cells between the upper and lower epidermis, B and C. Well developed root-tips after 14 d and 15 d culture, respectively, D. A longer root with root hairs after 16 days in culture (Gurel and Wren 1995a).

that callus-induction frequency from the cotyledon region adjacent to the hypocotyl (hypo-cotyledon) was higher than that from the distal region. Also, Nourouzi et al. (2005) reported that cells near the main leaf vein were more responsive for transformation and subsequent regeneration. Another critical factor is the physiological age of the donor tissue. Several studies have shown that younger tissues are more responsive than older tissues (Gurel 1991, Gurel and Wren 1995b, Kuykendall et al. 2003).

It has been reported by several groups that 2,3,5-triiodobenzoic acid (TIBA), an anti-auxin, has a stimulatory effect on the development of callus, shoots and somatic embryos (Abe et al. 1991, Kulshreshtha and Coutts 1997). Even, pretreating detached sugar beet leaves with TIBA was shown to inhibit root development from the cut surface of leaf petiole bases, while non-treated leaf cuttings could produce roots in the absence of auxin (Miedema et al. 1980). TIBA is known to inhibit polar transport of auxins in plants, which leads to loss of apical dominance and promotion of axillary shoot growth. However, it is not known whether effects of TIBA are due to its hormone-like activity, which induces cell division and/or differentiation, or to its effects on the relative concentrations of endogenous auxin in the excised tissues.

Several pretreatments of the cultured explants or their donor tissues improved the regeneration efficiency in sugar beet: (i) culturing seedlings on medium containing 6-benzylaminopurine (BAP) (Gurel et al. 2003a), thidiazuron (TDZ) (Gurel et al. 2003b) or TIBA combined with either BAP (Jacq et al. 1993) or L-proline (Moghaddam et al. 2000) and (ii) preconditioning greenhouse-grown sugar beet plants with TIBA (Roussy et al. 1996). Zhang et al. (2001b) preconditioned tissues in the dark to induce somatic embryo regeneration. When Jacq et al. (1993) tested several preconditioning media on organogenic callus induction, pretreatment with BAP or TIBA alone was insufficient but very productive when used in combination. Cold pretreatment of callus tissue seemed to have a conflicting effect on the subsequent regeneration success. For example, De Greef and Jacobs (1979) demonstrated that callus pretreatment at 4°C for 3 to 9 weeks had no effect on shoot formation from callus while Gurel et al. (2001) found higher shoot regeneration from callus after two-week preculture at 4°C. On the other hand, earlier reports indicated that multiple hormonal sequences were necessary for the development of somatic embryos from callus (Tetu et al. 1987), however, later studies proved otherwise, as single-step protocols using supplemental vitamins and amino acids (Freytag et al. 1988) were shown to result in highly effective regeneration (Kubalakova 1990, Tsai and Saunders 1999a,b, Ivic and Smigocki 2003).

Following earlier studies on the effects of a single-stage pretreatment of three sugar beet breeding lines by germinating seedlings on a medium containing different concentrations of BA (Gurel et al. 2003a) or TDZ (Gurel et al. 2003b), the same group of workers also tested the effects of a two-stage hormonal pretreatment (i.e., first pretreatment during germination and second pretreatment during growth stage) on adventitious shoot regeneration capacity from two different types of explants (leaf and petiole) isolated from two sugar beet breeding lines (Fig. 4) (Gurel et al. 2011). It was shown that the explants isolated from untreated seedlings (i.e., germinated and further grown on medium containing no plant growth regulators—PGRs) produced either no shoots at all (as in lamina explants) or in very few numbers with low frequencies of organogenic explants (as in petiole explants). This observation was more evident when an overall comparison of different germination media containing either no PGRs, 0.5 mg/l BA or TIBA was made irrespective of the hormonal compositions of growth and regeneration media; both types of explant and breeding lines produced significantly less numbers of shoots per explant with lower frequencies of organogenic explants when cultured on regeneration medium containing either no PGRs or 0.5 mg/l TIBA. However, those germination media supplemented with 0.5 mg/l BA were much more productive for shoot development than the media containing 0.5 mg/l TIBA. As suggested in a recent study on quince (*Cydonia oblonga* Mill.), it might be expected that the effects of a growth regulator pretreatment on directing cell competence may differ according to the degree of cell differentiation, thus producing different regenerating responses (D'Onofrio and Morini 2006). This assumption would be in agreement with the competence-determination theory of Christianson and Warnick (1985), who postulated that organogenesis proceeds through three sequential stages; (i) the acquisition of competence

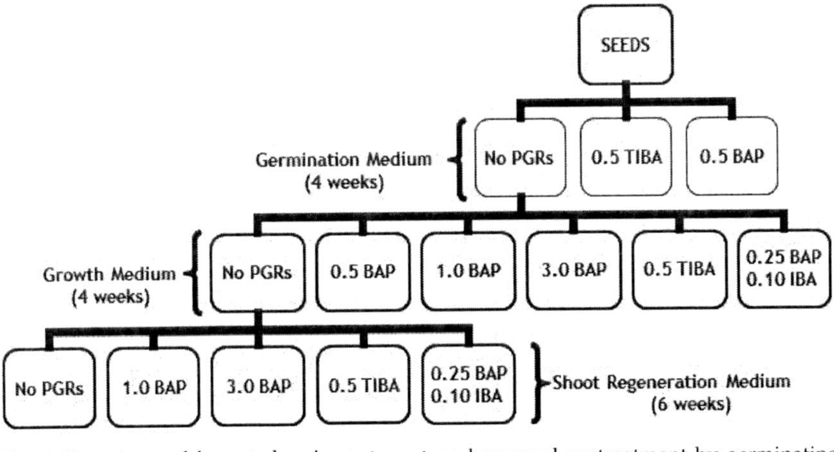

Fig. 4. Experimental layout showing a two-stage hormonal pretreatment by germinating seeds on three different germination media for four weeks and then further growing on six different growth media before culturing the explants excised from the pretreated seedlings on five different regeneration media for 6 weeks (Gurel et al. 2011).

(which is believed to be predetermined at an early stage of development) to respond to a particular inductive signal, (ii) induction, and (iii) morphogenic differentiation and development. In accordance with this theory, we may suggest that the sugar beet tissues obtained from seedlings germinated and then further grown on a medium containing BA might have gained a kind of competence at an earlier stage of development, so that their cells can be more readily directed into shoot induction.

In the same study (Gurel et al. 2011), when six different growth media irrespective of the germination and regeneration media were compared (Fig. 4), it was clear that those media containing no PGRs were the least effective while those supplemented with a combination of 0.25 mg/l BA and 0.1 mg/l IBA were the most productive. On the combined medium (0.25 mg/l BA and 0.10 mg/l IBA), petiole explants of the line EK345 produced the highest mean number of shoots per explant (4.3 shoots) and frequency of organogenic explants (54.2%). On the other hand, the comparison of five different shoot regeneration media (Fig. 4) irrespective of the hormonal compositions of germination and growth media revealed that the combination of BA (0.25 mg/l) and IBA (0.10 mg/l) were the most prolific treatment for shoot regeneration from either type of explant of both breeding lines; mean numbers of shoots per explant ranging from 2.0 to 3.4 and mean frequencies of organogenic explants from 20.3 to 55.4% (Gurel et al. 2011).

In an attempt to find a correlation between (i) the capacity for adventitious root organogenesis, (ii) the extent of necrosis or blackening (largely caused by polymerization of phenolic compounds) and (iii) levels of polyphenol oxidases of cultured sugar beet leaf lamina explants, it was found that, although a relationship between rooting capacity and degree of blackening could be established but no correlation with the level of phenolic contents was evident (Gurel 1997). Also, it was recently reported that reducing sucrose content from 3 to 1% in the culture medium significantly reduced phenolic content of leaf discs, thus decreasing necrosis during culture and resulting in higher rates of shoot regeneration (Yildiz et al. 2007).

The composition of the culture media is important in determining the morphogenetic pathway. A cytokinin, usually BAP, and an auxin, mostly IAA, NAA or 2,4-D, are normally included in the primary culture medium. Then, after callus formation has occurred, a lower auxin to cytokinin ratio is needed for shoot induction and a high ratio for rooting in the subsequent media. TIBA has been included in some cultures, but reports of TIBA's effect on callus formation and organ induction are conflicting. The growth of callus from cotyledons was inhibited by TIBA (Miedema et al. 1980), but growth of callus from embryos and bud formation from callus derived from various explants was promoted (Tetu et al. 1987). Genotypic variation is also important in relation to callus production and subsequent organ formation, some genotypes being more amenable to organogenesis than others (see the Section 9 later in this section).

In addition to the very commonly used MS (Murashige and Skoog 1962), several other medium formulations have been used for *in vitro* regeneration studies in sugar, including B5 (Gamborg et al. 1968) and PGo (Negrutiu et al. 1975, although often cited as De Greef and Jacobs 1979) and MS-RV medium (Freytag et al. 1988). The MS-RV medium, a modified MS medium with additional ten vitamins and six amino acids, was used by several groups (Catlin 1990, Owens and Wozniak 1991, Owens and Eberts 1992, Tenning et al. 1992, Munthali et al. 1996) and reported to be much more effective than other formulations.

Types of calluses and their hormone-autonomous status

Two types of calluses developed from different explants of sugar beet have been repeatedly described; (i) white-friable callus with large-sized cells and (ii) green-compact callus with small-sized cells (Fig. 5) (Saunders and Daub 1984, Nakashima et al. 1988, Gurel et al. 2001). It has been reported that there is a relationship between the callus type and organogenic potential; white-friable callus commonly producing roots and shoots (Konwar and Coutts 1990, Gurel et al. 2001, Shimamoto et al. 1993), while green-compact callus having no organogenic capacity at all (Ritchie et al. 1989, Tetu et al. 1987). Using electron microscopy, it was recently shown that cells from the two types of calluses had clear differences in terms of cell structure and organellar contents; cells from nonembryogenic callus being more likely to be polyploid, with rough endoplasmic reticulum, polysomes, poly-nucleoli and incomplete cell walls while cells from embryogenic callus had vacuolated cytoplasm, normal cell walls and a high ratio of nuclear to cytoplasmic area.

De Greef and Jacobs (1979) subsequently identified regenerative callus (which was later termed and widely accepted as *hormone-autonomous callus* or *habituated callus*) since such a callus was able to continue growing on medium containing no plant growth regulators at all. Soon after De Greef and Jacobs (1979), another group of workers have also reported a hormone-autonomous (habituated) callus which did not require plant growth regulators for further growth and subsequent regeneration (Van Geyt and Jacobs 1985, Saunders and Shin 1986, Masuda et al. 1988). Cytokinin habituation was reported to occurs spontaneously (Van Geyt and Jacobs 1985), while auxin habituation appeared only after sequential lowering of external hormone concentrations. Also, detailed biochemical studies showed that organogenic and nonorganogenic habituated cells differ in terms of several parameters, including level of peroxidase activity and

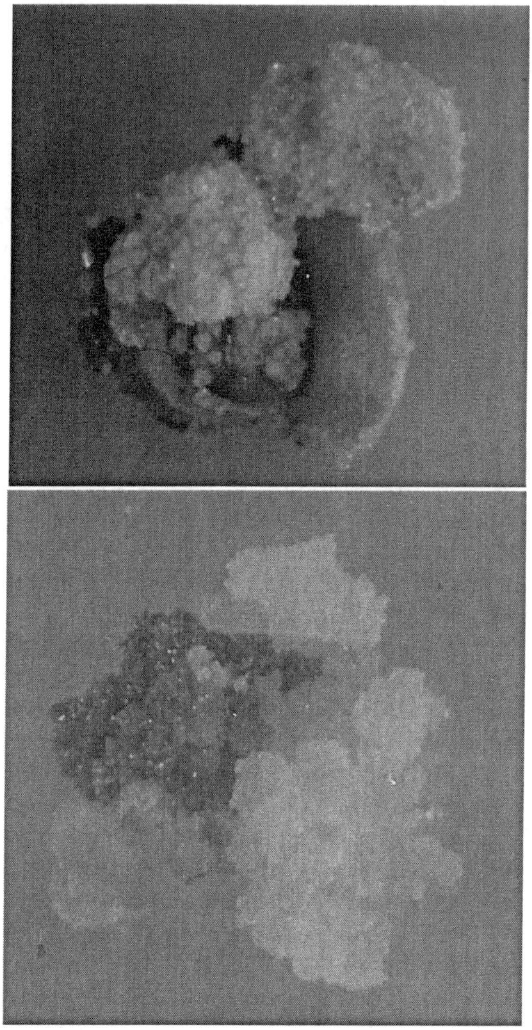

Fig. 5. Green and compact callus with small-sized cells (top) and white and friable callus with large-sized cells.

Color image of this figure appears in the color plate section at the end of the book.

inhibitors, auxin content and protectors (Kevers et al. 1981a,b), calcium-controlled secretion of peroxidases by suspension cells (Gaspar et al. 1983, Kevers et al. 1982, 1983) and polyamine and ethylene metabolism (Bisbis et al. 2000).

Suspension Cultures

The essential cell material in most plant cell culture systems is known as callus, which can be introduced into a similarly complex but liquid medium and agitated so that the cells disperse throughout the liquid to form a cell suspension culture. Such cells are in theory also totipotent and they should also have a potential to synthesize any of the compounds normally associated with the whole plant. When leaf explants of sugar beet are inoculated into culture medium and cultured on a reciprocal shaker, single cells begin to be released from the cut surfaces of the leaf pieces after 6 days (Masuda et al. 1988, Gurel et al. 2002a). The metabolic heterogeneity of cell suspension cultures is far less as compared to callus cultures. As far as morphology is concerned, cell suspension cultures often tend to form cell aggregates rather than single cells in liquid media. Cell suspension cultures have the same growth characteristics as callus cultures, although the cycle is shortened because of increased growth rate. During the incubation period, the biomass of the suspension cultures increase due to cell division and cell enlargement. As new cells are formed they are dispersed into the liquid medium and become clusters and aggregates. Cells in suspension can exhibit much higher rates of cell division than do cells in callus culture. Thus, cell suspension offer advantages when rapid cell division or many cell generations are desired, or when a more uniform treatment application is required such as during cell selection procedures (Philips et al. 1995).

The techniques of plant cell culture facilitate the rapid production of variant cell lines via selection procedures similar to those employed in microbial systems. These variant cell lines are useful in research into the genetics and biochemistry of plant cells and in biotechnology for production of a new plant varieties and secondary metabolites. Rapidly growing, fine suspension cultures or friable calluses are generally the most suitable for selection purposes. Sugar beet is generally known to be more sensitive to residual herbicides than many other crops. Flexibility of herbicide use could be achieved by obtaining herbicide-resistant sugar beet cultivars from somatic cell selection (Saunders et al. 1992). Where it is possible to regenerate plants from variant cells, selection techniques have potential for the production of crop varieties with new characteristics such as salt-tolerance (Freytag et al. 1990), cold-tolerance, disease-resistance and metal-tolerance (Cresswell 1995). The diverse group of compounds known as plant secondary metabolites includes many compounds with pharmaceutical activity (e.g., morphine and vinvristine), fragrances, pigments, latex, enzymes and carbohydrates (Hunter and Kilby 1990). Sugar beet cells are used for obtaining the pigment betalanins (Weller and Lasure 1981), betanin (Hunter and Kilby 1990), betaxanthin and betacyanin (Hamill et al. 1986, Girod and Zryd 1991). The red and yellow pigments produced by beet root are important to food industry as color additive (Weller and Lasure 1981). Cell suspensions have also proven to be excellent starting materials for the isolation of protoplasts to be used in a wide range of applications including cell fusion and genetic manipulation studies (Hall 1991).

Regeneration of Haploid Plants from Gametic Cells

Generation of haploid plants permits the production of a series of homozygous breeding lines from a given genotype in a rather short period of time. In sugar beet breeding, pure breeding lines can be obtained also by conventional inbreeding and back-crossing which takes no less than six years. Induction and detection of recessive mutations can also be achieved by haploid plant production technology. Following the classical work of Guha and Maheshwari (1964) in *Datura inoxia*, haploid plant production has been used achieved in many plant species including sugar beet (Fig. 6), and the regenerated haploids/double haploids were mostly employed for breeding studies.

In sugar beet (*Beta vulgaris* L.), many different approaches have been suggested for the induction of haploid and dihaploid plants. Classical techniques such as natural polyembryony, pollen irradiation and crosses between diploid and tetraploid lines or with wild species yielded very low numbers of haploids (Yuce 1973, De Jong and De Bock 1978, Pedersen and Keimer 1996). Androgenesis and gynogenesis are the two *in vitro* approaches used for haploid plant production. However, although attempted by several workers, androgenesis has been unsuccessful, producing only callus from microspores (Van Geyt et al. 1985, Speckman et al. 1986), callus and roots from anthers (Banba and Tanabe 1972, Welander 1974, Rogozinska and Goska 1976) and diploid plants from anther-derived callus (Rogozinska et al. 1977, Rogozinska and Goska 1982). The most suitable stage of microspore cells for callus proliferation is late meiosis to early uninucleate (Van Geyt et al. 1985).

In contrast to the androgenesis, the gynogenetic pathway has been found to be much more successful in sugar beet. Hosemans and Bossoutrot (1983) were the first to report haploid plant production from culturing unfertilized ovules isolated from male-sterile plants. Later, induction of haploid plants from unfertilized ovules of male-fertile genotypes was also demonstrated by the same group (Bossoutrot and Hosemans 1985) and many others (Van Geyt et al. 1987, Svirshchevskaya et al. 1993, Hansen et al. 1994, 1995, 1998, 2000).

Doubled haploids/dihaploids (which have a complete homozygous nature) can be produced through chromosome doubling by treating regenerated haploids with anti-mitotic agents, most often by including colchicine in the culture medium, either during multiplication (Bossoutrot

Fig. 6. A. An isolated ovule (round, glossy structure in the center), B. A haploid plant (n=9) developing from ovule, C. A rooted haploid plant (n=9), D. A doubled-haploid plant (2n=18) tranferred to soil.

and Hosemans 1985, Ragot and Steen 1992, Gurel et al. 2000, 2003c) or directly at the ovule culture stage (Hansen et al. 1994). Colchicine is known to inhibit mitosis in a wide variety of plant and animal cells by interfering with the structure of the mitotic spindle, but it is highly toxic and often induces chimeric plants (Rao and Suprassana 1996). Several other agents (i.e., amipro-phosmethyl (APM), pronamide, trifluralin and oryzalin) were also tested (Hansen et al. 1998, 2000), and APM was found the most effective of these four agents. In a comparative study, trifluralin was shown to be as effective as colchicine, even at much lower concentrations (Gurel et al. 2000). Therefore, trifluralin is strongly recommended as it is much cheaper than colchicine as well as being less toxic (Zhao et al. 1996).

Most haploid plants from unfertilized ovules were shown to develop through direct embryogenesis. It was reported by Ferrant and Bouharmont (1994) that viable gynogenetic embryos originated only from the egg cell. In other species, besides egg cells, synergids and antipodals are also capable of undergoing embryogenesis or callus formation (Mukhambetzhanov 1997). Regeneration of more than one haploid sugar beet plant from ovule callus been reported by Galatowitsch and Smith (1990). They placed the ovules onto the medium in the dark conditions at 27°C in order to obtain callus and than transferred to the light (16h/8h light/dark) at 25°C to promote shoot development. Haploid plants can also be obtained by using ovary rather than ovule culture. However, ovary culture was reported to be more successful than ovule culture when callus from diploid cells of ovary tissue was removed and ovules were transferred to auxin-free medium with 0.5% charcoal (Van Geyt et al. 1987). On the other hand, it was demonstared that culturing ovaries in the dark for two weeks before transferring to light/dark conditions produced significantly more diploid shoots from ovary-derived callus than those kept continuously in the dark (Gurel and Gurel 1998). Galatowitsch and Smith (1990) reported that adventitious shoot regeneration originated mostly from diploid cells of the ovary tissue, unless they were not spontaneously doubled haploids, as also reported by others (D'Halluin and Keimer 1986, Wang et al. 1991).

The ploidy level of different regenerants can be confirmed by various methods such as chromosome counting (Magnusson et al. 1985, D'Halluin and Keimer 1986; also see Fig. 7), chloroplast counting (Doctrinal et al. 1989; also see Fig. 7), leaf morphology (Galatowitsch and Smith 1990), isozyme analysis (Van Geyt et al. 1987) and flow cytometry (Alemanno and Guilderdoni 1994).

As in all types of tissue cultures studies in sugar beet, genotypic variation is a serious problem for the overall success of haploid plant production from unfertilized ovules (Doctrinal et al. 1989, Lux et al. 1990, Hansen et al. 1995, Gurel et al. 2000). Physical and chemical properties of the culture medium are also critical factors. Addition of charcoal increased frequency of embryo formation (D'Halluin and Keimer 1986, Gurel et al. 2000), but inclusion of $AgNO_3$ decreased or completely inhibited their formation (Gurel et al. 2000). Áthough earlier studies reported no effect of cold pretreatment of flower buds on the induction of haploid embryos (D'Halluin and Keimer 1986), later studies suggested otherwise (Gurel et al. 2000). Physical properties of the culture medium used during the chromosome doubling process were also important. Significantly higher rates (29.1%) were reported when haploid shoots were immersed in liquid medium versus culturing on agarose-solidified (20.7%) or agar-solidified medium (15.4%) (Gurel et al. 2000).

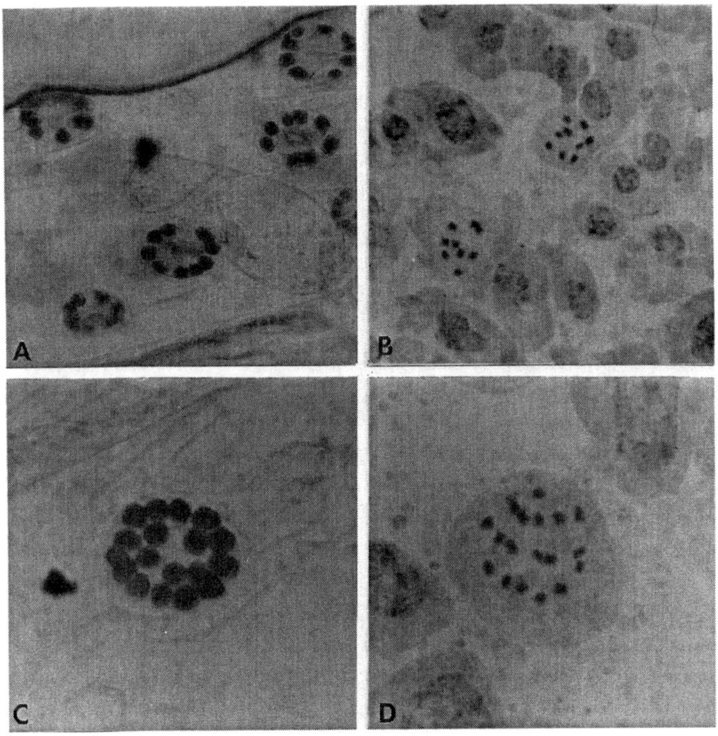

Fig. 7. A. Chloroplasts of a stoma cell of a haploid (n=9) plant, B. Chromosomes of leaf mesophyll cell of a haploid (n=9) plant, C. Chloroplasts of a stoma cell of a double-haploid (n=18) plant, D. Chromosomes of leaf mesophyll cell of a double-haploid (n=18) plant.

Protoplast Isolation and Culture

Protoplasts are able to divide and form cell colonies, which can regenerate into plants *in vitro*. Following electroporation or chemical treatments (e.g., polyethylene glycol), protoplasts fuse and directly take up plasmids or naked DNA fragments. These properties make protoplasts important, not only as alternatives for plant regeneration, but also for somatic hybridization of cells from distantly related plants and for direct gene transfer.

For an efficient protoplast isolation from cell suspensions or leaf and petiole tissues of a wide range of species including sugar beet, certain cell wall-digesting and middle lamella-digesting enzymes (i.e., cellulase, pectinase, cellulycin, cellulase Onozuka R10, macerozyme R10 and driselase) (Schlangstedt et al. 1994) and some sugars (i.e., mannitol, sorbitol or glucose as osmoticums) (Majewska-Sawka et al. 1994) have been widely used. Medium osmolality is greatly influential in successful protoplast isolation and the inclusion of n-propylgallate (nPG) at a concentration of 0.1 mM proved to be critical not only for the isolation but also for the subsequent culture of beet protoplasts (Krens et al. 1990). Prior to enzyme incubation, keeping the chopped leaf tissue in a medium with a high calcium concentration for 4–6 h was also found beneficial (Hall et al. 1993b).

For isolating protoplasts, greenhouse grown plants exhibited not only the problem of obtaining sterile isolations but also a subsequent poor protoplast survival rate (Hall et al. 1993b). Aseptically germinated seedlings proved to be a very reliable source of mesophyll protoplast (Krens et al. 1990). However, shoot cultures are the most favoured choice as they provides a uniform and reusable supply of leaf material for which sterility can be guaranteed. Growth on hormone-free medium is preferred as this leads to essentially normal plant morphology. However, for some lines (e.g., subspecies *maritima* lines) sustained growth was not possible in the absence of cytokinin (Hall et al. 1993b). Pedersen et al. (1993) stated that only protoplasts isolated from *in vitro* grown material are capable of sustained division.

The most commonly used source of plant protoplast is the leaf tissue since it allows the isolation of a large number of uniform cells (Schlangstedt et al. 1994). Also, because the mesophyll cells are loosely arranged, the enzymes have an easy access to the cell wall. The conditions under which the donor plants have been grown may be also important for the success of protoplast isolation in many species. However, when isolating culturable protoplast from leaf tissue is difficult (e.g., in cereals), suspension cultures can be a reliable source of protoplast. Majewska-Sawka et al. (1994) isolated sugar beet protoplasts from hypocotyl-derived suspension cells. Although leaf tissue is known as a source of totipotent protoplasts, colony formation in cultures of such protoplasts is extremely poor (Krens et al. 1990). Petioles can be also used as an alternative tissue source for protoplast isolation. While leaf protoplasts cultured under the most optimal conditions yielded plating efficiencies of only approximately 0.5%, the ratio for petiole protoplasts was 12–15%.

After the first report of protoplast isolation from sugar beet (Li et al. 1976), attempts were made to achieve plant regeneration from protoplasts isolated from cell suspensions, leaf mesophyll or petiole cells. However, most efforts with protoplast-derived callus resulted in either callus formation only (Szabados and Gaggero 1985, Ford-Lloyd and Bhat 1986, Majewska et al. 1990, 1994, Pedersen et al. 1993, Ma et al. 1995, Gurel et al. 2002b) or root formation (Schlangstedt et al. 1992). The first successful plant regeneration from sugar beet mesophyll protoplasts came from efforts to transfer cytoplasmic male sterility by asymmetric cell fusion (Krens et al. 1990). Use of a lipoxygenase inhibitor, n-propylgallate (nPG), appeared necessary as nPG reduced ethylene production by protoplasts significantly (Krens et al. 1994). Lenzner et al. (1995) also established a plant regeneration system from mesophyll-derived protoplasts; however, only friable callus yielded shoots (Lenzner et al. 1995). Low plating efficiency was problematic (Hall and Krens 1988), despite attempts to improve it by, e.g., immobilizing cells in calcium alginate (Hall et al. 1993a, 1994, Lenzner et al. 1995), embedding cells in agarose (Lindsey and Jones 1989), and incorporating anti-oxidative agents (Krens et al. 1990, 1994) or polyamines in the culture medium (Majewska-Sawka et al. 1997, Jazdzewska et al. 2000). When Hall et al. (1993a) investigated the effects of NaCl, feeder cells and the embedding of protoplasts in Ca-alginate, they found that the use of NaCl in all instances proved to be detrimental to protoplast development, the other two treatments having no clear effects. However, the same group (Hall et al. 1993b) also reported that embedding cells in Ca-alginate prior to culture resulted in an enhancement of a compact and non-regenerable callus production. The plating efficiency increased up to 3.7% when the protoplasts were given a cold pretreatment in a salt solution prior to embedding in alginate (Schlangstedt et al. 1994). Also, using MS medium solidified by 1.2% agarose was found superior, thus immobilizing the protoplast in agarose proved to be essential for obtaining sustained protoplast division and reproducible colony formation (Majewska-Sawka et al. 1994). Strong genotypic variation and somaclonal variation were also evident (Lenzner et al. 1995, Jazdzewska et al. 2000).

Although limited progress has been achieved, sugar beet is still regarded as a recalcitrant species for protoplast regeneration (Gurel et al. 2002c). Recently, however, using computer-assisted microscopy to follow divisions of heterogeneous cell populations from a recalcitrant variety, Hall et al. (1995, 1996b) showed that cells producing friable, regenerable callus were of stomatal guard-cell origin, indicating totipotency of the highly specialized guard cells. A method to enrich protoplast preparations from stomatal guard cells was developed by blending de-ribbed leaf material and purifying by gradient centrifugation; this led to cell division frequencies of approximately 50% compared to 0.01 to 0.5% for unfractionated leaf protoplasts. The protocol was reported to be highly reproducible and genotype-independent and it has since been exploited for sugar beet transformation (for a detailed review, see Gurel et al. 2008). Presently it is the only practical approach to use protoplasts from sugar beet (Hall et al. 1996a, 1997, Hall 1998, Krens et al. 1998, Wisniewska et al. 2002). Highly efficient plant regeneration was also reported from protoplasts that were isolated from friable callus developed from etiolated hypocotyl explants (Dovzhenko and Koop 2003). The method appears simpler, faster and more efficient than that of Hall et al. (1996a, 1997); however, the callus-based approach has not been reported to be effective for transformation.

Recently, Wisniewska and Majewska-Sawka (2007) reported beneficial effects of arabinogalactan protein (AGP)-rich extracts, isolated from culture medium of embryogenic and nonembryogenic suspension cells, on development and differentiation of sugar beet guard cell protoplasts and on shoot regeneration capacity of callus developed from guard cell protoplasts. AGPs, a class of proteoglycans implicated in development and differentiation of cells and tissues both *in planta* and *in vitro*, were effective for *in vitro* development of many species when isolated from organogenic or embryogenic cultures. AGPs appear to contribute to intra- and intercellular signaling (Wisniewska and Majewska-Sawka 2007, 2008).

Protoplast Fusion (Somatic Hybridization)

There are many wild relatives of beet which possess desirable characters such as disease- and pest-resistance. Owing to limited or poor crossability between *Beta vulgaris* L. and other *Beta* species, the use of interspecific hybridization in beet breeding has been limited. Techniques of proptoplast culture and fusion have the potential to produce somatic hybrids between different species of this genus (Krens et al. 1990). The ability to change a plant cell cytoplasm, e.g., from a cytoplasmic male sterile (CMS) to a male fertile (MF) type, or vice versa, would greately assist the breeding programmes in, for example, allowing the creation of the perfect maintainer line or permitting the production of F_1 hybrid seed. While such a goal in beet could be achived, at least with related lines via conventional breeding, the time required to carry out the necessary backcrosses to regain the original nuclear genome is very long in this biennial species. Furthermore, in beet, a potentially high-risk situation exists regarding the very limited cytoplasmic genetic variation, which has resulted from the almost exclusive use of a single CMS cytoplasm for breeding and seed production purposes. This problem could be significantly resolved through the rapid introduction and assesment of some of the potentially new CMS cytoplasms identified recently (Hall et al. 1993b).

The transformation techniques are valuable to genetic engineers because they allow the introduction of specific genes into plants. At the same time, this general strategy offers little immediate hope for the manipulation of polygenic traits, even if all the component genes influencing a quantitative character could be identified and isolated. A large amount of DNA would have to be transferred and the individual genes presumably would have to be expressed in a coordinated manner. Protoplast technology has provided an alternative approach to this problem, with the help of the non-sexual fusion of both nucleated protoplasts and enucleate cytoplasts, to generate hybrid cells with novel combinations of both nuclear and cytoplasmic genes.

Although fragile in comparison with an intact cell wall, protoplast membrane still has a relatively high degree of structural stability (Warren

1996). Cell-to-cell fusion is possible using either electrical fusion or chemical fusions induced by (i) sodium nitrate, (ii) high pH and high Ca, (iii) polyethylene glycol (PEG) or (iv) a combination of methods (ii) and (iii) (Bhat et al. 1985). Protoplasts are first brought into contact, after which their plasma membranes can be induced to fuse together to form ultimately a single cell. In sugar beet, high fusion frequencies can be obtained using electrical methods although, for the achivement of binary fusion products (involving just two cells), there is a clear optima. Under optimal conditions (100 V/cm AC alignment, 1 Mhz; 1500 V/cm DC pulse, 2x50 μ sec), up to 15% binary fusion products per 100 aligned cells is obtainable. However, for certain lines (i.e., subspecies *maritima* genotypes) electrofusion techniques had a definite detrimental effect on subsequent cell development. In contrast, PEG fusion techniques had a stimulatory effect on plating efficiencies of these genotypes. The choice of fusion method must be, therefore, made only after taking the genotypes to be used into account. Using PEG method, approximately 5% of the cells available for plating out is found to be heterofusion products (Hall et al. 1993b). When PEG is diluted at high pH in high concentrations, a relatively high frequency of fused protoplasts is obtained (Keller and Melchers 1973).

The identification and selection of hybrids from a background of like parental (homocaryotic) fusion products can be carried out previously *in vitro*, by metabolic complementation, fluorescence-activated cell sorting or identification of molecular markers or at a whole plant level where morphological, biochemical and molecular markers of hybridity can be employed (Hall et al. 1993a). To increase efficiency of asymmetric cell fusions, techniques were developed to produce cytoplasts from suspension culture-derived protoplasts (Van Ark et al. 1992). In 2007, using confocal laser scanning microscopy and flow cytometry, Famelaer et al. (2007) investigated synchronization and micro-nucleation in sugar beet by correlating DNA content with the size of the micronuclei and the micro-protoplasts; however, it was non-linear, affected by a degree of DNA condensation, total amount of DNA and presence of cytoplasm.

Preservation of Genetic Resources

Given the negative impacts of genetic variation, preservation of plant genetic resources as somatic tissues has been widely exploited. Culturing vegetative tissues was performed *in vitro* by keeping it at low temperature for a short- and mid-term storage or freezing in liquid nitrogen for a long-term storage. By keeping clonally propagated plantlets at 12°C, Hussey and Hepher (1978) decreased subculturing frequencies to 18-week intervals. *In vitro* regenerated shoots were preserved at low light intensity for one year at 2 or 5°C (Miedema 1982a). Although low temperatures are useful for conserving breeding materials (e.g., CMS lines, inbred O-type lines, tetraploid pollinators) for years, maintenance of vegetative materials is laborious and time-consuming and also bears the risk of losing valuable materials due to contamination or human error (Vandenbussche et al. 2000). To minimize losses, cryopreservation was developed for sugar beet using either shoot-tips (Braun 1988, Vandenbussche and De Proft 1996, 1998, Vandenbussche et al. 1999, 2000) or root-tips from transgenic hairy root cultures (Benson and Hamill 1991).

Cryopreservation of shoot-tips leads to long-term storage, maximal stability of phenotypic and genotypic characteristics of stored germplasm, and minimal storage space and maintenance requirements (Ding et al. 2008). Braun (1988) first showed applicability of freezing shoot-tip meristems excised from clonally propagated sugar beet shoots of floral buds. Slow-freezing involved preculturing isolated shoot-tips on agar medium with TIBA and 5% dimethylsulfoxide (DMSO), a cryoprotectant, transferring to liquid medium with two cryoprotectants (sorbitol and DMSO), and freezing at –40°C before storing in liquid nitrogen. A frequency of 58.4% regrowth occurred when frozen meristems were thawed at 39°C. Improved protocols were developed that employ encapsulation-dehydration, combined with (Vandenbussche et al. 2000) or without (Vandenbussche and De Proft 1996, 1998) prior vitrification. Best results were obtained with encapsulation-dehydration, following three-step preculture and dehydration (Vandenbussche and De Proft 1996). Survival rates increased to 45 to 90% when shoot-tips were excised from cold-acclimated *in vitro* shoots (Vandenbussche and De Proft 1998), which appears to enhance tolerance to dehydration and freezing by inducing physiological changes. Plants from cold-acclimated shoot-tips have more sucrose, D-glucose and D-fructose and fatty acids (Vandenbussche et al. 1999); desaturation of membrane lipids also increased. The optimized encapsulation-dehydration method was, however, lengthy and laborious, i.e., one-day pretreatment, three-day preculture and seven-hour dehydration. These disadvantages were largely overcome with a combination of vitrification and cryopreservation to preserve shoot-tips (Vandenbussche et al. 2000), enabling survival rates of up to 100% and faster regrowth. The combination of vitrification and cryopreservation of shoot-tips from field-grown plants was used to remove pathogens, especially viruses (Ding et al. 2008).

Cryopreservation methods for root-tips from hairy root cultures of transgenic sugar beet were used to study effects of culture age, pre-growth, cryoprotection, freezing rate and post-freeze culture conditions (Benson and Hamill 1991). Although ca. 80% of root-tip meristems were viable after freezing, based on fluorescein diacetate staining, less than 20% of roots regenerated. Growth rates, secondary metabolite production and T-DNA structure, however, were unchanged after cryopreservation.

Somaclonal Variation and *in vitro* Cell Selection

Somaclonal variation has been exploited in breeding programmes to identify desirable characteristics in several species that are not readily available within compatible germplasm sources. It has been more successful in crops with limited genetic bases, like many ornamental species where spontaneous or induced somaclonal variation provides a source of novel variation for plant improvement (Davey et al. 2007). In sugar beet, somaclonal variation was commonly observed in plants resulting from indirect regeneration from callus derived from either seedling tissues (i.e., leaf lamina, petiole or hypocotyl explants) (Brears et al. 1989, Yu 1989) or from protoplasts (Jazdzewska et al. 2000). Somaclonal variation among the regenerants directly developed from explant tissues was also reported (Harms et al. 1983, Dikalova et al. 1993, Zhong et al. 1993b). However, somaclonal variation among sugar beet plants obtained via meristem proliferation has not been reported.

Somaclonal variation can be a visually observed phenotype (i.e., alterations in morphology, especially in the leaf) (Saunders 1982, Saunders and Mahoney 1982, Freytag et al. 1988) or a molecular change, as in isozyme analyses, RFLP and RAPD (i.e., structural changes in nuclear and mitochondrial genomes) (Brears et al. 1989, Levall et al. 1994, Munthali et al. 1996, Dikalova et al. 1993, Jazdzewska et al. 1995, Sadoch et al. 2000). Polymorphism frequencies determined by RAPDs (Munthali et al. 1996) were similar to those seen in isozyme and RFLP analyses (Sabir et al. 1992, Levall et al. 1994).

Somaclonal variation can occur through genetic (heritable) (Jacq et al. 1992) or epigenetic (nonheritable) means, as observed when regenerants were monitored morphologically and cytologically after three cycles of petiole culture (Detrez et al. 1989). Somaclonal variation is more often induced when the cultures were kept in culture for longer periods; 96% of regenerated plantlets being true-to-type diploids after a short callus phase while it was 83% after a long phase (Jacq et al. 1992). Extent of variability also seems to depend on explant source used for regeneration; variation, based on chromosome number, was highest in plants regenerated from protoplasts and lowest in those from calluses (Steen et al. 1986). Thus, several factors, e.g., explant type, culture conditions and duration, impact the frequency, extent and nature (heritable or nonheritable) of spontaneously occurring somaclonal variation (Saunders et al. 1990).

Artificial induction of somaclonal variation can also occur following certain *in vitro* cell selections, including those focused on UV-tolerance (Levall and Bornman 1993, Levall et al. 1994), salt-tolerance (Pua and Thorpe 1986, Freytag et al. 1990, Yang et al. 2004), herbicide-tolerance (Saunders et al. 1992, Hart et al. 1993, Wright and Penner 1998a,b, Wright et al. 1998) and reduced nitrogenous impurities (Saunders et al. 1997). Levall and Bornman (1993) exposed calluses to UV radiation, regenerated shoots from surviving calli and screened for tolerance to UV radiation. Unselected plants had significantly greater damage following UV radiation compared to selected plants. Subsequently, regenerated plants from UV-treated calluses were studied by RFLP analysis to determine the extent of somaclonal variation at the DNA level. Cytological analyses of somaclones revealed considerable ploidy variation, while DNA methylation differences were not significant (Levall et al. 1994).

To select for somaclonal variants with multiple salt-tolerances, Freytag et al. (1990) regenerated plants from petiole explants cultured on medium with five different salts. Surviving shoots were exposed to salt in three subsequent challenges, during which the number of petioles developing tolerant shoots steadily increased from 5 to 80%. Selected regenerants grew in soil containing 0.1% (w/w) salt, with no detectable morphological differences compared to control plants grown in salt-free soil. The multiple salt-tolerant plants might have come from single mutant cells, as previously suggested (Detrez et al. 1988). Yang et al. (2004) obtained salt-tolerant plants from multiple bud clumps following culture of immature inflorescence tips exposed to γ-radiation and then selected with NaCl.

Numerous attempts to exploit somaclonal variation were made to obtain herbicide-tolerant clones (Saunders et al. 1992, Wright and Penner 1998a,b, Wright et al. 1998). Identification of a monogenic dominant, sulfonylurea-tolerant line (REL-1) was achieved when unmutagenized suspension cell clusters were exposed to 2.8 nM concentrations of chlorsulfuron, an acetolactate synthase (ALS)-inhibiting herbicide (Saunders et al. 1992). Shoots from tolerant colonies survived 2.8 nM chlorsulfuron, a concentration that killed shoots from nontolerant colonies of the same line. REL-1 was also used to develop lines tolerant to another group of ALS-inhibiting herbicides, imidazolinones (Wright and Penner 1998a). The imidazolinone- and sulfonylurea-tolerant lines, obtained following cell selection, were studied for inheritance and cross-tolerance characteristics (Wright et al. 1998, Wright and Penner 1998b). Presence of a single-copy ALS gene in diploid regenerants was believed to limit the ability to stack tolerance traits in the same plant by traditional breeding.

Genotypic Variation

Certain genotypes of sugar beet are more amenable than others to direct and indirect organogenesis (Gurel et al. 2001, 2003a), to somatic embryogenesis (Saunders and Tsai 1999), to haploid plant production from unfertilized ovules (Doctrinal et al. 1989, Lux et al. 1990, Hansen et al. 1995, Gurel et al. 2000), to protoplast isolation, culture and regeneration (Lenzner et al. 1995, Gurel et al. 2002b, Jazdzewska et al. 2000), and to clonal propagation from pre-existing meristems (Mezei et al. 1990, Sabir and Ford-Lloyd 1991, Hayakawa et al. 1994).

Large *inter-variety* differences were observed among ten different commercial varieties when variability of callus development and root organogenesis was investigated at different levels, i.e., variety, plant and organ/leaf (Gurel 1997). Significant *inter-plant* (plant-to-plant) and *intra-plant* (*inter-leaf*) variability occurred in a variety even at similar physiological ages, but explants from different positions in the leaf lamina behaved similarly. Strong intra-plant variation in callusing and shooting within a cultivar (genotype population) was also observed (Doley and Saunders 1989, Ivic-Haymes and Smigocki 2005). Since sugar beet is highly heterozygous, it is understandable that considerable inter-plant variation exists, even with phenotypically similar individuals. Thus, genotypic differences persisting within a variety are likely responsible for inter-plant variability, which accounts for plants of the variety, being categorized as good-, poor- or non-rooters *in vitro* (Gurel 1997). Intra-plant variability is more difficult to explain.

A higher correlation was expected in explant behavior from different leaves of the same plant but that is not observed. Means exist to cope with variability in sugar beet. One is to use large numbers of explant replicates from many individual plants within a single line/population (Doley and Saunders 1989, Ivic-Haymes and Smigocki 2005). But the laboriousness of plant tissue culture, coupled with the need to standardize production of parental plants, often restricts feasibility of this approach. Material can also be sorted prior to culture to eliminate those with low organogenic or embryogenic potential. To facilitate this approach, attempts were made to correlate blackening of leaf lamina discs on culture medium and the inability to produce roots with levels of naturally-occurring phenolics. The latter was measured through polyphenol oxidase (PPO) activity with a highly sensitive, small-scale, reproducible method (Gurel and Wren 1995b). Although a two-fold variation in levels of enzyme activity was observed, no relationship was seen between PPO activity and rooting capacity, or between PPO activity and degree of blackening; a weak negative correlation was found between degree of blackening and rooting capacity (Gurel 1997). Other factors, e.g., nature and concentration of phenolic substrates rather than enzyme activity, might be more critical in determining regenerative capacity.

References

Abe, J., H. Nakashima, K. Mitsui, T. Mikami and Y. Shimamoto. 1991. Tissue culture response of *Beta* germplasm, callus induction and plant regeneration. Plant Cell Tiss. Org. Cult. 27: 123–127.

Alemanno, L. and E. Guiderdoni. 1994. Increased doubled haploid plant regeneration from rice (*Oryza sativa* L.) anthers cultured on colchicine-supplemented media. Plant Cell Rep. 13(8): 432–436.

Atanassov, A.I. 1980. Method for continuous bud formation in tissue cultures of sugar beet (*Beta vulgaris* L). Z. Pflanzenzuchtung 84: 23–29.

Atanassov, A.I. 1986. Sugar beet. *In*: Handbook of Plant Cell Culture. D.A. Evans, W.R. Sharpe and P.V. Ammirato (eds.). Macmillan Publishing Co., New York, pp. 652–680.

Banba, H. and H. Tanabe. 1972. On anther culture in sugar beet. Bull. Sugar Beet Res. 14: 9–16.

Bekheet, S.A. and M.E. Solliman. 2007. An efficient transformation system for sugar beet mediated by *Agrobacterium tumefaciens*. J. Food Agr. Environ. 5: 193–197.

Benson, E.E. and J.D. Hamill. 1991. Cryopreservation and post freeze molecular and biosynthetic stability in transformed roots of *Beta vulgaris* and *Nicotiana rustica*. Plant Cell Tiss. Org. Cult. 24: 163–172.

Bhat, S.R., B.V. Ford-Lloyd and J.A. Callow. 1985. Isolation of protoplasts and regeneration of callus from suspension cultures of cultivated beets. Plant Cell Rep. 4: 348–350.

Bisbis, B., C. Kevers, J. Dommes and T. Gaspar. 2000. Interactions between polyamine and ethylene metabolisms in a hormone autonomous sugar beet callus. J. Plant Physiol. 157: 24–30.

Bossoutrot, D. and D. Hosemans. 1985. Gynogenesis in *Beta vulgaris* L.: From *in vitro* culture to the production of doubled haploids plants in soil. Plant Cell Rep. 4: 300–303.

Braun, A. 1988. Cryopreservation of sugar beet germplasm. Plant Cell Tiss. Org. Cult. 14: 161–168.

Brears, T., G.J. Curtis and D.M. Lonsdale. 1989. A specific rearrangement of mitochondrial DNA induced by tissue culture. Theor. Appl. Genet. 77: 620–624.

Catlin, D.W. 1990. The effect of antibiotics on the inhibition of callus induction and regeneration from cotyledons of sugar beet. Plant Cell Rep. 9: 285–288.

Christianson, M.L. and D.A. Warnick. 1985. Temporal requirement for phytohormone balance in the control of organogenesis. *In vitro*. Dev. Biol. 112: 494–497.

Coumans-Gilles, M.F., C. Kevers, M. Coumans, E. Ceulemans and T. Gaspar. 1981. Vegetative multiplication of sugar beet through *in vitro* culture of inflorescence pieces. Plant Cell Tiss. Org. Cult. 1: 93–101.

Cresswell, R. 1995. Improvement of plants via plant cell culture. *In*: Plant Cell, Tissue and Organ Culture. O.L. Gamborg and G.C. Phillips (eds.). Springer and Verlag, Heidelberg, pp. 101–123.

Davey, M., P. Anthony, K. Lowe, M. Castellanos, G. Ribas-Vargas and B. Power. 2007. Somatic cell technologies for genetic improvement of ornamental plants. Propag. Ornam. Plants 7: 122–128.

De Greef, W. and M. Jacobs. 1979. *In vitro* culture of the sugar beet: description of a cell line with high regeneration capacity. Plant Sci. Lett. 17: 55–61.

De Jong, J.H. and T.S.M. De Bock. 1978. Use of haploids of *Beta vulgaris* L. for the study of orcein and giemsa stained chromosomes. Euphytica 27: 41–47.

Detrez, C., R.S. Sangwan and B.S. Sangwan-Norrel. 1989. Phenotypic and karyotypic status of *Beta vulgaris* plant regenerated from direct organogenesis in petiole culture. Theor. Appl. Genet. 77: 462–468.

Detrez, C., T. Tetu, R.S. Sangwan and B.S. Sangman-Norrel. 1988. Direct organogenesis from petiole and thin cell layer explants in sugar beet cultured *in vitro*. J. Exp. Bot. 39: 917–926.

D'Halluin, K. and B. Keimer. 1986. Production of haploid sugar beets (*Beta vulgaris* L.) by ovule culture. *In*: Genetic Manipulation in Plant Breeding. W. Horn, C.J. Jensen, W. Odenbach and O. Schieder (eds.). Walter de Gruyter, Berlin, Germany, pp. 307–309.

Dikalova, A.E., N.A. Dudareva, M. Kubalakova and R. Salganik. I. 1993. Rearrangements in sugar beet mitochondrial DNA induced by cell suspension, callus cultures and regeneration. Theor. Appl. Genet. 86: 699–704.

Ding, F., S. Jin, N. Hong, Y. Zhong, Q. Cao, G. Yi and G. Wang. 2008. Vitrification-cryopreservation, an efficient method for eliminating *Candidatus Liberobacter asiaticus*, the citrus Huanglongbing pathogen, from *in vitro* adult shoot tips. Plant Cell Rep. 27: 241–250.

Doctrinal, M., R.S. Sangwan and B.S. Sangwan-Norreel. 1989. *In vitro* gynogenesis in *Beta vulgaris* L.: Effect of plant growth regulators, temperature, lines and season. Plant Cell Tiss. Org. Cult. 17: 1–12.

Doley, W.P. and J.W. Saunders. 1989. Hormone-free medium will support callus production and subsequent shoot regeneration from whole plant leaf explants in some sugar beet (*Beta vulgaris* L.) populations. Plant Cell Rep. 8: 222–225.

D'Onofrio, C. and S. Morini. 2006. Somatic embryo, adventitious root and shoot regeneration in *in vitro* grown quince leaves as influenced by treatments of different length with growth regulators. Sci. Hort-Amsterdam 107: 194–199.

Dovzhenko, A. and H.U. Koop. 2003. Sugar beet (*Beta vulgaris* L.): Shoot regeneration from callus and callus protoplasts. Planta 217: 374–381.

Draycott, A.P. 2006. Introduction. *In*: Sugar Beet. A.P. Draycott (ed.). Blackwell, Oxford, pp. 1–8.

Famelaer, I., H.A. Verhoeven, P. Dijkhuis and K.S. Ramulu. 2007. A study of the process of synchronization and micronucleation in *Beta vulgaris* and the monitoring of an isolation procedure for micro-nuclei and micro-protoplasts by confocal laser scanning microscopy and flow cytometry. Plant Cell Tiss. Org. Cult. 90: 169–179.

Ferrant, V. and J. Bouharmont. 1994. Origin of gynogenetic embryos of *Beta vulgaris* L. Sex. Plant Reprod. 7: 12–16.

Ford-Lloyd, B.V. and S. Bhat. 1986. Problems and prospects for the use of protoplasts in beet breeding. *In*: Genetic Manipulation in Plant Breeding. W. Horn, C.J. Jensen, W. Odenbach and O. Schieder (eds.). Walter de Gruyter, Berlin, Germany, pp. 437–440.

Freytag, A.H., S.C. Anand, A.P. Rao-Arelli and L.D. Owens. 1988. An improved medium for adventitious shoot formation and callus induction in *Beta vulgaris* L. *in vitro*. Plant Cell Rep. 7: 30–34.

Freytag, A.H., J.A. Wrather and A.W. Erichsen. 1990. Salt tolerant sugar beet progeny from tissue cultures challenged with multiple salts. Plant Cell Rep. 8: 647–650.

Galatowitsch, M.W. and G.A. Smith. 1990. Regeneration from unfertilized ovule callus of sugar beet (*Beta vulgaris* L.). Can. J. Plant Sci. 70: 83–89.

Gamborg, O.L., R.A. Miller and K. Ojima. 1968. Nutrient requirements of suspension cultures of soybean root cells. Exp. Cell Res. 50: 151–158.

Gaspar, T., C. Kevers, C. Penel and H. Greppin. 1983. Auxin control of calcium-mediated peroxidase secretion by auxin-dependent and auxinindependent sugar beet cells. Phytochemistry 22: 2657–2660.

Giord, P.A. and J.P. Zryd. 1991. Secondary metabolism in cultured red beet (*Beta vulgaris* L.) cells: Differential regulation of betaxanthin and betacyanin biosenthesis. Plant Cell Tiss. Org. Cult. 25: 1–12.

Grieve, T.M., K.M.A. Gartland and M.C. Elliot. 1997. Micropropagation of commercially important sugar beet cultivars. Plant Growth Regul. 21: 15–18.

Guha, S. and S.C. Maheshwari. 1964. *In vitro* production of embryos from anthers of *Datura*. Nature 204: 497–498.

Gurel, E. 1991. Callus Development and Organogenesis in Cultured Leaf Explants of Sugar Beet (*Beta vulgaris* L). Ph.D. Dissertation, University of Leeds, Leeds, England.

Gurel, E. 1997. Callus and root development from leaf explants of sugar beet (*Beta vulgaris* L.): Variability at variety, plant and organ level. Turk. J. Bot. 21: 131–136.

Gurel, E. and S. Gurel. 1998. Plant regeneration from unfertilized ovaries of sugar beet (*Beta vulgaris* L.) cultured *in vitro*. Turk. J. Bot. 22: 233–238.

Gurel, E. and M.J. Wren. 1995a. *In vitro* development from leaf explants of sugar beet (*Beta vulgaris* L.): Rhizogenesis and the effect of sequential exposure to auxin and cytokinin. Ann. Bot. 75: 31–38.

Gurel, E. and M.J. Wren. 1995b. Measuring polyphenol oxidase activity in small leaf discs of sugar beet (*Beta vulgaris* L.). Turk. J. Bot. 19: 497–502.

Gurel, S. 2000. *In vitro* Growth and Development of Sugar Beet (*Beta vulgaris* L.) from Callus, Ovule, Suspension Cells and Protoplasts. Ph.D. Dissertation, Middle East Technical University, Ankara, Turkey.

Gurel, S., E. Gurel and Z. Kaya. 2000. Doubled haploid plant production from unpollinated ovules of sugar beet (*Beta vulgaris* L.). Plant Cell Rep. 19: 1155–1159.

Gurel, S., E. Gurel and Z. Kaya. 2001. Callus development and indirect shoot regeneration from seedling explants of sugar beet (*Beta vulgaris* L.) cultured *in vitro*. Turk. J. Bot. 25: 25–33.

Gurel, S., E. Gurel and Z. Kaya. 2002a. Establishment of cell suspension cultures and plant regeneration in sugar beet (*Beta vulgaris* L.). Turk. J. Bot. 26: 197–205.

Gurel, E., S. Gurel and Z. Kaya. 2002b. Protoplast isolation and culture in sugar beet (*Beta vulgaris* L.). Plant Cell Biotechnol. Mol. Biol. 3&4: 11–20.

Gurel, S., E. Gurel and Z. Kaya. 2002c. Protoplast fusion in sugar beet (*Beta vulgaris* L.). Turk. J. Biol. 26: 163–170.

Gurel, S., E. Gurel, Z. Kaya, M. Erdal and E. Guler. 2003c. Effects of antimitotic agents on haploid plant production from unpollinated ovules of sugar beet (*Beta vulgaris* L.). Biotechnol. Biotech. Eq. 17: 97–101.

Gurel, E., E. Topal and S. Gurel. 2003a. The effect of pretreating seedlings with BAP on direct shoot regeneration from petiole explants of sugar beet (*Beta vulgaris* L.). Biotechnol. Biotech. Eq. 17: 89–96.

Gurel, S., E. Topal and E. Gurel. 2003b. The effect of pretreating seedlings with TDZ on direct shoot regeneration from petiole explants of sugar beet (*Beta vulgaris* L.). Asia Pas. J. Mol. Biol. Biotechnol. 11: 57–62.

Gurel, E., S. Gurel and P.G. Lemaux. 2008. Biotechnology applications for sugar beet. Crit. Rev. Plant Sci. 27: 1–33.

Gurel, S., M.C. Baloglu, E. Gurel, H.A. Oktem and M. Yucel. 2011. A two-stage pretreatment of seedlings improves adventitious shoot regeneration in sugar beet (*Beta vulgaris* L.). Plant Cell Tiss. Org. Cult. 106: 261–268.

Hall, R.D. 1991. The initiation and maintenance of plant cell suspension cultures. Plant Tissue Culture Manual, A3: 1-21. Kluwer Academic Publishers, Dordrecht.

Hall, R.D. 1998. Biotechnological applications for stomatal guard cells. J. Exp. Bot. 49(SI): 369–375.

Hall, R.D. and F.A. Krens. 1988. The production and electrofusion of *Beta* cytoplasts. *In*: Progress in Plant Protoplast Research. K.J. Puite, J.J.M. Dons, H.J. Huizing, A.J. Kool, M. Koornneef and F.A. Krens (eds.). Kluwer, Dordrecht, pp. 263–264.

Hall, R.D., C. Pederson and F.A. Krens. 1993a. Improvement of protoplast culture protocols for *Beta vulgaris* L. (sugar beet). Plant Cell Rep. 12: 339–342.

Hall, R.D., C. Pedersen and F.A. Krens. 1993b. Progress towards the development of a general somatic hybridization protocol for *Beta*. J. Sugar Beet Res. 30: 275–290.

Hall, R.D., G. Pedersen and F.A. Krens. 1994. Regeneration of plants from protoplasts of *Beta vulgaris* (sugar beet). *In*: Biotechnology in Agriculture and Forestry, Vol. 29, Plant Protoplasts and Genetic Engineering V. Y.P.S. Bajaj (ed.). Springer-Verlag, Berlin, pp. 16–37.

Hall, R.D., H.A. Verhoeven and F.A. Krens. 1995. Computer-assisted identification of protoplasts responsible for rare division events reveals guard-cell totipotency. Plant Physiol. 107: 1379–1386.

Hall, R.D., T. Riksen-Bruinsma, G. Weyens, M. Lefebvre, J.M. Dunwell and F.A. Krens. 1996a. Stomatal guard cells are totipotent. Plant Physiol. 112: 889–892.

Hall, R.D., T. Riksen-Bruinsma, G. Weyens, I.J. Rosquin, P.N. Denys, I.J. Evans, J.E. Lathouwers, M. Lefebvre, J.M. Dunwell, A. van Tunen and F.A. Krens. 1996b. A high efficiency technique for the generation of transgenic sugar beets from stomatal guard cells. Nature Biotechnol. 14: 1133–1138.

Hall, R.D., T. Riksen-Bruinsma, G. Weyens, M. Lefebvre, J.M. Dunwell, A. Van Tunen and F.A. Krens. 1997. Sugar beet guard cell protoplasts demonstrate a remarkable capacity for cell division enabling applications in stomatal physiology and molecular breeding. J. Exp. Bot. 48: 255–263.

Hamill, J.D., A.J. Parr, R.J. Robins and M.J.C. Rhodes. 1986. Secondary product formation by cultures of *Beta vulgaris* and *Nicotiana rustica* transformed with *Agrobacterium rhizogenes*. Plant Cell Rep. 5: 111–114.

Hansen, A.L., A. Gertz, M. Joersbo and S.B. Andersen. 1995. Short-duration colchicine treatment for *in vitro* chromosome doubling during ovule culture of *Beta vulgaris* L. Plant Breed. 114: 515–519.

Hansen, A.L., A. Gertz, M. Joersbo and S.B. Andersen. 1998. Antimicrotubule herbicides for *in vitro* chromosome doubling in *Beta vulgaris* L. ovule culture. Euphytica 101: 231–237.

Hansen, A.L., A. Gertz, M. Joersbo and S.B. Andersen. 2000. Chromosome doubling *in vitro* with amiprophos-methyl in *Beta vulgaris* ovule culture. Acta Agr. Scand. 50: 89–95.

Hansen, A.L., C. Plever, H.C. Pedersen, B. Keimer and S.B. Andersen. 1994. Efficient *in vitro* chromosome doubling during *Beta vulgaris* ovule culture. Plant Breed. 112: 89–95.

Harms, C.T., I. Baktir and J.J. Oertli. 1983. Clonal propagation *in vitro* of red beet (*Beta vulgaris* ssp.) by multiple adventitious shoot formation. Plant Cell Tiss. Org. Cult. 2: 93–102.

Hart, S., J. Saunders and D. Penner. 1993. Semidominant nature of monogenic sulfonurea herbicide resistance in sugarbeet *(Beta vulgaris)*. Weed Sci. 41: 317–324.

Hayakawa, H., Y. Shimamoto, S. Asano and T. Izuka. 1994. Development of shoot-base tissue on leaf blade for raising transgenic plants in beets. Proc. Jpn. Soc. Sugar Beet Technol. 43: 24–28.

Hisano, H., Y. Kimoto, H. Hayakawa, J. Takeichi, T. Domae, R. Hashimoto, J. Abe, S. Asano, A. Kanazawa and Y. Shimamoto. 2004. High frequency *Agrobacterium*-mediated transformation and plant regeneration via direct shoot formation from leaf explants in *Beta vulgaris* and *Beta maritima*. Plant Cell Rep. 22: 910–918.

Hooker, M.P. and M.W. Nabors. 1977. Callus initiation, growth, and organogenesis in sugar beet *(Beta vulgaris* L). Z. Pflanzenzuchtung 84: 237–246.

Hosemans, D. and D. Bossoutrot. 1983. Induction of haploid plants from *in vitro* culture of unpollinated beet (*Beta vulgaris* L.). Z. Pflanzenzuchtung 91: 74–77.

Hunter, C.S. and N.J. Kilby. 1990. Betanin production and release *in vitro* from suspension cultures of *Beta vulgaris*. *In*: Methods in Molecular Biology, Vol. 6, Plant Cell and Tissue Culture. J.W. Pollard and J.M. Walker (eds.). Humana Press, New Jersey, pp. 545–554.

Hussey, G. and A. Hepher. 1978. Clonal propagation of sugar beet plants and the formation of polyploids by tissue cultures. Ann. Bot. 42: 447–478.

Ivic, S.D. and A.C. Smigocki. 2003. Transformation of sugar beet cell suspension cultures. *In vitro* Cell. Dev. Biol.-Plant 39: 573–577.

Ivic-Haymes, S.D. and A.C. Smigocki. 2005. Identification of highly regenerative plants within sugar beet (*Beta vulgaris* L.) breeding lines for molecular breeding. *In vitro* Cell. Dev. Biol.-Plant 41: 483–488.

Jacq, B., T. Tetu, R.S. Sangwan, A. De Laat and B.S. Sangwan-Norreel. 1992. Plant regeneration from sugarbeet (*Beta vulgaris* L.) hypocotyls cultured *in vitro* and flow cytometric nuclear DNA analysis of regenerants. Plant Cell Rep. 11: 329–333.

Jacq, B., T. Tetu, R.S. Sangwan, A. De Laat and B.S. Sangwan-Norreel. 1993. Efficient production of uniform plants from cotyledon explants of sugarbeet (*Beta vulgaris* L.). Plant Breed. 110: 185–191.

Jazdzewska, E., A. Niklas and A. Majewska-Sawka. 1995. Progress towards sugar beet improvement through somatic hybridization. 1. Inactivation of nuclei and cytoplasm in donor and recipient protoplasts. Acta Soc. Bot. Pol. 64: 341–347.

Jazdzewska, E., Z. Sadoch, A. Niklas and A. Majewska-Sawka. 2000. Plant regeneration from sugar beet leaf protoplasts: analysis of shoots by DNA fingerprinting and restriction fragment length polymorphism. Can. J. Bot. 78: 10–18.

Keller, W.A. and G. Melchers. 1973. The effect of high pH and calcium on tobacco leaf protoplast fusion. Z. Naturforsch. 28: 737–741.

Kevers, C., M. Coumans, W. De Greef, M. Jacobs and T. Gaspar. 1981a. Organogenesis in habituated sugar beet callus: Auxin content and protectors, peroxidase pattern and inhibitors. Z. Pflanzenphysiol. 101: 79–87.

Kevers, C., M. Coumans, W. De Greef, M. Hofinger and T. Gaspar. 1981b. Habitation in sugar beet callus: Auxin content, auxin protectors, peroxidase pattern, and inhibitors. Physiol. Plant 51: 281–286.

Kevers, C., L. Sticher, C. Penel, H. Greppin and T. Gaspar. 1982. Calcium controlled peroxidase secretion by sugar beet cell suspensions in relation to habituation. Plant Growth Regul. 1: 61–66.

Kevers, C., L. Sticher, C. Penel, H. Greppin and T. Gaspar. 1983. The effect of ergosterol, ergocalciferol and cholecatciferol on calcium controlled peroxidase secretion by sugar beet cells. Physiol. Plant. 57: 17–20.

Konwar, B.K. 1994. *Agrobacterium tumefaciens*-mediated genetic transformation of sugar beet (*Beta vulgaris* L.). J. Plant Biochem. Biotechnol. 3: 37–41.

Konwar, B.K. and R.H.R. Coutts. 1990. Rapid regeneration of sugarbeet (*Beta vulgaris* L.) plants from *in vitro* cultures. *In*: Progress in Plant Cellular and Molecular Biology. H.J.I. Nijkamp, L.H.W. Van Der Plas and J. Van Aartrijk (eds.). Kluwer Academic Publishers, The Netherlands, pp. 114–118.

Krens, F.A., D. Jamar, L.C.P. Keizer and R.D. Hall. 1994. The effect of *n*-propyl gallate on the formation of ethylene during protoplast isolation in sugarbeet (*Beta vulgaris* L). J. Exp. Bot. 45: 1899–1901.

Krens, F.A., D. Jamar, G.J.A. Rouwendal and R.D. Hall. 1990. Transfer of cytoplasm from new *Beta* CMS sources to sugar beet by asymmetric fusion. I. Shoot regeneration from mesophyll protoplasts and characterization of regenerated plants. Theor. Appl. Genet. 79: 390–396.

Krens, F.A., H.A. Verhoeven, A.J. Van Tunen and R.D. Hall. 1998. The use of an automated cell tracking system to identify specific cell types competent for regeneration and transformation. *In vitro* Cell. Dev. Biol.-Plant 34: 81–86.

Kubalakova, M. 1990. Somatic embryogenesis and cytoplasmic sterility in *Beta vulgaris* L. var. Saccharifera. Biol. Plant. 32: 414–419.

Kulshreshtha, S. and R.H.A. Coutts. 1997. Direct somatic embryogenesis and plant regeneration from mature sugarbeet (*Beta vulgaris* L.) zygotic cotyledons. Plant Growth Regul. 22: 87–92.

Kuykendall, L.D., T.M. Stockett and J.W. Saunders. 2003. *Rhizobium radiobacter* conjugation and callus-independent shoot regeneration used to introduce the cercosporin export gene *cfp* from *Cercospora* into sugar beet (*Beta vulgaris* L.). Biotechnol. Lett. 25: 739–744.

Lenzner, S., K. Zoglauer and O. Schieder. 1995. Plant regeneration from protoplasts of sugar beet (*Beta vulgaris*). Physiol. Plant. 94: 342–350.

Levall, M.W., K. Bengtsson, N.O. Nilsson, A. Hjerdin and C. Hallden. 1994. Molecular characterization of UV-treated sugar-beet somaclones using RFLP markers. Physiol. Plant. 90: 216–220.

Levall, M.W. and J.F. Bornman. 1993. Selection *in vitro* for UV-tolerant sugarbeet (*Beta vulgaris*) somaclones. Physiol. Plant. 88: 37–43.

Li, S.N., B. Landova and Z. Landa. 1976. Isolation of protoplasts from sugar beet leaves. Biol. Plant. 18: 389–392.

Lichter, R. 1978. Restoration of male fertility in cytoplasmic male-sterile sugarbeet by heat-treatment and meristem culture. Z. Pflanzenzuchtung 81: 159–165.

Lindsey, K. and M.G.K. Jones. 1989. Stable transformation of sugar beet protoplasts by electroporation. Plant Cell Rep. 8: 71–74.

Lux, H., L. Herman and C. Wetzel. 1990. Production of haploid sugar beet *(Beta vulgaris* L.) by culturing unpollinated ovules. Plant Breed. 104: 177–183.

Ma, Y., H. Nakashima and Y. Shimamoto. 1995. Colony formation from protoplast of *Beta vulgaris* L. Proc. Jpn. Soc. Sugar Beet Technol. 37: 19–23.

Magnusson, E., E. Wremerth-Weich and C.H. Bornman. 1985. Haploidization of *Beta vulgaris* ovule culture. Hereditas Suppl. 3: 147–148.

Mahn, A., A. Matzk, C. Sautter and J. Schiemann. 1995. Transient gene expression in shoot apical meristems of sugarbeet seedlings after particle bombardment. J. Exp. Bot. 46: 1625–1628.

Majewska-Sawka, A., H. Nakashima and K. Mori. 1990. Protoplast isolation from mesophyll of *Beta* species. Proc. Jap. Soc. Beet Technol. 32: 129–133.

Majewska-Sawka, A., H. Nakashima and K. Mori. 1994. Isolation and culture of suspension-derived protoplasts of *Beta vulgaris* L. Biol. Plant. 36: 9–13.

Majewska-Sawka, A., A. Niklas and E. Jazdzewska. 1997. The effect of polyamines on the development of sugar beet protoplasts. Biol. Plant. 39: 561–567.

Margara, J. 1970. Neoformation de bourgeons *in vitro* chez la betterave sucrieve *(Beta vulgaris* L.). C. R. Acad. Sci. D. Nat. 270: 698–701.

Margara, J. 1977. La multiplication vegetative de la betterave *(Beta vulgaris* L.) en culture *in vitro.* C. R. Acad. Sci. D. Nat. 285: 1041–1044.

Masuda, H., R. Nakagawa and S. Sugawara. 1988. Hormone-autonomous suspension culture from leaf explants of sugar beets in liquid medium. Plant Cell Physiol. 29: 75–78.

Mezei, S., S. Jelaska and L. Kovacev. 1990. Vegetative propagation of sugar beet from floral ramets. J. Sugar Beet Res. 27: 90–96.

Miedema, P. 1982a. A tissue culture technique for vegetative propagation and low temperature preservation of *Beta vulgaris* L. Euphytica 31: 635–643.

Miedema, P. 1982b. Vegetative propagation of *Beta vulgaris* by leaf cuttings with axillary buds. Euphytica 31: 771–772.

Miedema, P. 1983. The effect of various cytokinin treatments on shoot initiation and shoot morphogenesis in *Beta vulgaris.* Acta Bot. Neerl. 32: 237

Miedema, P., P.J. Groot and J.H.M. Zuidgeest. 1980. Vegetative propagation of *Beta vulgaris* by leaf cuttings. Euphytica 29: 425–432.

Mikami, T., R. Sudoh, E. Nagao and T. Kinoshita. 1989a. Genotypic variation in the *in vitro* morphogenesis from leaf explants of *Beta vulgaris* L. and *B. maritima* L. Euphytica 40: 271–273.

Mikami, T., Y. Yanai and T. Kinoshita. 1989b. High frequency of organogenesis in leaflet culture of sugarbeet. Proc. Jpn. Soc. Sugar Beet Technol. 31: 145–150.

Mishutkina, Y.V. and A.K. Gaponenko. 2006. Sugar beet (*Beta vulgaris* L.) morphogenesis *in vitro*: Effect of phytohormone type and concentration in the culture medium, type of explants, and plant genotype on shoot regeneration frequency. Russian J. Genet. 42: 150–157.

Moghaddam, B.E., M. Mesbah and N. Yavari. 2000. The effect of *in planta* TIBA and proline treatment on somatic embryogenesis of sugar beet (*Beta vulgaris* L.). Euphytica 112: 151–156.

Mukhambetzhanov, S.K. 1997. Culture of nonfertilized female gametophytes *in vitro*. Plant Cell Tiss. Org. Cult. 48: 111–119.

Munthali, M.T., H.J. Newbury and B.V. Ford-Lloyd. 1996. The detection of somaclonal variants of beet using RAPD. Plant Cell Rep. 15: 474–478.

Murashige, T. and F. Skoog. 1962. A revised medium for rapid growth and bioassays with tobacco tissue cultures. Physiol. Plant. 15: 473–497.

Nakashima, H., N. Tabayashi, J. Abe and Y. Shimamoto. 1988. Tissue culture of sugar beet: Effects of genotypes on callus induction from seeds, callus proliferation and regeneration from callus. Proc. Jpn. Soc. Sugar Beet Technol. 30: 113–117.

Negrutiu, I., F. Beeftink and M. Jacobs. 1975. *Arabidopsis thaliana* as a model system in somatic cell genetics. 1. Cell and tissue culture. Plant Sci. Lett. 5: 293–304.

Norouzi, P., M.A. Malboobi, K. Zamani and B. Yazdi-Samadi. 2005. Using a competent tissue for efficient transformation of sugar beet *(Beta vulgaris* L.) *In vitro* Cell. Dev. Biol.-Plant 41:11–16.

Owens, L.D. and D.R. Eberts. 1992. Sugarbeet leaf disk culture: An improved procedure for inducing morphogenesis. Plant Cell Tiss. Org. Cult. 31: 195–201.

Owens, L.D. and C.A. Wozniak. 1991. Measurement and effects of gel matric potential and expressibility on production of morphogenic callus by cultured sugar-beet leaf-disks. Plant Cell Tiss. Org. Cult. 26: 127–133.

Pedersen, H.C., R.D. Hall and F.A. Krens. 1993. Petioles as the tissue source for isolation and culture of *Beta vulgaris* and *B. maritima* protoplasts. Plant Sci. 95: 89–97.

Pedersen, H.C. and B. Keimer. 1996. Haploidy in sugar beet (*Beta vulgaris* L.). *In*: *In vitro* Haploid Production in Higher Plants, Vol. 3, Important Selected Plants. S.M. Jain, S.K. Sopory and R.E. Veilleux (eds.). Kluwer Academic Publishers; Dordrecht; Netherlands, pp. 17–36.

Phillips, G.C., J.F. Hubstenberger and E.E. Hansen. 1995. Plant regeneration by organogenesis from callus and cell suspension cultures. *In*: Plant Cell, Tissue and Organ Culture. O.L. Gamborg and G.C. Phillips (eds.). Springer and Verlag, Heidelberg, pp. 67–78.

Pua, E.C. and T.A. Thorpe. 1986. Differential response of non-selected and Na2SO4-selected callus cultures of *Beta vulgaris* L. to salt stress. J. Plant Physiol. 123: 241–248.

Ragot, M. and P. Steen. 1992. Genetic and environmental effects on chromosome doubling of sugarbeet (*Beta vulgaris* L.) haploids. Euphytica 63: 233–237.

Rao, P.S. and P. Suprasanna. 1996. Methods to double haploid chromosome numbers. *In*: *In vitro* Haploid Plant Production in Higher Plants, Vol. 1, S.M. Jain, S.K. Sopory and R.E. Veilleus (eds.). Kluwer Academic Publishers, The Nedherlands, pp. 317–339.

Ritchie, G.A., K.C. Shoort and M.R. Davey. 1989. *In vitro* shoot regeneration from callus, leaf axils and petioles of sugar beet (*Beta vulgaris* L.). J. Exp. Bot. 40: 277–283.

Rogozinska, J.H. and M. Goska. 1976. Growth and differentiation of sugar beet anthers and callus. B. Acad. Pol. Sci. Biol. 24: 57–60.

Rogozinska, J.H. and M. Goska. 1978. Induction of differentiation and plant formation in isolated sugar beet leaves. B. Acad. Pol. Sci. Biol. 25: 343–345.

Rogozinska, J.H. and M. Goska. 1982. Attempts to induce haploids in anther culture of sugar, fodder and wild species of beet. Acta Soc. Bot. Pol. 51: 91–105.

Rogozinska, J.H., M. Goska and A. Kuzdowicz. 1977. Induction of plants from anthers of *Beta vulgaris* cultured *in vitro*. Acta Soc. Bot. Pol. 46: 471–479.

Roussy, I., F. Dubois, R.S. Sangwan and B.S. Sangwan-Norreel. 1996. *In planta* 2,3,5-triiodobenzoic acid treatment promotes high frequency and routine *in vitro* regeneration of sugarbeet (*Beta vulgaris* L.) plants. Plant Cell Rep. 16: 142–146.

Sabir, A.A. and B.V. Ford-Lloyd. 1991. Processing crop plant germplasm *in vitro* for mass production of regenerants: Acase study with beet. J. Biotechnol. 17: 257–268.

Sabir, A., H.J. Newbury, G. Todd, J. Catty and B.V. Ford-Lloyd. 1992. Determination of genetic stability using isozymes and RFLPs in beet plants regenerated *in vitro*. Theor. Appl. Genet. 84: 113–117.

Sadoch, Z., A. Majewska-Sawka, E. Jazdzewska and A. Niklas. 2000. Changes in sugar beet mitochondrial DNA induced during callus stage. Plant Breed. 119: 107–110.

Saunders, J.W. 1982. A flexible *in vitro* shoot culture propagation system for sugar beet that includes rapid floral induction of ramets. Crop Sci. 22: 1102–1105.

Saunders, J.W., G. Acquaah, K.A. Renner and W.P. Doley. 1992. Monogenic dominant sulfonylurea resistance in sugarbeet from somatic cell selection. Crop Sci. 32: 1357–1360.

Saunders, J.W. and M.E. Daub. 1984. Shoot regeneration from hormoneautonomous callus from shoot cultures of several sugar beet (*Beta vulgaris* L.) genotypes. Plant Sci. Lett. 34: 219–223.

Saunders, J.W., W.P. Doley, J.C. Theurer and M.H. Yu. 1990. Somaclonal variation in sugarbeet. *In*: Biotechnology in Agriculture and Forestry. Vol. 11, Somaclonal Variation in Crop Improvement. Y.P.S. Bajaj (ed.). Springer-Verlag, Berlin, pp. 465–490.

Saunders, J.W. and M. Mahoney. 1982. Benzyladenine induces foliar adventitious shoot formation on young plants of two sugar beet (*Beta vulgaris* L.) cultivars. Euphytica 31: 801–804.

Saunders, J.W. and K. Shin. 1986. Germplasm and physiologic effects on induction of high frequency hormone autonomous callus and subsequent shoot regeneration in sugarbeet. Crop Sci. 26: 1240–1245.

Saunders, J.W. and C.J. Tsai. 1999. Production of somatic embryos and shoots from sugarbeet callus: Effects of abscisic acid, other growth regulators, nitrogen source, sucrose concentration and genotype. *In vitro* Cell. Dev. Biol.-Plant 35: 18–24.

Saunders, J.W., C.J. Tsai and E. Samper. 1997. Evaluation of alternative nitrogen and carbon sources for sugarbeet suspension culture platings in development of cell selection schemes. *In vitro* Cell. Dev. Biol.-Plant 33: 56–61.

Schlangstedt, M., B. Hermans, K. Zoglauer and O. Schieder. 1992. Culture of sugar beet *(Beta vulgaris* L.) mesophyll protoplasts in alginate—callus formation and root organogenesis. J. Plant Physiol. 140: 339–344.

Schlangstedt, M., K. Zoglauer, S. Lenzner, B. Hermans and M. Jacobs. 1994. Improvement of sugar beet *(Beta vulgaris* L.) protoplast culture: leaf petioles as a protoplast source. J. Plant Physiol. 143: 227–233.

Shimamoto, Y., H. Hayakawa, J. Abe, H. Nakashima and T. Mikami. 1993. Callus induction and plant regeneration of *Beta* germplasm. J. Sugar Beet Res. 30: 317–319.

Skaracis, G.N. and M. McGrath. 2005. Molecular biology and biotechnology. *In*: Genetics and Breeding of Sugar Beet. E. Biancardi, L. Campbell, M. De Biaggi and G.N. Skaracis (eds.). Science Publishers Inc., Enfield, NH. pp. 221–286.

Speckman, G.J., J.P.C. Van Geyt and M. Jacobs. 1986. The induction of haploids of sugarbeet (*Beta vulgaris* L.) using anther and free pollen culture or ovule and ovary culture. *In*: Genetic Manipulation in Plant Breeding. W. Horn, C.J. Jensen, Odenbach and O. Schieder (eds.). Walter de Gruyter, Berlin, Germany, pp. 351–353.

Steen, P., B. Keimer, K. D'Halluin and H.C. Pedersen. 1986. Variability in plants of sugar beet (*Beta vulgaris* L.) regenerated from callus, cell-suspension and protoplasts. *In*: Genetic Manipulation in Plant Breeding. W. Horn, C.J. Jensen, W. Odenbach and O. Schieder (eds.). Walter de Gruyter, Berlin, Germany, pp. 633–635.

Svirshchevskaya, A.M., T.P. Kozyrevich and V.E. Bormotov. 1993. Haploids in the culture of unfertilized ovules in beet. Dokl. Akad. Nauk Belar. 37: 74–76.

Szabados, L. and C. Gaggero. 1985. Callus formation from protoplasts of a sugar beet cell suspension culture. Plant Cell Rep. 4: 195–198.

Tenning, P., E.W. Weich, U.B. Kjarsgaard, M.A. Lelu and M. Nihlgard. 1992. Somatic embryogenesis from zygotic embryos of sugar beet (*Beta vulgaris* L.). Plant Sci. 81: 103–109.

Tetu, T., R.S. Sangwan and B.S. Sangwan-Norreel. 1987. Hormonal control of organogenesis and somatic embryogenesis in *Beta vulgaris* callus. J. Exp. Bot. 38: 506–517.

Toldi, O., G. Gyulai, J. Kiss, I. Tarns and E. Balazs. 1996. Anti-auxin enhanced microshoot initiation and plant regeneration from epicotyl-originated thinlayer explants of sugar beet *(Beta vulgaris* L). Plant Cell Rep. 15: 851–854.

Tsai, C.J. and J.W. Saunders. 1995. Somatic embryos from callus of sugar beet biotechnology clone REL-1. J. Sugar Beet Res. 32: 215–227.

Tsai, C.J. and J.W. Saunders. 1999a. Evaluation of sole nitrogen sources for shoot and leaf disc cultures of sugarbeet. Plant Cell Tiss. Org. Cult. 59: 47–56.

Tsai, C.J. and J.W. Saunders. 1999b. Encapsulation, germination and conversion of somatic embryos in sugarbeet. J. Sugar Beet Res. 36: 11–32.

Turkish Sugar Co (2011) Sector Report, TSFAS, Ankara, Turkey. www.turkseker.gov.tr.

Van Ark, H.F., R.D. Hall, J. Creemersmolenaar and F.A. Krens. 1992. High yields of cytoplasts from protoplasts of *Lolium perenne* and *Beta vulgaris* using gradient centrifugation. Plant Cell Tiss. Org. Cult. 31: 223–232.

Vandenbussche, B. and M.P. De Proft. 1996. Cryopreservation of *in vitro* beet shoot tips using the encapsulation-dehydration technique: development of a basic protocol. Cryoletters 17: 137–140.

Vandenbussche, B. and M.P. De Proft. 1998. Cryopreservation of *in vitro* sugar beet shoot tips using the encapsulation-dehydration technique: influence of abscisic acid and cold acclimation. Plant Cell Rep. 17: 791–793.

Vandenbussche, B., S. Leuridan, V. Verdoodt, M. Gysemberg and M. De Proft. 1999. Changes in sugar content and fatty acid composition of *in vitro* sugar beet shoots after cold acclimation: influence on survival after cryopreservation. Plant Growth Regul. 28: 157–163.

Vandenbussche, B., G. Weyens and M. De Proft. 2000. Cryopreservation of *in vitro* sugar beet (*Beta vulgaris* L.) shoot tips by a vitrification technique. Plant Cell Rep. 19: 1064–1068.

Van Geyt, J. and M. Jacobs. 1985. Suspension culture of sugar beet (*Beta vulgaris* L). Induction and habitation of dedifferentiated and self-regenerating cell lines. Plant Cell Rep. 4: 66–69.

Van Geyt, J., G.J. Speckmann, K. D'Halluin and M. Jacobs. 1987. *In vitro* induction of haploid plants from unpollinated ovules and ovaries of the sugarbeet (*Beta vulgaris* L). Theor. Appl. Genet. 73: 920–925.

Wang, S., A. Hang and T. Tsuchiya. 1991. Chromosome studies of callus tissues and regenerated plants from an unfertilized ovule culture of sugar beet (*Beta vulgaris* L.). J. Genet. Breed. 45: 161–168.

Warren, G. 1996. Protoplast isolation and fusion. *In*: Plant Cell and Tissue Culture, A. Stafford and G. Warren (eds.). John Wiley and Sons, Chichester, UK, pp. 48–81.

Weller, T.A. and L.L. Lasure. 1981. Betalains in beetroot tissue culture. J. Food Sci. 47: 162–163.

Wisniewska, E. and A. Majewska-Sawka. 2007. Arabinogalactan-proteins stimulate the organogenesis of guard cell protoplasts-derived callus in sugar beet. Plant Cell Rep. 26: 1457–1467.

Wisniewska E. and A. Majewska-Sawka. 2008. The differences in cell wall composition in leaves and regenerating protoplasts of *Beta vulgaris* and *Nicotiana tabacum*. Biol. Plant 52: 634–641.

Wisniewska, E., A. Munster, H. Nakashima and A. Majewska-Sawka. 2002. Sugar beet guard cells as a source of efficiently regenerating protoplasts. Proc. Jpn. Soc. Sugar Beet Technol. 44: 1–6.

Wright, T.R. and D. Penner. 1998a. Cell selection and inheritance of imidazolinone resistance in sugar beet (*Beta vulgaris*). Theor. Appl. Genet. 96: 612–620.

Wright, T.R. and D. Penner. 1998b. *In vitro* and whole-plant magnitude and cross-resistance characterization of two imidazolinone-resistant sugarbeet (*Beta vulgaris*) somatic cell selections. Weed Sci. 46: 24–29.

Wright, T.R., N.F. Bascomb, S.F. Sturner and D. Penner. 1998. Biochemical mechanism and molecular basis for ALS-inhibiting herbicide resistance in sugarbeet (*Beta vulgaris*) somatic cell selections. Weed Sci. 46: 13–23.

Yang, A., L. Zhu, S. Zhao and J. Zhang. 2004. Induction of multiple bud clumps from inflorescence tips and regeneration of salt-tolerant plantlets in *Beta vulgaris*. Plant Cell Tiss. Org. Cult. 77: 29–34.

Yildiz, M., S. Onde and M. Ozgen. 2007. Sucrose effects on phenolic concentration and plant regeneration from sugarbeet leaf and petiole explants. J. Sugar Beet Res. 44: 1–15.

Yu, M.H. 1989. Callus induction and differentiation from leaf explants of different species of the genus *Beta*. Crop Sci. 29: 205–209.

Yuce, S. 1973. Haploidie bei der Zuckerrube. Ph.D. Dissertation, Justus Liebig University, Giessen, Germany.

Zhang, C.L., D.F. Chen, A.C. McCormac, N.W. Scott, M.C. Elliott and A. Slater. 2001a. Use of the GFP reporter as a vital marker for *Agrobacterium*-mediated transformation of sugar beet (*Beta vulgaris* L.). Mol. Biotechnol. 17: 109–117.

Zhang, C.L., D.F. Chen, M.C. Elliott and A. Slater. 2001b. Thidiazuron induced organogenesis and somatic embryogenesis in sugar beet (*Beta vulgaris* L.). *In vitro* Cell. Dev. Biol.-Plant 37: 305–310.

Zhang, C.L., D.F. Chen, M.C. Elliott and A. Slater. 2004. Efficient procedures for callus induction and adventitious shoot organogenesis in sugar beet (*Beta vulgaris* L.) breeding lines. *In vitro* Cell. Dev. Biol.-Plant 40: 475–481.

Zhang, C.L., D.C. Xu, X.C. Jiang, Y. Zhou, J. Cui, C.X. Zhang, D.F. Chen, M.R. Fowler, M.C. Elliott, N.W. Scott, A.M. Dewar and A. Slater. 2008. Genetic approaches to sustainable pest management in sugar beet (*Beta vulgaris*). Ann. Appl. Biol. 152: 143–156.

Zhao, J., D.H. Simmonds and W. Newcomb. 1996. High frequency production of doubled haploid plants of *Brassica napus* cv. Topas derived from colchicine-induced microspore embryogenesis without heat shock. Plant Cell Rep. 15: 668–671.

Zhong, Z., H.G. Smith and T.H. Thomas. 1993a. Micropropagation of wild beet (*Beta maritima*) from inflorescence pieces. Plant Growth Regul. 12: 53–57.

Zhong, Z., H.G. Smith and T.H. Thomas. 1993b. *In vitro* culture of petioles and intact leaves of sugar beet (*Beta vulgaris*). Plant Growth Regul. 12: 59–66.

7

Biology and Biotechnology of *Crocus*

Mahdi Ahouran

ABSTRACT

Crocus is a big genus of the Iridaceae family with a wide range of plants, a variety of colors and various applications. Plants of the *Crocus* genus are generally called saffron. Among these *C. sativus* has been at the center of attention as the most important saffron and the most expensive agricultural crop in the world from a very long past up to the present day. Many of the plants of this genus are also used as decorative plants in the parks and stone gardens. In view of the importance of certain plants of this genus and their sterility, tissue culture and micropropagation techniques can be employed as one of the considerable methods toward tackling the present problems of proliferation and traditional cultivation of *C. sativus* and the decorative species thereof as well as supplementing the projects of genetic breeding and production of new varieties. Mass proliferation of healthy corms that are free from pathogen, production of secondary metabolites with medicinal, spice and industrial values and change of the ploidy level with the aim of increasing the yields and removal of the sterility problem are among important applications of the tissue culture methods in saffron.

Keywords: *Crocus sativus*, polyploidy, micropropagation, organogenesis, somatic embryogenesis, saffron, genetic engineering

Department of Agricultural Biotechnology Engineering, College of Engineering and Technology, Imam Khomeini International University, Qazvin, Iran.
Email: mahdi.ahouran@gmail.com

Crocus Overview

Crocus is a big genus of the Iridaceae family that includes some 80 species with herb-like leaves and beautiful flowers in light and dark purple, white, multi-colorful and yellow. Plants of the *Crocus* genus are generally known as saffron however by saffron we usually mean *Crocus sativus* and its three-branch stigmas. *C. sativus* was the first saffron species cultivated by the human being which has been known as an economic and important plant for a long time. Saffron as the most expensive agricultural crop of the world is used as a flavoring and colorant in many cases (Fernandez 2007). This plant is also very valuable for medicinal purposes and has long been used in preparation of traditional medicines. Saffron is more or less cultivated in many countries including Iran, India, Spain, Turkey, Greece, France, Morocco, Egypt, China, Azerbaijan and the United Arab Emirates however Iran is presently considered as the biggest producer and exporter of saffron in the world (Winterhalter and Straubinger 2000, Rangahau 2003, Behdani et al. 2008).

History

Some researchers believe that cultivation and consumption of saffron dates back to 3500 years ago in Egypt and the Middle East. Archeological and historical sources also show that domestication of saffron has a long history and dates back to 1500 to 2500 B.C. (Plessner et al. 1989, Negbi 1999). Different opinions have been reported on the origin of saffron but most probably this plant is a native of Iran, Minor Asia and/or Greece whose cultivation has then spread in India, China, the Mediterranean region and Eastern Europe (Grilli Gaiola 2004). The name "saffron" has been derived from the Arabic word zaferan meaning yellow. Results of investigation of various sources reveal that the ancient civilizations of Iran primarily found out the value of this plant. Historical evidences attribute the primary use of saffron to the Achaemenian dynasty. Saffron was used for producing various types of fragrances, fragrant oils and baking bread at that age and was exported from Iran to Rome as an important product at the time of the Parthians. Moreover, according to the information contained in the manuscripts and documents of the traditional medicine, Iranian physicians were aware of the value of saffron in the preparation of simple and compound medicines. Saffron cultivation was spread by Arabs in Spain in the 10th century A.D. and then in Walden region in Essex in England in the 18th century A.D. This region is known as Saffron Walden today. According to one of the earlier historical evidences, saffron was at first applied in the ancient Egypt within a fragrant and relieving product. In ancient Greece saffron was considered as a precious product due to its coloring and sweet smelling characteristics. Hippocrates and other Greek physicians also used it as a medicine and prescribed it for treatment of smallpox, cholera, renal diseases and insomnia. It is important to know that saffron has had a constant presence as a coloring and flavoring condiment during a long period of 2500 years (Fernandez 2004).

Adaptation and distribution

Saffron is a plant which grows in dry and semi-torrid regions. Although it can grow in various climates, it grows more properly and with a better quality yield in regions with moderate weathers, dry summers and mild winters. The plant well tolerates temperature scales between –15°C in winter and +40° C in summer. Elevation of the saffron farmlands from the sea level differs from 800 meters to 2300 meters however it yields a more suitable product in the elevations of more than 1300 meters. The growth period of saffron depends on the regions and begins from early fall with flowering. The leaves grow in fall and winter up to the late spring. Flowering in some species is delayed until winter and spring.

The *Crocus* genus has more than 80 species that covers very extensive areas whose dispersion is from coordinates 10° W to 80° E in the Mediterranean region and west Asia and from coordinates 30° N to 50° N in the Irano-Touranian region. The most dispersion of the *Crocus* genus is seen in the European-Asian regions which begins from Morocco and Portugal in the west and goes toward the east to Russia, Kyrgyzstan and eventually to the Sinkiang Province in western China.

Taxonomy and botany

The taxonomic classification of saffron is as follows:

1. *Division*: Spermatophyta, 2. *Subdivision*: Angiospermae, 3. *Class*: Monocotyledonae, 4. *Sub-class*: Liliidae, 5. *Order*: Liliales, 6. *Family*: Iridaceae, 7. *Genus*: *Crocus*

The beautiful flowers of this genus are used as home-grown plants for decoration of gardens, stone gardens and parks. In proper climates the saffron flowers dehisce in bouquets producing a carpet of beautiful colors. The plants of the *Crocus* genus have somehow symmetrical and spherical corms with several sheath fibers that proliferate through producing new corms or seeds (in wild species with seeds), and among various species it is only *C. sativus* that proliferates annually through vegetative reproduction by means of the daughter corms, formed from the buds contained in the mother corm, because it is sterile and triploid (Fernandez 2004). The sheath of the corms plays a very important role in definition and diagnosis of the saffron species. The sheath fibers grow from the base of the corm and meet at the apex of the corm to prepare a protective sheath for the corm and apical bud. The number of apical buds in the corms differs from 1 to 4 depending on the size of the corm. The bigger and healthier the corm is, the more the number of apical buds there will be. The corm is in fact an underground stem rhizome and a rich source of starch. The corms in *C. sativus* are spherical and compressed and wide at the base and approximately 1–3 cm in diameter. The sheath of the corm consists of reticulated fibers and extends toward the apex of the corm up to about 5 cm above the plant. Growth of adventitious buds on the corm results in production of one to several corms on the old corm. The roots form at the beginning of the growth season (usually fall) and remains until the flowering stage in *C. sativus*. The root does not usually have a branch and the branch, if it exists, is rarely distinct in certain species. Bulb-like contractile roots sometimes form at the base of the growing corm and powerfully pull down the new corm into a proper depth of the soil. The leaves are sharp-tipped with a length of 30–40 centimeters. Surface of the leaves are darker and their thickness varies in different species. There is usually a middle line with silver or light yellow color on the upper (axial) surface which is an apparent feature of the saffron leaves. Differences such as presence of strengthening sclerenchyma, existence or non-existence of ciliate along with other apparent specifications can be useful in the diagnosis of species. Different species have different flowering time and the *Crocus* genus is divided to three groups of spring, fall and winter in terms of the flowering time. For example, flowers of *C. sativus* dehisce in fall in 1 to 3 numbers. Each flower grows on a short underground stem rhizome which is sometimes surrounded by infra leave membrane sheathes. The perianth is regular and the style is long and narrow, and the perianths are six in number that are arranged in two identical rows (rings). The perianth consists of three sepals and three petals in purple with darker streaks. Three stamens, usually narrow and spear-shaped, are on the edge of the perianth which split outwardly. The length of the filament is twice as much as the anther which is yellow in color. Examination of the features of the filament, anther and pollens are valuable in distinguishing the species. The ovary has three locules with two rows of ovules with each locule containing 9–15 ovules. The style is long and extended measuring about 7–10 cm in yellow which is divided to a three-branch red stigma in the end. Each stigma has a length of 2–3 cm.

Uses

The red stigma of *C. sativus* is the most precious agricultural crop in the world which is generally known as saffron. Saffron contains carbohydrates, minerals, vitamins, mucilage, fat, pigment, sweet fragrance, a little cinol and hydrozides by the name of picrocrocin, picrotin and crocin. The active ingredient of saffron is these three hydrozides that cause fragrance, flavor and sweet smell of the saffron.

Solomon and Carrar discovered crocin ($C_{44}H_{64}O_{24}$) which is the main element of the color of saffron for the first time in crystal form. Crocins are several types of unique carotenoids existing in nature that are easily soluble in water. This solubility in water is one of the reasons of extensive application of saffron as a colorant in food stuff and drugs as compared with other carotenoid compounds. As a result of hydrolysis, crocin produces crocetin ($C_{20}H_{24}O_4$) and glucose of which crocetin is not soluble in water (Moghaddasi 2010). Crocin oxidizes in the presence of light and oxygen and produces lesser color that is why it is recommended to keep saffron away from light and open air. The bitter taste of the saffron comes from a colorless glycoside by the name of picrocrocin ($C_{16}H_{26}O_7$). Picrocrocin is soluble in water and diluted alcohol and slightly soluble in ether and chloroform. This compound has a melting point of 156° C and changes into safranal under proper temperature and conditions. The fragrance and smell of saffron relate to the aroma of certain volatile oils and special essences that exist in this material. The oil essences are separated with distillation of the oil by water and in the presence of CO_2. This is an almost colorless fluid that causes the strong smell of saffron. This material is among the terpenes which are highly sensitive to oxidation and absorb oxygen easily to change into a thick brown fluid. This compound has a deterrent effect on the growth of microorganisms, while other saffron compounds do not have such a property. Although saffron is known as an important spice, it has many other foodstuff, pharmaceutical, dying, cosmetic and decorative applications.

Food industries

Use of saffron as a spice has a long history. This spice is a natural colorant which has attracted the attention of researchers because of having 150 different chemical compounds in its extract. In addition to the colorant property, other properties such as taste and fragrance have caused saffron to be used in many cases. One of its most evident applications of saffron is its use as a colorant and flavoring element for foodstuff, confectioneries and beverages. In many villages of northeastern Iran saffron is fixed like normal tea and offered to guests as an entertainment. This herbal product has many applications in ink making, illumination of color papers as well as book binding. However, in today's world saffron is mostly used as a precious spice and relish to give special flavor, color and taste to foods and drinks. According to the investigations of scientists the corms of the plants of *Crocus* genus contain glucose, minerals and amino acids such as alanine, glutamic acid/glutamine, glycine and serine. Corms of many species of *Crocus* genus that flower in fall, are used cooked or uncooked in some regions in Iran, Turkey and Syria. Petals of these plants are used as vegetables and leaves of the plants of *Crocus* genus are used in the production of a local cheese in Turkey by the name of Otlupeynir (Ozdemir and Kilinc 2008).

Medicinal and ornamental uses

Irrespective of its food status, saffron has been paid attention in medicinal point of view from the past to the present time (Abdullaev and Espinosa-Aguirre 2004). Traditional medicine of many countries have mentioned saffron as an important herbal medicine for treatment of many diseases including coughs, stomach disorders, colic, colds, insomnia, chronic uterine haemorrhage, femine disorder, scarlet fever, smallpox, asthma and cardiovascular disorders (Giaccio 1990, Winterhalter and Straubinger 2000, Abdullaev 2003). Saffron extract can be useful to heal neurodegenerative disorders accompanying memory (Zhang et al. 1994). In a number of studies saffron has been reported to have antinociceptive, anti-inflammatory, antiseizure and blood-pressure reducing effects in animal models (Hosseinzadeh and Younesi 2002, Hosseinzadeh and Khosravan 2002, Fatehi et al. 2003). Saffron has also been showed to act as a protective element in chromosomal damages (Premkumar et al. 2001). Among the identified compounds of saffron crocin and its similar compounds have the highest effect as curing properties of this plant (Escribano et al. 1996). Crocin is a very effective anti-oxidant that fights with oxidative stress in neurons (Ochiai et al. 2007). Anti-depression effects of the extracts of saffron petals and stigmas in mouse have also been reported (Karimi et al. 2001). Saffron extract is reported to have positive effects in patients who suffer from allergic asthma (Haggag et al. 2003). Today, saffron has shown its protective role against supraventricular arrhythmia (Khori et al. 2006). Hartwell (1982) reports that saffron extract was used in ancient time to cure different tumors and cancers and that many tumors of liver, spleen, kidney, stomach and uterus were treated by saffron medicinal products. Biochemical and medicinal studies on the elements forming saffron and examination of its anti-cancerous and anti-tumor properties are attracting the attention of scientists more than before. Some researchers have recently reported on the anti-cancerous, anti-tumor and anti-mutation properties of saffron and species of the *Crocus* genus. They have also pointed that extract of the herbs of this genus have a high cellular toxicity on different cancerous cells and can deter the synthesis of the nucleic acids in the malignant cancerous cells. Apart from the medicinal properties *Crocus* genus is attractive because of having very beautiful species, various colors and a high potential in the domain of decorative plants. Many plants of this genus are used as decorative plants in parks and stone gardens.

Reproductive characteristics

The *Crocus* genus has a wide range of chromosomes however most species of this genus except for *C. sativus* have a diploid with 2*n*=12, 2*n*=16, 2*n*=20 and 2*n*=24 chromosomes. Karasawa (1933) proposed the autotriploid or natural sterility of *C. sativus* and described its chromosomes as 2n=3X=24 which is divided into eight trivalents at the meiosis stage. Ghaffari (1986) too, concluded in his studies that *C. sativus* at the meiosis and metaphase stages produces gametes that may have 8–16 chromosomes. Such a condition results in a large number of cells to have an imbalanced number of chromosomes that is why distribution of chromosomes on the poles has been observed in the forms of 9–15, 10–14, 11–13 and 12–12. Moreover, a number of *C. sativus* chromosomes are laggard chromosomes which do not participate in the above-mentioned groupings. Since *C. sativus* is autotriploid and does not produce seed, there is no possibility of its inoculation with other types of saffron. Autotriploid not only sterilizes saffron but also acts as a deterrent factor against breeding of *C. sativus*.

Role of Tissue Culture in *Crocus*

Botanical biotechnology is in fact development of the science of botanical biology and an outcome of different sciences such as molecular biology, plant breeding, botanical biology, genetics, tissue culture, chemical engineering, plant pests and diseases and so on. The botanical biotechnology has presented extraordinary opportunities for the rapid scientific progress. The word "Biotechnology" is in fact referred to the scientific and targeted biological processes for production, alteration and application of materials. The most important object of biotechnology is to make changes in the structure of the genetic data of the plant and to stabilize these changes toward predicted goals. Culturing of plant tissue with a wide scope of different techniques can be employed as a powerful tool toward the goals such as plant breeding and preservation, plant proliferation and creation of the banks of botanical genes.

Producing disease-free plants

One of the primary goals of plant tissue culturing is to produce plants free from diseases because plant diseases influence all stages of the plant growth and eventually cause reduction of production. *C. sativus* as the most important plant of the *Crocus* genus is a perennial plant so that upon cultivation of the corm in the first year there is no need for its re-cultivation in the following years. The corms to be used for cultivation of *C. sativus* should be healthy and big and be selected from the farms with the age of 4 years and more. *C. sativus* corm can be cultivated from late May to early October. The best time for transfer of the corms from storage or mother farms to the new farms is late May. Research shows that storage of corms outside the soil injures the development process of the organogenesis which takes place in the corm buds in August. In the traditional cultivation method of *C. sativus*, corms should be pulled out from the soil at least every 5 years and new corms should be planted with an interval of 1 to 2 years. During the interval the farm is strengthened through cultivation of other crops and fertilization in order to be prepared for the cultivation of new corms. Whereas *C. sativus* is planted within a perennial farming (usually 8 years) saffron farmers pull out the corms every 7–8 years, something which can result in reduction of the production and its quality because the soil loses its nutrients after a while and whereas the number of corms increases every year there would be no sufficient space for the growth and nutrition of the plant. Since *C. sativus* is a triploid and sterile plant, it is usually proliferated by growing and multiplication of corms that is why it is infected by pathogenic agents such as bacteria, fungi and viruses. These agents remain active after harvesting the corms. Despite observation of hygienic procedures these pathogenic agents cause decay and necrosis of the corm and leave and often reduce or prevent the growth and flowering of the plant. Plants infected by fungus or bacterium can be treated by bactericide or fungicide compounds whereas treatment of the plants infected by virus does not prove to be successful. In view of such problems in these plants and other bulbous plants, tissue culture and micro-multiplication have been tried to be used for the mass-production free from pathogenic agents at a large scale.

Genetic engineering

Many existing hindrances in the way of plant breeding through traditional methods can be removed by using various biotechnological methods. In the event the *in vitro* culturing methods are well-matured, protoplast culture, protoplast fusion, anther culture and various types of gene transfer may be used to breed saffron. Fusion of plant protoplasts as one of the biotechnological methods would result in new somatic hybrids through a combination of nuclei or cytoplasmic genomes by overcoming sexual incompatibility between species. Protoplast fusion is the mixture of the protoplasts of two different genomes. When two protoplasts (cells whose walls are assimilated and only their plasma remain as a protector) are mixed, a very different condition of genetic materials occurs and this process is called "somatic hybridization". Somatic breeding was conducted for the first time by Carlson et al. (1972). They combined the protoplasts of two species of *Nicotiana glauca* and *Nicotiana langsdorffii* and achieved the first somatic hybrid plant. The somatic hybridization technique is an effective tool for genetic breeding of many valuable plants (Johnson and Veilleux 2001). Somatic hybridization by means of protoplast fusion is a method for confluence of the species that are incompatible and by using this method and by employing the germplasm of the wild species and increasing the genetic variation the features of the farm plants can be bred. Whereas *C. sativus* is a sterile plant and the classical and common breeding methods cannot be easily applied on it, tissue culture and micropropagation method can be a proper way to help its genetic breeding. As we know the useful section of *C. sativus* is its three-branch red stigma while its style and petal have a wide application, too. The utilizable product of *C. sativus* is generally its flower. Different tests have shown that an increase in the ploid level of plants would lead to the increase in the size of the flower and eventually the flower parts. An increase in the number of chromosome groups in plants often enlarges their size and enhances their consumption and marketing value. For example in tobacco, upon an increase in the number of chromosome groups from haploid upward, the size of the style grows bigger and cells, pollens, microsporocytes, and protective cells enlarge accordingly. In snapdragon too, the size of inflorescence, thickness of stem, flower size, pollen size and stomata of tetraploids are bigger than diploids. On the other hand, *C. sativus* is a triploid and sterile crop which is proliferated through a non-gamete method. Therefore, an increase in the ploid level of this plant can be a desirable mechanism toward further enlargement of the flower parts and removal of its sterility problem. A lot of studies have been conducted through polyploidy of *C. sativus* with different colichicine thicknesses in different time intervals however none of them have so far proven to be successful. Therefore, fusion of plant protoplasts as a powerful biotechnological tool can help researchers to solve this problem. Meanwhile, many species of the *Crocus* genus are decorative so we can take advantage of the protoplast fusion of these species to create a variety of colors and phenotype of these plants for further marketing. The advantages of employing methods of tissue culture and micropropagation of saffron can be generally summarized as follows: 1. Rapid and mass proliferation of new bred varieties or different echo types; 2. Production of corms free from pathogenic agents and growing healthy plants through cormogenesis, organogenesis, or somatic embryogenesis; 3. Somaclonal variation to achieve new genotypes and varieties; 4. Anther and pollen culture, protoplast preparation and somatic hybrid between *C. sativus* and wild and keen species to achieve superior varieties; 5. Fusion of *C. sativus* protoplasts to enhance the pliody level and achievement of fertile plants.

Micropropagation

Applications of the plant tissue culture is highly varied and numerous. Micropropagation by means of different skills of plant tissue culture is the most important and successful application in this respect. Micropropagation is the *in vitro* multiplication of the plant from organs, tissues, cells and protoplasts. Depending on the type of the plant and the existing conditions different sections of the plant like bud, root tip, leave, embryo, stamen and even individual cell can be used as an explant. Today, micropropagation has turned into an important industry with a special status in international trade. *Crocus* is a large genus with a wide range of various plants and colors. Since 1600 BC plants of this genus have been used in the manufacture of dyes, fragrances, medicines as well as decorative plants (Ozdemir et al. 2005). Due to the importance of seed production and sterility of certain species of this genus, micropropagation can be considered as a suitable method for mass production within a short time. Rapid multiplication of plants by means of micropropagation particularly in those groups of plants that are proliferated through traditional methods or their proliferation by means of these methods are not cost-effective, is the most suitable method. Micropropagation is also used for the plants which in addition to having the problem of non-gamete multiplication, their gamete multiplication too faces the problem of genetic instability, failure of producing seeds or producing unsuitable seeds for cultivation. Many farm plants that are proliferated through non-gamete method (through cutting, bulb or graft) contain bacterium, fungus and virus intra-structurally as a result of which infection is transferred from the mother plant to the next generation, causing quantitative and qualitative reduction of the yield. By using this method plants can be mass produced free from the pathogenic agents therefore healthy plants can be propagated and botanical germplasm can be exchanged safely. Moreover, contrary to the non-gamete multiplication methods which are limited to a particular season and condition, micropropagation allows plant production in the laboratory during different seasons of the year. This method is highly important in general propagation of bred cultivars and also as a supplement of the biotechnological methods and genetic engineering.

The current techniques of culturing a single cell and protoplast too, allow production of thousands of plants from a single cell within a relatively short period of time. On the other hand, we can take advantage of protoplast culture and protoplast fusion of the plants of this genus to transfer genes and create a variety in the morphology of the flower sections in decorative plants.

Organogenesis and somatic embryogenesis

Organogenesis and somatic embryogenesis are two different mechanisms in which explants are changed into a complete plant. Branches and roots in organogenesis form in response to the proper environmental condition, particularly the type and density of the regulators existing in the sample. The first definitions concerning the production of somatic embryo in the *in vitro* condition were proposed by Steward et al. (1958) and Reinert (1959) concerning carrot and somatic embryo is considered as the best method for marketing in the *in vitro* condition today. Somatic embryo is a process in which the diploid or haploid embryo cells are developed without annexation to gametes to produce structures similar to the somatic embryos. Somatic embryos are induced in two ways: 1. Induction is carried out directly from a tissue such as a leaf, stem, embryogenesis, protoplast or microspores; 2. Induction is carried out indirectly through an intermediate stage of cellular callus or suspension. Researches show that the genesis stages are similar for embryogenesis and somatic embryo. Embryos in both cases in dicotyledonous go through the globular, heart-shaped, torpedo, and cotyledonary stages and in monocotyledonous go through the globular, scutellar, and coleoptilar stages (Beyl 2010). Induction of somatic embryogenesis may occur either automatically or may need complicated protections. The embryo (non-embryo) stage in the explants cells may be induced in different ways such as exposure to growth regulators, pH shock, heat tension or treatment with different chemicals. Although growth regulators play an important role in the induction of somatic embryogenesis, there are other numerous factors that affect the condition of a particular tissue toward somatic embryogenesis. The type and density of the regulators of organogenesis and embryogenesis growth can occur simultaneously depending on the selected explants and culture medium.

In vitro studies on micropropagation of *Crocus*

Culture of the saffron tissue was first reported by Ding et al. (1979). They could produce buds by culturing superficial sections of the *Crocus sativus* corm. In 1981 these researchers used MS (Murashige and Skoog 1962) culture medium containing 2.4-D (2,4-Dichlorophenoxyacetic acid), KIN (Kinetin) and NAA (α-Naphthaleneacetic acid) hormones to induce calluses on the explants of the cultured corm and eventually succeeded in producing plantlets from the produced calluses. Many studies were conducted on different aspects of saffron afterwards. Among the studies conducted on callusogenesis, embryogenesis, and organogenesis of various saffron explants these cases may be mentioned: Isa and Ogasawara (1988) induced callusogenesis by using corm explants in MS medium containing different densities of NAA and BA (N6-Benzylaminopurine). They described environmental conditions such as darkness and temperatures below 20°C as being appropriate to increase the period of the callus growth. Milyaeva et al. (1998) used explants such as stigma, bud, torus, and leaf in addition to corm explants and were able to induce callusogenesis in MS medium through a different harmonic compound. They reported that corm explants possessed a higher morphogenesis capacity than other sections. Ilahi et al. (1987) could induce callus formation on the corm explants in ½ MS medium by using different densities of 2,4-D and BA hormones and coconut juice 2%. Chichiricco and Caiola (1982) conducted research on the potential of growing pollens in *C. sativus* and showed that although pollens kept their growth potential in the medium containing saccharine and chemical element bore, they hardly grow in the medium containing agar. This proved that although pollens grow in this condition, a tiny percent of them would be able to fertilize stigmas since sterility of *C. sativus* is due to cytological defects of the pollens. Chichiricco and Caiola (1987) showed that by culturing ovary explants in MS medium containing BA, 2,4-D and GA_3 (Gibberellic acid) hormones the ovules inflate and eventually produce Parthenocarpic fruits. Some studies have also been conducted in the area of organogenesis including that of Huang (1987) who used explants of leaf bases in N6 (Chu et al. 1975) medium containing 2 mg/l 2,4-D, 0.5 mg/l BA and 2 mg/l NAA, and extracted calluses that produced a large number of buds. After transferring these buds to the ½ MS medium containing 1 mg/l IAA (Indole-3-acetic acid) the buds grew up and produced young plantlets. Milyaeva et al. (1988) could induce calluses in the medium containing 2,4-D and BA and then transferred these calluses to the medium containing 0.5 mg/l 2,4-D and BA hormones where buds formed on the calluses. They showed that an increase in the amount of 2,4-D auxin led to the increase of calluses and as cytokinin of BA increased the buds came into existence. Isa and Ogasawara (1988) too, succeeded to direct the calluses induced in the medium containing NAA and BA toward formation of buds. Ebrahimzadeh et al. (1996) cultured the corm explants of *C. sativus* and four species namely *C. speciosus*, *C. cancellatus*, *C. almehensis* and *C. gilanicus* in MS and LS (Linsmaier and Skoog 1965) medium containing NAA, BA, KIN, Zeatin and 2,4-D hormones and showed that in addition to callus, different newly created organs like leaf structures, contractile and ordinary roots and newly created corms were forming on different explants. Majourhat et al. (2007) studied the growing buds of the *C. sativus* corms and caused an increase in the regeneration from the culture of the plant's meristem. In an effort to examine the reaction of these explants to different type and densities of cytokinins they were cultured in the media containing these growth regulators for a period of 6 weeks. They found that 5 mg/l produced

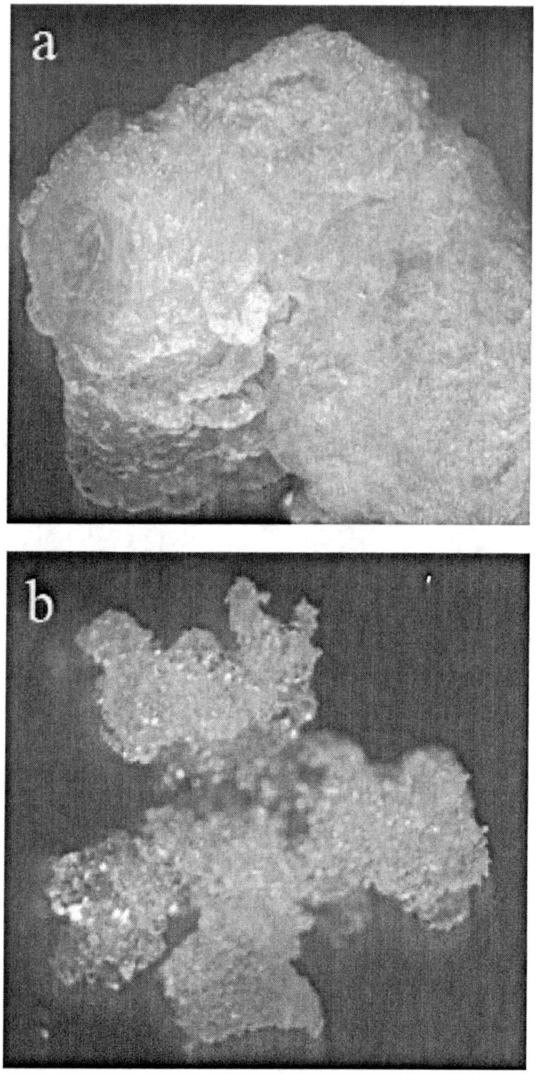

Fig. 1. Non-embryogenic callus: (a) Non-embryogenic callus of *C. sativus* (b) Non-embryogenic callus of *C. cancellatus*.

Color image of this figure appears in the color plate section at the end of the book.

greater and more vigorous shoots than either 2iP (2-Isopentenyladenine) or TDZ (Thidiazuron). They could regenerate saffron plantlets after several subcultures.

Fig. 2. Embryogenic callus of *C. sativus*.

Color image of this figure appears in the color plate section at the end of the book.

Some surveys have also taken place in the area of saffron corm formation in the *in vitro* condition. Homes et al. (1987) were able to produce very small corms on the surface of the corm explants in MS and B5 (Gamborg et al. 1968) media. Re-cultivation of the corms produced in the medium caused formation of new corms and buds. Plessner et al. (1990) cultured small and young corm explants as well as corm explants carrying apical buds and produced small corms proving that Zeatin hormone had an extraordinary effect on the growth of the new corms while being necessary for the growth of leaves. In 1993 Dhar and Sapru could induce calluses in the KIN and 2,4-D medium and upon transfer of these calluses to the MS medium containing NAA and KIN they were able to produce corm and flower-like structures. In 1994 Aguero and Tizio could produce a large number of buds by using marginal buds of the old corms in the MS media containing BA hormones with densities of 1 and 5 mg/l and nitrogenous compounds with a ½ density. The buds were then transferred to the MS medium containing 60 gr/l saccharine and free from hormone, which produced a number of corm explants. Milyaeva et al. in 1995 too, could use marginal buds and apical buds in a dormant condition to raise the regeneration coefficient of the *C. sativus* corms. In 1999 Piqueras et al. used the meristemic tissues of the corm in MS media containing high densities of BA (2 mg/l) and low densities of 2,4-D (0.1 mg/l) and conducted successful experiments proving that the medium was suitable for corm production. Homes et al. (1987) could produce micro-corm at the level of corm explants in MS and B5 media. Re-culturing of the produced corms in the culture medium resulted in the production of new corms and buds. Karaoglu et al. (2007) too, cultured corm pieces in the basic MS medium containing 2 mg/l BA and 0.5 mg/l NAA could induce the corm after six months. They kept the new corm explants in the temperature of 4°C for 5 weeks and then transferred them directly into the soil. Raja et al. (2007) placed leaf pieces in the MS medium containing 4 mg/l BA and 0.5 mg/l NAA and 9% saccharine and induced new corm explants. They also put the leaf pieces in an 1 mg/l each BA and 2,4-D medium and produced embryogenesis calluses. Transfer of these calluses to the medium containing 2.25 mg/l BA and 0.1 mg/l 2,4-D resulted in the formation of somatic embryos. After placing the matured embryos in the basic MS medium containing 20 mg/l GA_3 and 2 mg/l ABA (Abscisic acid) embryos began to sprout. Sharma et al. (2008) also cultured the buds derived from apical meristem in ½ MS medium containing 3 mg/l BA and 80 gr/l saccharine and induced new micro-corms. They also achieved the sprouting of the apical and marginal meristems of the produced micro-corms in the *in vitro* condition and in the basic MS medium containing 3 mg/l BA, 30 mg/l IBA (Indole-3-butyric acid) and 30 gr/l saccharine. Ahouran et al. (2012) cultured upper and lower explants of *C. cancellatus* and produced micro-corm. These researchers stated that both upper and lower sections were able to induce micro-corm although the upper section of the corm is far more potent than the lower section, and reported the maximum induction of the micro-corm in the MS medium containing 2 mg/l NAA and 1 mg/l BA. These researchers also described cold treatment as an effective factor for sprouting the new micro-corms. Studies have also been conducted on the formation of stigma-like structures as follows: Sano

Fig. 3. Direct micro-corms formation in *C. cancellatus*.

Color image of this figure appears in the color plate section at the end of the book.

and Himeno (1987) used ovary explants in LS and N (Nitsch 1972) culture media that contained KIN, BA, 2,4-D, NAA and IBA hormones and surveyed the formation of the stigma-like structures. They could produce two types of stigma-like structures; one in the form of direct and non-integrated filament on the ovary tissue and the other in the form of indirect and non-integrated filament of callus tissue in the aforementioned culture media with hormonal densities of 1.5 mg/l KIN, and 1 and 0.1 mg/l NAA and IBA. They also showed the qualitative similarity of the pigments of these structures with those of the natural varieties. The researchers examined the ingredients of the stigma-like structures in the culture medium and proved that these structures contain crocin and picrosin. Upon heat treatment of the stigma-like structures they were also able to detect safranal. The general results of their studies showed that these stigma-like structures resembled the natural varieties both from morphological and biochemical points of view. Himeno et al. (1988) proved through their analyses that among different sections of the *C. sativus* flower, carpel and placentas enjoy a high potential for the formation of the stigma-like structures. In 1988, other researchers including Koyama et al. (1988) too, induced formation of the stigma-like structures on the explants of stigma and style within LS medium containing NAA and BA hormones, and showed complete resemblance of the pigments produced in these structures and those of the natural stigmas by using HPLC. Hori et al. (1988) cultured pistils of *C. sativus* in an MS, White (1963) and NN (Nitsch and Nitsch 1969) media together with BA and NAA with the amount of 1 mg/l from each and produced callus. The resultant callus produced crocin, crocetin-di (monoglucosyl-diglucosyl) ester and crocetin-di (mono-glucosyl) ester. The calluses containing these pigments could be preserved in darkness for more than two years. Fakhari and Evans (1989) cultured different explants of *C. chrysanthus* Herbert in an MS medium containing different densities of NAA, BA and 2,4-D hormones and proved that among different explants of the flower only the ovary explants were able to produce stigma-like structures in 5, 10 mg/l BA densities. Exposure of these explants to light resulted in increase in the color of the stigma-like structures. The same researchers reported the formation of the stigma-like structures by using the *C. sativus* explants in the white culture medium containing Zeatin, NAA, 2,4-D and BA hormones in 1990. They also showed that culturing of the flower's explants in the medium containing different densities of 2,4-D and BA hormones induces the callus. Sarma et al. (1990) were other researchers who used stigma explants in MS medium containing NAA and BA hormones and produced stigma-like structures. Chemical analysis of these structures showed that they had more crocin and picrosin compounds than natural stigmas however no safranal was detected. In another study Visvanath el al. (1990) used explants of different sections of a flower in the MS medium containing different hormones and were able to produce red calluses and red petal and string-like structures. They used various spectrometric methods and chemical methods to show that the red calluses contained crocin pigments however contrary to the Sharma's theory they put the amount of these compounds as much as 6–11 times less than the natural amount. Nevertheless, the safranal amount in the petal structures was equal to that in the natural red and dry stigmas of saffron. Lu et al. (1992) examined factors like age and type of the explants and the effect of the exogenous hormones on the amount of regeneration of the stigma-like structures of saffron. These researches proved that young perianth and style explants have the most potential in production of stigma-like structures. Otsuka et al. (1992) cultured the flower style (connected to the perianth) and ovary in an altered MS medium (containing 5–12% saccharine) with different concentrations of auxin and cytokinin and produced stigma-like structures whose pigments were more than those produced in the common methods of tissue culture. Also Han and Zhang (1993) proposed a suitable culture medium for the sustainable induction of stigma-like structures on the explants of the pieces of *C. sativus* flower. Analysis with different chemical methods confirmed the presence of picrocrocin and safranal metabolites in these structures. Yongjiong et al. (1996) cultured parts of the flower organs including perianth and style of the *C. sativus* in the MS medium containing NAA and KIN hormones and produced red stigma-like structures. According to their results these structures were more induced in the perianth explants. In addition to the above-mentioned studies researches have also been conducted in the field of somatic embryogenesis in saffron. George et al. (1992) cultured meristemic parts of the *C. sativus* corms and produced calluses which were then transferred to the MS medium containing IAA, KIN, and ascorbic acid and produced globular embryos. These embryos were transferred to the ½ MS medium containing GA_3 which made the embryos to sprout. Transfer of these embryo buds to the solid ½ MS medium and by using NAA, KIN and ascorbic acid changed them into shoots. The same researchers in another experiment transferred the calluses produced from meristemic areas of the corm, to the MS medium containing IAA, KIN and ascorbic acid and produced globular embryos. These embryos changed into shoots in the liquid ½ MS medium containing 1 mg/l ABA. Ebrahimzadeh et al. (2000a) in their experiments succeeded in producing embryogenetic calluses in *C. sativus* by using the explants of the shoot meristem. These calluses went through complete stages of embryo maturity and were then transferred to the ½ MS medium containing GA_3 after maturity in the half-strength liquid MS basal medium-free from hormones. This procedure made the embryos sprout. Transfer of these embryo buds to the solid ½ MS medium containing NAA and BA resulted in the formation of complete plantlets with roots and shoots. In 1993 Dhar and Sapru induced

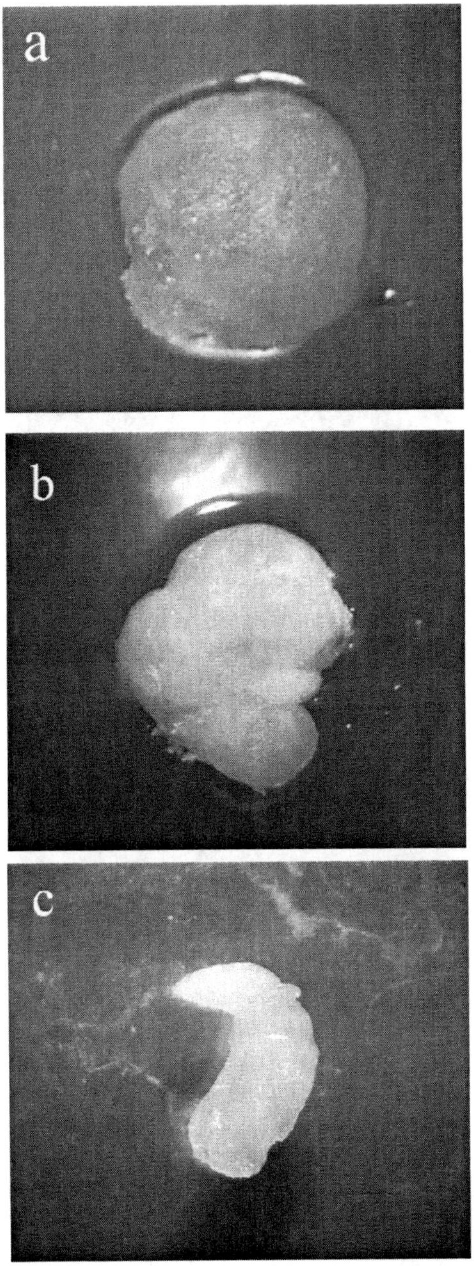

Fig. 6. Genesis stages of somatic embryogenesis in *C. sativus*: (a) globular stage (b) scutellar stage (c) coleoptilar stage.

Color image of this figure appears in the color plate section at the end of the book.

calluses in KIN and 2,4-D medium and by transferring these calluses to the MS medium containing NAA and KIN they were able to produce corms and flower-like structures. In another research Yonghua et al. (1996) examined the differences which existed in the organogenesis and various types of calluses produced from the saffron explants in order to achieve a rapid and effective method for micropropagation of *C. sativus*. Karamian (2004) surveyed somatic embryogenesis in four species of saffron and was able to induce the somatic embryogenesis by using the meristemic explants of stem and their culture in the LS media containing 4 mg/l NAA in combination with 4 mg/l BA as well as the medium containing 1 mg/l 2,4-D in combination with 4 mg/l KIN. The results showed that the somatic embryos in these four species do not grow congruently and concurrently. Embryos were induced to sprout by transferring calluses containing globular embryos to the ½ MS medium containing 1 mg/l ABA and then directed toward maturity by transferring to the ½ MS medium containing 25 mg/l GA_3 within a photoperiod of 16 hours of light and 8 hours of darkness under the temperature of 20°C. Blazquez et al. (2004) too, used corm explants and cultured them in the solid MS medium containing various hormones of 0.1 mg/l 2,4-D in combination with 0.5 mg/l BA in order to produce embryogenetic calluses and eventually used them to examine the morphological distinction and the role of the anti-oxidant enzymes. They showed a link between type, occurrence and expression of antioxidant enzymes (superoxide dismutases and catalase) and the stage of somatic embryogenesis. Darvishi (2007) produced embryogenetic calluses by using explants of the apical meristems in an MS medium containing 1 mg/l 2,4-D and 1 mg/l BA. During the research, the highest percentage of non-embryogenetic calluses was achieved by the treatment of 2 mg/l NAA and 2 mg/l BA. Sheibani et al. (2007) placed different parts of the *C. sativus* corms in different densities of TDZ and were able to induce embryogenetic calluses. Upon examining different densities they termed 0.5 mg/l TDZ as the most suitable density for induction of embryogenetic calluses and the number of globular embryos in each explant. Demeter et al. (2010) put pieces of corm bud explants of *C. heuffelianus* in the basic MS medium containing vitamins of B5 medium, 10 mg/l NAA and 1 mg/l BA and could induce embryogenetic calluses. Transfer of the embryos to a medium containing as much as 8 times of the density of the elements of the basic MS medium and vitamins of B5 medium, 0.05 mg/l NAA and 0.5 mg/l BA and under the conditions of 14 hours of light and 10 hours of darkness resulted in the regeneration of the plantlet. In addition to the above-mentioned studies, researches have been conducted in the area of culturing saffron protoplast with the following examples: Ebrahimzadeh et al. (2000b) were able to isolate protoplast from the produced embryogenetic calluses. The isolated protoplasts were immobilized in calcium alginate beads and

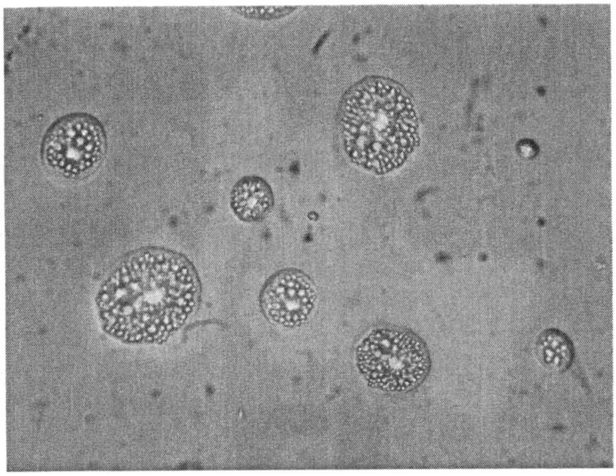

Fig.4. Isolation of protoplasts derived from embryogenic callus of *C. sativus*.

Color image of this figure appears in the color plate section at the end of the book.

then cultured in the liquid MS medium containing 0.5 mg/l each of the 2,4-D and BA hormones along with the nurse cells of the same species and were stored in darkness. Micro-calluses appeared on the surface of the beads after several weeks. Transfer of the beads to the solid MS medium containing 0.1 mg/l NAA and 1 mg/l BA resulted in the maturity and sprouting of the embryos. Karamian and Ebrahimzadeh (2001) were able to regenerate plantlets, from the protoplasts of *C. cancellatus*. By using the technique of immobilizing the protoplasts in calcium alginate beads these researchers could induce micro-calluses in the medium containing 2 mg/l KIN and 0.1 mg/l 2,4-D. As the micro-calluses were transferred to the solid ½ MS medium containing 0.2 mg/l KIN and 0.1 mg/l 2,4-D the embryos matured. Transfer of the matured embryos to the ½ MS medium containing 25 mg/l GA_3 and then to the medium containing 1 mg/l each of the NAA and BA hormones under the conditions of 16 hours of light and 8 hours of darkness with a temperature of 20°C resulted in their sprouting and developing to complete plantlets. Darvishi et al. (2006) analyzed the impact of the time of enzyme digestion on the viability of the protoplasts extracted from saffron's embryogenetic calluses and declared the 3-hour treatment of the enzyme digestive solution of 1% driselase, 1% cellulas R-10 and pectolyase y-23 with the viability of 98% as the most appropriate treatment time. Karamian (2007) used the method of immobilization of protoplasts in calcium alginate beads and could regenerate plantlets from the protoplasts produced from the tissue of the embryogenetic calluses on the pieces of the

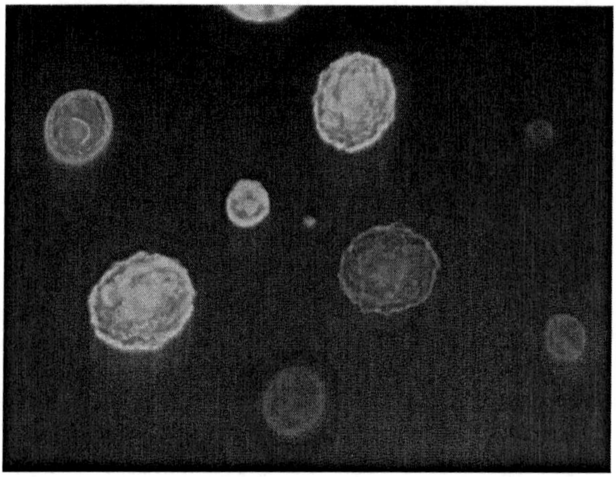

Fig. 5. *C. sativus* protoplasts after staining with FDA (Fluorescein diacetate).

Color image of this figure appears in the color plate section at the end of the book.

meristem explants of *C. pallasii* and made embryogenetic calluses with transferring of the embryogenetic calluses to the suitable culture media. Chaloushi et al. (2007) too, reported the induction of the non-embryogenetic calluses in treatment with 1 mg/l 2,4-D and 0.2 mg/l KIN after culturing the protoplasts of the farmed saffron and formation of embryogenetic calluses in the medium containing 0.1 mg/l 2,4-D and 0.2 mg/l KIN.

References

Abdullaev, F.I. 2003. *Crocus sativus* against cancer. Archives of Med. Res. 34: 354.

Abdullaev, F.I. and J.J. Espinosa-Aguirre. 2004. Biomedical properties of saffron and its potential use in cancer therapy and chemoprevention trials. Cancer Detect Prev. 28: 426–432.

Aguero, C. and R. Tizio. 1994. *In vitro* mass bulbification as a preliminary contribution to saffron (*Crocus sativus* L.) micropropagation. Biocell. 18: 55–63.

Ahouran, M., R. Hosseini and R. Zarghami. 2012. Corms as a source of explants for the successful Clonal propagation of *Crocus cancellatus*. J. Crop Sci. Biotech. 15: 47–51.

Behdani, M.A., A. Koocheki, P. Rezvani and M.J. Al-Ahmadi. 2008. Agro-ecological zoning and potential yield of saffron in Khorasan-Iran. J. Boil. Sci. 8: 298–305.

Beyl, C.A. 2010. PGRs and their use in micropropagation. *In*: R.N. Trigiano and D.J. Gray (eds.). Plant Tissue Culture, Development and Biotechnology. CRC Press, Boca Raton, FL, pp. 33–56.

Blazquez, S., E. Olmos, J.A. Hernandez, E. Hellin, J.A. Fernandez and A. Piqueras. 2004. Somatic embryogenesis in saffron (*Crocus sativus* L.): morphological differentiation and the role of the antioxidant enzymatic system. Acta Hortic. 650: 261–267.

Carlson, P.S., H.H. Smith and R.D. Dearing. 1972. Parasexual interspecific plant hybridization. Proc. Natl. Acad. Sci. USA 69: 2292–2294.

Chaloushi, B., R. Zarghami, C. Abd-Mishani, M. Omidi, Y.M. Agayev and B. Pakdaman-Sardood. 2007. Effects of different hormonal treatments on the callus production and plantlet regeneration in saffron (*Crocus sativus* L.). Pak. J. Biol. Sci. 10: 1625–1631.

Chichiricco, G. and M.G. Caiola. 1982. Germination and viability of the pollen of *Crocus sativus* L. Plant Biosystem. 116: 167–173.

Chichiricco, G. and M.G. Caiola. 1987. *In vitro* development of parthenocarpic fruits of *Crocus sativus* L. Plant Cell Tiss. Org. Cult. 11: 75–78.

Chu, C.C., C.C. Wang, C.S. Sun, C. Hsu, K.C. Yin, C.Y. Chu and F.Y. Bi. 1975. Establishment of an efficient medium for anther culture of rice through comparative experiments on the nitrogen source. Sci. Sin. 18: 659–668.

Darvishi, E., R. Zarghami, C.A. Mishani and M. Omidi. 2007. Effects of different hormone treatments on nonembryogenic and embryogenic callus induction and time-term enzyme treatments on number and viability of isolated protoplasts in saffron (*Crocus sativus* L.). Acta Hortic. 739: 279–284.

Darvishi, E., R. Zarghami, C.A. Mishani, M. Omidi and A. Sarkhosh. 2006. Investigation of the best time of enzyme treatment in order to isolate the protoplast from embryogenic callus to saffron (*Crocus sativus* L.). Biotechnology 5: 284–286.

Demeter, Z., G. Suranyi, V.A. Molnar, G. Sramko, D. Beyer, Z. Konya, G. Vasas, M. M-Hamvas and C. Mathe. 2010. Somatic embryogenesis and regeneration from shoot primordia of *Crocus heuffelianus*. Plant Cell Tiss. Org. Cult. 100: 349–353.

Ding, B.Z., S.H. Bai, Y. Wu and B.K. Wang. 1979. Preliminary report on tissue culture of corms of *Crocus sativus*. Acta. Bot. Sin. 21: 387.

Ding, B.Z., S.H. Bai, Y. Wu and X.P. Fang. 1981. Induction of callus and regeneration of plantlets from corms of *Crocus sativus* L. Acta. Bot. Sin. 23: 419–420.

Dhar, A.K. and R. Sapru. 1993. Studies on saffron in Kashmir III. *In vitro* production of corm and shoot-like structures. Ind. J. Genet. 53: 193–196.

Ebrahimzadeh, H., R. Karamian and M.R. Noori-Daloii. 1996. Regeneration of neoformant organs from organ parts of some *Crocus* species. J. Sci. I. R. Iran 7: 65–76.

Ebrahimzadeh, H., R. Karamian and M.R. Noori-Daloii. 2000a. Shoot regeneration from saffron protoplasts immobilizes in Ca-alginate beads. J. Sci. I. R. Iran 11: 69–72.

Ebrahimzadeh, H., R. Karamian and M.R. Noori-Daloii. 2000b. Somatic embryogenesis and regeneration of plantlet in saffron, *Crocus sativus* L. J. Sci. I. R. Iran 11: 169–173.

Escribano, J., G.L. Alonso, M. Coca-Prados and J.A. Fernandez. 1996. Crocin, safranal and picrocrocin from saffron (*Crocus sativus* L.) inhibit the growth of human cancer cells *in vitro*. Cancer Lett. 100: 23–30.

Fakhari, F. and P.K. Evans. 1989. Morphogenic potential of cultured explants of *Crocus chrysanthus* Herbert CV. E. P. Bowles. J. Exp. Bot. 40: 809–812.

Fakhari, F. and P.K. Evans. 1990. Morphogenic potential of cultured floral explants of *Crocus sativus* L. for the *in vitro* production of saffron. J. Exp. Bot. 41: 47–52.

Fatehi, M., T. Rashidabady and Z. Fatehi-Hassanabad. 2003. Effects of *Crocus sativus* petals extract on rat blood pressure and on response induced by electrical field stimulation in rat isolated vas deferens and guinea-pig ileum. J. Ethnopharmacology 84: 2–3, 199–203.

Fernandez, J.A. 2004. Biology, biotechnology and biomedicine of saffron. Plant Sci. 2: 127–159.

Fernandez, J.A. 2007. Genetic resources of saffron and allies (*Crocus* spp.). Acta Hortic. 739: 167–185.

Gamborg, O.L., R.A. Miller and K. Ojima. 1968. Nutrient requirements of suspension culture of soybean root cells. Ex. Cell. Res. 50: 151–158.

George, P.S., S. Visvanath, G.A. Ravishankar and L.V. Venkataraman. 1992. Tissue culture of saffron (*Crocus sativus* L.): Somatic embryogenesis and shoot regeneration. Food Biotechnol. 6: 217–223.

Giaccio, M. 1990. Components and features of saffron. Proceedings of the International Conference on Saffron, (ICS'90), Italy, pp. 135–148.

Ghaffari, S.M. 1986. Cytogenetic studies of cultivated *Crocus sativus*. (Iridaceae). Plant. Syst. Evol. 153. 199–204.

Grilli Caiola, M. 2004. Saffron reproductive biology. Acta Hortic. 650: 25–37.

Haggag, E.G., M.A. Abou-Moustafa, W. Boucher and T.C. Theoharides. 2003. The effect of a herbal water extract on histamine release from mast cells and on allergic asthma. J. Herb. Pharmcother. 3: 41–54.

Han, L.L and X.Y. Zhang. 1993. Morphogenesis of style-stigma-like structures from floral explants of *Crocus sativus* L. and indentification of the pigments. Acta Botanica Sinica. 35: 157–160.

Hartwell, J.L. 1982. Plants Used Against Cancer: A Survey. Quaterman Publications, Lawrence, CA.

Himeno, H., H. Matsushima and K. Sano. 1988. Scanning electron microscopic study on the *in vitro* organogenesis of saffron stigma and style-like structures. Plant Sci. 58: 93–101.

Homes, J., M. Legros and M. Jaziri. 1987. *In vitro* multiplication of *Crocus sativus* L. Acta Hortic. 212: 675–676.

Hori, H., K. Enomoto and M. Nakaya. 1988. Induction of callus from pistils of *Crocus sativus* L. and production of color compounds in the callus. Plant Tiss. Cult. Lett. 5: 72–77.

Hosseinzadeh, H. and H.M. Younesi. 2002. Antinociceptive and anti-inflammatory effects of *Crocus sativus* L. stigma and petal extracts. BMC. Pharmacol. 2: 7–7.

Hosseinzadeh, H. and V. Khosravan. 2002. Anticonvulsant effects of aqueous and ethanolic extracts of *Crocus sativus* L. stigmas in mice. Arch. Iran Med. 5: 44–47.

Huang, S.Y. 1987. A study on the tissue culture of *Crocus sativus* L. Plant Physiol. Commun. 6: 17–19.

Ilahi, I., M. Jabeen and N. Firdous. 1987. Morphogenesis with saffron tissue cultures. J. Plant Physiol. 128: 227–232.

Isa, T. and T. Ogasawara. 1988. Efficient regeneration from the callus of saffron (*Crocus sativus* L.). Jpn. J. Breed. 38: 371–374.

Johnson, A.A.T. and R.E. Veilleux. 2001. Somatic hybridization and applications in plant breeding. Plant Breed. Rev. 20: 167–225.

Karamian, R. 2004. Plantlet regeneration via somatic embryogenesis in four species of *Crocus*. Acta Hortic. 650: 253–259.

Karamian, R. 2007. Somatic embryogenesis and plant regeneration from protoplast culture of *Crocus pallasii* Subsp. *haussknechtii*. Pak. J. Biol. Sci. 10: 659–663.

Karamian, R. and H. Ebrahimzadeh. 2001. Plantlet regeneration from protoplast-derived embryogenic calli of *Crocus cancellatus*. Plant Cell Tiss. Org. Cult. 65: 115–121.

Karaoglu, C., S. Cocu, A. Ipek, I. Parmaksiz, S. Uranbey, E. Sarihan, N. Arslan, M.D. Kaya, C. Sancak, S. Ozcan, B. Gurbuz, S. Mirici, C. Er and K.M. Khawar. 2007. *In vitro* micropropagation of Saffron. Acta Hortic. 739: 223–227.

Karasawa, K. 1933. The triploidy of *Crocus sativus* L. and its high sterility. Jpn. J. Genet. 9: 6–8.

Karimi, G.R., H. Hosseinzadeh and P. Khaleghpanah. 2001. Study of antidepressant effect of aqueous and athanolic of *Crocus sativus* in mice. Iran. J. Basic. Med. Sci. 4: 11–15.

Khori, V., M. Nayebpour and A.R. Mansourian. 2006. The effect of aqueous extract of *Crocus sativus* on the electrophysiological properties of isolated perfused rabbit AV-node. Pak. J. Biol. Sci. 9: 2647–2651.

Koyama, A., Y. Ohmori, N. Fujioka, H. Miyagawa, K. Yamasaki and H. Kohda. 1988. Formation of stigma-like structures and pigment in cultured tissues of *Crocus sativus* L. Planta Med. 54: 375–376.

Linsmaier, E.M. and F. Skoog. 1965. Organic growth factor requirement of tobacco tissue culture. Physiol. Plant 18: 100–127.

Lu, W.L., X.R. Tong, Q. Zhang and W.W. Gao. 1992. Study on *in vitro* regeneration of style-stigma-like structure in *Crocus sativus* L. Acta Bot. Sin. 34: 251–256.

Majourhat, K., P. Martinez-Gomez, A. Piqueras and J.A. Fernandez. 2007. Enhanced plantlet regeneration from cultured meristems in sprouting buds of saffron corms. Acta Hortic. 739: 275–278.

Milyaeva, E.L., E.V. Komarova, N.S. Azizbekova, D.D. Akhundovam and R.G. Butenko (eds.). 1988. Features of morphogenesis in *Crocus sativus in vitro* culture. Biol. Kultiviruemykh Kletok I Biotekhnologiya 1: 146.

Milyaeva, E.L., N.S. Azizbekova, E.N. Komarova and D.D. Akhundova. 1995. *In vitro* of regenerant corms of saffron *crocus* (*Crocus sativus* L.). R. J. Plant Physiol. 42: 112–119.

Moghaddasi, M.S. 2010. Saffron chemicals and medicine usage. J. Med. Plants Res. 4: 427–430.

Murashige, T. and F. Skoog. 1962. A revised medium for rapid growth and bioassays with tobacco tissue cultures. Physiol. Plant 15: 473–497.

Negbi, M. 1999. Saffron-*Crocus sativus* L. Harwood Acad. Pub, Amesterdam, The Netherlands.

Nitsch, J.P. 1972. Haploid plant from pollen. Z. Pflanzenzuchtz 67: 3–18.

Nitsch, J.P. and C. Nitsch. 1969. Haploid plants from pollen grains. Sci. 163: 85–87.

Ochiai, T., H. Shimeno, K. Mishima, K. Iwasaki, M. Fujiwara, H. Tanaka, Y. Shoyama, A. Toda, R. Eyanagi and S. Soeda. 2007. Protective effects of carotenoids from saffron on neuronal injury *in vitro* and *in vivo*. Biochim. Biophys. Acta: General Subjects 1770: 578–584.

Otsuka, M., H. Saimoto, Y. Murata and M. Kawashima. 1992. Method for producing saffron stigma-like tissue and method for producing useful components from saffron stigma-like tissue. United State- Patent, US 5085995, 8 pp. A 28. 08. 89-US- 399037. P 04. 02. 92.

Ozdemir, C. and M. Kilinc. 2008. Morphology and anatomy of three subsp. of *Crocus speciosus* Bieb. Bangl. J. Bot. 37: 97–103.

Ozdemir, C., P. Baran and Y. Akyol. 2005. The morphology and anatomy of *Crocus flavus* Weston subsp. *flavus* (Iridaceae). Turk. J. Bot. 30: 175–180.

Piqueras, A., B.H. Han, J. Escribano, C. Rubio, E. Hellin and J.A. Fernandez. 1999. Development of cormogenic nodules and microcorms by tissue culture, a new tool for the multiplication and genetic improvement of saffron. Agronomie 19: 603–610.

Plessner, O., M. Negbi, M. Ziv and D. Basker. 1989. Effects of temperature on the flowering of the saffron crocus (*Crocus sativus* L.): Induction of hysteranthy. Israel. J. Bot. 38: 1–7.

Plessner, O., M. Zive and M. Negbi. 1990. *In vitro* corm production in the saffron *Crocus* (*Crocus sativus* L.). Plant Cell Tiss. Org. Cult. 20: 89–94.

Premkumar, K., S.K. Abraham, S.T. Santhiya, P.M. Gopinath and A. Ramesh. 2001. Inhibition of genotoxicity by saffron (*Crocus sativus* L.) in mice. Drug Chem. Toxicol. 24: 421–428.

Raja, W., G. Zaffer and S.A Wani. 2007. *In vitro* microcorm formation in saffron (*Crocus sativus* L.). Acta Hortic. 739: 291–296.

Rangahau, M.K. 2003. Growing Saffron—The world's most expensive spice. Crop and Food Research 20: 1–4.

Reinert, J. 1959. Untersuchungen uber die Morphogenese an Gewebekulturen. Berichte der Deutschen Botanischen Gesellschaft 71: 15.

Sano, K. and H. Himeno. 1987. *In vitro* proliferation of saffron (*Crocus sativus* L.) stigma. Plant Cell Tiss. Org. Cult. 11: 159–166.

Sarma, K.S., K. Maesato, T. Hara and Y. Sonoda. 1990. *In vitro* production of stigma-like structure from stigma explants of *Crocus sativus* L. J. Exp. Bot. 41: 745–748.

Sheibani, M., S.H. Nemati, G.H. Davarinejad, A.V. Azghandi and A.A. Habashi. 2007. Induction of somatic embryogenesis in saffron using thidiazuron (TDZ). Acta Hortic. 739: 259–268.

Sharma, K.D., R. Rathour, R. Sharma, S. Goel, T.R. Sharma and B.M. Singh. 2008. *In vitro* cormlet development in *Crocus sativus*. Biologia Plantarum 52: 709–712.

Steward, F.C., M.O. Mapes and K. Mears. 1958. Growth and organized development of cultured cells. II: Organization in cultures grown from freely suspended cells. Am. J. Bot. 45: 505–708.

Visvanath, S., G.A. Ravishankar and L.V. Venkataraman. 1990. Induction of crocin, crocetin, picrocrocin, and safranal synthesis in callus cultures of saffron *Crocus sativus* L. Biotechnol. Appl. Biochem. 12: 336–340.

White, P.R. 1963. The Cultivation of Plant and Animal Cells. Ronald Press, New York.

Winterhalter, P. and M. Straubinger. 2000. Saffron. Renewed interest in an ancient spice. Food Rev. Int. 165: 39–59.

Yonghua, Y., G. Zibiao and X. Fenghui. 1996. Differences on organogenesis of callus types in saffron. J. Plant Resour. Environ. 5: 63–64.

Yongjiong, J., Ch. Fang, L. Honghui, C. Youlong, L. Ying and W. Shui. 1996. Induction of style stigma like structure and regeneration of plantlets from corm of *Crocus sativus in vitro*. Sichuan Daxue Xuebao (Ziran Kexueban). 33: 747–750.

Zhang, Y., Y. Shoyama, M. Sugiura and H. Saito. 1994. Effects of *Crocus sativus* L. on the ethanol-induced impairment of passive avoidance performances in mice. Biol. Pharm. Bull. 17: 217–221.

8

Environmental and Life Cycle Cost Assessment of Cassava Ethanol

Shabbir H. Gheewala, Sébastien Bonnet* and *Thapat Silalertruksa*

ABSTRACT

Biofuels have been extensively promoted by many countries worldwide as a means to contribute to improved national energy security via reduced dependency on imported fossil fuels. They are also widely perceived as "green fuels" providing environmental benefits that are particularly emphasized for climate change mitigation through reduced greenhouse gas (GHG) emissions. However, the rapid increase in the production and use of ethanol and biodiesel over the last decade has been accompanied with rising concerns over the environmental benefits and cost competitiveness they may offer. In this chapter the case of cassava ethanol is taken as an example to address such concerns with a particular focus on Thailand where biofuels have been actively promoted over the last 15 years. To maintain the GHG performance and cost competitiveness of cassava ethanol as compared to its fossil

The Joint Graduate School of Energy and Environment (JGSEE), King Mongkut's University of Technology Thonburi (KMUTT), Bangkok, Thailand; and Center for Energy Technology and Environment, Ministry of Education, Bangkok, Thailand.
Emails: shabbir_g @jgsee.kmutt.ac.th; thapat.sil@kmutt.ac.th
* Corresponding author: shabbir_g @jgsee.kmutt.ac.th

equivalent gasoline, careful considerations are required over a range of parameters including, but not limited to, agricultural intensification of cassava production, land use expansion/land use change, enhanced productivity and co-products utilization, etc. Life-cycle greenhouse gas emissions and life-cycle costing studies have been conducted to evaluate the influence of these parameters and help maintain the benefits from biofuels.

Challenges of Biofuels

At present, bioethanol and biodiesel are the two main liquid biofuels being used worldwide. They have been attached significance as alternative fuels for transportation to mitigate the dependency on petroleum derivatives which have several constraints in their long-term utilization, including volatile prices and insecurity of supply (Silalertruksa and Gheewala 2010, Bell et al. 2011). It is also anticipated that the use of local feedstocks will provide benefits to the rural economy by stabilizing the prices of certain agricultural produce. Biofuels are also anticipated to help in mitigating climate change as they may release fewer greenhouse gases (GHGs) than their fossil energy counterparts (Nguyen et al. 2007a,b, Pleanjai et al. 2009a,b). The emissions of greenhouse gases (GHGs), in particular carbon dioxide, are expected to decrease when replacing fossil fuels with biofuels because the latter are derived from plant materials. However, this only due to the carbon neutrality of the CO_2 released from combustion of the biofuel. When the whole life-cycle is considered, the GHG emissions from feedstock cultivation (particularly from the manufacture and application of nitrogen fertilizers), feedstock processing and transportation between the various life-cycle phases are outside the scope of carbon neutrality for biofuels. Also, more recently, several studies have shown that land-use changes and management caused by increased growing demand for feedstocks could substantially increase GHG emissions from the changes of carbon stock which could outweigh the GHG benefits as compared to fossil fuels. It may also contribute to the loss of biodiversity and ecosystem services especially in the case of deforestation (Silalertruksa and Gheewala 2012). As biofuels are intensively promoted worldwide and their demand is substantially increasing, to guarantee they can help mitigate global warming impact when replacing petroleum fuels, GHG emissions reduction is set as one of the principal criteria for environmental sustainability of biofuels (Silalertruksa and Gheewala 2012).

Environmental Sustainability of Cassava Ethanol

Biofuels in Thailand

Thailand has been promoting the use of biofuels, i.e., biodiesel from oil palm and bioethanol from cane molasses and cassava, for the transportation sector for over a decade. The ministry of Energy has set production targets for bioethanol for 2011, 2016 and 2022 at 2.96, 6.2 and 9.0 million litres/day, respectively (DEDE 2009). Ethanol is currently being produced from sugar-cane molasses and cassava. Ethanol production directly from sugar-cane juice is also planned for the future. Sugar cane and cassava are both well-established agricultural products and important for the domestic market as well as export. Although ethanol is predominantly produced out of sugarcane, the Thai government is increasingly promoting cassava as a major feedstock for ethanol production. This is because cassava, a starch-accumulating tuber crop with up to 30% of fermentables, seems more advantageous to be used for the alcohol fuel industry. First, cassava can adapt well to a wide range of growing conditions and requires minimal inputs. Second, unlike sugar-based distilleries that are seasonally operated, cassava-based ethanol plants can run year round, thanks to the crop's unbound time window for growing and harvesting, plus its capability to be stored as dried chips (Nguyen et al. 2007c). A 10% blend of ethanol with gasoline, E10, has been available for several years since 2005 and has well-established usage; prices of E10 are maintained lower than gasoline through government incentives to encourage its use. Since 2008, E20 (a 20% blend of ethanol with gasoline) has been available once again with government incentives to maintain an attractive price. Many automobile companies are producing vehicles which can use E20. E85 (an 85% blend of ethanol with gasoline) has also been introduced since late 2008 on a very limited scale, although vehicles that can use this blend are not yet readily available in Thailand (Gheewala 2012).

Environmental performance of cassava ethanol

Life-cycle assessment

For assessing the environmental performance of cassava ethanol in a holistic approach, life cycle assessment (LCA) can be used as a useful environmental sustainability assessment tool. It has been widely used to assess environmental performance of renewable technologies and also bioenergy systems such as bioethanol and biodiesel. The crucial advantage of a life-cycle approach is that all burdens from raw material extraction through production, to use and disposal, will be accounted for. Even though LCA does not only focus on GHG impact assessment, many other impact categories can be considered as described in the ISO standards (e.g., acidification, eutrophication, toxicity, etc.), it is especially useful for evaluating the GHG performance of transportation biofuels in fair comparison with conventional petroleum fuels. This is because in such a approach, LCA enables to consider the entire life-cycle of a biofuel rather than just the combustion phase in a vehicle.

In the following sections, the life cycle GHG emissions of cassava ethanol are assessed based on the production system depicted in Fig. 1. The total GHG emissions are from the various life cycle stages such as land use change, cultivation and harvesting, feedstock processing, bioethanol conversion, and final use for transportation in vehicles. The overall GHG performance of the cassava ethanol system is analyzed in comparison with that of a conventional petroleum fuel system. For a fair assessment of the two systems, since fuels are used as a source of energy and the energy content of ethanol (21.2 MJ/L ethanol) and gasoline (32.4 MJ/L gasoline) differ, the GHG performance is estimated on the basis of one Megajoule (Nguyen 2007a).

Cassava ethanol production

Production of cassava ethanol for transport in Thailand is composed of five main stages (1) planting, (2) chips processing, (3) chips conversion to ethanol, (4) blending of ethanol with gasoline to produce gasohol and all (5) transport activities. A short description of each stage is provided below:

- *Cassava plantation*: This stage consists of land preparation, stake preparation, planting, treatment and harvesting.
- *Dried chips processing*: This stage consists in chopping fresh roots, drying by sunlight and packing cassava chips before handing over to other industries.
- *Ethanol conversion*: The cassava-ethanol plant consists of four main sub-processes: milling, mixing and liquefaction, fermentation, distillation and molecular sieve dehydration. Ethanol 99.5% concentration is the product from this conversion process.
- *Blending*: Blending ethanol with gasoline serves two purposes: (1) extends the gasoline supplies, and (2) improves the quality of gasoline by increasing the octane level. In Thailand, 10% ethanol was previously blended with gasoline (E10) as octane booster to replace methyl tertiary butyl ether (MTBE). Later, E20 and E85 were introduced in the market respectively in order to reduce dependency on imported crude oil.

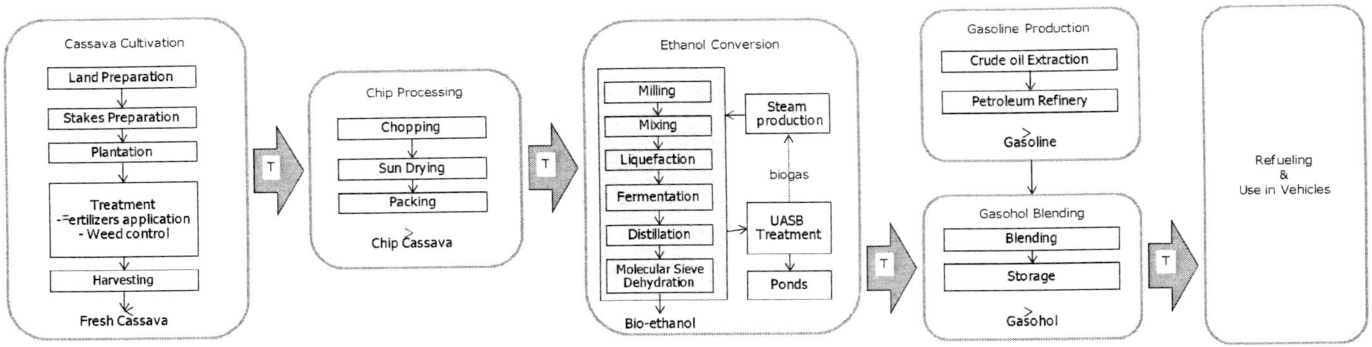

Fig. 1. Cassava ethanol production system.

- *Transport*: Various types of transport facilities are involved between each stage through different distances. For instance, truck and pick-up are widely used for transport of fresh cassava roots from fields to drying floor and ethanol producers.

Life-cycle GHG emissions of cassava ethanol

The GHG emissions of cassava ethanol are considered for several production scenarios in which co-products utilization are considered to evaluate the GHG benefits they may provide to the overall production system. The scenarios considered are:

- *Base case scenario*: Existing cassava ethanol system where cassava ethanol plant uses imported coal as fuel to produce steam and electricity is supplied from the national grid system. Distillery spent wash, the high organic wastewater discharged from ethanol conversion stage, is sent to aerobic ponds without waste recycling systems.
- *Scenario 1*: An upflow anaerobic sludge blanket (UASB) system for biogas recovery from spent wash is included as an additional component to the base case scenario. This biogas is used in the boiler to substitute imported coal. The GHG emission credits from coal displacement are accounted for in the GHG balance.
- *Scenario 2*: The production of dry distillers grains with solubles (DDGS) from spent wash is added to the base case scenario. Animal feed (soybean meal) is the product substituted by DDGS. Each kilogram of DDGS is considered to substitute 0.78 kg of soybean meal.
- *Scenario 3*: Rice husk is used as fuel in place of imported coal for the base case scenario.
- *Scenario 4*: Combination of scenarios 1 and 3.
- *Scenario 5*: Combination of scenarios 2 and 3.

In addition to the above, for each of the scenarios looked at, land use change considerations are also included. These involve the conversion of tropical forest and/or grassland to crop land for cassava production.

The results of the GHG assessment for each of the 6 scenarios including direct land use change (LUC) are shown in Table 1.

GHG emissions for cassava ethanol are in the range 27–91 gCO_2eq per MJ ethanol. The wide range in GHG emissions observed between the various scenarios stem from the differences in the production environment. As ethanol conversion is an energy intensive process, the use of fossil fuels such as coal to produce steam and electricity result in significantly higher GHG emissions than systems where bioenergy is used (scenario 1 and 3).

Looking further into the details of the GHG emissions for the base scenario for each of the life cycle phases leading to cassava ethanol production, their share contribution to total GHG emissions are as shown in Fig. 2.

It is clear that the ethanol conversion stage is by far the highest contributor to the overall life-cycle GHG emissions at 74%. This is due to the steam boiler in the ethanol plant which is using sub-bituminous coal as fuel to produce steam and the electricity consumption of the plant from the grid system which is mainly produced from natural gas.

It is also noticed from Table 1 that the inclusion of direct LUC can have substantial effect on the overall GHG emissions of the ethanol system considered. This could challenge its environmental competitiveness as compared to gasoline. The conversion of tropical forest and/or grassland to cropland can possibly increase GHG emissions three to nine times as compared to the cases where LUC is excluded. The conversion of tropical forest to cropland results in the highest GHG emissions as a result of the

Table 1. GHG emissions of cassava ethanol system (gCO_2eq per MJ cassava ethanol).

		Including LUC	
	Excluding LUC	Forest – to – crop land	Grassland – to – crop land
Base scenario	91	313	127
Scenario 1	82	304	118
Scenario 2	81	303	117
Scenario 3	37	258	73
Scenario 4	28	250	64
Scenario 5	27	249	63

Based on: Silalertruksa and Gheewala (2011)

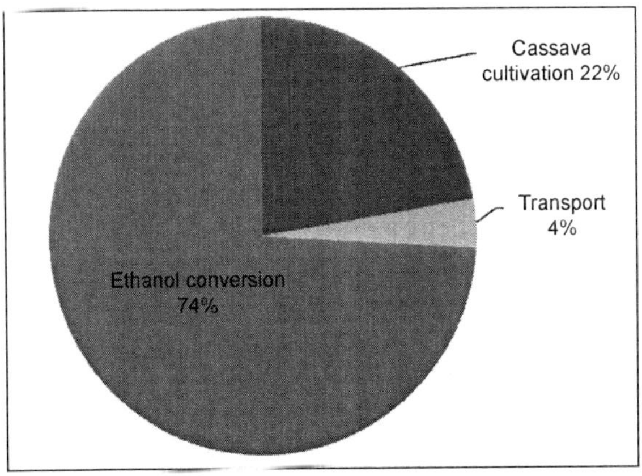

Fig. 2. Environmental impacts of cassava ethanol production (based on: Silalertruksa and Gheewala 2009).

loss of carbon stock in the above and below-ground biomass and non-CO_2 emissions from burning biomass as part of the first clearance of land, totaling about 14.5 tonnes CO_2-eq ha^{-1} yr^{-1}. In addition, soil carbon stock change from this direct LUC also creates GHG emissions of about 2.1 tonnes CO_2-eq ha^{-1} yr^{-1}. This is in contrast to the case of converting grassland to cropland for which GHG emissions originate mainly from soil carbon stock changes, i.e., 1.9 tonnes CO_2-eq ha^{-1} yr^{-1}, while the losses of carbon stocks in the above and below-ground biomass are just about 0.8 tonne CO_2-eq ha^{-1} yr^{-1}. Such findings indicate that the carbon debt from converting grassland to cropland could be paid off by the GHG credits of bioethanol within 10 years for the base case scenario of cassava ethanol if we consider that overall GHG emissions for a conventional gasoline system are about 2.92 kg CO_2-eq per liter gasoline or 90 g CO_2-eq per MJ. This payback period is an important consideration when dealing with environmental performance of biofuels indicating the important influence land use change may play when included in the GHG emission assessment.

So far LUC induced by domestic demand for bioethanol in Thailand has not been much of a focus as bioethanol production mainly relied on the surplus feedstocks obtained from the existing plantation areas. However, it would need to be considered if the approved ethanol plants were operated and the government target of 9 million liters ethanol per day was achieved. The government policy emphasizes increasing feedstocks production by improving yields instead of expanding plantation areas. However, the demand-supply of feedstocks estimated by the Department of Alternative Energy Development and Efficiency (DEDE) shows that there was 2.6 million tonnes of surplus cassava for producing ethanol in 2011 equivalent to 1.2 million liter of ethanol per day; however, the total installed capacity of cassava ethanol plants in Thailand is expected to be much higher at 4 million liters per day. Thus, LUC for cassava production will become an important factor in the long run.

Life Cycle Cost of Cassava Ethanol

Life-cycle cost

Life-cycle cost (LCC) is an economic method for determining all costs arising over an entire project's or product's life cycle from raw material acquisition, installation, operation, maintenance, to final disposal (Fuller and Petersen 1996, Luo et al. 2009). For the LCC of cassava ethanol production, all costs arising in the production system, i.e., cassava cultivation, chip processing, ethanol conversion and all transport activities need to be considered. A steady-state cost model can be assumed for the LCC analysis and with only production costs considered.

The LCC assessment of cassava ethanol is illustrated for the case of Thailand. It includes production data such as materials, chemicals and energy consumption, and their costs. These data are considered over past few years to reflect and investigate ranges of variation in the production costs of ethanol. For comparative purposes the cost performance of cassava ethanol can be compared with that of gasoline. This comparison can be performed in terms of ex-refinery price and on the basis of one litre of gasoline equivalent. All costs arising from cassava planting to ethanol production and transportation to refinery are to be included. A traditional way to compare ethanol and gasoline is to assume that one litre of ethanol produces the same performance as 0.65 litre of gasoline (Yoosin and Sorpipatana 2007, Henke et al. 2005). This factor comes from the ratio of the different calorific values of bio-ethanol (21.2 MJ per litre) and gasoline (32.4 MJ per litre). However, this factor is not yet the consensus for use as the substitution ratio between ethanol and gasoline (Henke et al. 2005). Another way for determining the substitution ratio between ethanol and gasoline is based on the real fuel economy in a motor vehicle engine (Nguyen et al. 2007b, Nguyen and Gheewala 2008). For instance, in Thailand, the PTT Research and Technology Institute, has performed test of various car models running on gasoline (ULG95) and gasohol E10. The results based on a 1.6 L Toyota vehicle show that the fuel economy running on ULG95 and E10 is 13.46 km per litre and 13.31 km per litre, respectively. Fuel economy comparison reveals that 1 L of gasohol is equal to 0.989 L of gasoline. The difference of 0.011 L is due to the 10% of ethanol blended in gasoline. The substitution ratio between ethanol (in E10 form) and gasoline in a motor vehicle engine would therefore be 1:0.89 based on fuel economy, instead of 1:0.65 based on energy content.

Estimation of life cycle cost

The life-cycle cost of cassava ethanol is the sum of all expenses in materials, energy, labor, transportation, depreciation of fixed assets, maintenance, and miscellaneous costs in all main-processes leading to cassava ethanol production. The costs of main-processes can be categorized into five groups, i.e., (1) cassava planting costs, (2) dry chips processing costs, (3) ethanol conversion costs, (4) gasohol production costs and (5) all transportation costs which take place in the system.

Cassava farming costs

The statistic of cassava farming costs in Thailand are available from the Office of Agricultural Economics (OAE). For the LCC assessment detailed below, production cost data for the years 2006, 2007 and 2008 have been retrieved as shown in Table 2 (OAE 2006, OAE 2007, OAE 2008). Fixed costs are the costs associated with land rental, depreciation and tax. Variable costs involve all operating costs, i.e., labor costs, material costs and miscellaneous costs, e.g., reparation and interest. Based on production data and costs, cassava farming costs can be classified into five categories, i.e., (1) land preparation, including costs of fuel, and hiring tractors and drivers; (2) hand planting including the cost of planting materials (cassava stems); (3) chemicals (fertilizers and herbicides); (4) harvesting; and (5) transportation of chemicals, stem cuttings and harvesting. These contribute to about 11%, 15%, 35%, 18% and 21% of the total farming costs, respectively (Nguyen 2007, Nguyen et al. 2008).

Table 2. Cassava farming costs.

		2006	2007	2008
Yield	Tonnes per ha	21.1	22.9	21.6
Total Production costs	THB per tonne roots	843	824	880
Fixed costs	THB per tonne roots	99	91	92
Variable costs	THB per tonne roots	744	733	788
Sale prices of fresh cassava	THB per tonne roots	1290	1180	1730
Farmer incomes	THB per tonne roots	447	356	850
	THB per ha	2,794	2,225	5,312

Source: Bonnet et al. 2009

Chips processing costs

Production costs of dried chips processing consist of raw material cost (fresh cassava roots), fuel used in the tractor to transfer roots into the hopper of the chopping machine and transfer of chopped roots on a large floor for drying by sunlight. Another cost is from the electricity used in the chopping machine. Today, the cassava chips market has expanded substantially as cassava chips can be used as feedstock to produce various chemicals (Sriroth and Piyachomkwan 2008). Its price can therefore easily vary depending on various market factors. However, ethanol producers generally have also their own facilities to produce dried chips. Dried chips processing is referred to as a process step taking place at the ethanol production plant. Ethanol producers will directly buy roots from farmers or suppliers. Only chips processing costs are considered in the analysis. The obtained results would differ from those using the market price of cassava chips for estimation of the ethanol production cost. Based on Nguyen et al. (2008), chips processing cost amounts to 152 THB per tonne of dry chips.

Ethanol conversion costs

The ex-refinery price of ethanol is a summation of ethanol conversion costs, profit margin of ethanol producer and transportation costs for delivery of ethanol to petroleum refinery for blending. The ethanol conversion costs include costs of raw material (cassava dry chips), utilities (e.g., energy and water), chemicals, repair, and maintenance, wages and salary, depreciation, fiscal charges, selling expenses, miscellaneous and profit margin. This information is available in Nguyen et al. (2008). The results of the assessment reveal that raw material cost (cassava chips) represents about 55% of the ethanol ex-refinery price. The share of other contributors to conversion costs are shown in Fig. 3.

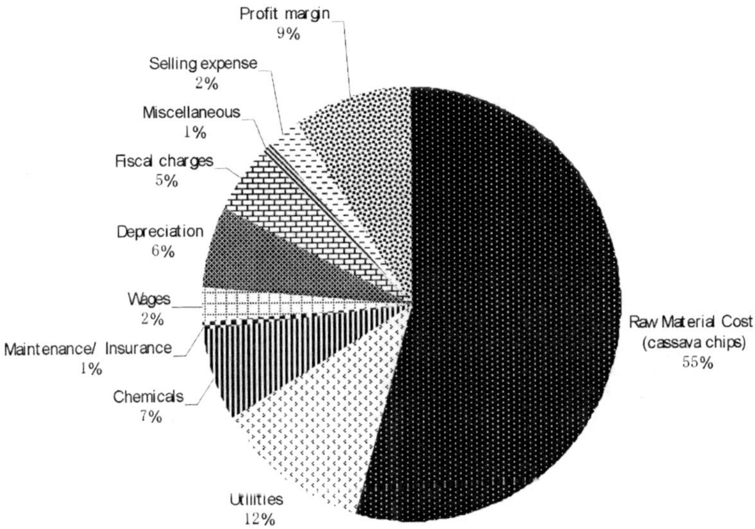

Fig. 3. Breakdown of ethanol conversion costs.

Comparison between gasoline and ethanol

As mentioned earlier, the cost performance of ethanol and gasoline can be compared based on their respective ex-refinery price (excluding taxes) and per litre of gasoline equivalent. The conversion to gasoline equivalent is based on the substitution ratio between ethanol and gasoline according to fuel economy. The substitution ratios between ethanol (in E10, E20 and E85 form) and gasoline (ULG95) in a motor vehicle engine are shown in Table 3. Based on the prices of cassava feedstock and gasoline during 2006–2008, the possible minimum, maximum and average ex-refinery prices of cassava ethanol and gasoline are shown in Fig. 4 and Fig. 5. It is

Table 3. Fuel economy for gasohol and gasoline.

Testing results	Fuel economy (km per litre)	Substitution ratio between ethanol (in gasohol forms) and gasoline	Source
Gasoline (ULG95)	13.46	1:0.89	Nguyen and Gheewala (2008)
E10	13.31		
Gasoline (ULG95)	13.19	1:0.78	Orbital Engine Company (2003)
E20	12.61		
Gasoline (ULG95)	13.26	1:0.67	Reading et al. (2002), Kelly et al. (1996)
E85	9.51		

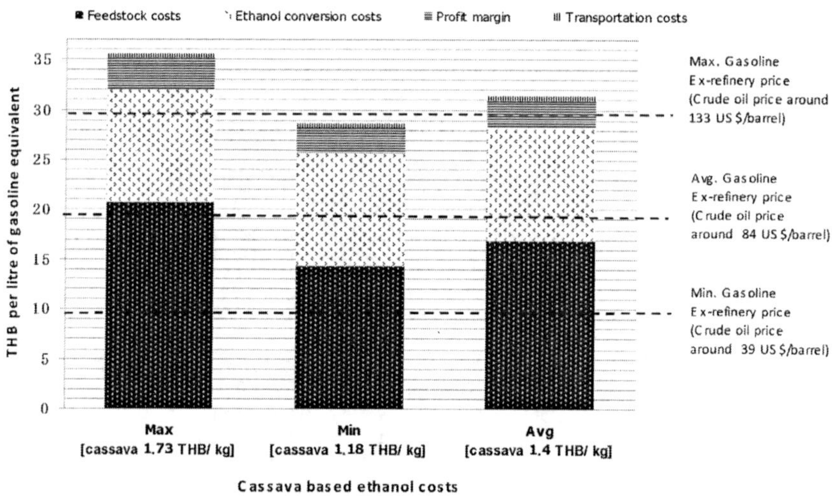

Fig. 4. Comparisons between ex-refinery price of cassava ethanol and gasoline (ULG95) on the basis of one litre gasoline equivalent (using substitution ratios based on energy content).

observed that the production costs of cassava ethanol are much higher than that of gasoline in particular for higher percentages of ethanol blends (Fig. 6). Hence for E10, the most popular gasohol available on the Thai market, it is noticed that the production cost of cassava ethanol has almost always been more expensive than that of gasoline except in cases when the crude oil price exceeded 110 US$ per barrel (during the economic crisis in 2008). The cost of cassava feedstock is the major contributor to total cost, ranging from 49 to 58% depending on its market price. The cost of ethanol conversion is the second major contributor to total cost, especially costs for utilities, e.g., fuel, electricity and water.

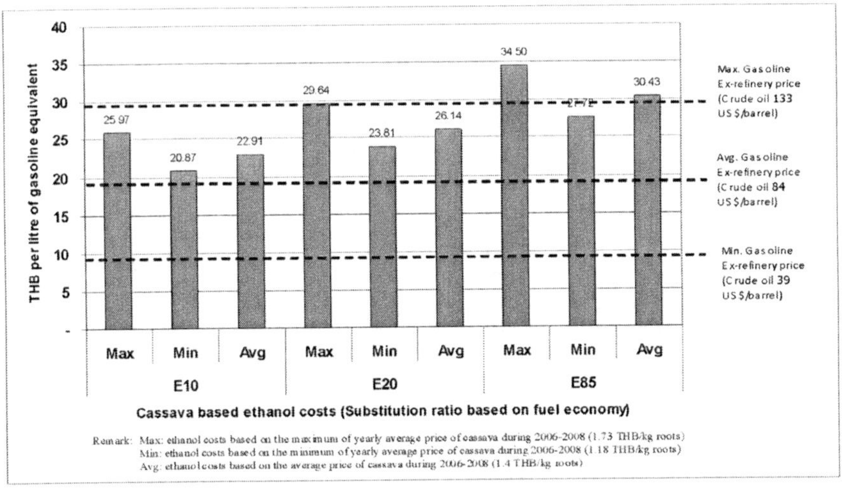

Fig. 5. Comparison between ex-refinery price of cassava ethanol (for various blends of gasohol) and gasoline (ULG95) on the basis of one litre gasoline equivalent (using substitution ratios based on fuel economy).

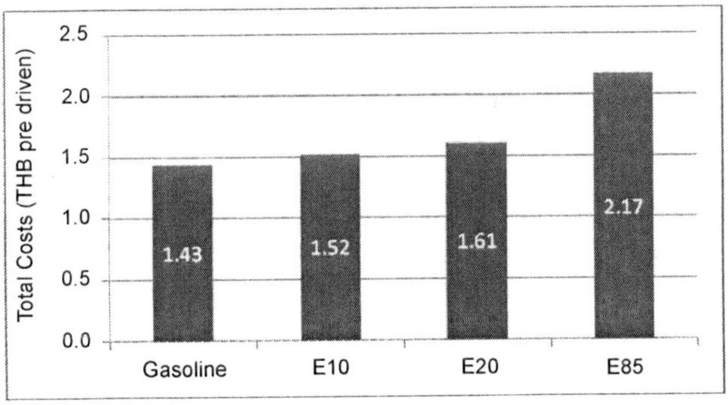

Fig. 6. Comparison of total costs per km driven distance between conventional gasoline and various gasohol blends (based on an average cassava feedstock price of 1.4 THB per kg fresh roots).

From the above results, it is clear that crude oil price is a key factor in influencing the success of any policy relating to the promotion of bioethanol in the transport sector. If the crude oil price rises up to a certain benchmark (133 USD per barrel), it would contribute to easing the introduction of the bioethanol program in Thailand as lesser subsidies would be needed to enhance the competitiveness of bioethanol as compared to conventional gasoline. However, if the crude oil price drops below that benchmark and falls below 84 USD per barrel, the competitiveness of bioethanol would be hindered as compared to gasoline even for the lowest cassava feedstock price (1.18 THB per kg roots).

Optimizing the cost-competitiveness of cassava ethanol

The LCC results show that the production cost of cassava ethanol in Thailand cannot compete with gasoline when compared in terms of gasoline equivalent. This is a common problem faced when promoting biofuels as substitute to petrol fuel not only for cassava based ethanol but also other biofuels in countries when subsidies are removed (Worldwatch Institute 2007, Nguyen 2007). However, this issue depends on prices of crude oil and energy crops. In May 2008, the crude oil price rose drastically up to 133 US$ per barrel, enabling cassava ethanol to compete with gasoline. However, few months later in the same year, the crude oil price dropped dramatically to around 39 US$ per barrel, which directly affected the competitiveness of bioethanol and therefore the bioethanol industry and renewable energy policy targets set by the Thai government for biofuels.

Another problem relating to the uncompetitive price of ethanol as compared to gasoline is the difficulty in expanding the production and use of cassava ethanol to achieve the government policy targets related to the use of higher percentages of gasohol blends such as E20 and E85. Consumers usually buy products for the lowest price if they can serve the same utility. A driver is interested in achieving a driving distance at the cheapest cost. As shown in Fig. 6, for the total costs per kilometer of achieved driven distance in passenger car, the cost per kilometer of E10 is close to that of gasoline. Also E20 can possibly compete with gasoline on the market. However, there is a large gap between the production costs per kilometer of E85 and gasoline at about 0.75 THB per km. For introducing E85 in the market, it is imperative therefore to set policies enabling to effectively promote biofuel, including tax exemption or blending mandates during the initial implementation period of the E85 policy. The above only considers issues related to higher production costs for higher blends of biofuels. However, other aspects such as providing support to the automobile industry for producing flex-fuel vehicles (FFVs) and to refineries for producing and selling E85 need also to be considered.

Besides the policy supporting tools such as tax exemption or blending mandates, measures for reducing ethanol production costs are also required to sustain the bioethanol industry in Thailand. This especially the case for situations in which cassava based ethanol is promoted to achieve nationwide biofuel production policy targets. As shown by the LCC analysis, the production costs of ethanol are mainly shared by raw material costs 55% and ethanol conversion costs 36%. Based on the ethanol policy in Thailand, the government has targeted that the root price of cassava should stabilize to about 1.50 THB per kg to benefit both the farmers and the industrial sector (Sriroth and Piyachomkwan 2008). In fact, cassava price varies as a result of many factors which are difficult to control. During 2006–2008 the market price of cassava root ranged between 1.12–2.23 THB per kg. Therefore, reduction of conversion costs by increasing the efficiency of material and energy use should be promoted by policy makers instead of attempting to control the market price of cassava. However, it is still important trying to maintain the price of cassava root above 1.5 THB per kg in order to maintain farmer's incentives in planting cassava and therefore avoiding feedstock shortage. Three possible approaches to reduce production costs are (1) improvement of conversion efficiency (from 7.5 to 6 kg roots per litre ethanol), (2) improvement of waste utilization (biogas from wastewater treatment used as fuel for fermentation process, Dry Distillers Grains with Soluble (DDGS) from slop of cassava and CO_2 recovery from fermentation) and (3) improvement of cassava yield (from 22 to 31 tonnes per ha based on government policy target) during the plantation stage. From such improvements the ethanol production costs could be further reduced as shown in Table 4.

Table 4. Ethanol production costs after implementing the improvement measures.

	Cassava price		
	Average (2006–2008)	**Policy target**	**Range (2006–2008)**
Cassava fresh root price (THB per kg)	1.40	1.50	1.12–2.23
Ethanol production costs (THB per litre)	20.45	21.20	18.14–26.47
Ethanol production costs saving & co-products credits from proposed options (THB per litre ethanol)			
- Improvement of conversion efficiency	2.10	2.50	2.50
- Improvement of waste utilization	1.10		
Improvement of cassava yield via improved agricultural practices (THB per litre ethanol)			
- improvement of cassava yield [1]	- Reduction in cassava production cost 0.28 THB/kg - Ethanol conversion cost would decrease 1.68 THB/L ethanol (@ efficiency 6 kg roots/L ethanol)		
	1.68	1.68	1.68
Ethanol production costs after implementing all the measures (THB per litre ethanol)	15.53	15.88	13.62–20.31
Ethanol production costs after implementing the measures (THB per litre gasoline equivalent)[2]	23.89	24.43	20.95–31.25
Range of gasoline (ULG95) Ex-refinery price during 2006–2008 (THB per litre)	Max: 29.61; Avg: 19.31; Min: 9.43		

Remarks:

[1] Several good practices for increasing yield, i.e., development of new varieties, selection of appropriate varieties, good plantation period, improving soil quality by using organic fertilizers, good practices in land preparation, plantation and harvesting, and weed control.

[2] Ethanol is assumed to contain 0.65 the energy of a litre of gasoline.

Note: 1THB = 30 USD

References

Bell, D.R., T. Silalertruksa, S.H. Gheewala and R. Kamens. 2011. The net cost of biofuels in Thailand: An economic analysis. Energ. Policy 39(2): 834–843.

Bonnet, S., S.H. Gheewala and T. Silalertruksa. 2009. Societal cost of biofuels in Thailand. Project report. King Mongkut's University of Technology Thonburi, Bangkok, Thailand.

Department of Alternative Energy Development and Efficiency (DEDE). 2009. Thailand's renewable energy and its energy future: Opportunities and challenges. Final Draft Report. Ministry of Energy. Available at: www.dede.go.th.

Fuller, S.K. and S.R. Petersen. 1996. Life-cycle costing manual for the Federal Energy Management Program. U.S. Department of Energy, Washington.

Gheewala, S.H. 2012. Food, fuel and climate change: policy performance and prospects for biofuels in Thailand. *In:* RAP Publication 2012/14 on: Sustainable Bioenergy in Asia: Improving residence to high food prices and climate change. FAO, ISBN 978-92-5-107291-2.

Henke, J.M., G. Klepper and N. Schmitz. 2005. Tax emption for biofuels in Germany: Is bio-ethanol really an option for climate policy? Energy 30: 2617–2635.

Kelly, K.J., B.K. Bailey, T.C. Coburn, W. Clark and P. Lissiuk. 1996. Federal test procedure emissions test results from ethanol variable-fuel vehicle Chevrolet Luminas. Proceedings of International Spring Fuels and Lubricants Meeting. Dearborn, MI: Society for Automotive Engineers.

Luo, L., E. van der Voet and G. Huppes. 2009. Life cycle assessment and life cycle costing of bioethanol from sugarcane in Brazil. Renew. Sust. Energ. Rev. 13(6-7): 1613–1619.

Nguyen, T.L.T., S.H. Gheewala and S. Garivait. 2007a. Fossil energy savings and GHG mitigation potentials of ethanol as a gasoline substitute in Thailand. Energ. Policy 35(10): 5195–5205.

Nguyen, T.L.T., S.H. Gheewala and S. Garivait. 2007b. Energy balance and GHG—abatement cost of cassava utilization for fuel ethanol in Thailand. Energ. Policy 35(9): 4585–4596.

Nguyen, T.L.T., S.H. Gheewala and S. Garivait. 2007c. Full Chain Energy Analysis of Fuel Ethanol from Cassava in Thailand. Environ. Sci. Technol. 41: 4135–4142.

Nguyen, T.L.T. 2007. Life-cycle assessment of bio-ethanol as an alternative transportation fuel in Thailand. Ph.D. dissertation. The Joint Graduate School of Energy and Environment, King Mongkut's University of Technology Thonburi, Bangkok, Thailand.

Nguyen, T.L.T.S. Bonnet and A. Bangviwat. 2008. Societal Costs of biofuels. Case study report. The Joint Graduate School of Energy and Environment, King Mongkut's University of Technology Thonburi, Bangkok, Thailand.

Nguyen, T.L.T. and S.H. Gheewala. 2008. Life cycle assessment of fuel ethanol from cassava in Thailand. Int. J. Life Cycle Ass. 13(2): 147–154.

OAE. 2006. Report on agriculture economics: Cassava. Office of Agricultural Economics. Ministry of Agriculture and Cooperatives, Bangkok, Thailand. Available at: www.oae.go.th.

OAE. 2007. Agricultural statistics of Thailand. Center for Agricultural Information. Office of Agricultural Economics. Ministry of Agriculture and Cooperatives, Bangkok, Thailand. Available at: www.oae.go.th.

OAE. 2008. The agricultural economics in 2008 and outlook for 2009. Office of Agricultural Economics. Ministry of Agriculture and Cooperatives, Bangkok, Thailand. Available at: www.oae.go.th.

Orbital Engine Company. 2003. Report to environment Australia. A testing based assessment to determine impacts of a 20% ethanol gasoline fuel blend on the Australian passenger vehicle fleet. March 2003.

Pleanjai, S., S.H. Gheewala and S. Garivait. 2009a. Greenhouse gas emissions from production and use of palm methyl ester in Thailand. International Journal of Global Warming 1(4): 418–431.

Pleanjai, S., S.H. Gheewala and S. Garivait. 2009b. Greenhouse gas emissions from production and use of used cooking oil methyl ester as transport fuel in Thailand, J. Clean. Prod. 17(9): 873–876.

Reading, A.H., J.O.W. Norris, E.A. Feest and E.L. Payne. 2002. Ethanol emissions testing. AEAT Unclassified. E&E/DDSE/02/021.

Silalertruksa, T. and S.H. Gheewala. 2009. Environmental sustainability assessment of bio-ethanol production in Thailand, Energy 34: 1933–1946.

Silalertruksa, T. and S.H. Gheewala. 2010. Security of feedstocks supply for future bio-ethanol production in Thailand. Energ. Policy 34: 7475–7486.

Silalertruksa, T. and S.H. Gheewala. 2011. Long-term bioethanol system and its implications on GHG emissions: A case study of Thailand, environ. Sci. Technol. 45(11): 4920–4928.

Silalertruksa, T. and S.H. Gheewala. 2012. Food, fuel and climate change: Is palm-based biodiesel a sustainable option for Thailand? J. Ind. Ecol. 16(4): 541–551.

Sriroth, K. and K. Piyachomkwan. 2007. Cassava ethanol technology and growth in Thailand. Project report.

Worldwatch Institute. 2007. Biofuels for transport: Global potential and implication for sustainable energy and agriculture.

Yoosin, S.C. and C. Sorpipatana. 2007. A Study of Ethanol production cost for gasoline substitution in Thailand and its competitiveness. Thammasat Int. J. Sc. Tech. 12(1): January–March 2007.

Section B

Flowering Plants

9

Genetic Transformation and Metabolic Engineering of *Lilium*

Pejman Azadi,[1,*] *Valentine Otang Ntui*,[2] *Dong Poh Chin*[2] and *Masahiro Mii*[2,*]

ABSTRACT

Lilies (species and cultivars of the genus *Lilium*) are economically important ornamental plants because of their large and attractive flowers. The establishment of an efficient transformation system of lilies is desired for their genetic improvement with respect to ornamental and marketable qualities such as novel flower color and plant form as well as benefits for commercial cultivation such as virus resistance and environmental stress tolerance. In *Lilium*, genetic transformation has been achieved by electroporation, and biolistic and *Agrobacterium*-mediated transformation. However, the transformation efficiency varies according to the methods, species or cultivars. In the initial trials, only the biolistic method was used for *Lilium* genetic transformation, but eventually, *Agrobacterium*-mediated method has been incorporated as a

[1] Tissue Culture and Gene Transformation Department, Agricultural Biotechnology Research Institute of Iran (ABRII), P. O. Box 31535-1897. Karaj, Iran.

[2] Laboratory of Plant Cell Technology, Graduate School of Horticulture, Chiba University, 648 Matsudo, Matsudo City, Chiba 271-8510, Japan.

* Corresponding authors: azadip22@gmail.com; mim@faculty.chiba-u.jp

successful approach for the genetic transformation of lilies in addition to a large number of monocots with better understanding of the process of monocot transformation, and availability of superior constructs, strains and vectors. This article summarizes the reported gene-transfer methods for *Lilium*, with an emphasis on the recent progress in introducing new traits into lily species in our laboratory.

Keywords: *Agrobacterium tumefaciens*, Biolistic, Electroporation, Gene transformation, *Lilium*, Metabolic engineering

Introduction

Lilium is a monocotyledonous genus of Liliaceae containing many species and cultivars with high ornamental values. Lilies are in high demand in the floriculture industry both as cut and potted flowers and are ranked amongst the top 10 flowers in the export market. The turnover of lilies in FloraHolland flower auction in 2008 was 159 million Euros with a ranking of 4th among cut flowers (FloraHolland Key figures 2008). The genus *Lilium* comprises more than 100 species (McRae 1998) and is classified into six sections (Smyth et al. 1989). Lilies are indigenous to Europe (c. 12 species), Asia (50–60 species) and North America (c. 24 species), while some are native to the tropics at higher elevations (Beattie and Whittle 1993). The three commercially important divisions of lily include Easter lily (*L. longiflorum*), Asiatic hybrid and Oriental hybrid.

Although major efforts in lily transformation have been directed toward the introduction of foreign genes for transient expression by direct gene delivery, methods such as particle bombardment (Nishihara et al. 1993, Wilmink et al. 1995, Tsuchiya et al. 1996, Watad et al. 1998, Lipsky et al. 2002, Irifune et al. 2003, Ahn et al. 2004, Cohen et al. 2004, Kamo and Han 2008) and electroporation (Miyoshi et al. 1995), the possibility of using *Agrobacterium* for *Lilium* transformation was first reported by Cohen and Meredith (1992). They co-cultured bulb-scale sections of *L. longiflorum* 'Nellie White' with *A. tumefaciens*-C58, which induced a small tumor-like swelling. Analysis of the callus homogenates revealed the expression of T-DNA, which encodes a gene for nopaline synthesis, indicating transformation of the *Lilium* tissue. They demonstrated that *L. longiflorum* 'Nellie White' is a host of *Agrobacterium* based on the formation of tumors after the inoculation. A few years later, Langeveld et al. (1995) also reported that the lily 'Harmony' was confirmed to be a host of *Agrobacterium* after the inoculation of stem internodes with the wild-type C58 strain. Compared to particle bombardment, *Agrobacterium* mediated transformation usually has a higher efficiency and results in a greater proportion of stable, low copy number of transgene, facilitating the stable inheritance of the transgene (Travella et al. 2005, Zhang et al. 2005).

Here we review the studies on genetic transformation of lily (Table 1), and present the achievements in our laboratory on *Lilium* transformation with improved transformation efficiency. Also, we describe metabolic engineering of *Lilium* using our improved protocol.

Gene Transformation System

Plant genetic transformation is performed using a wide range of methods such as *Agrobacterium*-mediated transformation, particle bombardment (biolistic), and microinjection, chemical (PEG) treatment and electroporation of protoplasts. Transformation using *Agrobacterium* and biolistic are currently the most extensively used methods.

Electroporation

Electroporation has been used to transfer genes into the pollen of lily to characterize the promoters of different genes (Tabata et al. 1993, Miyoshi et al. 1995, Pelkonen 2005). Using lily pollen protoplasts, Miyoshi et al. (1995) reported the direct transfer of genes into pollen by electroporation. Only pollen protoplasts electroporated with the pollen-specific Zml3 and LAT52 promoters showed apparent expression of GUS. The authors, however, reported that GUS activity in lily pollen protoplasts that had been electroporated with the Zml3-GUS construct was far higher, about 800-fold, than in the case of the LAT52-GUS construct. The frequency of blue-staining pollen protoplasts was about 70% of the total protoplasts electroporated with GUS gene driven by the Zml3 promoter (Miyoshi et al. 1995). Consequently, they indicated that the Zml3 promoter, derived from monocotyledonous maize, was more active in the monocotyledonous lily pollen than the LAT52 promoter from the dicotyledonous tomato. Although the coding region of the Zml3 gene from maize includes sequences homologous to the LAT52 gene from tomato, the respective upstream promoter regions are very different (Hamilton et al. 1989, McCormick 1991). Also, it was reported that the LAT52 promoter was more active in dicotyledonous tobacco pollen than in monocotyledonous lily pollen (Plegt et al. 1992, Nishihara et al. 1993). Although plant regeneration from protoplasts has been reported in *L.* × *formolongi* (Mii et al. 1994) and somatic hybrids were successfully produced between Oriental hybrid lilies ('Acapulco' and 'Shirotae') and *L.* × *formolongi* 'Hakucho' (Horita et al. 2003), there are no reports on genetic transformation by electroporation in these cultivars.

Table 1. Genetic transformation studies in *Lilium* spp.

Species	Explant	Co-cultivation medium	Gene	Method, Strain-plasmid	Reference
Lilium longiflorum 'Nellie White'	Calli from bulb scale	LS	GUS	A.t. C58/pTiC58 etc.	Cohen and Meredith 1992
L. longiflorum 'Gelria'	Pollen	—	GUS	Biolistic pCPO1.2'GUS etc.	Plegt et al. 1992
L. longiflorum, *L. speciosum*	Protoplast from calli and young anther, microsporocytes	—	GUS	Elecroporation pBI221 etc.	Tabata et al. 1993
L. longiflorum 'Georgia'	Pollen	—	GUS	Biolistic pLAT52-7, pBI221	Nishihara et al. 1993
L. longiflorum 'Harmony'	Stem node	—	GUS	A.t. C58/pTiC58/ p35SGUSINT etc.	Langeveld et al.1995
L. longiflorum 'Georgia'	Pollen protoplast	—	GUS	Elecroporation pZM13 etc.	Miyoshi et al. 1995
L. longiflorum 'Snow Queen'	Leaf	—	GUS	Biolistic pActl-F etc.	Wilmink et al. 1995
L. × formolongi 'Kitadake'; *L. dauricum; L. japonicum*	Bulb scales and immature embryos	—	GUS	Biolistic pREX-GUS etc.	Tsuchiya et al. 1996
L. longiflorum 'Snow Queen'	Bulblet scale-derived calluses	—	GUS, PAT	Biolistic pPHP6517, pPHP3953	Watad et al. 1998
Lilium formosanum	Calli from Bulbscale-derived	MS	GUS, NPTII, HPT	A.t. EHA101/pIG121-Hm	Suzuki et al. 1998
L. longiflorum 'Snow Queen'	Calli from Bulbscale-derived	—	GUS, PAT CMV	Biolistic pUBQ3genGUS, p35Sac, pSAN101	Lipsky et al. 2002

L. longiflorum 'Snow Queen'	Friable embryogenic calli	—	rolA+B+C	A.t. LBA4404/pBin19	Mercuri et al. 2003
L. × *fomolongi* 'Miyazono'	Calli from bulb scale	—	GUS, PAT	Biolistic pAct1-F, pDM302	Irifune et al. 2003
L. longiflorum cv. Osnat	Calli from bulb scale	—	GUS, NPTII	Biolistic pUBQ3genGUS, pCAMBIA2301	Cohen et al. 2004
L. longiflorum (Nellie White×Ace)	Calli from bulb scale	—	GUS, PAT, CMV	Biolistic pDM302 etc.	Ahn et al. 2004
Lilium 'Acapulco'	Filament–deriverd calli	NH4NO3-free MS medium	GUS, NPTII, HPT	A.t. EHA101/pIG121-Hm	Hoshi et al. 2004
L. longiflorum 'Georgia'	Filament-derived calluses	NH4NO3-free MS medium	GUS, NPTII, HPT	A.t. EHA101/pIG121-Hm	Hoshi et al. 2005
L. longiflorum	Bulblet scales	—	GUS, SUP and PAT	Biolistic pB G5, pB- SUP	Benedito et al. 2005
Lilium davidii 'Duchartre'	Pollen	1/10 MS	GUS, NPTII, HPT	A.t. EHA101/pIG121Hm LBA4404/pTOK233; etc.	Dong et al. 2007
L. longiflorum 'Georgia'	Pollen	PGM	GUS, NPTII	A.t. LBA4404/pBI121	Kim et al. 2007
Lilium longiflorum × *L. formosanum*	Bulb scales	NH4NO3-free MS medium	Zm401, NPTII	A.t. LBA4404/pBI121	Qiu-Hua et al. 2008
L. × *fomolongi* 'Akasu'	Meristematic nodular calli	MS	GUS, NPTII, HPT	A.t. EHA101/pIG121-Hm	Ogaki et al. 2008
L. longiflorum 'Nellie White'	Compact calli from bulb scale	—	GUS, bar, NPTII	Biolistic pDM327, pCAMBIA2301	Kamo and Han 2008

Table 1. contd....

Table 1. contd.

Species	Explant	Co-cultivation medium	Gene	Method, Strain-plasmid	Reference
Lilium longiflorum 'Tiepao'	Direct and indirect shoot from nodal stem	MS	GUS, NPTII	A.t. AGL1/pCAS04	Liu et al. 2011
L. × formolongi; L. longiflorum 'Acapulco'	Meristematic nodular calli	Modified MS	GUS, NPTII, HPT	A.t. EHA101/pIG121-Hm	Azadi et al. 2010a
L.× formolongi	Meristematic nodular calli	Modified MS	7 enzyme gene (carotenoid pathway), HPT	A.t. EHA101/pCrtZW-N8idi-crtEBIY	Azadi et al. 2010b
L. longiflorum 'Acapulco'	Meristematic nodular calli	Modified MS	CMV2-GDD, NPTII, HPT	A.t. EHA101/pEKH2-gdd	Azadi et al. 2011
Lanzhou lily (*Lilium davidii* var. unicolor)	scale leaves	—	GLDH, GUS, HPT	Biolistic pCAMBIA-GLDH	Shi et al. 2012
Multiple lily cultivars	Calli from filaments, styles and bulb scales	NH4NO3-free MS medium	GUS, HPT	A.t. AGL0/pCAMBIA1301 + GUS; pMF2 + GUS	Wang et al. 2012

LS: Linsmaier and Skoog
A.t.: *Agrobacterium tumefaciens*
PAT: phosphinothricin-*N*-acetyltransferase
SUP: SUPERMAN is a zinc-finger transcription factor
GLDH: L-galactono-1, 4-lactone dehydrogenase
ZM401: a pollen-specific gene of maize

Biolistic-mediated transformation

Another important method in plant genetic transformation is the biolistic (particle bombardment). It is mostly used to transform non-host *Agrobacterium* species, such as monocots and conifers, and other recalcitrant species. This method can be applied to many different tissues, but as with other transformation methods, the genotypic make up of the plant strongly affects the transformation efficiency (Aronen et al. 1995, Häggman and Aronen 1998). Technically, gold or tungsten particles coated with objective gene constructs are introduced into the cells of the target plant tissue by shooting with high pressure. After the bombardment, the tissues can be cultured normally in tissue culture media with a suitable selection procedure. As a selective marker, antibiotic or herbicide resistance genes are commonly used. In addition, some reporter genes, such as β-glucuronidase gene (GUS), can be used to confirm successful transformation (Jefferson et al. 1987). This method has also been successfully applied to some lily species to achieve transient and stable expression of reporter gene in pollen, bulb scales, callus and immature embryos (Nishihara et al. 1993, Tsuchiya et al. 1996, Watad et al. 1998). In lily, the highest frequency of GUS expressing pollen was 0.74% when mature pollen grains were bombarded with pLAT52-7 (Nishihara et al. 1993).

Wilmink et al. (1995) evaluated the activities of several promoters isolated from monocot species such as maize *Adhl-based pEmu*, rice *Actl* and maize *Ubi* promoters, coupled to the β-glucuronidase (*gus*) reporter gene by transient GUS gene expression upon particle bombardment of tissues of tobacco, rice, tulip, lily and leek. The results indicated that activities of these monocot promoters were much lower in the Liliaceous plants including lily than in rice. Although monocot promoters generally did not significantly exceed the activity of the CaMV 35S promoter, the *Actl* and *Ubil* promoters resulted in higher transient expression than the 35S promoter in lily leaves.

For efficient selection of cells expressing transferred genes following particle bombardment, the effects of different promoters on the transient expression of the β-glucuronidase gene (*gusA*) were estimated in bulb scales and immature embryos of lily (Tsuchiya et al. 1996). When four plasmids having the *gusA* gene under the control of different promoters were introduced by particle bombardment, the patterns of transient expression in the bulb scales showed differences among three *Lilium* species, *L.* × *formolongi*, *L. dauricum* and *L. japonicum*. A high level of transient expression in lily tissues was observed when the rice *Act1* promoter and the modified 35S promoter with an intron of the phaseolin gene (pREX-GUS) were used.

Transgenic lily plants produced by bioloistic method were reported by Lipsky et al. (2002) who showed the recovery of plants after bombardment of morphogenic primary callus of *L. longiflorum* 'Snow Queen' with plasmids containing *uidA* reporter and *PAT* selectable marker genes, and the plasmid pSAN101 containing the CMV defective replicase gene, followed by selection with bialaphos. Irifune et al. (2003) also successfully produced transgenic bialaphos-resistant plants of *L.* × *formolongi* displaying PAT activity using *bar* gene under the control of rice actin promoter (pDM302 plasmid), after optimization of gene delivery condition in calli and bombardment with pAct1-F-coated gold. The yield of transgenic plants was 2.6% approximately. Transgenic *L. longiflorum* plants, transformed by biolistic method using a vector containing *SUP* gene (SUPERMAN is a zinc-finger transcription factor involved in flower development) driven by the petunia FBP1 promoter and *PAT* gene under the control of CaMV35S promoter showed *in vitro* resistance to herbicide in the culture but no alteration for flower morphology was observed compared to wild type plants (Benedito et al. 2005).

Kamo and Han (2008) reported a successful and efficient protocol for lily transformation. At first, they developed a reliable selection system by subjecting non-transformed calli and *in vitro* plants to different selective agents such as phosphinothricin, kanamycin, geneticin, paromomycin, and hygromycin on MS medium. The non-transformed calli were highly sensitive to 0.5 mgL^{-1} phosphinothricin, 25 mgL^{-1} geneticin, and 5 mgL^{-1} hygromycin. Based on their sensitivity results, they bombarded callus with either the *bar-uidA* fusion gene under the control of CaMV 35S promoter or *npt II* or *uidA* under the control of CaMV 35S promoter, and obtained transgenic 'Nellie White' fertile plants. When the transgenic plants were raised to T1, 41% of the progeny showed stable integration of the *bar* gene. Expression of the *uidA* gene was demonstrated in leaves of transformed plant lines by specific activity of GUS and the presence of the RNA transcript (Kamo and Han 2008).

Agrobacterium-mediated Transformation

Characteristics of *Agrobacterium*-mediated transformation in monocots

Compared to biolistic techniques, *Agrobacterium*-mediated transformation generally offers potentially a low copy number of the transgene (Hiei et al. 1994). It had generally been recognized that multiple copies of the transgene with tandem or inverted repeats easily induce DNA methylation and trigger homology-dependent gene silencing (Matzke and Matzke 1995, Spiker and Thompson 1996). Because of these reasons, *Agrobacterium*-mediated transformation is well established in dicotyledonous plants since most of them are susceptible to the species of this bacterial genus, and a variety of transgenic plants with different uses have been produced. However, *Agrobacterium*-mediated transformation of monocotyledonous plants was extremely difficult, and reliable transformation methods were initially absent since monocots are not the natural hosts of *A. tumefaciens* (De Cleene and De Ley 1976). Consequently, only the direct delivery methods were used for monocots and the first transgenic rice, maize and wheat were produced by the biolistic methods (Gordon-Kamm et al. 1990, Christou et al. 1991, Vasil et al. 1992). With better understanding of the mechanisms involved in the transformation of plants with *Agrobacterium* infection, the first successful *Agrobacterium*-mediated transformation method in monocots was developed by Hiei et al. (1994) in rice with the use of acetosyringon for inducing the infection. Moreover, with the development of superior bacterial strains and vector constructs, *Agrobacterium*-mediated method has gained popularity in monocots and a large number of transgenic plants have successfully been produced in monocots and other non-natural host plants (Sood et al. 2011). In ornamental Liliaceous plants, *Chlorophytum capense* was the first species in the Liliaceae reported to be susceptible to *Agrobacterium tumefaciens* infection (Hooykaas-Van Slogteren et al. 1984), and transgenic plants have already been produced in the species of some genera such as *Agapanthus* (Suzuki et al. 2001), *Muscari* (Suzuki and Nakano 2002), *Ornithogalum* (van Emmenes et al. 2008) and *Tricyrtis* (Adachi et al. 2005).

Transformation of Somatic Tissues of *Lilium*

Genetic transformation of *Lilium* through *Agrobacterium*-mediated method was first examined by Langeveld et al. (1995), who tested the effect of three different *Agrobacterium* strains (*A. tumefaciens* strain C58, and *A. rhizogenes* strains 1855 and ATCC 15834) on the transformation of 'Harmony'. Consequently, only the nopaline strain C58 was able to transform stem node tissues of Harmony. Suzuki et al. (1998) also tried to transform *L. formosanum* using *A. tumefaciens* strain EHA101/pIG121Hm by examining the effect of several selective agents. Although transient expression of the *gus* gene in co-cultivated calli was observed, neither transformed tissues nor plants were obtained.

Successful result on the production of transgenic lily plants by *Agrobacterium*-mediated method was first reported by Mercuri et al. (2003). In this study, *L. longiflorum* 'Snow Queen' was transformed (0.8% efficiency) with *rol A, B* and *C* genes by infecting friable embryogenic calli with *A. tumefaciens* strain LBA4404, and co-cultured for seven days after infection. The *rol ABC* transformants displayed increased rooting ability with root branching. Some of the transgenic lilies exhibited a bushy and disorderly habit with six or seven stems, whereas others presented a desirable aspect with one or two stems and a severe reduction in internode length. Although some transgenic plants flowered earlier, this was not a general trend. These new phenotypes may be considered as innovative products. On the other hand, Hoshi et al. (2004) found that a combined treatment with scratching the calli with sandpaper prior to inoculation and using NH_4NO_3-free medium for co-cultivation of the Oriental hybrid lily 'Acapulco' gave 3% transformation efficiency. This method was also confirmed to be effective for the transformation of *L. longiflorum* 'Georgia' (Hoshi et al. 2005). Using NH_4NO_3-free medium for co-cultivation also resulted in a successful integration of *Zm401* gene (a maize pollen-specific gene) by *Agrobacterium* strain LBA4404 containing binary vector pBI121 in *L. longiflorum* × *L. formosanum* (Qiu-Hua et al. 2008). Liu et al. (2011) reported a high rate of *L. longiflorum* transformation via *Agrobacterium* for direct and indirect shoot regeneration from the nodal stem explants using MS salts and vitamins with 49.1 mg/L acetosyringone (AS) as co cultivation medium. However, most recent report using NH_4NO_3-free medium for co-cultivation in eight lily cultivars transformed with *A. tumefaciens* strain AGL0 carrying binary vectors pCAMBIA1301+GUS or pMF2 + GUS resulted in a low rate of transformation (1.4% as the maximum) (Wang et al. 2012). This could be due to the differences in genotypes used.

Under these circumstances, we tried to determine the reasons for the difficulty to transform lilies by *Agrobacterium*-mediated method. We used meristematic nodular callus culture of *L.* × *formolongi* 'Akasu', which was established using the method described previously (Mii et al. 1994). In this lily transformation, we found a drastic drop of pH in the medium and inhibition of *Agrobacterium* growth probably through the unknown substances secreted from the calli, suggesting that the low pH was one of the factors militating against *Agrobacterium* transformation of *Lilium* (Ogaki et al. 2008). Therefore, the effects of 2-(*N*-morpholino) ethanesulfonic acid (MES) on the maintenance of medium pH and transformation efficiency were examined by adding it at different concentrations to both incubation and co-cultivation medium as a buffer agent. As a result, both transient GUS expression and hygromycin-resistant transgenic calli were only observed when MES was applied in these media. The highest frequency of transient GUS expression was obtained at 10 mM, which favored the slight overgrowth of *Agrobacterium*. Consequently, transgenic plants were efficiently obtained by the addition of MES to the infection and co-cultivation medium (Ogaki et al. 2008).

Further improvement of lily transformation was achieved by modifying the composition of the media for *Agrobacterium* inoculation and co-cultivation (Azadi et al. 2010a). In this study, the effect of three types of modified MS media used for inoculation and co-cultivation on transformation was assessed by inoculating calli of *L.* × *formolongi* with *A. tumefaciens*. In full strength medium, GUS expression was observed at low level and no hygromycin-resistant callus was obtained after 4 months of culture on selection medium. Interestingly, a dramatic increase in transient GUS expression (72.5 GUS spots per 100 mg calli) and the highest frequency of transformation (25.4%) was observed when Murashige and Skoog (MS) medium lacking KH_2PO_4, NH_4NO_3, KNO_3, $CaCl_2$ was used (Fig. 1). Using the medium lacking all elements (containing only distilled water, sucrose and agar) decreased the number of GUS spots and the transformation efficiency compared to the medium lacking four macro elements (Azadi et al. 2010a). A favorable effect of this medium was also observed in three *Lilium* cultivars 'Acapulco', 'Casa Blanca' and 'Red Ruby', all of which showed significant increase in transient GUS expression. Using full component MS medium, neither GUS expression nor hygromycin-resistant calli production was observed in 'Acapulco'. The highest number of GUS spots (100.5 per 100 mg calli) and the highest frequency of hygromycin-resistant calli (23.3%) were scored in 'Acapulco' when the medium lacking KH_2PO_4, NH_4NO_3, KNO_3 and $CaCl_2$, was used for inoculation and co-cultivation. These values are approximately eight times higher than the result reported by Hoshi et al. (2004) for the same cultivar using NH_4NO_3-free medium. The visible proliferation of *Agrobacterium* cells around the calli during co-cultivation

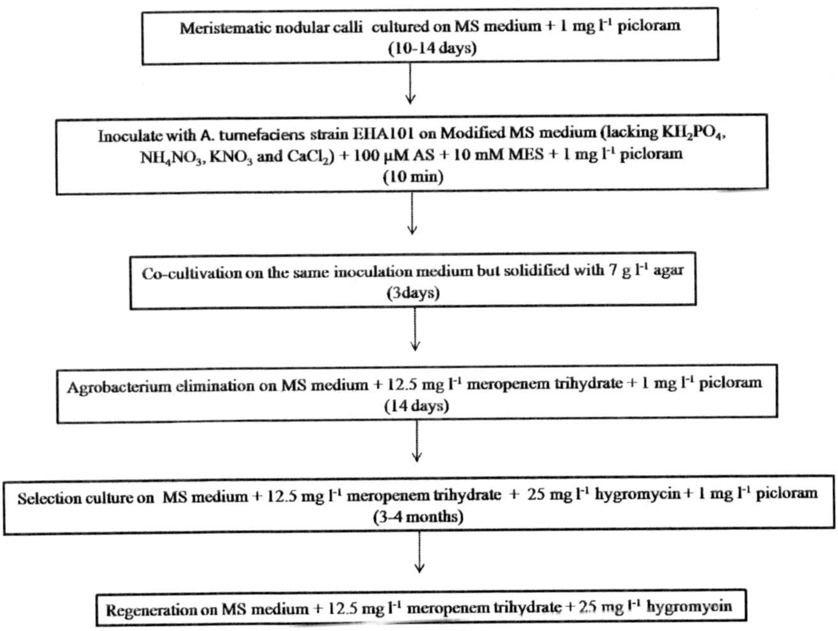

Fig. 1. The complete Protocol for *Agrobacterium*–mediated transformation of *Lilium* × *formolongi* 'Akasu'.

increased in all three cultivars when the medium lacking four macro elements was used. No appreciable change was observed in the calli by bacterial overgrowth in 'Acapulco' and *L.* × *formolongi*, whereas more than 50% of the calli showed partial necrosis in 'Red Ruby', and browning in 'Casa Blanca', respectively. These results suggest that they might be more sensitive to the bacterial overgrowth, resulting in the death of tissues including transformed cells. The amount of *Agrobacterium* that proliferated over and around the calli during co-cultivation increased as some elements were removed, suggesting some physiological changes of the calli caused by removing some macro elements, which gave the inhibitory effect on proliferation of *Agrobacterium*. The results also indicate that the period of co-cultivation should be optimized for sensitive cultivars such as 'Casa Blanca' and 'Red Ruby' in order to prevent bacterial overgrowth.

Transformation of Pollen with *Agrobacterium*-mediated Method

For the genetic transformation of plants, pollen has been considered as an important target material since it can be efficiently used to obtain transgenic progenies through fertilization events (Aziz and Machray 2003, Hess and Dressler 1989, Morikawa and Nishihara 1997). Other than biolistic and electroporation methods as described above, transformation of lily pollen has also been tried with *Agrobacterium*-mediated methods. Transient β-glucuronidase expression in lily (*L. longiflorum* L.) pollen via wounding-assisted *Agrobacterium*-mediated transformation has been reported (Kim et al. 2007). Physical wounding to lily pollen grains prior to the agroinfiltration process is considered a useful option for successful lily pollen transformation according to the analysis of histochemical and fluorometric assay for GUS enzyme activity and of cDNA blot hybridization for GUS mRNA detection.

In pollen transformation of David lily (*L. davidii*), Dong et al. (2007) found that only germinating pollen were efficiently transformed with two *Agrobacterium* strains, whereas no successful result was obtained in the case of non-germinating pollen. In this study, they obtained extremely high transformation efficiency (92.7%) by optimizing medium composition and concentration, combination of bacterial strains and plasmids, *Agrobacterium* cell density and duration of co-cultivation. For example, 10 times diluted MS medium was more suitable for the co-cultivation medium compared with the other dilution of MS medium and LB medium. Despite these pollen transformation studies, there have been no successful results on the production of transgenic plants using transformed pollen.

Introduction of Useful Genes into *Lilium*

Using the efficient protocol developed by Azadi et al. (2010a), a cultivar of *L. × formolongi* was genetically engineered for ketocarotenoid biosynthesis by transformation with the plasmid pCrtZW-N8idi-crtEBIY containing seven key enzyme genes under the regulation of 35S CaMV promoter. Probably due to the large size of the vector, the very low efficiency of transformation (0.25%) was observed. In the transformants, ketocarotenoids were detected in both calli and leaves, which showed a strong orange colour. In transgenic calli, the total amount of carotenoids (133.3 μg/g fresh weight) was 26.1-fold higher than in the wild-type calli. Unexpectedly, after several subcultures, green leaves started appearing from some of the six months old plantlets. By continuing the subculture and multiplication of the transgenic orange plantlets, new green bulblets emerged nine months after transfer to regeneration medium, which later produced completely green plantlets. HPLC-PDA analysis was used to determine the carotenoid profiles of the calli and leaves in wild type and transgenic *L. × formolongi* (orange and green plantlets). The leaves of transgenic green plantlets were collected from the first and second generations of micropropagated plantlets.

The wild type calli contained only phytoene, lutein, β-carotene and α-carotene, whereas the transgenic calli showed some novel peaks such as β-cryptoxanthin, lycopene, echinenone, canthaxanthin, 3′-hydroxyechinenone, 3-hydroxyechinenone and astaxanthin beside these compounds. Interestingly, in the leaves of transgenic green plantlets, only five novel peaks were detected. Astaxanthin, an important ketocarotenoid, was not detected in the leaves of transgenic green plantlets. Also, photosynthetic efficiency of the transgenic orange leaves was significantly low compared to that of the transgenic green leaves, which was however slightly lower than the wild type. The bulbs of transgenic lily plants are now in growth chamber (Fig. 2), and it is expected to bear flowers with orange to yellow colorations (Azadi et al. 2010b).

Using the same protocol, a cultivar of Oriental Hybrid, 'Acapulco' was transformed with *A. tumefaciens* strain EHA101 containing pEKH2-gdd which harbored a defective cucumber mosaic virus (CMV) replicase gene (CMV2-GDD) construct. Of the 300 cultured calli of 'Acapulco', 110 Hm resistant calli were generated. A high efficiency of transformation (36%) was observed using the plasmid pEKH2-gdd. Four lines were analyzed for gene expression and resistance to CMV-O strain. Expression of the CMV2-GDD gene in the transgenic plants was confirmed by reverse transcription PCR (RT-PCR). When these four lines were mechanically inoculated with CMV-O, no signal of coat protein (CP) messages was detected in the newly produced leaves of two transgenic lines. Dot-immunobinding assay (DIBA)

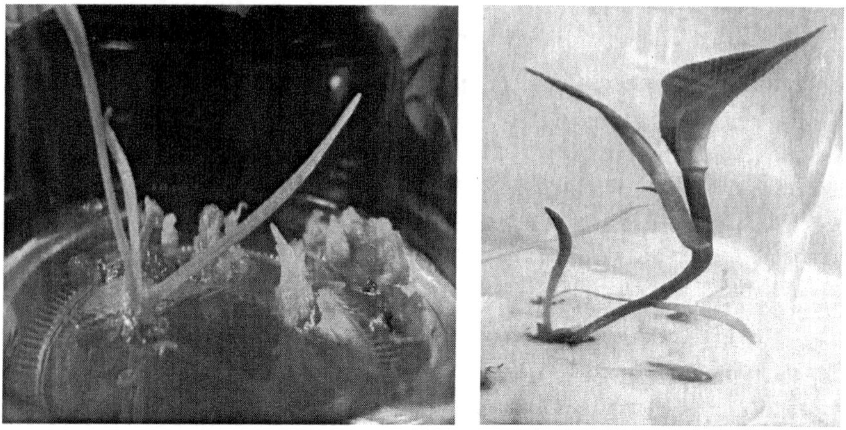

Fig. 2. Production of transgenic orange calli and regeneration of transgenic orange plants and expression of carotenoid gene cassettes in *Lilium* × *formolongi* (left). Transgenic plant from meristematic nodular calli 15 months after transfer to regeneration medium (right).

Color image of this figure appears in the color plate section at the end of the book.

of CP was performed to examine the presence of the CMV in the newly produced leaves of the challenged plants. Results obtained were similar to those observed in RT-PCR of the CP messages. This implies that the two lines showed increased levels of resistance to CMV, and that CMV-GDD replicase gene is an effective construct that has protection against CMV in *Lilium* (Azadi et al. 2011).

To increase the levels of vitamin C in Lanzhou lilies (*Lilium davidii* var. *unicolor*) which is an edible lily, genetic transformation mediated by particle bombardment was carried out in this lily variety. Using the vector pCAMBIA-GLDH, containing the *L-galactono-1, 4- lactone dehydrogenase* gene from apple, the levels of reduced ascorbate and total ascorbate (reduced ascorbate+oxidized ascorbate) in transgenic plants were increased two to seven times compared to non-transgenic controls. The efficiency of transformation was about 1.3% in this study (Shi et al. 2012).

Conclusion

Genetic transformation systems of *Lilium* has successfully been established through biolistic as well as *Agrobacterium*-mediated methods during the last ten years, and several useful transgenic plants have also been produced as described above. However, these transformation systems have been applied to limited genotypes and are still difficult to apply to a wide range of lily species and genotypes. Therefore, transformation of lilies by *Agrobacterium*-

mediated method is still regarded as a difficult task. It is therefore necessary to develop more widely applicable methods by elucidating the factors involved in the transformation of recalcitrant genotypes such as 'Casa Blanca' (Azadi et al. 2010a).

References

Adachi, Y., S. Mori and M. Nakano. 2005. *Agrobacterium*-mediated production of transgenic plants in *Tricyrtis hirta* (Liliaceae). Acta Hort. 673: 415–419.

Ahn, B.J., Y.H. Joung and K.K. Kamo. 2004. Transgenic plants of Easter lily (*Lilium longiflorum*) with phosphinothricin resistance. J. Plant. Biotech. 6: 9–13.

Aronen, T., A. Hohtola, H. Laukkanen and H. Häggman. 1995. Seasonal changes in the transient expression of a 35S CaMV-GUS gene construct introduced into Scots pine buds. Tree Physiol. 15: 65–70.

Azadi, P., D.P. Chin, K. Kuroda, R.S. Khan and M. Mii. 2010a. Macro elements in inoculation and co-cultivation medium strongly affect the efficiency of *Agrobacterium*-mediated transformation in *Lilium*. Plant Cell Tissue Organ. Cult. 101: 201–209.

Azadi, P., V.O. Ntui, D.P. Chin, I. Nakamura, M. Fujisawa, H. Harada, N. Misawa and M. Mii. 2010b. Metabolic engineering of *Lilium* × *formolongi* using multiple genes of the carotenoid biosynthesis pathway. Plant Biotechnology Reports 4: 269–280.

Azadi, P., O.N. Valentaine, H. Supaporn, S.K. Raham, D.P. Chin, I. Nakamura and M. Mii. 2011. Increased resistance to cucumber mosaic virus (CMV) in *Lilium* transformed with a defective CMV replicase gene. Biotechnol. Lett. 33: 1249–1255.

Aziz, N. and G.C. Machray. 2003. Efficient male germ line transformation for transgenic tobacco production without selection. Plant Molecular Biology 51: 203–211.

Beattie, R.B. and M.J. Whittle. 1993. Doppler and fetal growth retardation. Arch. Dis. Child 69: 271–273.

Benedito, V.A., B.C.E. Kronenburg-van der Ven, J.M. Tuyl, G.C. Angenent and F.A. Krens. 2005. Transformation of *Lilium longiflorum* via particle bombardment and generation of herbicide resistant plants. Crop Breed Appl. Biotechnol. 5: 259–264.

Cheng, M., B. Lowe, T. Michael, M. Spencer, X. Ye and C. Armstrong. 2004. Factors influencing *Agrobacterium*-mediated transformation of monocotyledonous species. *In vitro* Cell Dev. Biol.-Plant. 40: 31–45.

Christou, P., T.L. Ford and M. Kofron. 1991. Production of transgenic rice (*Oryza sativa* L.) plants from agronomically important indica and japonica varieties via electric discharge particle acceleration of exogenous DNA into immature zygotic embryos. Bio. Technology 9: 957–962.

Cohen, A., A. Lipsky, T. Arazi, A. Ion, R. Stav, D. Sandler-Ziv, C. Pintea, R. Barg, Y. Salts, S. Shabtai, V. Gaba and A. Gera. 2004. Efficient genetic transformation of *Lilium longiflorum* and *Ornithogalum dubium* by particle acceleration followed by prolonged selection in liquid medium. Acta Hort. 651: 131–138.

Cohen, A. and C.O. Meredith. 1992. *Agrobacterium*-mediated transformation of *Lilium*. Acta Hort. 325: 611–618.

De Cleene, M. and J. De Ley. 1976. The host range of crown gall. Bot. Rev. 42: 389–466.

Dong, W., Y.F. Mao and W. Li. 2007. Factors influencing T-DNA transferring into pollen of lily *in vitro*. Russian Journal of Plant Physiology 54: 420–425.

Gordon-Kamm, W.J., T.M. Spencer, M.L. Mangano et al. 1990. Transformation of maize cells and regeneration of fertile transgenic plants. The Plant Cell 2: 603–618.

Häggman, H. and T. Aronen. 1998. Transgene expression in regenerating cotyledons and embryogenic cultures of Scots pine. J. Exp. Bot. 49: 1147–1156.

Hamilton, D.A., D.M. Bashe, J.R. Stinson and J.P. Mascarenhas. 1989. Characterization of a pollen-specific genomic clone from maize. Sex Plant Reprod. 2: 208–212.

Hess, D. and K. Dressler. 1989. Tumor transformation of *Petunia hybrida* via pollen culture with *Agrobacterium tumefaciens*. Bot. Acta 102: 202–207.

Hiei, Y., S. Ohta, T. Komari and T. Kumashiro. 1994. Efficient transformation of rice (*Oryza sativa* L.) mediated by *Agrobacterium* and sequence analysis of the boundaries of the T-DNA. Plant J. 6: 271–282.

Hooykaas-VanSlogteren, G.M.S., P.J.J. Hooykaas and R.A. Schilperoort. 1984. Expression of Ti plasmid genes in monocotyledonous plants infected with *Agrobacterium*. Nature 311: 763–764.

Horita, M., H. Morohashi and F. Komai. 2003. Production of fertile somatic hybrid plants between Oriental hybrid lily and *Lilium x formolongi*. Planta 217: 597–601.

Hoshi, Y., M. Kondo, H. Kobayashi, S. Mori and M. Nakano. 2005. *Agrobacterium*- mediated transformation of *Lilium longiflorum*. Acta Hort. 673: 543–547.

Hoshi, Y., M. Kondo, S. Mori, Y. Adachi, M. Nakano and H. Kobayashi. 2004. Production of transgenic lily plants by *Agrobacterium*-mediated transformation. Plant Cell Rep. 22: 359–364.

Irifune, K., Y. Morimoto and M. Uchihama. 2003. Production of herbicide resistant transgenic lily plants by particle bombardment. J. Jpn. Soc. Hor. Sci. 72: 511–516.

Jefferson, R.A., T.A. Kavanagh and M.W. Bevan. 1987. GUS-fusions: glucuronidase as a sensitive and versatile gene fusion marker in higher plants. EMBO J. 6: 3901–3907.

Kamo, K. and B.H. Han. 2008. Biolistic-mediated Transformation of *Lilium longifiorum* cv. Nellie White. Hort. Science 43: 1864–1869.

Kim, S., D. Shin and H.S. Park. 2007. Transient β-glucuronidase expression in lily (*Lilium longiflorum* L.) pollen via wounding-assisted *Agrobacterium*-mediated transformation. Biotechnol. Lett. 29: 965–969.

Langeveld, S.A., M.M. Gerrits, A.F.L.M. Derks, P.M. Boonekamp and J.F. Bol. 1995. Transformation of lily by *Agrobacterium*. Euphytica 85: 97–100.

Lipsky, A., A. Cohen, V. Gaba, K. Kamo, A. Gera and A. Watad. 2002. Transformation of *Lilium longiflorum* plants for cucumber mosaic virus resistance by particle bombardment. Acta Hort. 568: 209–214.

Liu, J., J. Zhang, B. Xu, C. Jia, J. Zhang, G. Tan and Z. Jin. 2011. Regeneration and production of transgenic *Lilium longiflorum* via *Agrobacterium tumefaciens*. *In vitro* Cell Dev. Biol. Plant 47: 348–356.

Matzke, M.A. and A.J.M. Matzke. 1995. How and why do plants inactivate homologous (trans) genes ? Plant Physiol. 107: 679–685.

McCormick, S. 1991. Molecular analysis of male gametogenesis in plants. Trends Genet. 7: 298–303.

McRae, E.A. 1998. Lilies. A Guide for growers and collectors. Timber Press, Portland, Oregon. USA. 392 pp.

Mercuri, A., L.D. Benedetti, S. Bruna, R. Bregliano, C. Bianchini, G. Foglia and T. Schiva. 2003. *Agrobacterium*-mediated transformation with rol genes of *Lilium longiflorum* Thunb. Acta Hort. 612: 129–136.

Mii, M., Y. Yuzawa, H. Suetomi, T. Motegi and T. Godo. 1994. Fertile plant regeneration from protoplasts of a seed-propagated cultivar of *Lilium* × *formolongi* by utilizing meristematic nodular cell clumps. Plant Sci. 100: 221–226.

Miyoshi, H., T. Usami and I. Tanaka. 1995. High level of GUS gene expression driven by pollen-specific promoters in electroporated lily pollen protoplasts. Sex Plant Rep. 8: 205–209.

Morikawa, H. and M. Nishihara. 1997. Use of pollen in gene transfer. *In:* K.R. Shivanna and V.K. Sawhney (eds.). Pollen Biotechnology for Crop Production and Improvement. Cambridge University Press, Cambridge, pp. 423–437.

Nishihara, M., M. Ito , I. Tanaka, M. Kyo, K. Ono, K. Irifune and M. Morikawa. 1993. Expression of the β glucuronidase gene in pollen of lily (*Lilium longiflorum*), tobacco (*Nicotiana tabacum*), *Nicotiana rustica*, and peony (*Paeonia lactiflora*) by particle bombardment. Plant Physiol. 102: 357–361.

Ogaki, M., Y. Furuichi, K. Kuroda, D.P. Chin, Y. Ogawa and M. Mii. 2008. Importance of co-cultivation medium pH for successful *Agrobacterium*-mediated transformation of *Lilium × formolongi*. Plant Cell Rep. 27: 699–705.

Pelkonen, V.P. 2005. Biotechnological approaches in lily (*Lilium*) production. Ph.D. thesis. Faculty of Science, Department of Biology, University of Oulu, Oulu. Finland.

Plegt, L.M., B.C.E. Ven, R.J. Bino, T.P.M. Salm and A.J. van Tunen. 1992. Introduction and differential use of various promoters in pollen grains of *Nicotiana glutinosa* and *Lilium longiflorum*. Plant Cell Rep. 11: 20–24.

Qiu-Hua, L., H. Bo, T. Zheng, M. Chao, G. Ai-Nong, Y. Jing-Juan and G. Jun-Ping. 2008. Establishment of regeneration system and transformation of Zm401 gene in *Lilium longiflorum × L. formosanum*. Chin. J. Agr. Biotechnol. 5: 113–119.

Shi, S., F. Ma, Y. Li, F. Feng and Z. Shang. 2012. Overexpression of L-galactono-1, 4-lactone dehydrogenase (GLDH) in Lanzhou lily (*Lilium davidii* var. unicolor) via particle bombardment-mediated transformation. *In vitro* Cell Dev. Biol. Plant 48: 1–6.

Smyth, D.R., K. Kongsuwan and S. Wisudharomn. 1989. A survey of C-band patterns in chromosomes of *Lilium* (Liliaceae). Plant. Syst. Evol. 163: 53–69.

Sood, P., A. Bhattacharya and A. Sood. 2011. Problems and possibilities of monocot transformation. Biologia Plantarum 55: 1–15.

Spiker, S. and W.F. Thompson. 1996. Nuclear matrix attachment regions and transgene expression in plants. Plant Physiol. 110: 15–21.

Suzuki, S. and M. Nakano. 2002. *Agrobacterium*-mediated production of transgenic plants *Muscari armeniacum* Leichtl. ex Bak. Plant Cell Rep. 20: 835–841.

Suzuki, S., Y. Niimi, T. Sakakibara, K. Hosokawa, S. Yamamura and M. Nakano. 1998. Effects of several antibiotics and bialaphos on the growth and organ formation of *Lilium formosanum* calli and transient expression of the gusA gene after co-cultivation with *Agrobacterium tumefaciens*. Plant Biotechnol. 15: 213–216.

Suzuki, S., K. Supaibulwatana, M. Mii and M. Nakano. 2001. Production of transgenic plants of the Liliaceous ornamental plant *Agapanthus praecox* ssp. *orientalis* (Leighton) Leighton via *Agrobacterium*-mediated transformation of embryogenic calli. Plant Sci. 161: 89–97.

Tabata, S., S. Sato, Y. Watanabe, M. Yamamoto and Y. Hotta. 1993. Evidence of meiosis-specific regulation of gene expression in lily microsporocytes. Plant Sci. 89: 31–41.

Travella, S., S.M. Ross, J. Harden, C. Everett, J.W. Snape and W.A.A. Harwood. 2005. Comparison of transgenic barley lines produced by particle bombardment and *Agrobacterium*-mediated techniques. Plant Cell Rep. 23: 780–789.

Tsuchiya, T., S. Takumi and T. Shimada. 1996. Transient expression of a reporter gene in bulbscales and immature embryos of three *Lilium* species is affected by 5 upstream sequences and culture conditions. Physiol. Plant 98: 699–704.

Van Emmenes, L., A. Veale, A. Cohen and T. Arazi. 2008. *Agrobacterium*-mediated transformation of the bulbous flower *Ornithogalum*. Acta Hort. 766: 420–425.

Vasil, V., A.M. Castillo, M.E. Fromm and I.K. Vasil. 1992. Herbicide resistant fertile transgenic wheat plants obtained by microprojectile bombardment of regenerable embryogenic callus. Bio Technology 10: 667–674.

Wang, Y., B. van Kronenburg, T. Menzel, C. Maliepaard, X. Shen and F. Krens. 2012. Regeneration and *Agrobacterium*-mediated transformation of multiple lily cultivars Plant Cell Tiss. Organ Cult. 111: 113–122.

Watad, A.A., D.J. Yun, T. Matsumoto, X. Niu, Y. Wu, A.K. Kononowicz, R.A. Bressan and P.M. Hasegawa. 1998. Microprojectile bombardment-mediated transformation of *Lilium longiflorum*. Plant Cell Rep. 17: 262–267.

Wilmink, A., B.C.E. van de Ven and J.J.M. Dons. 1995. Activity of constitutive promoters in various species from the Liliaceae. Plant Mol. Biol. 28: 949–955.

Zhang, Y.S., X.Y. Yin, A.F. Yang, G.S. Li and J.R. Zhang. 2005. Stability of inheritance of transgenes in maize (*Zea mays* L.) lines produced using different transformation methods. Euphytica 144: 11–22.

10

Molecular Markers in Caladium

Development, Characterization and Applications

Zhanao Deng

ABSTRACT

Caladium is an important ornamental aroid. The development and characterization of molecular markers in caladium are still quite limited compared to some other bulbous plants; yet, the available AFLP, TRAP, particularly EST-SSR, markers have become a powerful tool for genetic, taxonomical, and breeding studies of caladium. AFLP and EST-SSR markers have allowed reliable differentiation of commercial caladium cultivars and the development of a DNA fingerprint database for these cultivars. Analysis of caladium DNA fingerprints has resulted in a good understanding of the genetic similarity among caladium cultivars and an assembly of core caladium germ plasm for preservation. Applications of this tool also have resulted in a significant amount of new information about genetic diversity and relationships among caladium species. Examples of discoveries that have resulted from such applications include molecular evidences for *Caladium humboldtii* as a distinct species (not a chromosomal race), the existence of vast differences between

University of Florida/IFAS, Environmental Horticulture Department, Gulf Coast Research and Education Center, 14625 County Road 672, Wimauma, FL 33598, USA.
Email: zdeng@ufl.edu

Caladium lindenii and other caladium species and the potential need to reclassify this species as a new genus, the most likely origin of cultivated caladiums (*Caladium* × *hortulanum*), etc. It is expected that molecular markers will be utilized a lot more frequently in caladium in the future for cultivar fingerprinting, registration and protection, taxonomical studies, gene tagging, and genome mapping.

Keywords: AFLP marker, *Caladium*, *Caladium bicolor*, *Caladium* × *hortulanum*, *Caladium humboldtii*, *Caladium lindenii*, fingerprinting, genetic diversity, genetic relationship, microsatellite marker, SSR marker, TRAP marker

Introduction

Caladium (*Caladium* Vent.) is a member of the family Araceae, placed in the tribe Caladieae in the subfamily Aroideae (Mayo et al. 1997). The genus is indigenous to South and Central Americas (Grayum 1986, Mayo et al. 1997). The number of species in this genus is still a matter of discussion, from seven (Madison 1981) to 17 (Croat 1994) depending on the taxonomists who worked on this genus. Birdsey (1951) proposed to refer to the cultivated caladiums as *Caladium* × *hortulanum* Birdsey. In some classifications, however, *C.*× *hortulanum* is considered synonymous with *Caladium bicolor* (Aiton) Vent. (http://theplantlist.org/). For simplicity and for the new insights gained from molecular marker analysis described below, the name *C.* × *hortulanum* will be used in this chapter to refer to the fancy- or lance-leaved caladium cultivars.

In the floriculture trade, caladiums are grown as pot or hanging basket plants or planted in the landscape as accent or border plants. Their ornamental values come mainly from colorful and variably-shaped leaves (Fig. 1). Commercial caladium plants generally are propagated through tuber division and production. Most of the caladium tubers used in the world is produced in central Florida (Bell et al. 1998, Deng et al. 2008). It is estimated that the annual farm-gate value of caladium tubers produced in Florida is about $15 million (Rosskopf et al. 2005). Currently there are approximately 100 caladium cultivars in commercial production; some 50 cultivars account for more than 95% of the field used for caladium tuber production (Bell et al. 1998, Deng et al. 2008). The great majority of these cultivars result from sexual hybridization among existing cultivars, breeding lines and/or species (Deng 2012).

Molecular markers are specific fragments of DNA that can be identified within the whole genome of an organism (Agarwal et al. 2008). Differences among individuals at molecular marker loci represent polymorphisms among the individuals at the DNA level. Over the past 25 years, numerous

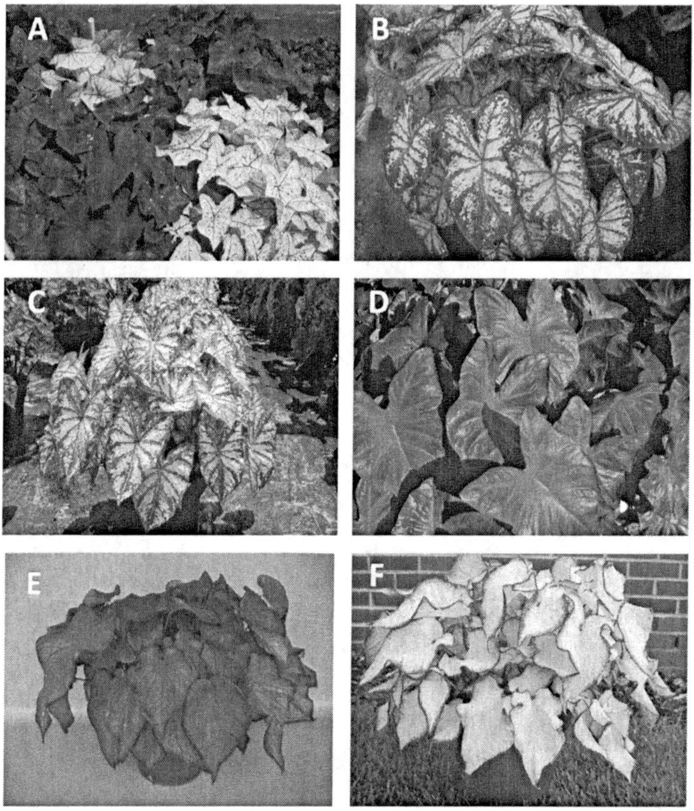

Fig. 1. Caladium plants in the landscape or containers. A: multiple cultivars grown in the landscape; B and C: 'Tapestry' and 'Florida Blizzard' caladium grown in containers; D: 'Royal Flush' caladium in the landscape; E and F: 'Fairytale Princess' and 'White Butterfly' caladium grown in containers.

Color image of this figure appears in the color plate section at the end of the book.

technologies have been developed to identify molecular markers, characterize them, and relate the identified DNA polymorphisms with phenotypic variations in morphology, physiology, biochemistry, production, product quality, disease resistance, and/or stress tolerance. In nowadays plant sciences and biotechnology, molecular markers have found wide applications in many research fields and have become an indispensable tool for many plant scientists including biotechnologists (Agarwal et al. 2008). The same is true in caladium. This chapter will review the development, characterization and applications of molecular markers in caladium, one of the most important ornamental aroids in the world.

Development and Characterization of Molecular Markers in Caladium

Among the numerous molecular marker technologies that have been described in the literature, three have been applied to caladium, including the amplified fragment length polymorphism (AFLP) technique, the microsatellite or simple sequence repeat (SSR) marker system, and the target region amplification polymorphism (TRAP) method.

AFLP markers

The AFLP marker technology was developed in the early 1990s (Vos et al. 1995). It is based on restriction enzyme digestion and PCR amplification of genomic DNA to reveal DNA polymorphisms and produce molecular markers. Two obvious and important advantages of this marker development technique are no requirement of any prior sequence information and high multiplex ratios. These advantages may have become irrelevant in plants whose genome or transcriptome has been sequenced, but they are still very important for use in caladium (and many other bulbous plants) whose genome is not sequenced and little sequence information is available in public databases.

Loh et al. (1999, 2000) applied the AFLP marker system to caladium. In their studies, caladium genomic DNA was digested with two endonucleases, *Eco*RI and *Mse*I, a six base- and a four-base cutter, respectively. The digested DNA fragments were then ligated to two oligonucleotide adapters (*Eco*RI and *Mse*I adapters) to provide priming sites for polymerase chain reaction (PCR) amplification of the DNA fragments. PCR amplification was conducted in two steps, i.e., pre-selective and selective amplification, respectively. In the pre-selective amplification, primers used were complementary to the adapter sequences and the restriction sites and contained one selective nucleotide at the 3′ ends. In the following selective amplification, primers contained the pre-selective primer sequences and two additional selective nucleotides at the 3′ ends. During these steps, only DNA fragments with end sequences matching the selective nucleotides of the primers could be amplified. In other word, a subset of the caladium restriction fragments was selectively amplified in a given PCR reaction. The amplified DNA fragments were resolved on 6% denaturing polyacrylamide gels and visualized by silver staining.

Lohet al. (1999) tested 17 primer combinations on seven caladium cultivars, six from *C. bicolor* and one from *Caladium schomburgkii* Schott. Each AFLP primer combination amplified as many as 110 fragments detectable by silver staining and revealed five to 24 unique markers. Together, the 17 primer combinations produced 173 unique markers among the seven caladium cultivars. The amplified fragments ranged from 50 base pairs to 600 base pairs. Subsequently, Loh et al. (2000) applied the AFLP marker system to four caladium species and obtained as many as 235 polymorphic DNA bands with three primer combinations. These data show that AFLP markers or fingerprints produced in caladium are reliable and reproducible and can reveal high levels of DNA polymorphism in caladium.

TRAP markers

The TRAP marker system is another PCR-based technique (Hu and Vick 2003). Deng et al. (2007) used this marker technique in caladium. In their study, caladium genomic DNA was amplified directly with pairs of "fixed" and "arbitrary" primers, initially at a low annealing temperature (35°C) for 5 cycles and then at a more stringent annealing temperature (50°C) for 35 cycles. The amplified products are separated in a 6.5% polyacrylamide sequencing gel on a LI-COR DNA analyzer and visualized by infrared laser scans. The fixed primers were designed from sunflower or lettuce expressed sequence tags (ESTs) in the public Compositae Genome Project database (http://compgenomics.ucdavis.edu/), whereas the "arbitrary" primers consisted three to four selective nucleotides at the 3′ end, four to six nucleotides with AT- or GC-rich regions (the core), and some filler sequences at the 5′ end. The arbitrary primers were labeled with the infrared (IR) dye 700 or IR dye 800 for DNA fragment detection on the LI-COR DNA analyzer.

Deng et al. (2007) used four fixed primers and four arbitrary primers in eight primer combinations. The eight primer pairs generated 297 clear-cut DNA bands in 59 caladium accessions. The number of DNA fragments amplified by each primer combination varied from 13 to 60, with an average of 37 fragments per primer combination. This TRAP marker analysis did not involve restriction enzyme digestion or adaptor linking, thus it was easier, quicker, and less expensive to perform than the AFLP marker analysis. Since the arbitrary primer sequences and the fixed primer sequences were not designed from caladium sequences, it was expected that the primer sequences could find only imperfect matches in the caladium genome. This may explain why the TRAP system amplified fewer DNA bands than the AFLP system did in caladium. Because of the same reason, reproducibility was a concern for molecular markers so generated. Nevertheless, duplicate samples from three cultivars shared 97.2% of the DNA fragments, suggesting that TRAP markers could achieve an acceptable level of reproducibility in caladium.

SSR markers

SSR sequences are abundant, widely present, and hyper-variable in plant genomes; molecular markers based on SSR variations have offered a number of advantages over other types of markers for applications in plant breeding and genetic studies (Varshney et al. 2005). SSR markers have good potential to detect multiple co-dominant alleles among cultivars, thus able to reveal high levels of genetic polymorphism for gene or genome mapping and to provide strong discriminatory power for cultivar identification. SSR markers may be transferable across species and ideal for anchoring and aligning genetic linkage maps, conducting comparative genetic mapping, and revealing species relationships, etc.

Gong and Deng (2009) attempted to transfer SSR markers from taro [*Colocasia esculenta* (L.) Schott] to caladium. When 43 taro SSR markers (Mace and Godwin 2002) were tested on several caladium cultivars, four of the markers amplified weak DNA bands in caladium. Subsequently 22 pairs of primers were designed from taro DNA sequences found in the GenBank, and one pair of the primers amplified products in caladium. Their results indicate that it was difficult and inefficient to apply taro SSR markers to caladium, even though the two aroids belong to the same tribe.

To develop caladium-specific SSR markers, Gong and Deng (2011) fragmented the genomic DNA of the caladium cultivar Florida Sweetheart, enriched the SSR-containing DNA fragments with biotinylatedoligo $(GA)_{15}$, constructed a small genomic library (~1,200 clones) from the enriched DNA, and sequenced 480 clones from the library. Three hundred and seventy sequences from the sequenced clones contained SSRs with $\geq$ 5 repeats; 173 sequences were unique, contained $\geq$ 7 repeats of a dinucleotide or $\geq$ 5 repeats of a trinucleotide motif, and were suitable for primer designing. A majority (163) of the 173 sequences contained the GA/TC motif, identical to the biotinylatedoligo used in the enrichment procedure. Also, a majority (143) of these 163 GA/TC microsatellites contained perfect repeats, and 20 contained compound motifs (GA/TC and CA/TG). Among perfect GA/TC SSRs, 45% of them contained 5–10 repeats, 49% had 11 to 25 repeats, and a small percentage (6%) consisted of more than 25 repeats.

When 114 pairs of primers were tested on seven caladium cultivars, 99 pairs of the primers amplified clear PCR products of expected sizes, yielding a success rate of 86.8%. These SSR markers also showed high levels of transferability to other caladium species: 100% to *C. bicolor*, 98.3% to *C. schomburgkii*, 86.7% to *Caladium steudneriifolium* Engl., 85.0% to *Caladium picturatum* K. Koch & C.D. Bouché, 70% to *Caladium humboldtii* (Raf.) Schott, 70% to *Caladium clavatum* Hett., Bogner & J. Boos (or *C. bicolor* var. *rubricundum*), and 51.7% to *Caladium marmoratum* Mathieu ex K. Koch.

Applications of Molecular Markers in Caladium

As mentioned above, molecular markers have been used in many ways in plant science and biotechnology (Agarwal et al. 2008). So far the application of molecular markers in caladium has been mainly in cultivar fingerprinting and differentiation and analysis of the genetic relatedness among cultivars and/or species. Other applications of molecular markers in caladium are expected to be reported in the near future.

Fingerprinting of caladium cultivars

Traditionally, plant and foliar characteristics are the main attributes used for caladium cultivar description and differentiation. It is known that the expression of many plant and foliar traits in caladium varies with plant developmental stages and growing environments (Ahmed et al. 2004). There are quite a number of cases where these attributes are not able to provide enough discriminatory power. For example, a few caladium cultivars were found to have different levels of resistance to Pythium root rot but to be indistinguishable morphologically (Deng et al. 2005). In recent years, there has been a surge of new caladium plant patent applications in the U.S. Patent and Trademark database (http://www.uspto.gov/). It is becoming increasingly difficult to differentiate some of the new caladium cultivars solely based on plant, foliar and floral attributes. There is a need for application of DNA fingerprinting techniques to caladium cultivars for plant patent application and protection.

Loh et al. (1999) was the first to apply AFLP-based DNA fingerprinting to caladium cultivars. They identified 10 to 30 DNA bands specific for each of the six *C. bicolor* cultivars ('E.O. Orpet', 'Itacapus', 'Jequitinhonha', 'Porto Novo', The Thing', and an unnamed strap-leaf cultivar) and 40 DNA bands specific for *C. schomburgkii* 'Changjur'. Their results showed that AFLP markers were very useful for distinguishing caladium cultivars.

Deng et al. (2007) used TRAP markers to fingerprint 45 caladium cultivars. Sixty-three DNA bands generated by eight primer combinations were polymorphic among these cultivars. All 45 cultivars in the study could be differentiated based on the profile of these polymorphic DNA bands. To test the discriminatory power of these markers, three pairs of cultivars that are morphologically highly similar ('Aaron' and 'Gray Ghost', 'Rosebud' and 'Rosebud Bear Hollow' and 'Candidum' and 'Candidum Junior') were included in the test. The first and the second pair of cultivars were indistinguishable based on plant morphology. TRAP marker analysis revealed differences at two DNA bands within these pairs of cultivars. 'Candidum Junior' was believed to be a sport mutation of 'Candidum' (Wilfret 1991). TRAP marker analysis detected differences at six DNA bands. Recently, Gong and Deng (2011) have shown that SSR markers are even more powerful in discriminating caladium cultivars. Fingerprinting with 99 SSR markers revealed differences at 53 (68.8%) out of 77 DNA bands between 'Aaron' and 'Gray Ghost' and differences at 28 (41.8%) out of 67 DNA bands between 'Candidum' and 'Candidum Junior'. A DNA fingerprint database based on these SSR markers is under construction for all major commercial caladium cultivars (Z. D., unpublished data).

Jaccard similarity coefficients (JSC) between caladium cultivars

In molecular marker analysis, JSC represents the proportion of DNA bands shared by two samples. This coefficient or index has been widely used for measuring and comparing the genetic similarity of plant cultivars, populations, or species. Recently it has been proposed to use this coefficient as a main criterion for identifying essentially derived cultivars in agricultural and horticultural crops (van Eeuwijk and Law 2004). Deng et al. (2007) analyzed the distribution of JSC among 45 caladium cultivars (990 pairwise comparisons) based on the DNA profile generated by TRAP markers. Their results indicate that the similarity coefficient between caladium cultivar pairs ranged from 0.691 to 0.982, with the average similarity coefficient of 0.801 (Fig. 2).

Five pairs of caladium cultivars, out of 990 pairs, had the similarity coefficient above 0.900: 'Carolyn Whorton' and 'Florida Elise' (0.911), 'Red Flash' and 'Red Frill' (0.911), 'Candidum' and 'Candidum Junior' (0.945), 'Rosebud' and 'Rosebud Bear Hollow' (0.963), and 'Aaron' and 'Gray Ghost' (0.982). 'Florida Elise' resulted from backcrossing with 'Carolyn Whorton' (Wilfret 1991); 'Rosebud Bear Hollow' was selected out asexually propagated 'Rosebud'. Also it was speculated that 'Candidum Junior' was a field mutation of 'Candidum' (Wilfret 1991), a cultivar that was developed by Bleu in France around 1880 (Hayward 1950) and is still a major commercial cultivar. These pieces of information seem to be in agreement with the similarity analysis: 'Florida Elise', 'Rosebud Bear Hollow', and 'Candidum Junior' are essentially derived cultivars. Information about the possible origin of 'Red Flash' and 'Red Frill', or 'Aaron' and 'Gray Ghost' is not available in the literature. The high similarity coefficients between these pairs of cultivars may indicate that one cultivar in each pair may be essentially derived from the other in the pair.

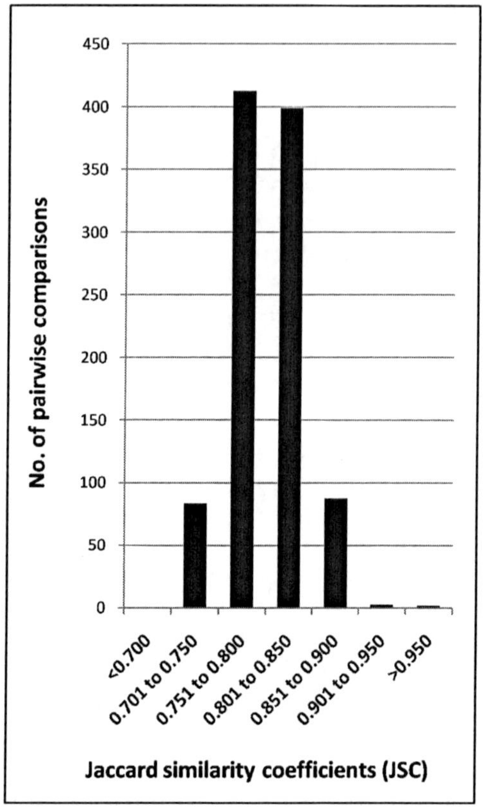

Fig. 2. Distribution of the Jaccard similarity coefficients among 45 caladium cultivars based on TRAP marker analysis.

Genetic relatedness among cultivars

Caladium cultivars are often separated into one of eight groups based on leaf shape (fancy, lance or strap) and leaf impact color (white, red, pink, or multi-colored) (Wilfret and Hurner 1982, Bell et al. 1998). However, little information is available in the literature about the relatedness among these groups and most of the caladium cultivars in commercial production and landscape use. Deng et al. (2007) analyzed the genetic relatedness of 45 caladium cultivars across the eight cultivar groups using TRAP markers. Clustering analysis, based on 141 TRAP markers, Jaccard similarity coefficients, and the unweighted pair group method of arithmetic averages (UPGMA), separated the cultivars into several clusters. However, none of these clusters corresponded to any cultivar group, and few of the clusters were confirmed by bootstrapping. Subsequently a principal-coordinate analysis (PCoA) was performed on these cultivars; again no specific grouping was observed in the PCoA plot. It remains to be found out whether the lack of specific grouping is due to extensive hybridization among the cultivars or insufficiency of the TRAP markers. A new study has been initiated to use recently developed SSR markers to understand the genetic relatedness of these cultivars. Results from this study should shed new light on this subject. A better understanding of the genetic relatedness of caladium cultivars might help choose caladium cultivars to be maintained in the core collection of caladium germplasm. This knowledge also will provide valuable guidance for choosing caladium cultivars as breeding parents to be used in sexual hybridization breeding.

Genetic relationships among *caladium* species

Previously morphological and anatomical examination was the main source of knowledge about the relationships among caladium species. Recent application of molecular marker analysis has shed new light on this subject. Loh et al. (2000) used AFLP markers to analyze the genetic relationships of four caladium species accession. Their data showed that *C. bicolor* and *C. schomburgkii* were highly similar and had a JSC of 0.737. *Caladium humboldtii* was less similar to *C. bicolor*, with a JSC of 0.326. *Caladium lindenii* (André) Madison was very different from *C. bicolor*, *C. schomburgkii*, and *C. humboldtii*; its JSC with these species was 0.184, 0.163, and 0.151, respectively.

Deng et al. (2007) investigated the genetic relationships of 14 caladium accessions from six species using 297 TRAP markers. These species accessions, maintained at the Marie Selby Botanic Gardens in Sarasota, Florida, were collected from at least five countries in South and Central Americas between 1977 and 2002. Two of the three *C. bicolor* accessions (Bi3A and Bi3B) were highly similar to each other, having a JSC of 0.900. But one *C. bicolor* accession (Bi7B) was quite dissimilar from Bi3A and Bi3B accessions (JSC of 0.248 and 0.261, respectively). These data seem to indicate that some significant divergence has evolved within *C. bicolor*, which might have resulted in the formation of different varieties within this species.

Plants of *C. humboldtii* are dwarf and have miniature leaves yet have a leaf coloration pattern similar to the leaf coloration pattern in some *C. bicolor* plants. This similarity led Madison (1981) to speculate that *C. humboldtii* might be a chromosomal race of *C. bicolor*. Loh et al. (2000) observed a vast difference between the two species in AFLP profiles and concluded that *C. humboldtii* would be a distinct and separate species. Deng et al. (2007) came to a similar conclusion based on TRAP marker analysis of three *C. humboldtii* accessions. *Caladium humboldtii* showed high levels of similarity among them (JSC of 0.941 to 0.990) but low levels of similarity with other caladium species (JSC of 0.079 to 0.416).

Molecular marker analysis also raised an interesting question about the status of *C. lindenii*. Previously this species was classified as a member of the genus *Xanthosoma*, but it was transferred to the genus *Caladium* based on its similarity with *Caladium* in pollen shedding pattern (Madison 1981). Pollen is shed as tetrads in *Xanthosoma*, whereas pollen shed as monads in *Caladium*. Loh et al. (2000) observed that *C. lindenii* did not share any of the *Xanthosoma*-specific AFLP markers, so it was not a *Xanthosoma* species. In an UPGMA cluster analysis, *C. lindenii* was clustered only distantly with *C. bicolor*, *C. schomburgkii*, or *C. humboldtii*. Its JSC with these three species was rather low (0.160 to 0.184). Similar genetic relationships between *C. lindenii* and other caladium species were observed by Deng et al. (2007) in TRAP marker-based analysis: *C. lindenii* showed extremely low JSC (0.060–0.148) with accessions from *C. bicolor*, *C. humboldtii*, *C. marmoratum*, *C. picturatum*, *C. schomburghkii*, or *C. steudneriifolium*. In PCoA based on TRAP marker fingerprints, *C. lindenii* was separated far from other caladium species. Recently, SSR marker analysis provided additional information about the relationship between *C. lindenii* and other caladium species. Gong and Deng (2011) observed a steep drop in SSR marker transferability from *C.* × *hortulanum* to *C. lindenii* compared to the transferability to *C. bicolor*, *C. humboldtii*, *C. marmoratum*, *C. picturatum*, *C. schomburgkii*, and *C. steudneriifolium* (25% vs. 51.7 to 100%). Furthermore, some of the DNA fragments amplified from *C. lindenii* by the transferable SSR marker primers were beyond the DNA fragment size range observed in other caladium species. These results from molecular marker analyses, plus the existence of vast differences in plant and foliar characteristics, suggest the need to reclassify *C. lindenii* in Araceae, probably a new genus status for *C. lindenii*.

Genetic relationship between caladium cultivars and species

Previously it was speculated that cultivated caladiums (*C.* × *hortunalum*) might have resulted from hybridization among *C. bicolor*, *C. marmoratum*, *C. picturatum*, and/or *C. schomburgkii* (Haward 1950, Birdsey 1951, Wilfret 1993), based on morphological comparisons and similarities. Deng et al. (2007) used TRAP marker data to examine the relationship between 45 caladium cultivars and seven accessions of *C. bicolor*, *C. marmoratum*, *C. picturatum*, and *C. schomburgkii*. In both UPGMA cluster analysis and PCoA, *C. marmaratum*, *C. picturatum*, and two accessions of *C. bicolor* were placed rather far away from caladium cultivars, and only one accession of *C. bicolor* (Bi7B) and one *C. schomburgkii* accession (Shm) were clustered close to the cultivars (Fig. 3). The cultivars' similarity to Bi7B and Shm was high, with JSC being 0.531 to 0.771. These results seem to suggest that

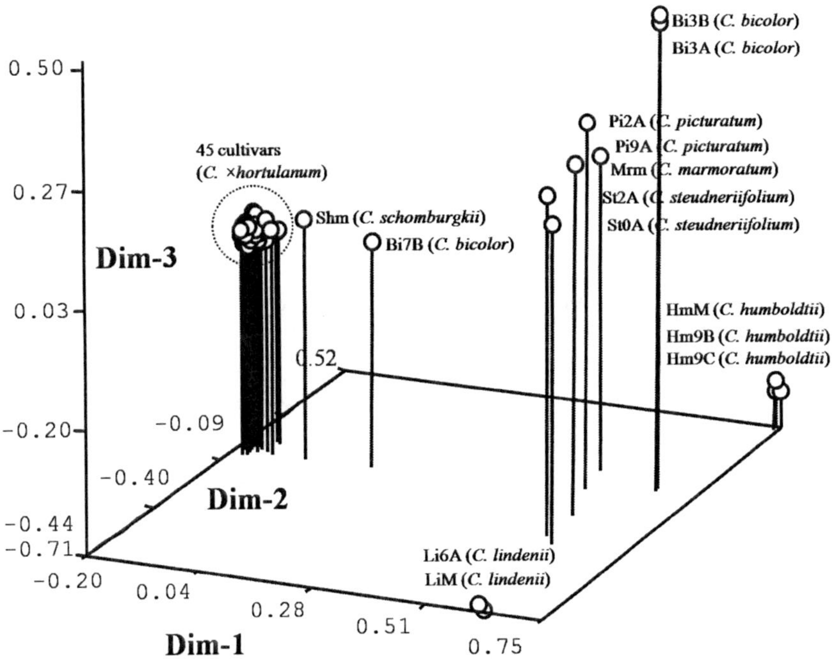

Fig. 3. Separation of 45 *Caladium* × *hortunalum* cultivars and 14 caladium species accessions in a principal coordinate analysis plot based on 297 TRAP markers (Deng et al. 2007).

most caladium cultivars, if not all, might have derived from some *C. bicolor* accessions and *C. schomburgkii*, but not from *C. marmoratum* or *C. picturatum*. Research is under way to apply newly developed SSR markers to deepen our understanding of the relationship between different caladium species and cultivars and unveil the origin of caladium cultivars.

Conclusion

The development and characterization of molecular markers in caladium are still quite limited compared to the amount of work that has been done in some other bulbous plants. Yet, the available molecular markers have provided a powerful tool for conducting genetic, taxonomical, and breeding studies in caladium. Applications of these markers have resulted in a significant amount of new information about genetic diversity and relationships among caladium cultivars and species. It is expected that there will be a lot more applications of molecular markers in caladium cultivar differentiation, registration and protection, taxonomical studies, gene tagging, and genome mapping.

Acknowledgements

The author would like to thank Dr. Hu and Dr. Gong for their molecular marker work in caladium.

References

Agarwal, M., N. Shrivastava and H. Padh. 2008. Advances in molecular marker techniques and their applications in plant sciences. Plant Cell Reports 27: 617–631.

Ahmed, E.U., T. Hayashi and S. Yazawa. 2004. Leaf color stability during plant development as an index of leaf color variation among micropropagated *Caladium*. HortScience 39: 328–332.

Bell, M.L., G.J. Wilfret and D.A. Devoll. 1998. Survey of caladium tuber producers for acreage of cultivars grown. Proceedings of the Florida State Horticultural Society 111: 32–34.

Birdsey, M.R. 1951. The Cultivated Aroids. Gillick Press, Berkeley, California.

Croat, T. 1994. Taxonomic status of neotropical Araceae. Aroideana 17: 33–60.

Deng, Z. 2012. Caladium breeding and genetics: Recent advances. *In*: J.A. Teixeira da Silva (ed.). Floriculture and Ornamental Biotechnology 6 (Special Issues 1). Global Science Books, London, UK, pp. 53–61.

Deng, Z., F. Goktepe and B.K. Harbaugh. 2007. Assessment of genetic diversity and relationships among caladium cultivars and species using molecular markers. Journal of the American Society for Horticultural Science 132: 219–229.

Deng, Z., B.K. Harbaugh, R.O. Kelly, T. Seijo and R.J. McGovern. 2005. Pythium root rot resistance in commercial caladium cultivars. HortScience 40: 549–552.

Deng, Z., B.K. Harbaugh, R.K. Schoellhorn and R.C. Andrew. 2008. 2003 Survey of the Florida caladium tuber production industry. Available online: http://edis.ifas.ufl.edu/ep258/.

Gong, L. and Z. Deng. 2009. Transferring SSR markers across genus for caladium breeding and genetic analysis. HortScience 44: 1153–1154.

Gong, L. and Z. Deng. 2011. Development and characterization of microsatellite markers for caladium (*Caladium* Vent.). Plant Breeding 130: 591–595.

Grayum, M.H. 1986. New taxa of *Caladium*, *Chlorospatha*, and *Xanthosoma* (Araceae: Colocasioideae) from southern Central America and northwestern Colombia. Annals of the Missouri Botanical Garden 73: 462–474.

Hayward, W. 1950. Fancy-leaved caladiums. Plant Life 6: 131–142.

Hu, J. and B.A. Vick. 2003. Target region amplification polymorphism: A novel marker technique for plant genotyping. Plant Molecular Biology Reporter 21: 1–6.

Loh, J.P., R. Kiew, A. Hay, A. Kee, L.H. Gan and Y.-Y. Gan. 2000. Intergeneric and interspecific relationships in Araceae tribe Caladieae and development of molecular markers using amplified fragment length polymorphism (AFLP). Annals of Botany 85: 371–378.

Loh, J.P., R. Kiew, A. Kee, L.H. Gan and Y.-Y. Gan. 1999. Amplified fragment length polymorphism (AFLP) provides molecular markers for the identification of *Caladium bicolor* cultivars. Annals of Botany 84: 155–161.

Mace, E.S. and I.D. Godwin. 2002. Development and characterization of polymorphic microsatellite markers in taro (*Colocasia esculenta*). Genome 45: 823–832.

Madison, M. 1981. Notes on *Caladium* (Araceae) and its allies. Selbyana 5: 342–277.

Mayo, S.J. and J. Bogner. 1988. A new species of *Caladium* (*Araceae*) with notes on generic delimitation in the *Colocasioideae-Caladieae*. Willdenowia 18: 231–242.

Mayo, S.J., J. Bogner and P.C. Boyce. 1997. The Genera of Araceae. Royal Botanical Gardens, Kew, Richmond, Surrey, UK.

Rosskopf, E.N., D.O. Chellemi, N. Kokalis-Burelle and G.T. Church. 2005. Alternatives to methyl bromide: a Florida perspective. Online APSnet Features. doi: 10.1094/APSnetFeature/2005-0605.

van Eeuwijk, F.A. and J.R. Law. 2004. Statistical aspects of essential derivation, with illustrations based on lettuce and barley. Euphytica 137: 129–137.

Varshney, R.K., A. Graner and M.E. Sorrells. 2005. Genic microsatellite markers in plants: features and applications. Trends in Biotechnology 23: 48–55.

Vos, P., R. Rogers, M. Bleeker, M. Reijans, T. van de Lee, M. Hornes, A. Fritjers, J. Pot, J. Peleman, M. Kuiper and M. Zabeau. 1995. AFLP: a new concept for DNA fingerprinting. Nucleic Acids Research 23: 4407–4414.

Wilfret, G.J. 1991. Florida Elise, a pink caladium for landscape and containers. University of Florida, Florida Agricultural Experiment Station Circular S-375, pp 6.

Wilfret, G.J. 1993. Caladium, *In*: A. de Hertogh and M. LeNard (eds.). The Physiology of Flower Bulbs, Elsevier, Amsterdam, The Netherlands, pp. 239–247.

Wilfret, G.J. and G.T. Hurner Jr. 1982. A survey of caladium cultivars grown in Florida and their characteristics as potted plants. Proceedings of the Florida State Horticultural Society 95: 190–194.

11

Genetic Control of Anthocyanin Synthesis in Dahlia (*Dahlia variabilis*)

Sho Ohno, Ayumi Deguchi* and *Munetaka Hosokawa*

ABSTRACT

In this chapter, we shall describe the genetic control of flower color in dahlia (*Dahlia variabilis*). Dahlia is one of the most popular ornamental tuberous plants in the world because of its wide range of flower color, such as purple, red, yellow, ivory white, bicolor and black. These colors are mainly derived from the accumulation of three major compounds, anthocyanin, butein and flavone. The mechanisms of pigment biosynthetic pathway, for example, anthocyanin synthesis, flower color intensity, bicolor and black coloring are described by each flower color. Finally, for future molecular breeding of dahlia, the target genes of each pigments or flower colors are summarized.

Keywords: Anthocyanin, bHLH, Black Flower, Butein, Chalcone Synthase, Cyanidin, Dahlia, Flavone, Flavone Synthase, Flower Color, Molecular Breeding

Laboratory of Vegetable and Ornamental Horticulture, Graduate School of Agriculture, Kyoto University, Sakyo-ku, Kyoto 606-8502, Japan.
* Corresponding author: sohno@kais.kyoto-u.ac.jp

Introduction

Dahlia (*Dahlia variabilis*) is one of the most popular tuberous plants in the world because of its varied colors of flower. In dahlia petals, three flavonoid pigments have been identified as contributing to flower color: anthocyanin for purple, flavone for pale yellow, and chalcone for yellow. Differences in anthocyanin quantities in petals, resulting in black to pink variation, are important factors in the large diversity of dahlia colors (Fig. 1).

Fig. 1. Dahlia flowers used in our experiment. a: "Michael J," b: "MJOr," c: "MJY," d: "Super Girl," e: "Yukino," g: "Cupid," f: "Evelyn Rumbold," h: "Jyunn-ai," i: "Saffron," j: "Hakuba," k: "Hakuyo," l: "Yuino," m: "Matsuribayashi," n: "Kazusa-shiranami," o: "Ori2," p: "Fidalgo Blacky," q: "Ms. Noir," r: "Kokucho," s: "Black Cat," and t: "Kokucho" purple mutant.

Color image of this figure appears in the color plate section at the end of the book.

The genes of flavonoid biosynthesis have been determined by many previous studies using flowers or fruits. However, we have found few reports of the application of this information to flower or fruit breeding. Dahlia is thought to be an autoallooctoploid with $2n = 8x = 64$ (Gatt et al. 1998), and the large diversity of flower traits is the product of complex expressions of several alleles and/or loci. Dahlias are easily propagated vegetatively from cuttings or tubers, so that a newly produced seedling plant can immediately become a new cultivar. Dahlia breeding by crossing requires an extremely large field. If molecular markers for flower color selection are available, we will be able to conduct a selection for each flower color. Elaborate genetic studies for determining alleles or loci contributing to each flower color will be needed for the development of such genetic markers. In this section, we shall introduce our recent studies on genes controlling each flower color, flower color intensities, and the bicolor trait.

Flower color and pigments in dahlia

The range of flower color in dahlia is enormous. According to the Royal Horticultural Society, dahlia colors may be classified into 11 groups: (white, yellow, orange, bronze, flame: red or dark red, pink, lilac, lavender or mauve, purple: wines or violets blends, bicolored, and variegated). Including cultivars that do not fall into this classification, such as black cultivars, it can be said that dahlia is one of the most diversified floricultural species in the world.

The dahlia flower pigments have been studied for a long time, and three major compounds contribute to flower color variation. The first is anthocyanin, which is well known as a purple or red pigment. In dahlia, cyanin (a glycoside of cyanidin) and pelargonin (a glycoside of pelargonidin) accumulate in petals. These anthocyanins are modified with malonylation and finally accumulate as malonyl glycosides. For example, pelargonidin 3-(6″-malonylglucoside)-5-glucoside and pelargonidin 3-(6″-malonylglucoside)-5-malonylglucoside accumulate in "Biddenham Strawberry" and cyanidin 3-(6″-malonylglucoside)-5-glucoside and cyanidin 3-(6″-malonylglucoside)-5-malonylglucoside accumulate in "Shandy" (Takeda et al. 1986).

The second compound is flavone, which is observed in cultivars of almost all colors except for black or the white areas of petals in bicolor cultivars. In dahlia, two flavones, apigenin and luteolin, have been detected and these too are modified with glycosylation. Apigenin 4′, 7-glucoside and luteolin 7-diglucoside have been detected in "Dandy" (Nordström and Swain 1953). The last compound is butein (6′-deoxychalcone), a species of chalcone named after *Butea frondosa*. Butein is not widely distributed and is detected in limited genera including dahlia (Price 1939), *Cosmos sulphureus*

(Geissman 1942), and *Coreopsis grandiflora* (Geissman 1943). In dahlia, butein is accumulated as butein 4′-malonylsophoroside and butein 4′-malonylglucoside (Harborne et al. 1990). In addition to these three principal compounds, aurone and flavonol are believed to be present. Combinations of these pigments account for the wide range of flower colors. The major colors and their pigment compositions are shown in Table 1.

Table 1. Major flower colors and pigment compositions.

Flower color	Anthocyanin	Butein	Flavone
Ivory white	−	−	○
Yellow	−	○	○
Pink	+	−	○
Purple	++	−	○
Deep purple	+++	−	○
Orange	+	○	○
Red	++	○	○
Black	+++	−	−

+: low anthocyanin, ++: medium anthocyanin, +++: high anthocyanin

General regulation of flavonoid synthesis pathway

All pigments in dahlia flowers are synthesized by the flavonoid biosynthetic pathway, the best known secondary metabolite pathway in plants. Anthocyanidin, the aglycone of anthocyanin, is formed in the anthocyanin synthesis pathway by the condensation of three molecules of malonyl-CoA with one molecule of 4-coumaroyl-CoA. The enzymes involved in this pathway are chalcone synthase (CHS), chalcone isomerase (CHI), flavanone 3-hydroxylase (F3H), dihydroflavonol 4-reductase (DFR), and anthocyanidin synthase (ANS). Generally, anthocyanidins are modified by glucosyltransferase (3GT, 5GT), malonyl transferase (MT), and/or acyl transferase (AT) and are accumulated as anthocyanin (Grotewold 2006, Tanaka et al. 2008). Flavones are synthesized from flavanone by flavone synthase (FNS). In the anthocyanidin and flavone synthetic pathway, cyanidin and luteolin are synthesized when flavonoid 3′-hydroxylase (F3′H) is active, whereas pelargonidin and apigenin are synthesized when F3′H is inactive. Butein is synthesized with three molecules of malonyl-CoA and one molecule of 4-coumaroyl-CoA, the same substrate as that of tetrahydroxychalcone (chalcone), is converted to trihydroxychalcone (isoliquiritigenin) by CHS and chalcone reductase (CHR) (Bomati et al. 2005) and then hydroxylated by chalcone 3-hydroxylase (CH3H). The simplified flavonoid synthetic pathway in dahlia is shown in Fig. 2.

Some genes encoding the abovementioned enzymes involved in the anthocyanin synthesis pathway are regulated by transcription factors such as basic helix–loop–helix (bHLH), R2R3-MYB, and WD40 repeats (WDR); these three proteins function as transcription factors by forming complexes

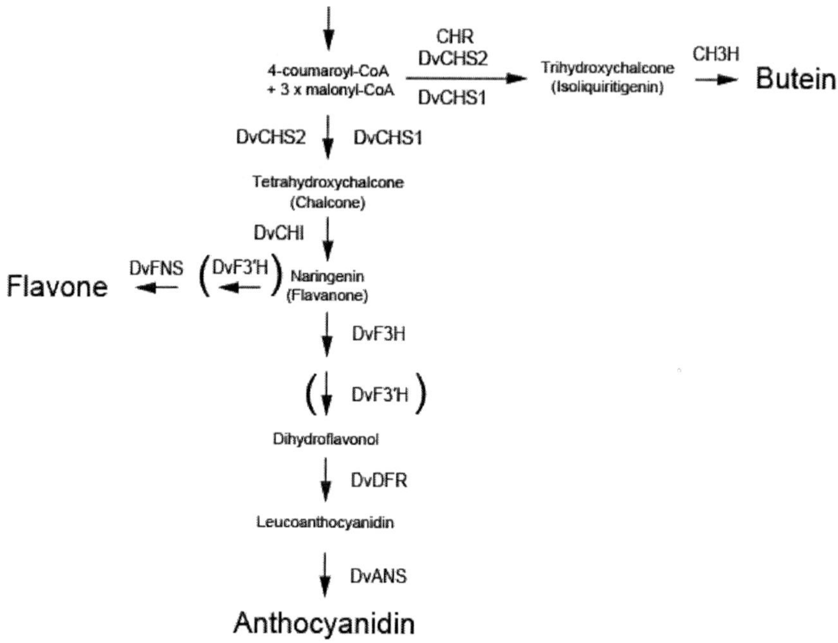

Fig. 2. The flavonoid synthesis pathway in dahlia. The simplified biosynthesis pathway of three major compounds: anthocyanidin, flavone, and butein in dahlia. Abbreviations: ANS, anthocyanidin synthase; CH3H, chalcone 3-hydroxylase; CHI, chalcone isomerase; CHR, chalcone reductase; CHS, chalcone synthase; DFR, dihydroflavonol 4-reductase; F3H, flavanone 3-hydroxylase; F3'H, flavonoid 3'-hydroxylase; FNS, flavone synthase.

or acting alone (Koes et al. 2005, Hichri et al. 2011). Thus not only structural genes in the flavonoid synthetic pathway but also transcription factors can be targets for molecular breeding.

Cyanic and acyanic cultivars

Cyanic cultivars have anthocyanins in their petals, while acyanic cultivars do not. This character is explained by a bHLH transcription factor, *DvIVS*, hence, *DvIVS* is targeted for molecular breeding. Flower color intensities are explained by the genotype of the promoter region in *DvIVS*, which is reflected in the expression level.

For molecular breeding, especially for marker-assisted selection, it is essential to determine the key gene regulating a flower color. However, as noted previously, owing to the high polyploidy of dahlia, it is not easy to determine the key gene because a single loss-of-function mutation is hidden by genetic redundancy. Accordingly, to reveal the gene regulating anthocyanin synthesis in dahlia we used a method called transposon tagging. Some cultivars have variegations in their petals, which are usually caused by transposition of transposable elements (TE). The uncolored parts correspond to the silent state owing to TE insertion and the colored parts to the active state from TE deletion. In our study (Ohno et al. 2011a), two lateral mutants MJOr (Fig. 1B) and MJY (Fig. 1C) that spontaneously occurred in "Michael J" (Fig. 1A), which has yellow ray florets with orange variegation, were used. MJOr, a bud mutant producing orange ray florets, accumulates anthocyanins, flavones, and butein, while MJY, another mutant producing yellow ray florets, accumulates flavones and butein. The lack of anthocyanin in MJY is believed to be due to TE insertion in a gene associated with anthocyanin synthesis. Because the genetic background is expected to be the same except for the candidate gene regulating anthocyanin biosynthesis, these mutants are very suitable for comparative genetic analysis.

Expression levels of *DvCHS1*, *DvF3H*, *DvDFR*, and *DvANS* were lower in MJY than in MJOr at all developmental stages, and the expression level of *DvIVS* was lower in MJY than in MJOr. This result suggested that *DvIVS* is a transcription factor that activates anthocyanin biosynthesis genes including *DvCHS1*, *DvF3H*, *DvDFR*, and *DvANS*. A 5.4-kb insertion sequence, which is a CACTA superfamily transposable element and is named *transposable element of Dahlia variabilis 1* (*Tdv1*) was found in the fourth intron of the *DvIVS* gene in MJY. In MJOr, *DvIVS* expression is followed by activation of *DvCHS1*, *DvF3H*, *DvDFR*, and *DvANS* and synthesis of anthocyanidin, flavone, and butein. In MJY, because of *Tdv1* insertion, only a truncated *DvIVS* mRNA is transcribed and transcriptional regulation cannot occur. Because transcriptions of *DvCHS2* and *DvCHI* are not affected by *DvIVS*, accumulation of flavone and butein were observed in MJY. Thus MJOr produces orange ray florets owing to the accumulation of anthocyanin, flavone, and butein, while MJY produces yellow ray florets owing to the accumulation of flavone and butein. From these results, we may infer that a bHLH transcription factor, *DvIVS*, is involved in anthocyanin synthesis in dahlia and that the regulations of anthocyanin and flavone/butein synthesis are different in dahlia (Fig. 3).

Regulation of flower color intensity

In many cases, flower color intensity can be explained by the amounts of pigments, but in some cases, other factors such as the structure of epidermal cells or cellular pH also contribute to flower color (Mol et al. 1998). Accordingly, we first aimed to determine which factor is most important for flower color intensity. In dahlia, there are a series of cultivars with different color intensities, for example, purple to pink or red to orange. We investigated the factors determining flower color intensity in dahlia (Ohno et al. 2013).

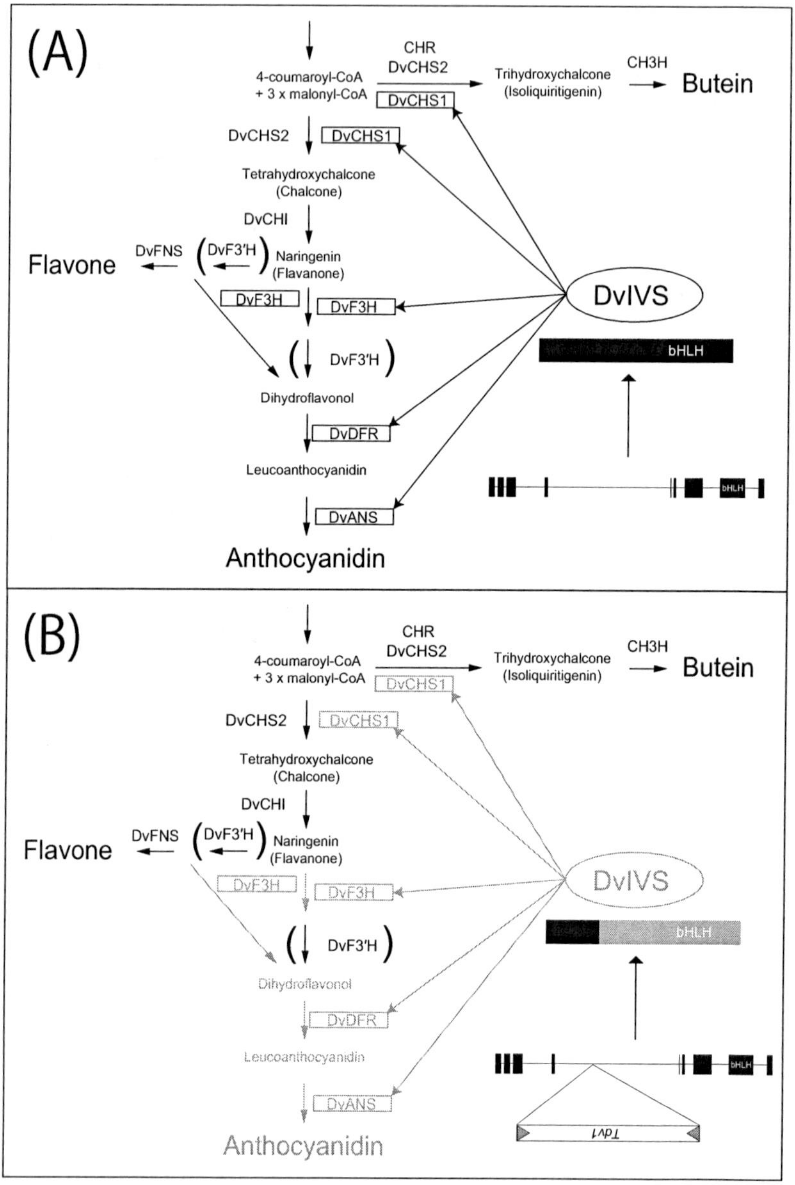

Fig. 3 A bHLH transcription factor *DvIVS* regulates anthocyanin synthesis in dahlia. (A) The model for MJOr. *DvIVS* is expressed, after which *DvCHS1*, *DvF3H*, *DvDFR*, and *DvANS* are activated, and anthocyanidin, flavone, and butein are synthesized. (B) The model for MJY. Because of the *Tdv1* insertion, only truncated *DvIVS* mRNA is expressed and transcriptional regulation cannot occur. Although *DvCHS1*, *DvF3H*, *DvDFR*, and *DvANS* are not activated, *DvCHS2* and *DvCHI* are not affected by *DvIVS*, resulting in accumulation of flavone and butein.

To test the factors determining flower intensity, we chose 12 cultivars with different color intensities: deep purple ["Super Girl" (Fig. 1D) and "Yukino" (Fig. 1E)], purple ["Cupid" (Fig. 1F), "Evelyn Rumbold" (Fig. 1G) and "Atom"], pink ["Magokoro," "Jyunn-ai" (Fig. 1H) and "Saffron" (Fig. 1I)] and ivory white cultivars ["Gitts Attention," "Zannsetsu," "Hakuba" (Fig. 1J) and "Hakuyo" (Fig. 1K)]. Deep purple, purple, and pink cultivars accumulate anthocyanins and flavones in their petals, while ivory white cultivars accumulate only flavones.

To describe flower color objectively, we measured color components as CIE $L^*a^*b^*$ coordinates. This enabled us to describe differences in petal colors as quantitative numerical values. Here, L^* indicates lightness, from 0 (black) to 100 (white). A positive a^* value indicates redness and a negative a^* value greenness. A positive b^* value indicates yellowness and a negative b^* value blueness. Chroma (c^*), purity or saturation of the color, was calculated as follows: $c^* = (a^{*2} + b^{*2})^{1/2}$. The L^* value can be used as an indicator of flower color intensity. When the L^* and c^* values of twelve cultivars were plotted onto coordinate axes, four color groups were clearly separated. Ivory white cultivars had the highest L^* and lowest c^* values, whereas deep purple cultivars had the lowest L^* and the highest c^* values. Pink and purple cultivars were plotted between deep purple and ivory white cultivars. Anthocyanin content was highest in deep purple cultivars, medium in purple cultivars, low in pink cultivars, and null in ivory white cultivars. An inverse relationship was observed between L^* value and anthocyanin content. All cultivars accumulated pigments in the epidermal cells of the petals. No relationship was detected between L^* value and either cell shape or petal pH. These results suggested that anthocyanin content is a key factor determining flower color intensity in dahlia.

What then determines anthocyanin content? As described above, *DvIVS* regulates anthocyanin synthesis in dahlia. We accordingly hypothesized that higher expression of *DvIVS* led to higher anthocyanin accumulation. We examined the correlation between anthocyanin content and gene expression. The expression levels of genes under regulation of *DvIVS* (*DvCHS1*, *DvF3H*, *DvDFR*, and *DvANS*) showed high correlations with anthocyanin content and that of *DvIVS* also showed a high correlation with anthocyanin content ($r = 0.96$). We inferred that the expression level of *DvIVS* determines the anthocyanin content.

The expression level of a gene is usually influenced by its promoter region. When we examined the promoter region of *DvIVS*, three types of promoter regions were detected in four color groups. Deep purple cultivars have Type 1 promoter, purple cultivars have Types 1 and 2 and/or Type 3 promoters, and pink cultivars have Type 2 and/or Type 3 promoter, whereas except for "Hakuyo," ivory white cultivars do not have these promoters. Thus, we found an association between *DvIVS* expression levels and *DvIVS* promoter types, suggesting that the genotype of the *DvIVS* promoter region determines the *DvIVS* expression level. We assumed that the Type 1 promoter has strong activity, whereas Types 2 and 3 have weak activity. Thus, the combination of *DvIVS* genotypes will determine the flower color intensity in dahlia (Fig. 4). At least, Type 1 promoter is expected to contribute to deep flower color, given that 22 of 23 cultivars carrying the Type 1 promoter showed deep color (purple, red, and black) (Ohno et al., unpublished data). These findings will contribute to marker-assisted selection for cyanic cultivars.

Ivory white cultivars

Ivory white cultivars accumulate only flavones in their petals. This character results from the failure of synthesis of anthocyanins and butein. Accordingly, nonfunctional alleles of both *DvIVS* and *CHR* are targets for breeding ivory white cultivars (Fig. 5).

In dahlia, there are many ivory white cultivars with a slightly cream color in the petals. This color is clearly distinguished from the white areas of bicolor cultivars where no pigment is accumulated (See below). In the petals of ivory white cultivars, *DvIVS*, *DvCHS1*, *DvF3H*, *DvDFR*, and *DvANS* were not expressed (Ohno et al. 2011b). Thus, loss of *DvIVS* function is important for ivory white genomic background. In the promoter region of *DvIVS*, there were at least two variants of nonfunctional *DvIVS* allele. One was unidentified and thought to be carried in "Gitts Attention," "Zannsetsu," and "Hakuba," which do not have Types 1, 2, or 3 promoters of *DvIVS*. It is likely that deletion or some modification of promoter region of *DvIVS* leads to the absence of expression. The other variant was found in 'Hakuyo', which has a Type 2 promoter. This allele is expressed, given that the 5′ untranslated region of a *DvIVS* transcript was detected. But no full length transcript was detected, suggesting that a deletion or insertion in an exon or intron leads to the expression of truncated, nonfunctional transcripts. The analysis of the genomic structure of nonfunctional alleles is now in progress.

Yellow Cultivars

Yellow cultivars accumulate butein and flavones in their petals without anthocyanin accumulation. In yellow cultivars, the synthesis of isoliquiritigenin always co-occurs with butein synthesis, suggesting *CHR* or its transcription factor as a target for breeding yellow cultivars. The genes for the butein synthetic pathway in dahlia have not been isolated;

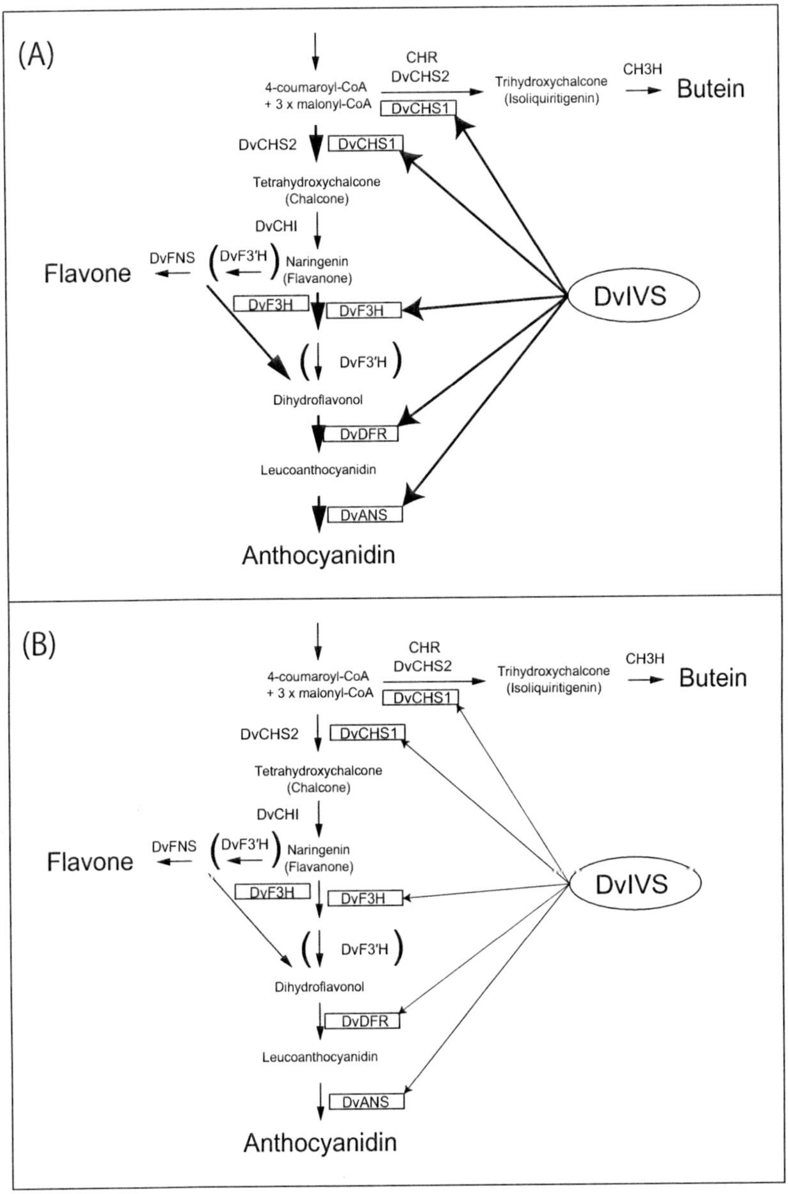

Fig. 4. Flower color intensity is regulated by the expression level of *DvIVS*. Flower color intensity is dependent on anthocyanin content in petals. The expression level of *DvIVS*, which may be determined by the genotype of its promoter region, shows a close correlation with the amount of anthocyanin. Heavy arrows indicate high and light arrows low expression levels. (A) In deep purple cultivars, *DvIVS* is abundantly expressed leading to accumulate high amount of anthocyanin. (B) In pink cultivars, *DvIVS* has lower expression, leading to lower anthocyanin accumulation.

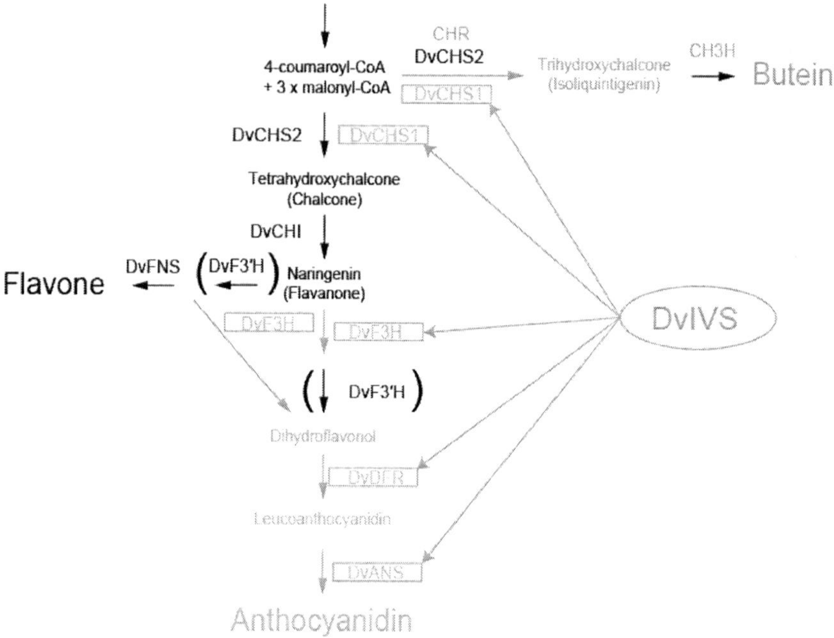

Fig. 5. Flavonoid synthetic pathway in ivory white cultivars. Ivory white cultivars accumulate only flavones in their petals. This character may be because of the absence of the function of *DvIVS* and *CHR*. *DvCHS2*, *DvCHI*, and *DvFNS* are expressed to synthesize flavones.

the inferred pathway from previous reports is shown in Fig. 6. Functional *CHR* (or its transcription factor) and nonfunctional *DvIVS* are targets for breeding yellow cultivars.

Bicolor Cultivars

Bicolor cultivars have white tips with colored petals. No pigment is accumulated in the white areas of these cultivars in contrast to flavone accumulation in ivory white cultivars. This is because post-transcriptional gene silencing (PTGS) of *CHS* genes occurs in the white areas of bicolor cultivars (Ohno et al. 2011b). PTGS is an RNA-silencing scheme mediated by small interfering RNA (siRNA).

No flavonoid derivatives were detected in the white areas of petals in bicolor cultivars, indicating that their white areas are pure white. In the white areas of petals in "Yuino" (Fig. 1 L), "Matsuribayashi" (Fig. 1M), "Kazusa-shiranami" (Fig. 1N), and "Ori2" (Fig. 1O), both *DvCHS1* and *DvCHS2*

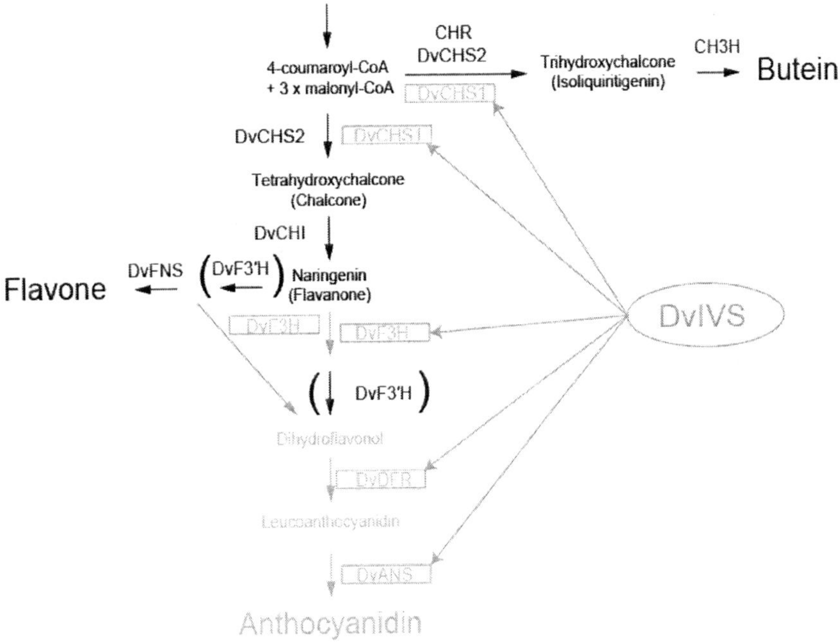

Fig. 6. Flavonoid synthetic pathway in yellow cultivars. Yellow cultivars accumulate butein and flavones in their petals. In the combination of nonfunctional *DvIVS* and functional *CHR*, anthocyanins are not synthesized but butein is synthesized.

were downregulated. CHS is an enzyme of the first step of the flavonoid synthetic pathway, which synthesizes chalcone (tetrahydroxychalcone) through the sequential condensation of one molecule of coumaroyl-CoA and three molecules of malonyl-CoA. Although "Yuino" retains two alleles for *DvCHS1* and *DvCHS2*, small RNAs were mapped onto all *CHS* transcripts (Fig. 7). Thus all *CHS* genes are silenced post-transcriptionally, resulting in the formation of white areas on petals (Fig. 8).

For breeding, the genetic feature causing PTGS of *CHS* is important. It is reported that two highly similar *glutelin* genes that form a tail-to-tail inverted repeat are a potent inducer of RNA silencing (Kusaba et al. 2003). With respect to *CHS* PTGS, inverted repeats of *CHS* in maize (*Zea mays*) (Della Vedova et al. 2005) and soybean (*Glycine max*) (Kasai et al. 2007) and a tandem repeat in petunia (*Petunia hybrida*) (Morita et al. 2012) are indicated as potent causes. Accordingly, we expect that bicolor cultivars too contain a specific *CHS* allele causing PTGS, and the identification of this genomic structure will contribute to breeding bicolor cultivars in dahlia.

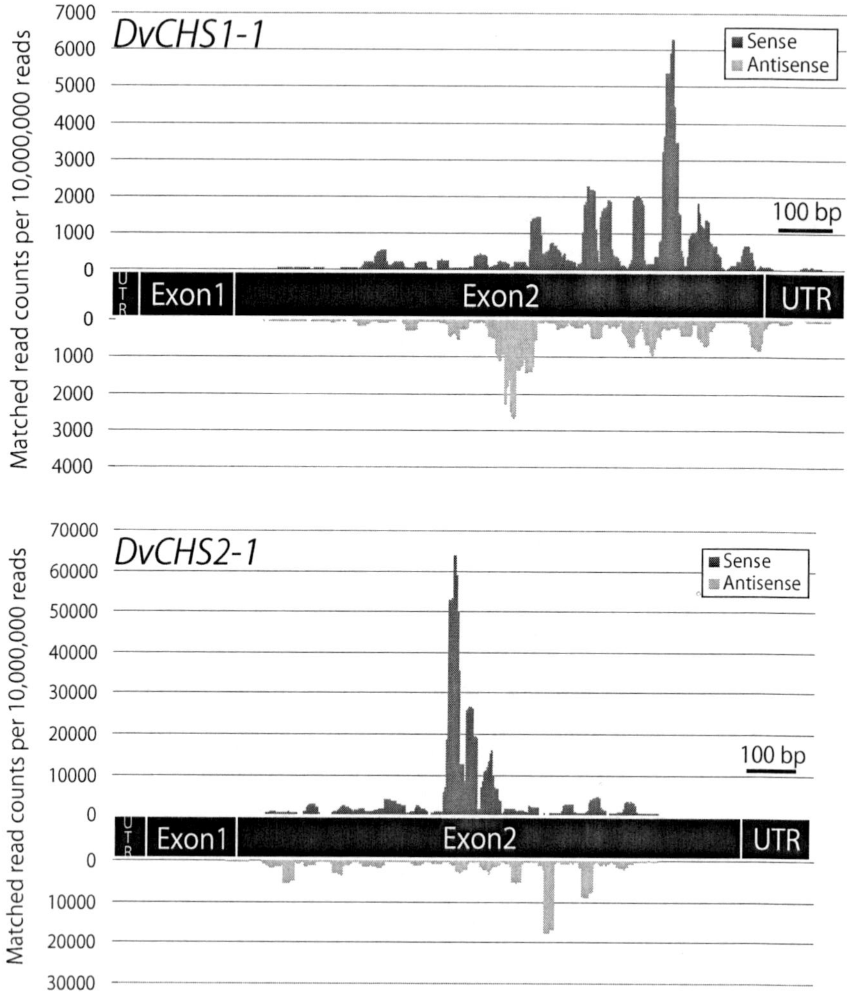

Fig. 7 Mapping of *CHS* small RNAs. 18–32 nt small RNAs with 100% match to the *DvCHS1-1* and *DvCHS2-1*genes were mapped on either the sense (black: above the X-axis) or antisense (gray: below the X-axis) strand. Number of matched reads is shown as per 10,000,000 reads.

Black Cultivars

Black coloring in flowers is an especially attractive trait, but this trait is rare and is present in few genera. The mechanism for black coloring has not been determined; therefore, breeding black cultivars requires large numbers of crosses. In our recent study (Deguchi et al. 2013), the mechanism for black coloring in dahlia was largely revealed. In the near future, marker-assisted breeding for selecting black cultivars will be available.

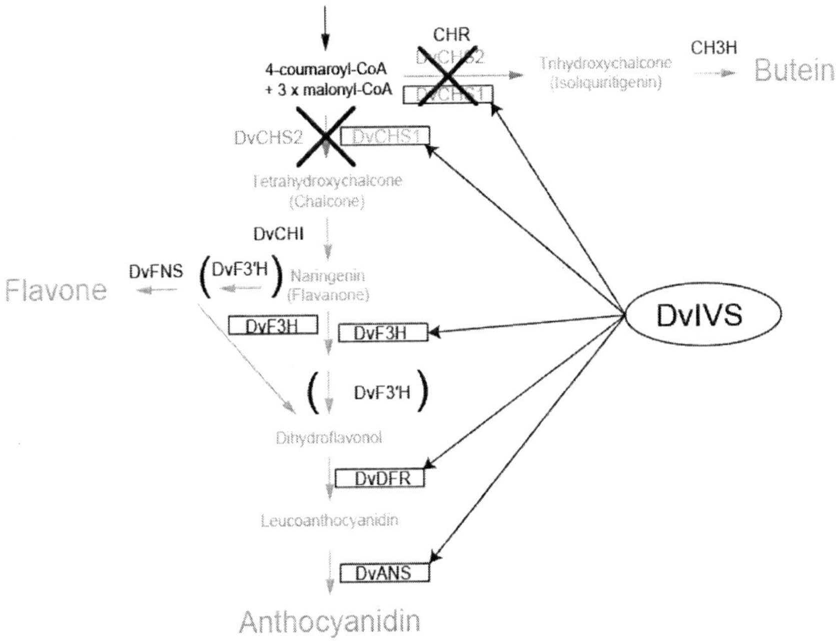

Fig. 8. Flavonoid synthetic pathway in the pure white areas of bicolor flowers. In pure white areas of bicolor cultivars, simultaneous post-transcriptional silencing of *DvCHS1* and *DvCHS2* occurs. CHS is the first step of flavonoid synthesis; hence, inactivation of CHS leads to an absence of flavonoids, resulting in the formation of pure white areas.

In dahlia, there are some black cultivars whose name includes "black" or an equivalent word for "black." Black cultivars have the lowest L^* value and have c^* values similar to those of purple or deep purple cultivars. Four black cultivars ["Fidalgo Blacky" (Fig. 1P), "Ms. Noir" (Fig. 1Q), "Kokucho" (Fig. 1R) and "Black Cat" (Fig. 1S)] accumulated comparatively high amounts of anthocyanin (anthocyanidin), high amounts of cyanidin, and no other flavonoids. But in fact two of the black cultivars, "Fidalgo Blacky" and "Black Cat," accumulated lower total anthocyanidin (pelargonidin and cyanidin) than the deep purple cultivars ("Yukino" and "Super Girl"). Thus, the black coloring of dahlia petals cannot be explained by only the total anthocyanidin amount. In a comparison between cultivars having almost the same amount of total anthocyanidins, the one with higher cyanidin contents had a lower L^* value. It was accordingly suggested that cyanidin contributes strongly to decreasing L^* value. However, given that the black cultivar "Kokucho" and the purple cultivar "Cupid" accumulated the same amount of cyanidin, "Kokucho" had a lower L^* value than "Cupid," black coloring could not be explained merely by cyanidin content. We accordingly concluded that not only total anthocyanidin but also cyanidin content is important for a black appearance. Considering that cyanidin has more contribution to a lower L^* value than pelargonidin, a simple model for black coloring of petals is shown in Fig. 9. In Fig. 9, the contribution to lowering L^* is calculated as "darkness units," each of which indicates the contribution of one molecule of anthocyanidin. The greater the sum of darkness units, the lower will be the L^* value. Under high total anthocyanidin and/or high cyanidin content, the sum of darkness units exceeds a hypothetical threshold and the petals appear black.

The quantification of pigments revealed another interesting result. A spontaneous lateral purple mutant of a black cultivar "Kokucho" (Fig. 1T) accumulated flavones and lower total anthocyanidin than the original "Kokucho." The decrease in anthocyanidin was nearly the same as the increase in flavones in the mutant. This observation may be accounted for by competition for substrates between flavone synthesis and anthocyanidin synthesis, whose pathways branch halfway along the anthocyanin pathway. The arrest of flavone synthesis could abolish the competition, resulting in high accumulation of total anthocyanidins. Furthermore, in the "Kokucho" mutant, the reduced amounts of both pelargonidin and cyanidin are nearly equal and only the amount of cyanidin was significantly lower than that in the original. This finding suggested that the absence of flavone could contribute to the black appearance via an increase of both cyanidin and total anthocyanidin amounts.

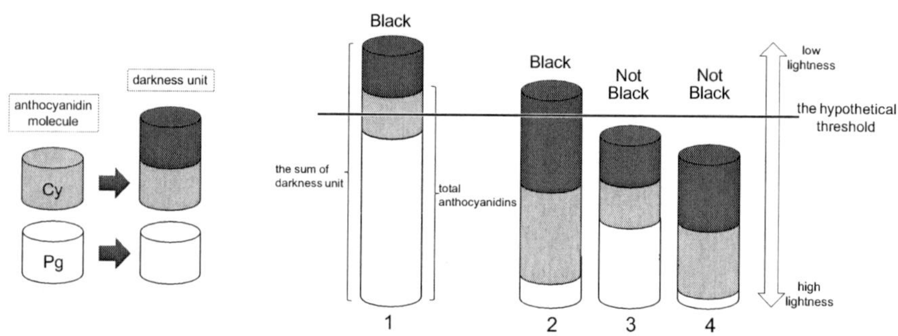

Fig. 9. A model for black coloring of petals. To explain black petal color we define the "darkness unit", which indicates the contribution of a molecule to lowering lightness (L^*). Cyanidin (Cy) molecules will contribute much more to lowering lightness than the same number of pelargonidin (Pg) molecules. For example, the darkness unit of cyanidin is considered to be twice that of pelargonidin (Saito and Mikanagi 1989). When the sum of darkness units approaches a hypothetical threshold, the flower color looks black to the eye. Four representative cases are shown. (1) Total anthocyanidins exceed the hypothetical threshold. (2) Cyanidin concentrations lead to exceeding the hypothetical threshold. (3), (4) The hypothetical threshold is not exceeded. White: pelargonidin content = pelargonidin "darkness units"; gray: cyanidin content, and gray + black: cyanidin "darkness units".

These black cultivars had the Type 1 (strong) promoter of *DvIVS* and its expression levels were comparatively high. However, the expression levels of the genes for anthocyanidin biosynthesis including *DvIVS* were not necessarily as high as those of deep purple cultivars containing the same amount of total anthocyanidin. The genes for modification and transportation, *Dv3GT*, *Dv3MT*, and *DvGST* were not highly expressed either. However, the expression level of *DvFNS* was significantly lower in all black cultivars than in deep purple and purple cultivars. Surprisingly, in black cultivars, the suppression of *DvFNS* occurred in a post-transcriptional manner, as determined by small RNA mapping. In petunia, artificial silencing of *FNS* or *FLS* led to increase of anthocyanins (Davies et al. 2003, Holton et al. 1993). Thus PTGS of *DvFNS* causing an absence of flavones could abolish the competition for substrates and induce high accumulation of total anthocyanidin and cyanidin, resulting in lowering L^* value to confer the black appearance in most black cultivars (Fig. 10).

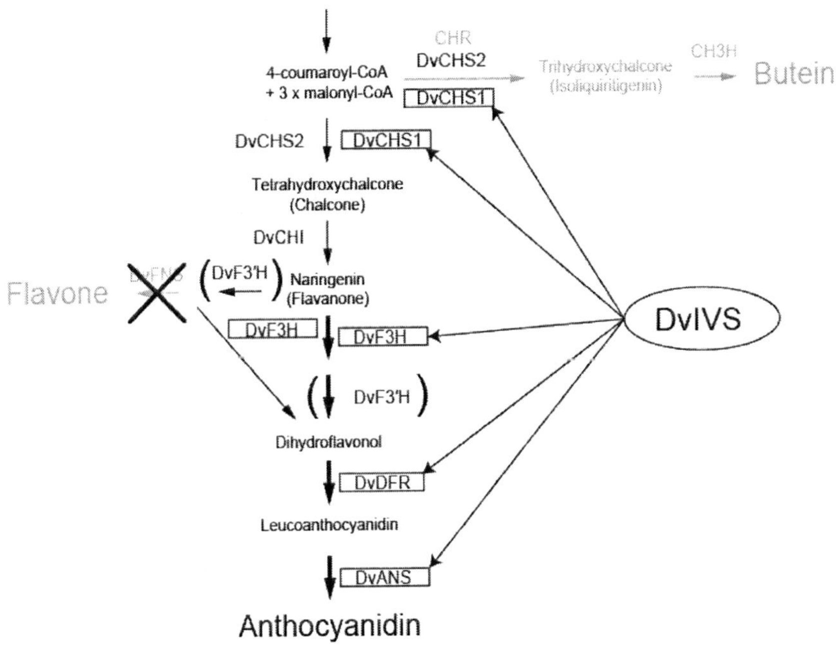

Fig. 10. The flavonoid synthetic pathway in most black cultivars. In most black cultivars, post-transcriptional gene silencing of *DvFNS* occurs. When *DvFNS* is suppressed, the competition for substrates between flavone and anthocyanidin synthesis is abolished. The substrates destined for flavone synthesis are then available for anthocyanidin synthesis. Because these cultivars have the Type 1 (strong) promoter of *DvIVS*, they can synthesize high amounts of anthocyanidin from the substrates including the newly available portion. With *DvFNS* suppression, increase of both the total amount of anthocyanidin and the cyanidin content result in lowering the L^* value to confer a black appearance.

It is worth mentioning that a few black cultivars accumulate both flavone and high amounts of cyanidin (Deguchi et al., unpublished data). This observation suggests that the silencing of *DvFNS* followed by no accumulation of flavones is not a prerequisite for black coloring and that (an)other mechanism(s) may be independently involved. However, in fact, almost all black dahlia cultivars undergo PTGS of *DvFNS*. Thus, the genomic background, probably an inverted or tandem repeat of *DvFNS*, that triggers the PTGS of *DvFNS* combined with a *DvIVS* Type 1 (strong) promoter will predict black cultivars with high probability. Work aimed at clarifying the genomic background causing the PTGS of *DvFNS* is currently in progress.

Molecular Breeding in Dahlia: Past and Future

In the 1920s and 1930s, Lawrence and colleagues proposed the following four elements of flower color inheritance in dahlia: (A) pale anthocyanin; (B) deep anthocyanin; (I) ivory flavones; and (Y) yellow flavone (butein) (Lawrence 1929, Lawrence and Scott-Honcrieff 1935). In their reports, B was dominant over A. Applying these factors to cultivars used in our experiment, pink cultivars may carry only A and purple cultivars may carry B. Thus, it is expected that the weakly functioning *DvIVS* allele may correspond to factor A, the strongly functioning *DvIVS* allele may correspond to B, and I may correspond to the nonfunctional *DvIVS* allele carried by ivory white cultivars. Yellow or red cultivars usually contain buteins as a yellow pigment, and cultivars without buteins contain neither butein nor its precursor, isoliquiritigenin. We have shown that anthocyanin synthesis and butein synthesis are mutually independent in dahlia; therefore, the Y element may correspond to the chalcone reductase gene or its transcription regulation factor.

We have shown that the flower color of dahlia is regulated with transcriptional regulation by *DvIVS* and PTGS of *DvCHS*s or *DvFNS*. These findings are essential for breeding programs of dahlia and are summarized below and in Table 2.

Table 2. Flower colors and corresponding inferred genomic backgrounds.

Flower color	*DvIVS* Strong promoter	*DvIVS* Weak promoter	*CHR*	*FNS* PTGS	*CHS* PTGS	*DvIVS-Tdv1*
Ivory white	–	–	–	–	–	–
Pink	–	o	–	–	–	–
Purple	o	o	–	–	–	–
Deep purple	o	–	–	–	–	–
Yellow	–	–	o	–	–	–
Orange	–	o	o	–	–	–
Red	o	o	o	–	–	–
Deep red	o	–	o	–	–	–
Black	o	o or -	–	o	–	–
Bicolor					o	
Variegation						o

Pigments

Anthocyanin	Anthocyanin synthesis is regulated by *DvIVS*. The genotype of *DvIVS* promoter region influences the amount of anthocyanin. Promoter genotypes may be classified into at least four types: high, medium, low, and none.
Butein	Genes for butein synthesis have not been isolated, but nonbutein-accumulating cultivars also do not accumulate isoliquiritigenin, suggesting that the butein synthesis pathway is stopped at isoliquiritigenin synthesis by CHS and CHR. Accordingly, *CHR* or its regulating gene may be targeted for breeding yellow cultivars.
Flavone	Flavones are synthesized in petals of almost all cultivars, except for black cultivars or the white parts of bicolor cultivars. The suppression of flavone synthesis in black cultivars is induced by *DvFNS* PTGS. A genomic structure leading to PTGS of *DvFNS* is predicted, and isolation of the responsible allele will contribute to selection of black cultivars.

Flower colors

Purple	High amount of anthocyanin with flavone. Strong *DvIVS* promoter is required.
Pink	Low amount of anthocyanin with flavone. Weak *DvIVS* promoter is required.
Red	High amount of anthocyanin with flavone and butein. Strong *DvIVS* promoter and functional *CHR* are required.
Orange	Low amount of anthocyanin with flavone and butein. Weak *DvIVS* promoter and functional *CHR* are required.
Ivory white	Flavone. Unfunctional *DvIVS* and *CHR* are required.
Bicolor	Bicolor phenotype is caused by *CHS* PTGS. The genomic structure triggering PTGS will be important for marker-assisted selection.
Black	High amount of anthocyanin without flavone. Strong *DvIVS* promoter and the genomic structure of *FNS* PTGS are needed.
Variegation	Variegation phenotype is usually caused by transposition of transposable elements in pigment synthesis genes. Variegation cultivars in dahlia are separated into two types: purple variegation on white petals and red variegation on

yellow petals. In addition, a study of "Michael J" showed that transposition of *Tdv1* in *DvIVS* is the cause of the variegation phenotype. Thus, it will be possible to design a DNA marker for variegation cultivars.

Breeding dahlia cultivars is currently limited by crossing, owing to its high polyploidy and self incompatibility. Broertjes and Ballego (1967) reported that mutants of flower color or shape were easily obtained by radiation-induced mutation, but this method can be adapted for only limited features. Thus, marker-assisted selection using DNA markers will be a very useful method. Next-generation sequencing will provide abundant genetic information, and identifying genetic markers in other polyploid plants will also soon be realistic.

References

Bomati, E.K., M.B. Austin, M.E. Bowman, R.A. Dixon and J.P. Noel. 2005. Structural elucidation of chalcone reductase and implications for deoxychalcone biosynthesis. J. Biol. Chem. 280: 30496–30503.

Broertjes, C. and J.M. Ballego. 1967. Mutation breeding of *Dahlia variabilis*. Euphytica 16: 171–176.

Davies, K.M., K.E. Schwinn, S.C. Deroles, D.G. Manson, D.H. Lewis, S.J. Bloor and J.M. Bradley. 2003. Enhancing anthocyanin production by altering competition for substrate between flavonol synthase and dihydroflavonol 4-reductase. Euphytica 131: 259–268.

Deguchi, A., S. Ohno, M. Hosokawa, F. Tatsuzawa and M. Doi. 2013. Endogenous post-transcriptional gene silencing of flavone synthase resulting in high accumulation of anthocyanins in black dahlia cultivars. Planta 237: 1325–1335.

Della Vedova, C.B., R. Lorbiecke, H. Kirsch, M.B. Schulte, K. Scheets, L.M. Borchert, B.E. Scheffler, U. Wienand, K.C. Cone and J.A. Birchler. 2005. The dominant inhibitory chalcone synthase allele *C2-Idf* (*inhibitor diffuse*) from *Zea mays* (L.) acts via an endogenous RNA silencing mechanism. Genetics 170: 1989–2002.

Gatt, M., H. Ding, K. Hammett and B. Murray. 1998. Polyploidy and evolution in wild and cultivated Dahlia species. Ann. Bot. 81: 647–656.

Geissman, T.A. 1942. Anthochlor pigments. III. The pigments of *Cosmos sulphureus*. J. Am. Chem. Soc. 64: 1704–1707.

Geissman, T.A. 1943. Anthochlor pigments. IV. The pigments of *Coreopsis grandiflora*, nutt. I. J. Am. Chem. Soc. 65: 677–683.

Grotewold, E. 2006. The genetics and biochemistry of floral pigments. Annu. Rev. Plant Biol. 57: 761–780.

Harborne, J.B., J. Greenham and J. Eagles. 1990. Malonylated chalcone glycosides in Dahlia. Phytochemistry 29: 2899–2900.

Hichri, I., F. Barrieu, J. Bogs, C. Kappel, S. Delrot and V. Lauvergeat. 2011. Recent advances in the transcriptional regulation of the flavonoid biosynthetic pathway. J. Exp. Bot. 62: 2465–2483.

Holton, T.A., F. Brugliera and Y. Tanaka. 1993. Cloning and expression of flavonol synthase from *Petunia hybrida*. Plant J. 4: 1003–1010.

Kasai, A., K. Kasai, S. Yumoto and M. Senda. 2007. Structural features of GmIRCHS, candidate of the I gene inhibiting seed coat pigmentation in soybean: Implications for inducing endogenous RNA silencing of chalcone synthase genes. Plant Mol. Biol. 64: 467–479.

Koes, R., W. Verweij and F. Quattrocchio. 2005. Flavonoids: A colorful model for the regulation and evolution of biochemical pathways. Trends Plant Sci. 10: 236–242.

Kusaba, M., K. Miyahara, S. Iida, H. Fukuoka, T. Takano, H. Sassa, M. Nishimura and T. Nishio. 2003. Low glutelin content1: A dominant mutation that suppresses the glutelin multigene family via RNA silencing in rice. Plant Cell 15: 1455–1467.

Lawrence, W.J.C. 1929. The genetics and cytology of Dahlia species. Journal of Genetics 21: 125–159.

Lawrence, W.J.C. and R. Scott-Honcrieff. 1935. The genetics and chemistry of flower colour in Dahlia: A new theory of specific pigmentation. Journal of Genetics 30: 155–226.

Mol, J., E. Grofewold and R. Koes. 1998. How genes paint flowers and seeds. Trends Plant Sci. 3: 212–217.

Morita, Y., R. Saito, Y. Ban, N. Tanikawa, K. Kuchitsu, T. Ando, M. Yoshikawa, Y. Habu, Y. Ozeki and M. Nakayama. 2012. Tandemly arranged *chalcone synthase A* genes contribute to the spatially regulated expression of siRNA and the natural bicolor floral phenotype in *Petunia hybrida*. Plant J. 70: 739–749.

Nordström, C.G. and T. Swain. 1953. The flavonoid glycosides of *Dahlia variabilis*. Part I. General introduction. Cyanidin, apigenin, and luteolin glycosides from the variety "Dandy". Journal of the Chemical Society 2764–2773.

Ohno, S., M. Hosokawa, A. Hoshino, Y. Kitamura, Y. Morita, K. Park, A. Nakashima, A. Deguchi, F. Tatsuzawa, M. Doi, S. Iida and S. Yazawa. 2011a. A bHLH transcription factor, *DvIVS*, is involved in regulation of anthocyanin synthesis in dahlia (*Dahlia variabilis*). J. Exp. Bot. 62: 5105–5116.

Ohno, S., M. Hosokawa, M. Kojima, Y. Kitamura, A. Hoshino, F. Tatsuzawa, M. Doi and S. Yazawa. 2011b. Simultaneous post-transcriptional gene silencing of two different chalcone synthase genes resulting in pure white flowers in the octoploid dahlia. Planta 234: 945–958.

Ohno, S., A. Deguchi, M. Hosokawa, F. Tatsuzawa and M. Doi. 2013. A basic helix-loop-helix transcription factor *DvIVS* determines flower color intensity in cyanic dahlia cultivars. Planta 238: 331–343.

Price, J.R. 1939. The yellow colouring matter of *Dahlia variabilis*. Journal of the Chemical Society 1017–1018.

Saito, N. and Y. Mikanagi. 1989. Sekishokushikiso to sekishokuka. Bio Hortic. 3: 49–62.

Takeda, K., J.B. Harborne and R. Self. 1986. Identification and distribution of malonated anthocyanins in plants of the compositae. Phytochemistry 25: 1337–1342.

Tanaka, Y., N. Sasaki and A. Ohmiya. 2008. Biosynthesis of plant pigments: Anthocyanins, betalains and carotenoids. Plant J. 54: 733–749.

12

Flowering in South African Iridaceae

Luise Ehrich

ABSTRACT

South African Iridaceae are a highly diverse and specialized group of bulbous plants and contain many genera with a great potential as a new floricultural crop. After a brief botanical introduction about the family in general and its South African genera, emphasis is placed on the description and illustration of the growth rhythms of Iridaceae species originating in the Cape Floral region. Thereafter, various aspects of the propagation and the cultivation of South African Iridaceae are presented and compared to the wellknown genus Freesia. Special attention is given to the different growth factors influencing the development and the physiology of these corm-bearing plants, especially their influence on flowering during cultivation. These factors are in particular: the applied storage, cultivation, temperature and light intensity regimes as well as the corm size used. Lastly, future prospects and the potential of some members of the South African Iridaceae that have been the subject of international research as new floricultural crops are discussed.

Keywords: Iridaceae, *Freesia, Gladiolus, Sparaxis, Tritonia, Watsonia,* inflorescence initiation, dormancy, corm, Cape Floral Region, new floricultural crops, South Africa, storage

P.O. BOX 12124, Garden Route Mall, George 6546, Western Cape, South Africa.
Email: luise_ehrich@hotmail.com

Introduction

The Iridaceae is one of the larger families of the Monocotyledonae, and is botanically recognised by a few critical features of the leaves and flowers. The flowers are characterized by their colourful perianth, or corolla, with three sepals each in the inner and outer whorl, as well as only having three stamens in the androecium and a mostly inferior ovary. The leaves are recognized by their sword-like shape as well as their edgewise orientation to the stem with both surfaces being unifacial.

Despite the current discussion amongst experts, about 66 genera and about 2000 species can be grouped into this family. This was reconfirmed by the investigation of the molecular sequences of several plastid DNA regions which endorsed that the Iridaceae constitute a single lineage (Goldblatt and Manning 2008).

The Iridaceae have a worldwide distribution, and amongst them are the well-known genera *Crocus*, *Iris* and *Tigridia*. However, the largest concentration of Iridaceae worldwide can be found in the Cape Floral Region (CFR) of South Africa. The CFR contains 27 genera and ca. 700 species of Iridaceae (Manning et al. 2002)—amongst them the well-known ornamental genera *Crocosmia*, *Freesia* and *Gladiolus*—and are characterized by a large diversity of flower colours (Niederwieser et al. 2002).

South African Iridaceae belong to the subfamily Aristeoideae (*Aristea*), Nivenioideae (*Nivenia, Klattia, Witsenia*) and Iridioideae (*Dietes, Bobartia, Ferraria, Moraea*), but the largest amount of genera belong to its sub-family Crocoideae (*Tritoniopsis, Lapeirousia, Pillansia, Thereianthus, Micranthus, Watsonia, Gladiolus, Malesphaerula, Crocosmia, Devia, Freesia, Xenoscapa, Radinosiphon, Romulea, Afrocrocus, Geissorhiza, Hesperantha, Babiana, Chasmanthe, Sparaxis, Tritonia, Ixia* and *Dierama*) (Goldblatt and Manning 2008). These genera have a wide flower colour range in shades of white, yellow, orange, red, pink, purple and blue and they always form corms, rhizomes or a woody caudex as their storage organ, but never the true bulbs typically displayed by the South American and Eurasian genera (Manning et al. 2002). For ease of reference and unless otherwise stated, further descriptions of the life-cycle and physiology of South African Iridaceae shall be for the findings regarding *Gladiolus* and *Freesia*, which are the two horticulturally and economically most important South African genera.

Growth rhythms of South African Iridaceae in the Cape Floral Region

Babiana, Chasmanthe, Freesia, Ixia, Sparaxis, Tritonia, Watsonia and many other South African Iridaceae genera originating from the CFR region in South Africa form corms that enable them to survive the hot and dry summers at dormancy stage, these sprout and grow during winter and flower in spring. Hence, the undoubtedly most impacting factor on their development is the availability of water and the ambient temperature.

The temperature decrease and the onset of winter rains evoke the Iridaceae's corms' sprouting and vegetative growth after dormancy, following which an inflorescence is formed by the shoot apex meristem (Manning et al. 2002). This life-cycle in correlation with temperature and rainfall of the CFR is illustrated in Fig. 1. Hence, unlike other South African geophyte species forming true bulbs (e.g., *Lachenalia* and *Ornithogalum dubium* Houtt), the inflorescence formation is only initiated within the apical meristem once the vegetative growth has commenced, but not during the (summer) dormancy (Rees 1989). This physiological pattern has been extensively studied (Ehrich 2007, Imanishi 1993). During the initiation of flowering, the apical meristem 'surrenders' its dominance and new apical buds are formed next to the inflorescence. These will remain inactive up to and throughout the dormancy and only become active in next year's growing cycle (Rees 1984).

Figures 2, 3 and 4 display *Sparaxis* corms for further illustration of the corm's yearly life-cycle. Contrary to other geophytes, the above listed corm forming genera replace these every year during the growing season and 'discharge' the corm formed in the previous year. In fact, the onset of the formation of the 'daughter corm' has been physiologically linked to

Fig. 1. Yearly life cycle in the winter rainfall region of the CFR of South Africa (Ehrich 2007).

Fig. 2. *Sparaxis* corms (Ehrich 2007).

Color image of this figure appears in the color plate section at the end of the book.

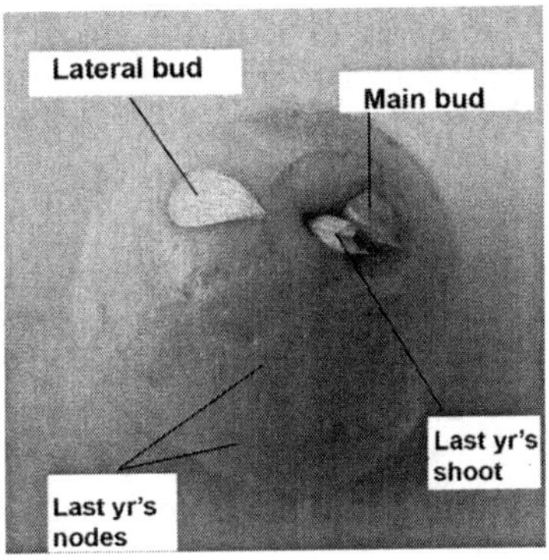

Fig. 3. *Sparaxis* corm in detail (Ehrich 2007).

Color image of this figure appears in the color plate section at the end of the book.

Fig. 4. *Sparaxis* mother and daughter corm illustrated (Ehrich 2007).

Color image of this figure appears in the color plate section at the end of the book.

the time point of flower initiation (Hartsema 1961). With the increase in temperatures, the lack of rainfall and the onset and ripening of seeds, these newly formed corms then again undergo dormancy until the next onset of winter rains and the decrease in ambient temperature. To be mentioned in this context should also be the formation of 'bulbils' (also called 'cormlets') alongside the daughter corm. These are produced by the lower axillary buds of the developing daughter corm and from those buds in the neck region of the mother corm, but are much smaller in size than the 'primary' daughter corm. They usually start developing after flowering and finish their development at the same time as the daughter corm, namely at the onset of dormancy (De Hertogh and Le Nard 1993).

Another factor which impacts on the natural development of South African Iridaceae in their natural habitat is the regularly occurring smoke in the Fynbos and Renosterveld vegetation types of the CFR. One of the main ingredients of the fire is ethylene (Halevy 1994) which has proven to have an effect on the physiology of commercially grown *Freesia* x *hybrida*. Lastly, the amount and the distribution of the annual rainfall significantly influence the duration of flowering as well as the number of individual flowers on the inflorescence (Struck 1994).

Commercial cultivation of South African Iridaceae

Commercial aspects

The two economically most important South African Iridaceae worldwide—especially for cut flower production—are *Freesia* and *Gladiolus*. Statistics from the International Association of Horticultural Producers (2011) show that the production area of freesias under glass in The Netherlands has dropped from 221 ha to 110 ha between 2000 and 2010. However, freesias are still one of the top sellers at the Dutch flower auctions, with 329.63 million stems sold in 2010 at an average price of 0.15 Euro per stem. Gladioli in contrast are one of the top selling cut flowers in the USA, with 89.68 million flowering stems being sold in 2010 at an average price of 0.19 Euro per stem. China is one of the largest producers of *Gladiolus*, with the production area covering 2891 ha in 2010 compared to only 722 ha in The Netherlands in the same year.

General aspects of cultivation

In the following paragraphs, the cultivation methods of freesias and factors affecting their physiology will be described. The cultivation of *Gladiolus* is not so readily comparable with South African Iridaceae originating in the CFR, since the modern hybrids have been influenced genetically by species originating outside of South Africa or originating from areas within South Africa with very different climatic conditions than the CFR. Although there is some variation between the different hybrids used in commercial horticulture, the following description should be seen a brief overview of the main cultivation methods applied.

Planting of freesia corms can happen throughout the year, although not all varieties are suitable for year round planting, since they react differently to the high temperatures during summer months as well as the low light intensities during winter months (Rupprecht 1988), henceforth an applied variety selection needs to be done by the grower. In general, flowering is achieved three to five months after planting, and then lasts for two to 3.5 months. Different pests and diseases can affect *Freesia* x *hybrida* during cultivation, amongst them are viruses (e.g., freesia leaf necrosis FLN), *Fusarium*, *Botrytis*, *Pseudomonas*, as well as aphids, thrips, mites and fungus gnats.

Corms are either produced in greenhouses (mainly in The Netherlands) or outdoors in frost-free climates. Soils should be within normal pH range (6.6 to 7.5), well drained but also water retaining, have low EC ranges and sterilized to prevent *Fusarium*. Corm and cut flower production can happen simultaneously. Harvest of corms usually happens about six weeks after flowering has ended. A more detailed description of commercial corm and cut flower production can be found in De Hertogh and Le Nard (1993).

In Ehrich (2007), common production features for the investigated South African *Freesia laxa* Ecklon, *Sparaxis* hybrids, *Tritonia deusta* Ker Gawler and *Tritonia securigera* Ker Gawler shared with other crops from the same area of origin (e.g., *Lachenalia*, spring flowering *Gladiolus* and to an extent also *Ornithogalum dubium*) were: Similar temperature requirements during storage, a pre-planting cooling of the corms/bulbs for 3 to 6 weeks at 9 to 17°C to enhance sprouting and shorten the time until flowering as well as cool temperatures during cultivation to prevent inflorescence abortion. Differences between species could be shown regarding their light requirements and their susceptibility towards pests during forcing and they require further, separate investigation for each species.

Methods of corm storage

The storage of *Freesia* x *hybrida* corms during their leafless stage is the determining factor which makes year-round production possible and guarantees the corms' sprouting upon planting. Long-term storage systems have made it possible to extend the leafless stage of the plants without impacting too much on the quality in respect of their performance upon planting.

In general, corms of *Freesia* x *hybrida* are subjected to a so-called "preparation" after harvest, i.e., storage of 10 to 17 weeks at 27 to 31°C and 60 to 70% relative humidity. This corresponds to the natural dormancy in the CFR which lasts for three to four months during summer. The overcoming of the dormancy guarantees the re-sprouting at planting. Horn (1996) reports that inflorescence initiation in freesia hybrids is only possible if the dormancy has been broken by a sufficiently long storage at 30°C. The exact duration of this "preparation" is dependent on variety as well as season in which the corms were harvested. Also, there have been results showing that ethylene treatments (2 days at 5 $\mu l\ l^{-1}$ after 4 weeks of storage at 30°C) can substitute 2 weeks of storage (Imanishi and Fortanier 1983) as well as guaranteeing very early sprouting (Horn 1996).

However, if long-term storage is planned before the next planting, the corms will be first subjected to a long cold storage at 1 to 3°C for up to 11 months after harvest. During that time and due to the very low temperatures, only on average 1.5% corm mass is lost due to respiration, although differences exist between varieties (Gilbertson-Ferriss et al. 1981a). At these low temperatures, the corms are kept inactive (hence the low respiration rate) and their dormant state is not changed. They can be removed from the cold storage at any time to be submitted to the above described "preparation" to overcome dormancy in preparation for planting. This long-term storage system can extend the period from harvest to planting to up to 16 months (Rupprecht 1988). For corms stored at 2°C for 6 months, Imanishi and Berghoef (1986) found that a three week long storage at 30°C and a consecutive ethylene treatment for three hours were sufficient to break dormancy.

Long-term storage of *Freesia* x *hybrida* corms is also possible when submitting the corms to storage at 13 to 15°C after four to six weeks storage at 30°C. During this storage, an interesting physiological phenomenon, namely 'pupation' of the corms, occurs. This physiological phenomenon denotes a process in which the original corm is replaced by a daughter corm without any aerial vegetative growth (Imanishi 1993). The newly formed corms are as dormant as the original mother corm and then need to be subjected to a "preparation" again for their dormancy to be successfully broken. Corms can thereby be stored for ten to twelve months. However, the pupation costs the corm up to 40% of its mass when stored for 9 months, making this type of storage unattractive to the commercial grower. The picture below in Fig. 5 illustrates the pupation of the corm.

Fig. 5. Pupation of *Tritonia deusta* corm (Ehrich 2007).

Color image of this figure appears in the color plate section at the end of the book.

Leaf and flower formation and associated meristem changes

Once the corms of South African Iridaceae have been released from dormancy, the corms sprout and undergo a period of strong vegetative growth, before the apical meristem undergoes transformation to the adult state initiating an inflorescence. Thompson et al. (2011) found the inflorescence initiation of *Watsonia* coinciding with the emergence of the second leaf. In this process, the apical meristem surrenders its dominance and new apical buds are formed laterally next to the initiated inflorescence. These are inactive throughout the subsequent dormancy of the corm (after flowering has been completed) and only become active once the corms sprout again after the next dormancy.

During its inflorescence initiation, the apical meristem undergoes a total of eight different initiation stages in *Freesia* x *hybrida* (De Hertogh and Le Nard 1993). Standard abbreviations for the stages of inflorescence are:

a) Stage I (vegetative): The flat shaped apical meristem differentiates leaf primordia.
b) Stage II (generative): The apical meristem takes on a more conical shape.
c) Stage Pr-Br: First flower primordia visible. Flowers bearing bracts, specialized leaves.
d) Stage Bo: The inner bract is initiated opposite the outer bract.
e) Stage A: Three stamen primordia appear.
f) Stage P_1: 1st whorl of perianth is formed.
g) Stage P_2: 2nd whorl of perianth is formed.
h) Stage G: The formation of the gynoecium is completed.

Propagation

Most South African Iridaceae can be propagated generatively (through seeds) or vegetatively by using the bulbils (cormlets) that form alongside the daughter corm. Usually, flowering is achieved earlier when using bulbils, since these are bigger in size than the corm of seedlings, but exceptions exist, e.g., *Freesia* x *hybrida*, where both seedlings and bulbils can reach flowering age in the second season. Another advantage of using the bulbils consists in the fact that these are clonal copies of the mother corm, and this is important for propagation of specific varieties. However, these have to undergo dormancy breaking treatments on the daughter corms before being able to be planted again. Seed propagation has obviously got its own important place when breeding new varieties, however it should be noted here that freesias are essentially self-incompatible.

Kapczynska and Piskornik (2003) showed that even 27 months after seed harvest, the germination rate was still more than 60% in *Sparaxis*, with the larger seeds germinating better than the smaller seeds. Of course, the temperature at which the seeds are being stored is important in this regard. For example, in Kapczynska and Piskornik (2003), the authors could show that seeds of *Sparaxis* stored at 10°C and 15°C displayed better germination rates that those seeds stored at 5°C. In Ascough et al. (2006), the authors clearly prove that there is a distinction in the optimal germination temperatures between those *Watsonia* species originating in the summer rainfall area of South Africa (germinating best at 25°C) and those species originating from the winter rainfall area (germinating best at 10°C), which the authors attribute to the occurrence of rainfall at these season specific temperatures in the original habitats of the respective species.

Freesia x *hybrida* is commonly propagated via bulbils, since these can reach flowering size within the next season (De Hertogh and Le Nard 1993) as well as reproducing the variety clonally. Tissue culture is applied when virus or disease-free plant material is required. It can be achieved through direct and indirect somatic embryogenesis as well as organogenesis (Wang 2006).

Successful clonal propagation as well as induced polyploidy through tissue culture has however also been achieved for more unusual and "new" South African Iridaceae crops, such as, e.g., *Babiana*, *Ixia*, *Romulea*, *Sparaxis* and *Watsonia*. With restricted plant material available for trialling and commercial introduction, tissue culture is often the only way to rapidly produce larger volumes of plant material.

In Thompson et al. (2010) and Ascough et al. (2008) *in vitro* regeneration and multiplication was easily achievable for *Watsonia* and the authors were hopeful to achieve desirable polyploidy for the improvement of the ornamental value (more compact growth habit and larger flowers) of the investigated six species of this genus. Explant survival was higher and more uniform with oryzalin treatment compared to that with colchicine. Ewald (1993) reported an improvement of the ornamental value of *Freesia laxa* through polyploidization, especially an increased flower diameter.

Sutter (1986) reported the successful *in vitro* propagation of *Ixia viridiflora* Lam., with many buds being produced on transversely-cut halves of the corms when using a Linsmaier-Skoog medium containing benzyladenine. These buds elongated into shoots which in turn formed corms at the base of the shoots when they were transferred to a medium containing naphthaleneacetic acid.

Swart et al. (2012) also achieved viable corms in tissue culture of the attractive and rare South African *Romulea sabululosa* Schltr. ex Beg. The heaviest corms were produced at 15°C on a MS (Murashige and Skoog) medium supplemented with 6% sucrose and shoots formed *in vitro* multiplied best on a medium with 2.5 μM *meta*-topolin riboside (*m*TR).

Factors Affecting Flowering

Corm size

The inflorescence formation is not "supported" by the reserves of the "mother corm", but through the assimilation during the active growth period, which is influenced by ambient growing conditions. However, the size (and age) of the corms directly influences the vigour of the sprouting of the corms, depending on whether they can be classified as juvenile or adult. This means that through the sprouting vigour of the corm relying on its reserves, it is determined whether the corm "is able" to initiate a flower, meaning that for corms below a certain minimum size (and age), i.e., which are still juvenile, an inflorescence is hardly ever initiated (Rees 1989b, Halevy 1990). For corms above this minimum size, flowering should occur in most cases, since the corms have become adult (Rees 1989b). For corm sizes in between, flowering is uncertain, since it is not clear whether juvenility has been overcome or not (Dole 2003). Adult corm size varies according to plant species and variety and are commonly indicated by circumference in cm to indicate whether these are considered marketable and able to produce flowers.

Additionally, the number of inflorescences as well as their quality is influenced by the size of the adult corm (Halevy 1990). Thompson et al. (2011) illustrated the varying differences in the relationship between corm size and flowering: Of *Watsonia angusta* Ker Gawler, even the smallest corms were able to flower, although they produced fewer flowers than those plants originating from larger corms. In opposition, most flowers per plant were recorded on the plants having been grown from the smallest corms in *Watsonia tabularis* Mathews & Bolus. Investigations of *Crocus sativus* L. revealed that that the minimum corm size able to flower is dependent on the field conditions prevalent during the year when corms were produced (Han 2001). Thompson et al. (2011) also mention the necessity of *Gladiolus* to have matured a critical number of leaves after which only flowering is possible. This is in contrast to *Freesia* x *hybrida* and the investigated *Watsonia* species of Thompson et al. (2011) where corm size has less influence on flowering than the environmental conditions prevailing during cultivation.

Pre-planting storage regimes

Berghoef and Zevenbergen (1990) also report that the extension of the activation of freesia corms used for the cultivation of potted freesias from three weeks to six or seven weeks at 13°C before planting managed to reduce plant height by 30 to 35 cm, which can reduce the use of necessary growth regulators (see section 4.7 below). However, this temperature treatment also led to a reduced number of flowers per inflorescence and a decrease in vase life down to 7 to 10 days. Ehrich (2007) could confirm this height reducing effect for *Freesia laxa* (height reduction of 5.5cm) as well as showing an increase in the number flowering plants compared to the control plants exposed to only 3 weeks at 13°C before planting (45% versus 13% during the spring batch at high ambient temperatures and 74% versus 43% during the autumn batch at cooler ambient temperatures). For *Tritonia deusta* this treatment led to a 4 to 6 week earlier inflorescence formation in the apical meristem and to a three week advanced anthesis, although only less than 20% of these plants flowered, probably caused by the growing conditions favouring flower abortion after inflorescence formation, namely the prevailing high temperatures during the European spring/early summer as well as low light intensities during the European autumn (Ehrich 2007) during trialing. Plant quality of *Sparaxis* hybrids and *Tritonia securigera* was greatly reduced by subjecting the corms to a longer storage at 13°C before planting, as this treatment lead to irregular and poor growth (Ehrich 2007).

Wulster and Gianfangna (1991) could also demonstrate a reduced plant height, an earlier anthesis as well as reduced shoot and inflorescence length by submitting the corms of *Freesia* x *hybrida* to a four-week long storage at 5°C after preparation. This height reducing effect was variety dependent and was displayed more strongly in naturally tall growing varieties compared to naturally more compact growing varieties. However, this treatment did not show any significant effects in Ehrich (2007) for the investigated *Freesia laxa*, *Sparaxis* hybrids as well as *Tritonia deusta* and *Tritonia securigera*. Gonzalez et al. (1998) could show a 20 and 11 days earlier anthesis for *Gladiolus tristis* L. (a species also originating in the CFR) corms stored for 3 and 6 weeks at

5°C before planting, with the flowering period increasing by 35 days after the longer storage period.

Another interesting set of results was presented by Ehrich (2007). The long-term storage (3.5 months) at 2°C for the batch planted in the European autumn was carried out after the corms of *Freesia laxa*, *Sparaxis* hybrids, *Tritonia deusta* and *Tritonia securigera* had already undergone a 3 to 4 month warm (> 20°C) storage in South Africa, and so this temperature treatment could not be fully compared with the cold storage carried out commercially for freesia corms directly after harvest and before the high temperature treatment as described by Imanishi (1993). After planting, the effect of the storage at 2°C differed among the selected species and hybrids. A portion of the corms subjected to the 2°C-storage of *Tritonia deusta* and *Sparaxis* hybrids pupated after planting, an unexpected result. The remainder of the corms sprouted and grew very poorly. The plants of *Tritonia securigera* subjected to the same treatment regime displayed an unacceptable growth vigour compared to the control plants, although no pupation was observed. In comparison, the reaction to the cold storage of *Freesia laxa* was most surprising. The plants whose corms were subjected to the 2°C storage before planting had significantly reduced plant height (ca. 10 cm), inflorescences formed 8 weeks earlier, anthesis was 7 weeks earlier, and there was a statistically significant increase in the number of inflorescences per plant compared to control plants. Furthermore, almost 100% flowering was achieved among the plants subjected to a 2°C storage and just over 40% in the control.

In Thompson et al. (2011), cold exposure (4 and 10°C) resulted in 100% sprouting of all the investigated *Watsonia* species, i.e., *Watsonia borbonica* (Pourr.) Goldblatt, *W. pillansii* Bolus and *W. tabularis* at the end of their dormancy. Only a partial release of dormancy was achieved when submitting the corms to 20°C instead (*W. borbonica* and *W. tabularis*). Also, Thompson et al. (2011) could link exposure to the above cooler temperatures to secondary shoot emergence and thereby an increase in total leaf number.

Light intensity

The prevalent high light intensity levels in South Africa result in a requirement for high light intensities during cultivation for *Freesia* x *hybrida* in Europe and North America, especially during the months of low ambient light intensities, such as autumn and winter, in order to achieve adequate product quality for the market (Rupprecht 1988). Horn (1996) states that the effect of light intensity is lower than the effect of temperature. However, low light intensities negatively impact on inflorescence formation and also vegetative growth. If light intensities are too low, flower abortion can occur in which the flower buds dry off (De Hertogh and Le Nard 1993). Supplemental lighting during winter months resulted in an increased number of inflorescences, slight advancement of anthesis as well as improved corm production, inflorescence quality and flowering as the light intensity applied increased (Doorduin 1992). However, the author also noted that 24 hr of supplemental lighting decreased plant quality in return. Berghoef and Zevenbergen (1990) managed to reduce plant height of *Freesia* x *hybrida* by 20 to 30 cm when cultivating the plants at 280 or 500 $W\ h^{-1}\ m^{-2}\ d^{-1}$ compared to plants grown under 105 $W\ h^{-1}\ m^{-2}\ d^{-1}$.

In Ehrich (2007), plants of *Freesia laxa*, *Sparaxis* hybrids, *Tritonia deusta* and *Tritonia securigera* grown in spring/early summer under naturally high light intensity and high temperatures during these months (over 22°C during the day almost daily for longer than two months) were more compact that those grown under a naturally cooler temperature regime and a lower light intensity in the European autumn months. Moreover, during the lower temperatures of the European autumn in Ehrich (2007), high light intensities achieved through additional lighting led to a statistically significant increase in plant height for *Tritonia securigera* and *Sparaxis* hybrids (both ca. 7 cm) as well as in the number of flower primordia per inflorescence for both *Tritonia* species (ca. 2 to 7) and *Sparaxis* hybrids (ca. 1), but these effects were not observed in *Freesia laxa*.

Besides these effects, the naturally low light intensities of the non-irradiated treatments during the winter months of the batch planted during the European autumn in Ehrich (2007) also led to flower abortion or a delay in flowering of *Sparaxis* hybrids, *Tritonia deusta* and *Tritonia securigera*, similar to the reaction of *Freesia* hybrids (Imanishi 1993) as well as *Gladiolus* (Halevy 1975) in this regard. Surprisingly, the described low light intensities had no negative effect on *Freesia laxa*, as can be confirmed by a similar or higher percentage of flowering plants of the non-irradiated plants of the control plants.

Thompson et al. (2011) also confirmed that reduced light intensities experienced by *Watsonia* up to the third leaf stage after sprouting increased plant mortality and induced flower abortion in the surviving plants despite the plants having been exposed to inductive temperatures. Thompson et al. (2011) relate this to the competition between the reproductive and the vegetative processes within the plant and reason that with a lack in photosynthetic activity none of these processes can be properly sustained. Interestingly, Thompson et al. (2011) could also demonstrate that after the third leaf stage, a significantly decreased light intensity did not impact as heavily anymore on the ability of the investigated *Watsonia* species to produce flower spikes, but still decreased the plant quality substantially.

Day length

The influence of day length on flower induction of *Freesia* x *hybrida* is discussed controversially in the relevant literature. It is considered less important than temperature (Horn 1996). Gilbertson-Ferriss (1989) reports the stimulation of flower induction through short day and increased plant growth during long day, but both effects decrease with an increase in cultivation temperature. Under short day conditions, the number of flower primordia per inflorescence is increased as well as the length of the inflorescence. During long day conditions, anthesis is advanced by 6 to 14 days, however the quality of the inflorescences decreases. Berghoef and Zevenbergen (1990) note that an 8 h short day reduces leaf length by 10 to 15% compared to a 16 h long day. Horn (1996) confirms that short days (8 to 10 h) promote flower induction and inflorescence differentiation. Long days prolong the phase of vegetative growth and lead to a high number of leaves before flower induction. Hence, Horn (1996) declares *Freesia* x *hybrida* to be a facultative (qualitative) short day plant. This is confirmed by Rünger (1976), however, the latter author has a different opinion on the influence of cultivation temperatures compare to the above mentioned Gilbertson-Ferriss (1989). Rünger (1976) states that with low cultivation temperatures, *Freesia* x *hybrida* hardly displays any reaction to daylength.

Thompson et al. (2011) report that the individual flowers of *Watsonia* matured under short day conditions were less intense in colour as well as facing downwards (instead of upwards), and thereby reducing the ornamental value of the plants, compared with those matured under long day conditions. Also, Thompson et al. (2011) categorized *Watsonia borbonica* and *Watsonia tabularis* as obligate long day or day neutral plants, since no or only poor flowering was achieved under short day conditions applied post-vernalization. In opposition, *Watsonia pillansii* flowered under long as well as short day conditions at cool temperatures, but its flowering was inhibited at higher temperatures. The authors concluded that *W. pillansii* is day neutral, but that the effects of temperature and photoperiod on flowering are complementary. However, flowering was achieved in all three Watsonia species when short day conditions were only applied after the third leaf stage of the plants.

Temperature

The active growth of *Freesia* x *hybrida* outside of its dormancy can be divided into four phases (Horn 1996), for which each a different optimal temperature regime applies. The first phase extends over three weeks until inflorescence initiation and comprises the vegetative development, i.e., the differentiation of shoots and leaves, which is promoted by temperatures of

15°C and above. In the second phase, temperatures between 12 and 15°C should be applied and maintained, since then inflorescence initiation is optimal. In the third phase, i.e., the phase after inflorescence initiation until anthesis, temperatures of 12 to 21°C increase plant height, number of flowers per inflorescence as well as corm mass. Phase 4 comprises anthesis and subsequent initiation of dormancy.

If dormancy has been adequately broken in *Freesia* x *hybrida*, inflorescences can be initiated earliest two weeks after planting. The temperature window for inflorescence initiation is between 5 and 20°C, but the optimum is at 12 to 15°C (Horn 1996). At temperatures over 20°C inflorescence initiation is suppressed or significantly delayed, and lateral shoot formation is also reduced (Rupprecht 1988).

Gilbertson-Ferriss (1981b) describe the relationship between the number of leaves and the associated photosynthetic active leaf surface in stating that a high number of leaves has a positive effect on the development of the inflorescences and the corms and that henceforth the temperatures during cultivation should not be decreased too soon to induce flower initiation.

Strong temperature fluctuations shortly after inflorescence initiation lead to a physiological phenomenon termed "gladiolus bloom", in which the flower spike as well as the individual flowers are formed with an abnormal shape. Cultivation temperatures which are kept too high throughout day and night after inflorescence initiation lead to flower abortion as well as thumbing (too great a distance between individual flowers on the flower spike) (Hirai and Mori 1996). Interestingly, Gilbertson-Ferriss (1981c) reported that *Freesia* x *hybrida* can tolerate cultivation at 24°C and still produce a marketable product at the same cultivation duration if for at least 8 h per day (at night or during daytime) the cultivation temperature can be reduced to 13°C.

In Ehrich (2007), the inflorescence differentiation in the terminal bud, which was microscopically monitored and determined, was temperature independent for the investigated *Freesia laxa*, *Tritonia deusta* and *Sparaxis* hybrids, and was even accomplished in the mentioned high temperatures for the batch planted during the European spring. This can be compared to the physiological reaction of *Freesia* hybrids, for which 12°C is optimal for flower initiation and differentiation, but for which higher average temperatures (up to 20°C according to Gilbertson-Ferris 1989) will delay, but not inhibit these processes (Imanishi 1993). Only for *Tritonia securigera,* a cool temperature regime (17°C day and 13°C night or continuous 13°C) was found to be required for the formation of a floral meristem. This can be compared to findings of Ascough et al. (2006) for several species of *Watsonia*, for which the authors showed facultative or obligate vernalization requirements for successful flowering. Similarly, in Thompson et al. (2011), *Watsonia pillansii* demonstrated an obligate vernalization requirement and *Watsonia borbonica* and *Watsonia tabularis* a facultative vernalization requirement.

However, for all three investigated *Watsonia* species of Thompson et al. (2011), cooler temperatures advanced flowering and maximized the number of plants which flowered in the trials. Where plants were subjected to elevated temperatures above 25°C thereafter, flower blasting was observed. Interestingly, the authors found that those plants which aborted their flowers ultimately produced more cormels, indirectly making these a sink for surplus assimilates. The authors proceed to conclude that since non-flowering *Watsonia* never fail to produce daughter corms, inflorescence differentiation is seemingly more sensitive to growth factor stresses than daughter corm initiation.

Similarly, in Ehrich (2007), continuous high temperatures after inflorescence initiation in the apical meristem resulted in its abortion within the corm or shortly before emergence of all chosen species investigated. *Freesia laxa* demonstrated to be the least sensitive in that regard with the highest percentage of flowering plants compared to the other investigated species/hybrids during the batch planted during the European spring (values up to 49% compared to values of only 0 to 33%). This reaction is also known for *Freesia* x *hybrida* when subjected to high temperatures during the period of flower development (Imanishi 1993). In Ehrich (2007), the lower growing temperatures realized in the batch planted during the European autumn led to a higher percentage of flowering plants of all investigated species and hybrids and were essential to achieve desirable plant qualities (attractive foliage and a satisfactory number of flowering plants per treatment).

In Ehrich (2007), plants grown at continuous 20°C or above (day and night) demonstrated that cultivation at these temperatures led to a delay or inhibition of flowering for all investigated species/hybrids, confirming the results recorded of the control plants planted during the European spring due to the ambient high temperatures. Plants in Ehrich (2007) grown in a climatic chamber at continuous 13°C in comparison displayed 70 to 100% flowering in both batches of all species/hybrids, although their number of inflorescences per plant as well as their number of flower primordia per inflorescence was usually lower than plants grown under supplemental lighting during the European autumn with day/night temperatures being 17/13°C, i.e., higher temperatures during the day.

Fertilization

Freesia x *hybrida* is generally classified as salt sensitive (Rupprecht 1988). The concentration of liquid feeding solutions or when using hydroponic growing systems should not be higher than 0.2%. Rupprecht (1988) advises to keep the ratio between the elementar nutrients at 1 (N) : 0.35 (P) : 1.8 (K), with the concentration being lower during cultivation in winter compared to cultivation during summer in order to avoid elongation. Growing in for the corms upon planting is much better when media are not pre-enriched with fertilizer. Once the plants are in full vegetative growth, nutrients can then be given through top dressing or liquid fertilization. Freesias are also classified as fluoride sensitive, therefore the appropriate phosphorous fertilizer as well as perlite needs to be selected. Otherwise, freesias can suffer from so called "leaf scorch" (De Hertogh and Le Nard 1993), especially if the pH is too low, which increases fluoride solubility.

In the results of Ehrich (2007) weekly fertigation (Kristalon Weiss [15:5:30] at 0.2%) as well as the use of a controlled release fertilizer (Plantacote Depot 8M at 3.5 g l^{-1} growing medium) was as well tolerated for *Freesia laxa*, *Sparaxis* hybrids, *Tritonia deusta* and *Tritonia securigera* as was the application of Osmocote Plus 5–6M (at 5 to 6 g per litre growing medium) in Startek (2002) for pot freesias. In Thompson et al. (2011) the four *Watsonia* species investigated responded well to a weekly fertigation (Chemicult at 2 g l^{-1}) with 6.5 N : 2.7 P : 13 K as well as Ca, Mg, S and trace elements.

Growth regulators

Wulster et al. (1989) state that growth regulation is necessary for potted freesias if varieties are used that are otherwise grown in cut flower production, since they are usually not sufficiently compact. The growth regulator advised by the International Bulb Centre is Ancymidol (30 mg l^{-1}), of which the efficacy is said to decrease at increasing cultivation temperatures (Wulster et al. 1989). De Hertogh and Milks (1990) prefer Paclobutrazol, since Ancymidol proved to be phytotoxic for 40% of the investigated freesia varieties. They advise to soak freesia corms before planting in a 200 ppm solution for one hour or apply 5 mg per 15 cm pot with 5 to 6 corms each at 4 cm leaf length. Wulster and Ombrello (2000) achieved attractive and marketable plants of *Ixia* hybrids by using Paclobutrazol as a preplant corm soak, a post-emergent drench or a post-emergent spray in combination with a 2 to 4 week pre-plant storage of corms at 7°C and an 18°C day/10°C night cultivation temperature. The soaking of individual *Watsonia* corms of different species in a 200 ml solution of Paclobutrazol to deliver 0.5, 1 and 2 mg of its active ingredient to the ready to plant corms produced non-flowering plants versus producing marketable plants with flowers when using a post-planting Paclobutrazol soil-drench that delivered 5, 10 and 20 mg of the active ingredient at 15 cm leaf height to the investigated species in Ascough et al. (2006).

The exposure of freesia corms to a four week storage at 5°C after preparation but prior to planting in Wulster and Gianfangna (1991) described in paragraph 4.2 could reduce plant height by more than 50% when combined with an Ancymidol treatment. The plants originating from the 5°C treatment without Ancymidol treatment displayed a 20 days earlier anthesis than those not submitted to this 5°C storage but having had an Ancymidol treatment.

Startek and Zurawik (2005) also showed the efficacy of Etephon (125 to 150 mg l^{-1}) on *Freesia hybrida*. Chlormequat (CCC) and Daminozid were shown not be effective for growth regulation in freesias (Berghoef and Zevenbergen 1990).

In Ehrich (2007), the application of the growth retardant Chlormequat (CCC) did not show a great effect at the concentration used (0.1 and 0.2% solution at leaf length of 3 to 7 cm) for *Freesia laxa*, *Sparaxis* hybrids as well as *Tritonia deusta* and *Tritonia securigera*. Noteworthy were obvious signs of more vigorous growth of the *Sparaxis* hybrids plants (visually determined only), which require further investigation. Furthermore, a slightly earlier anthesis date was recorded for the CCC-treated *Tritonia deusta* plants. These results were similar to Halevy and Shillo (1970) and Bhattacharjee (1984) who found an earlier anthesis date and promotion of growth of the related genus *Gladiolus*.

Startek (2002) could display a good growth habit and easy cultivation for the 'Easy Pot' freesia range specifically bred for cultivation in pots. This should be the route taken considering the prohibition of many of the above mentioned growth regulators in many European countries to date.

Future Prospects: Market Potentials and 'New Floricultural Crops'

From the huge reservoir of potential for floricultural crops in the Cape Floral Region (CFR) (Van Wyk and Smith 2001), its outstanding variety of geophytes, namely 16% of the total species, form a largely untouched plant group in horticulture, amongst them the diverse species of Iridaceae.

Exporting the dormant corms of commercially grown Iridaceae from South Africa and leaving them in their South African growth rhythm in order to force them during the European winter offers two interesting prospects. Firstly, the European pot plant assortment during that season is very limited and needs diversification. Secondly, originating from a Mediterranean climate, the chosen species and hybrids have low temperature requirements for forcing in greenhouses and their production is therefore energy efficient compared to crops originating in more tropical climates, e.g., *Euphorbia pulcherrima*. The ability to force the chosen species and hybrids as pot plants, detailed investigations on their growth rhythms and growth requirements are needed in order to provide information on their physiology and on how their cultivation can be regulated and optimised to eventually determine their potential as new pot plants (Rees 1992, Dole 2003).

For example, in Ehrich (2007), due to the limited information regarding the physiology and cultivation of the investigated species/hybrids, the different trial treatments were based on and the results compared to findings for *Freesia* x *hybrida*. Detailed results have been mentioned in the preceding paragraphs. In summary, the results of Ehrich (2007) showed that the export of dormant corms of *Freesia laxa*, *Sparaxis* hybrids as well as two *Tritonia* species from South Africa for trials in Germany proved to be uncomplicated and their dormancy could be further maintained in subsequent storage in Berlin at temperatures above 20°C. Temperature was found to be the main criterion to successfully realise flowering after planting. The species varied in their sensitivity, but generally cultivation at 13°C at night was essential, with temperatures of 17°C and above possible during the day. During the Central European summer months, inflorescences in the terminal bud failed to completely develop or flower primordia were aborted within the corm due to the high ambient temperatures present during the trials. A reduction in plant height and enhanced flowering could be achieved for some species by specific storage regimes and the application of a growth regulator. During naturally low light conditions in the European winter months, supplemental lighting was essential to produce plants of adequate quality for both *Tritonia* species and *Sparaxis* hybrids, but was not required for *Freesia laxa*.

But in summary, the investigated species/hybrids clearly displayed a great potential for an energy saving production system as well as for diversifying the autumn and winter pot plant assortment in Europe. The different physiological reactions of the investigated species/hybrids indicate that some of them share these with *Freesia* x *hybrida* also originating in the CFR, but others have to be individually determined for each new species.

Acknowledgements

The author would like to thank PD Dr. Heiner Grüneberg of Humboldt University in Berlin, Germany as well as Mandy and Paul Fick from New Plant Nursery in South Africa for their continued support during her research over several years as well as the subsequent writing of this manuscript.

References

Ascough, G.D., N.P. Mtshali, D.I. Thompson, N.O. Anderson, J.E. Erwin and J. Van Staden. 2006. Watsonia. *In*: J.A. Teixere da Silva (ed.). Floriculture, Ornamental and Plant Biotechnology, Vol. IV, Global Science Books, UK, pp. 559–605.

Ascough, G.D., J. Van Staden and J.E. Erwin. 2008. Effectiveness of Colchicine and Oryzalin at inducing Polyploidy in *Watsonia lepida* N.E. Brown. Hort. Science 43(7): 2248–2251.

Bhattarcharjee, S.K. 1984. The effect of growth regulating chemicals on *Gladiolus*. Gartenbauwissenschaft 49: 103–106.

Berghoef, J. and A.P. Zevenbergen. 1990. The effect of precooling environmental factors and growth-regulating substances on plant height of *Freesia* as pot plant. Acta Horticulturae 266: 251–257.

De Hertogh, A.A. and R. Milks. 1990. Forcing Dutch-grown freesias as potted plants in the U.S. and Canada. Acta Horticulturae 266: 115–122.

De Hertogh, A.A. and M. Le Nard (eds.). 1993. The physiology of flower bulbs. Elsevier, Amsterdam/London/New York/Tokyo.

Dole, J.M. 2003. Research approaches for determining cold requirements for forcing and flowering of geophytes. Hort. Science 38(3): 341–346.

Doorduin, J.C. 1992. Effects of photosynthetic lighting on *Freesia* grown for winter-flowering. Acta Horticulturae 325: 85–90.

Ehrich, L. 2007. Untersuchungen zur Eignung ausgewaehlter neuer Zierpflanzen aus Suedafrika fuer de Export und die weiterfuehrende Kultur unter mitteleuropaeischen Bedingungen. Ph.D. Thesis, Humboldt University of Berlin, Germany. (An English summary of this thesis can be found *In:* L. Ehrich, C. Ulrichs and H. Grueneberg. 2009. Factors Influencing Flowering of Different South African Iridaceae. Hort. Science 44(6): 1792–1795).

Ewald, A. 1993. *In vitro*-Polyploidisierung von *Anomatheca laxa*. Gartenbauwissenschaft 3: 120–124.

Gilbertson-Ferriss, T.L., M.L. Brenner and H.F. Wilkins. 1981a. Effects of storage temperatures on endogenous growth substances and shoot emergence in *Freesia hybrida* corms. Journal of the American Society for Horticultural Science 106: 455–460.

Gilbertson-Ferriss, T.L., M.L. Brenner and H.F. Wilkins. 1981b. Corm and shoot development of *Freesia hybrida* 'Super Emerald Mixture' as related to endogenous abscisic acid and indoleacetic acid levels in the developing corm. Journal of the American Society for Horticultural Science 106: 463–466

Gilbertson-Ferriss, T.L., M.L. Brenner and R. Hoberg. 1981c. Influence of alternating day and night temperature on flowering of *Freesia hybrida*. Journal of the American Society for Horticultural Science 106: 466–469.

Gilbertson-Ferriss, T.L. 1989. *Freesia* x *hybrida*. *In*: A.H. Halevy (ed.). Handbook of flowering, Vol. 3, CRC Press, Boca Raton, Florida, pp. 34–37.

Goldblatt, P. and J.C. Manning. 2008. The Iris Family—Natural History and Classification. Timber Press, Portland/London.

Gonzalez, A., S. Bañón, J.A. Fernández, J.A. Franco, J.L. Casas and J. Ochoa. 1998. Flowering responses of *Gladiolus tristis* (L.) after exposing corms to cold treatment. Scientia Horticulturae 74: 279–284.

Halevy, A.H. and R. Shillo. 1970. Promotion of growth and flowering and increase in content of endogenous gibberellins in *Gladiolus* plants treated with the growth retardant CCC. Physiologia Plantarum 23: 820–827.

Halevy, A.H. 1975. Light energy flux and distribution of assimilates as factors controlling the flowering of flower crops. Proc. 19th International Horticultural Congress 4: 125–134.

Halevy, A.H. 1990. Recent advances in control of flowering and growth habit of geophytes. Acta Horticulturae 266: 35–42.

Halevy, A.H. 1994. The use of plant bioregulators in ornamental crops. Acta Horticulturae 394: 37–43.

Han, S.S. 2001. Flowering of three species of Brodiaea in relation to bulb size and source. Scientia Horticulturae 91: 349–355.

Hartsema, A.M. 1961. Influences of temperatures on flower formation and flowering of bulbous and tuberous plants. *In*: W. Ruhland (ed.). Handbook of Plant Physiology, Springer. New York 16: 123–167.

Hirai, H., G. Mori, K. Tozai, C. Kubota, K. Fujiwara, Y. Ibaraki and S. Sase. 1996. Forcing culture of *Freesia* and Dutch *Iris* using spot cooling system. Acta Horticulturae 440: 262–267.

Horn, W. 1996. Zierpflanzenbau. Blackwell, Berlin/Vienna.

Imanishi, H. and E.J. Fortanier. 1983. Effects of exposing *Freesia* corms to ethylene or to smoke on dormancy-breaking and flowering. Scientia Horticulturae 18: 381–389.

Imanishi, H. and J. Berghoef. 1986. Some factors affecting dormancy breaking by ethylene in *Freesia* corms. Acta Horticulturae 177: 637–640.

Imanishi, H. 1993. Freesia. *In:* A.A. De Hertogh and M. Le Nard (eds.). The Physiology of Flower Bulbs. Elsevier, Amsterdam, pp. 285–296.

Kapczyńska, A. and M. Piskornik. 2003. Generative propagation of harlequin flower (*Sparaxis tricolor* (Curt) Ker. Gawl). Folia Horticulturae 15(2): 173–184.

Manning, J.C., P. Goldblatt and D. Snijman. 2002. The color encyclopedia of Cape bulbs. Timber Press, Oregon/Cambridge.

Niederwieser, J.G., R. Kleynhans and F.L. Hancke. 2002. Development of a new flower bulb crop in South Africa. Acta Horticulturae 570: 67–73.

Rees, A.R. 1984. Dormancy, flowering and periodicity in ornamental bulbs. The Plantsman 6: 33–41.

Rees, A.R. 1989. Ornamental bulbous plants. *In*: A.H. Halevy (ed.). Handbook of flowering, Vol. 1, CRC Press, Boca Raton/Florida, pp. 259–267.

Rees, A.R. 1992. Ornamental bulbs, corms and tubers. CAB-Publishing International, Wallingford/UK.

Rünger, W. 1976. Licht und Temperatur im Zierpflanzenbau. 3rd edn., Paul Parey, Berlin/Hamburg.

Rupprecht, H. 1988. Die Freesie. VEB Dt. Landwirtschaftsverlag, Berlin.

Startek, L. 2002. Growth dynamics and decorative value of 'Easy Pot' potted *Freesia* depending on the growing conditions. Acta Horticulturae 570: 385–390.

Startek, L. and P. Zurawik. 2005. Effect of ethephon on 'Easy Pot' *Freesia*. Acta Horticulturae 673: 617–623.

Struck, M. 1994. Flowering phenology in the arid winter rainfall region of Southern Africa. Bothalia 24(1): 77–90.

Sutter, E.G. 1986. Micropropagation of *Ixia viridifolia* and a *Gladiolus* x *Homoglossum* hybrid. Scientia Horticulturae 29: 181–189.

Swart, P.A., M.G. Kulkarni, M.W. Bairu, J.F. Finnie and J. Van Staden. 2012. Micropropagation of *Romulea sabulosa* Schltr. ex Beg.—A potential ornamental plant. Scientia Horticulturae 135: 151–156.

Thompson, D.I., N.O. Anderson and J. Van Staden. 2010. Colchincine-induced Somatic Polyploids from *in vitro*-germinated Seeds of South African *Watsonia* species. Hort. Science 45(9): 1398–1404.

Thompson, D.I., N.P. Mtshali, G.D. Ascough, J.E. Erwin and J. Van Staden. 2011. Flowering Control in Watsonia: Effects of corm size, temperature, photoperiod and irradiance. Scientia Horticulturae 129: 493–502.

Van Wyk, A.E. and G.F. Smith. 2001. Regions of floristic endemism in Southern Africa. Umdaus Press, Hatfield/South Africa.

Wang, L. 2006. Freesia. *In*: N.O. Anderson (ed.). Flower Breeding and Genetics. Springer. The Netherlands, pp. 665–693.

Wulster, G.J., S. Cartwright and T.J. Gianfagna. 1989. The effects of greenhouse temperature and ancymidol concentration on height and flowering time of *Freesia hybrida* grown as container plant. Acta Horticulturae 252: 97–103.

Wulster, G.J. and T.J. Gianfagna. 1991. *Freesia hybrida* respond to ancymidol, cold storage of corms, and greenhouse temperatures. Hort. Science 26(10): 1276–1278.

Wulster, G.J. and T.M. Ombrello. 2000. Control of Height and Flowering of Ixia Hybrids as Container Plants. Hort. Science 35(6): 1087–1088.

13

In vitro Regeneration of *Alstroemeria*: A Review

Amir Ghaffar Shahriari,[1,*] *Abdolreza Bagheri*[1] and *Zolfaghar Shahriari*[2]

ABSTRACT

Having beautiful and durable flowers, *Alstroemeria* is considered among the most important cut-flowers in the global market. Commercial-scale propagation of *Alstroemeria* by traditional methods such as rhizome splitting or seed planting is confronted by a number of limitations, and *in vitro* micropropagation is therefore of considerable importance. This will require optimization of different factors of both physical (light and temperature) and chemical (growth regulators, carbohydrates, mineral salts, vitamins, agar, and the pH of the culture medium) nature, as well as issues concerning plant material and technical considerations (explant type and excision). This chapter describes the protocol for *Alstroemeria* micropropagation and reviews the impact of physic-chemical factors on its regeneration and micropropagation.

Keywords: *Alstroemeria*, *in vitro* culture, micropropagation

[1] Ferdowsi University of Mashhad, Department of Plant Biotechnology and Breeding, Mashhad, Iran.

[2] Shiraz University, Department of crop production and Plant Breeding. Shiraz, Iran.

* Corresponding author: shahriari.ag@gmail.com.

Introduction

Alstroemeria (Alstroemeriaceae), commonly called the Peruvian Lily or Lily of the Incas or Parrot Lily is a South American genus of about 50 species of flowering plants. *Alstroemeria* is named after the Swedish botanist Klas von Alstroemer, who was a pupil of the great botanical classifier Linnaeus. Beautiful and durable flowers of Alstroemeria (Fig. 1) place it among the important and desirable ornamentals on the global market. Floriculture industry is fast emerging as a commercial venture throughout the world with increasing per capita consumption of flowers due to globalization and its effect on income generation. Rising demand has resulted in new production centres in Latin America, Africa and Asia, which were traditionally in USA, Japan, the Netherlands and Columbia. *Alstroemeria* spp. production has increased recently in Iran due to its demand as a cut flower. Traditionally, this plant is propagated by rhizome splitting, but this method of commercial propagation is confronted by viral contamination, low propagation rates and seasonal limitations. Seed propagation is also inefficient due to germination problems and segregation of traits in the progeny (Khaleghi et al. 2008). In order to overcome these problems and produce virus-free *Alstroemeria* plants, *in vitro* methods have attracted much attention. The global significance and market position of this flower on the other hand necessitate the application of novel mass propagation techniques for its commercial production.

Fig. 1. Alsteromeria Flowers; Beautiful and durable cut flowers.

Color image of this figure appears in the color plate section at the end of the book.

In vitro Culture of *Alstroemeria*

Commercial propagation of this plant currently makes use of tissue culture techniques due to the limitations of rhizome splitting and the problems associated with seed-mediated propagation. Shoot-tips (Pedraz Santos et al. 2005), shoot nodes (Lin et al. 2000), rhizome buds (Khaleghi et al. 2008) and mature and immature embryos (Akustu and Sato 2002) have been used as explants for *Alstroemeria* micropropagation.

Factors Influencing *Alstroemeria in vitro* Culture

Chemical factors

Plant growth regulators

The impacts of different auxins and cytokinins during different stages of *in vitro* plantlet regeneration have been reported (Pierik et al. 1977, Chiari and Bridgen 1996, Podwyszynska et al. 1997, Podwyszynska 1998, Chiari and Mark 2000, Lin et al. 2000, Akustu and Sato 2002).

Auxins

In a study of *Alstroemeria* hybrid micropropagation, Pierik et al. (1997) tested culture media with different auxins for root induction in plantlets which had been regenerated in cytokinin-containing media, and found that NAA at a concentration of 0.5 mg/l induced higher rooting compared to IBA. Lin et al. (2000) cultured *Alstroemeria* shoot segments on a medium containing 4 mg/l 2,4-D and 0.5–1 mg/l BA for induction of embryogenic calli. After 10 weeks, 1 percent of the explants produced callus at their nodes. In another experiment, NAA (0.5 mg/l) was used to induce rooting on stems directly regenerated from shoot nodes of *Alstroemeria* (Lin et al. 2000). Podwyszynska et al. (1997) studied the impact of different auxins (NAA, IAA and IBA) together with growth retardants during the rooting phase. The results indicated that among the different auxin treatments, NAA was the most efficient, and growth retardants such as paclobutrazol and flurprimidol in combination with NAA considerably enhanced rooting (with 100 percent of the rhizomes producing roots) and application of the growth regulator combined with a growth retardant (such as Paclobutrazol) prohibited elongation of the roots (Podwyszynska et al. 1997). Most of the reports concerning root induction in micropropagated *Alstroemeria* plantlets mention the application of different auxins, all of which specifying NAA as the most efficient treatment (Han et al. 1994, Gabryszewska 1996, Podwyszynska et al. 1998, Kristiansen et al. 1999, Chiari and Mark 2002).

Cytokinins

In a study by Chiari and Bridgen (1996), all the three *Alstroemeria* genotypes cultured on mediums containing 9, 18 or 36 μmol BA proved significantly different from controls (i.e., no BA included) with respect to bud number, shoot number and fresh weight; but no difference was observed among BA levels concerning the mentioned traits (Chiari and Bridgen 1996).

In a study of *Alstroemeria* meristem culture, modified MS medium containing 0.7 μmol /l gibberellic acid or 8.9 μmol /l BA or a combination of both was used for apical and lateral rhizome meristem and apical shoot meristem cultures, with the highest rate of rhizome regeneration obtained with 8.9 μmol BA (Chiari and Mark 2002). Lin et al. (1997) used MS medium containing 10 μmol TDZ and 0.5 μmol IBA for regenerating plantlets from shoot nodes, which were subsequently transferred onto an MS medium containing 2.2 μmol BA. After three weeks of culture on this regeneration medium, the buds emerged between petiole and shoot node and after two weeks the first leaf and a main shoot developed. Plantlets were subcultured for two months in this medium until normal plantlets with rhizome tips and roots were formed (Lin et al. 1997). BAP at a concentration of 0.5 mg/l has been used for rhizome induction at the base of plantlets obtained from seeds of lines in MS medium. After two months of subculture on this medium, a number of rhizomes formed on the plantlets (Lin et al. 2000).

The influence of cytokinins such as BAP, 2iP and kinetin at different concentrations—alone or in combination with growth retardants—was studied on *Alstroemeria* micropropagation, with the highest number and the lowest length of shoots being obtained with 6 mg/l BAP (Podwyszynska et al. 1998). Han et al. (1994) cultured rhizome buds of hybrid *Alstroemeria* cultivars on MS medium containing different growth regulators. Among the cytokinins tested (2iP, BA and kinetin), BA at a concentration of 1–2 mg/l together with an auxin (NAA or IAA at 0.2–0.3 mg/l) brought about the best results in terms of regenerating shoots and rhizomes, compared to the other cytokinins. BA at 0.5–1 mg/l in combination with 4 mg/l 2,4-D has been used for callus induction from *Alstroemeria* stems, with explans being subcultured bi-weekly for two months on this medium.

After production of soft calli, the cultures were transferred onto MS medium with different levels of BA (0, 0.1, 0.5 and 2 mg/l) for induction of somatic embryos, among which the medium with 2 mg/l BA produced the highest number of pre-embryos (184 per 0.5-gram samples of transferred soft callus). This treatment also resulted in the highest rate of transformation of pre-embryos to somatic embryos (33%). The same medium was used for regeneration of plants from somatic embryos. The results showed that embryos cultured in BA-free MS medium produce no stems. The highest number of shoots (8 shoots) and roots (8 roots) were obtained on a medium with 0.5 mg/l BA and in a BA-free medium, respectively (Lin et al. 2000).

Akutsu and Sato (2002) reported induction of embryogenic calli from ovules cultured on ½ MS salts with 0.5 mg/l BA and 0.5 mg/l NAA, and regeneration of plantlets from pre-embryos or secondary embryos on a solid medium with ½ MS salts and 0–2 mg/l BA (with the highest number of plantlets obtained from the 2 mg/l BA treatment). The published literature (Han et al. 1994, Gabryszewska 1996, Podwyszynska et al. 1998, Kristiansen et al. 1999, Chiari and Mark 2002, Shahriari et al. 2012) generally suggests that among auxins, NAA is the most efficient with respect to root induction. 2,4-D is another auxin which mainly plays a part in callus induction and results in production of soft and fragile calli when applied at high concentrations (Lin et al. 2000). Among cytokinins, BA has been the most important growth regulator for *in vitro Alstroemeria* regeneration, inducing callus and somatic embryos or regenerating shoots, depending on explant type and the time of application (Han et al. 1994, Akustu and Sato 2002).

Carbohydrates

In most of the reports, 3–4% sucrose has been added to the medium used for micropropagation of *Alstroemeria* (Buitendijk et al. 1992, Chiari and Bridgen 1996, Lin et al. 1997, Pierik et al. 1997, Podwyszynska et al. 1997, Podwyszynska et al. 1998, Chiari and Mark 2000, Lin et al. 2000, Chiari and Mark 2002).

Gabryszewska (1996) studied the influence of sucrose concentration on *in vitro* growth and development of *Alstroemeria* cv Zebra. Sucrose was added at concentrations of 0, 10, 20, 30, 40, 50, 60, 70 and 80 g/l. The highest shoot number was obtained with 60 or 80 g/l sucrose, and 20 or 30 g/l sucrose resulted in regeneration of the longest shoots. No rooting was observed on rhizomes cultured on mediums containing 0 or 10 g/l sucrose. The highest sucrose level (80 g/l) also totally prevented rooting (Gabryszewska 1996). The literature generally indicates that the best sucrose concentration for *in vitro Alstroemeria* culture is 20 to 80 g/l (Buitendijk et al. 1992, Chiari and Bridgen 1996, Lin et al. 1997, Pierik et al. 1997, Podwyszynska et al. 1997, Podwyszynska et al. 1998, Chiari and Mark 2000, Lin et al. 2000, Chiari and Mark 2002). Sucrose concentrations lower than 20 and higher than 80 g/l inhibit growth and will suppress rooting (Gabryszewska 1996).

Mineral salts

In most cases, *Alstroemeria* has been cultured on MS basal medium (Han et al. 1994, Chiari and Bridgen 1996, Gabryszewska 1996, Lin et al. 1997, Pierik et al. 1997, Bakken et al. 1999, Lin et al. 2000, Chiari and Mark 2002). Lin et al. (2000) used a ½ MS medium with 10 g/l sucrose and 2.2 g/l gelrite for germination of *Alstroemeria* seeds. The seeds germinated after two months and produced 1 cm long stems which were subsequently used to induce callus and somatic embryos on MS medium with different concentrations of growth regulators. Bakken et al. (1999) used a medium with MS micro salts and vitamins, ½ concentration of MS macro salts and 400 mg/l casein hydrolysate for embryo rescue of triploid *Alstroemeria* plants. Akutsu and Sato (2002) used a liquid medium with ½ MS salts containing 0.5 mg/l NAA and 0.5 mg/l BA for induction of embryogenic calli. They obtained embryo-like bodies on a solid medium with ½ MS salts and without growth regulators. MS medium has been used for regeneration of *Alstroemeria* from rhizome explants (Khaleghi et al. 2008), but seed germination (Lin et al. 2000), embryo rescue (Bakken et al. 1999) and callus induction (Akutsu and Sato 2002) have been achieved on mediums with 1/2 MS mineral salts.

Vitamins

Most of the reports concerning *Alstroemeria in vitro* culture mention the application of MS vitamins including 100 mg/l myo-inositol, 2 mg/l glycine, 0.1 Mg/l thiamine and 0.5 mg/l of both nicotinic acid and pyridoxine (Gabryszewska 1996). Pierik et al. (2002) used vitamin B_1 alone for *in vitro* vegetative propagation of *Alstroemeria* hybrids. A medium with MS salts plus 9.4 µM pantothenic acid, 1.2 µM thiamine and 0.5 mM myo-inositol has been also used successfully (Chiari and Mark 2002). MS vitamins can be considered appropriate for *in vitro* regeneration of *Alstroemeria* (Gabryszewska 1996), although application of other vitamin formulations has also been reported (Pierik et al. 2002, Chiari and Mark 2002).

Agar

Alstroemeria has been generally cultured on mediums with 0.8% agar (Lin et al. 2000). Pierik and coworkers (1997) used a medium with 0.7% difco-baco agar for *in vitro* culture of rhizome buds of *Alstroemeria* hybrids.

pH of the culture medium

For *Alstroemeria* tissue culture, pH of the medium has generally been adjusted to 5.6–6 (Bakken and Bavre 1999, Chiari and Mark 2000, Akustu and Sato 2002).

Environmental factors

Light

In a study on *Alstroemeria* regeneration from callus, cultures were kept under continuous light of cold white fluorescent bulbs with a light capacity of 50 µmol/m2/s (Akutsu and Sato 2002). Lin et al. (2000), studying micropropagation of *Alstroemeria* shoot buds, placed the explants cultured on shoot induction as well as regeneration mediums under dark conditions, and those cultured on rhizome induction medium under 12 h light. The influence of different light intensities (0–40 mmol/m2/s) on *Alstroemeria* cultures was studied, with the results indicating that higher light intensities result in lower propagation rates and that there is an interrelationship between light intensity and sucrose concentration. The number of aerial shoots increased with increasing light at 3% sucrose concentration, whereas light intensity had no influence on the number of aerial shoots in 5–7% sucrose concentrations; and the case was totally opposite for the root induction medium (Chrisiansen et al. 1999). Chiari and Mark (2002) cultured rhizome explants of *Alstroemeria* under 16 h light with an intensity of 50 µmol /m2/s, and the rhizome buds regenerated an average of one new tip every three weeks.

Bond et al. (1993) studied the influence of light on *in vitro* growth of *Alstroemeria* rhizome. The results indicated that decreasing radiation from 20 to 5 watts and an increase of day length from 8 to 20 hours did not have a significant impact on the number of roots, lateral rhizomes and aerial shoots per explant, with the best results obtained with 15°C, 5 watts/m2 radiation, and a day length of 8 hours. Pierik et al. (1997) kept the rhizome explants under 8 h light conditions; and placed the cultures under 16 h light conditions for rooting. In a study on *Alstroemeria* meristem culture for virus elimination, the meristems were cultured under 16 h light with an intensity of 50 µmol /m2/s with weekly subcultures (Chiari and Mark 2002). Podwyszynska (1977) placed the cultured rhizomes under 16 h light of 55 µmol/m^2/s for micropropagation of the Juanita hybrid. *In vitro* germination of *Alstroemeria* seeds was achieved under 12 h day length (Lin et al. 2000). Han et al. (1994) reported the highest rate of shoot induction (an average of 6.8 shoots/explant) on rhizome explants to be achieved under continuous 3000 lux light, while rhizome induction occurred under totally dark conditions.

Temperature

Alstroemeria in vitro cultures are generally reported to be kept under a temperature of 18°C (Lin et al. 1997, Chiari and Mark 2000, Lin et al. 2000).

Pierik et al. (1997) placed the cultured rhizomes of *Alstroemeria* hybrids in 21°C; whereas Podwyszynska et al. (1997) kept rhizome cultures of the Juanita hybrid under 25°C conditions. Gabryszewska (1996), in a study on the influence of temperature on *in vitro* growth and development of *Alstroemeria* cv Zebra, kept the cultures at two temperatures (17 and 25 °C) and reported that regeneration of lateral rhizomes and aerial shoots was better induced under the 25°C treatment. Bond and Alderson (1993) reported a significant increase in the number of lateral rhizomes and aerial shoots per explant by an elevation of temperature from 9 to 16°C, but there was no influence on the number of regenerated roots.

Plant material

Explant type

Rhizome buds possess the greatest potential for *Alstroemeria* regeneration (Fig. 2a). Application of other explants such as segments of vegetative shoots, shoot nodes, ovules, and mature and immature embryos have also been reported (Lin et al. 1997). Lin et al. (2000) reported callus induction from shoot segments of two tetraploid *Alstroemeria* lines and subsequent regeneration of plantlets from the calli. The fragile, embryogenic calli were cultured on MS basal medium containing different BA concentrations (0.5, 1 and 2 mg/l). After 6 months, plantlets were regenerated from the obtained pre-embryos.

Apical shoot meristem as well as apical and lateral rhizome meristems of *Alstroemeria* were cultured for production of virus-free plants, with the highest regeneration rate being that of lateral rhizome meristem on MS medium containing 2 mg/l BA; and the plantlets regenerated from lateral

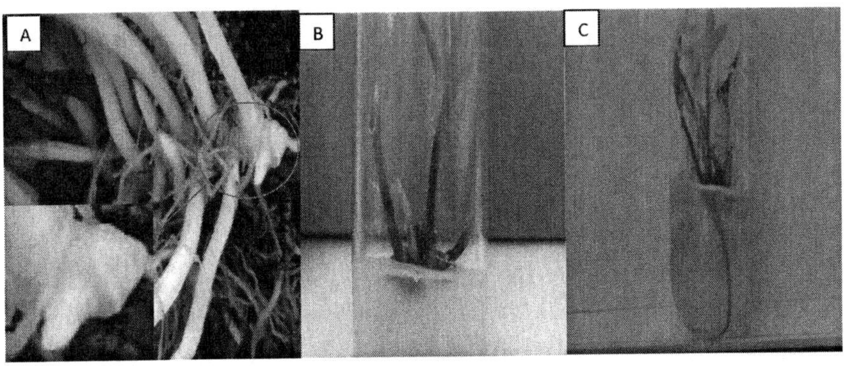

Fig. 2. Explants preparation and *in vitro* growth (A) Preparation of rhizome-tip explants. (B) Explants growth and shoot emergence in the 5th week. (C) Rooting in the 7th week.

Color image of this figure appears in the color plate section at the end of the book.

shoot meristems did not produce rhizomes and were eventually lost (Chiari and Mark 2002). Shoot nodes are a good source of explant as they are less contaminated compared to sub-soil organs. Lin et al. (1997, 2000) used shoot nodes as explants for direct shoot regeneration. This method includes adventitious stems growing from shoot epidermis at the region between leaf and stem tissues which will develop into a full plant. Explants were kept for 10 days on the shoot induction medium under dark conditions and then transferred onto regeneration medium. Subcultures were maintained every 3 weeks for 2 months until regenerated stems with rhizomes were normally developed. The best mediums for induction and for regeneration were MS with 10 μmol TDZ and 0.5 μmol IBA, and with 2.2 μmol BAP, respectively. The average regeneration rate of leaf bud explants is reported to be 88% (Lin et al. 1997, 2000).

Ovules From intra-specific hybrids of *Alstroemeria palgrana* (cv Alba and Magenta) were cultured on solid MS medium without growth regulators. After two months, 2 gram of the produced callus was cultured as cell suspension in MS medium containing 1 mg/l picloram. Fragile, embryogenic callus was then obtained using solid and liquid mediums with ½ MS salts and 0.5 mg/l of both NAA and BA. The callus produced pre-embryos when transferred to a hormone-free medium with ½ MS salts, which eventually regenerated plantlets, either directly or via secondary embryos, on a solid medium with ½ MS salts. Organogenesis was achieved at a high rate on a solid medium with ½ MS salts and 0.5–1 mg/l NAA and 0.5 mg/l kinetin. 90 plantlets were regenerated from 1 g weight of callus during 8 months by organogenesis, whereas embryogenesis from the same weight of callus resulted in regeneration of approximately 450 plantlets during three months. Somatic embryogenesis is thus more suited than organogenesis according to the shorter time and higher rate of regeneration (Akutsu and Sato 2002).

The influence of excision orientation

Excision of the explant on the medium will influence the development of lateral stems and induction of adventituous roots (Nobre 1994). Rhizome buds of *Alstroemeria* cv Sweet Laura and A30, excised either horizontally or vertically, were cultured on MS medium containing 0.8 mg/l BA, together with full buds as controls. The vertically excised and the control explants produced more (2.23 and 2.38 respectively) stems compared to the vertically excised explants (which produced 1.95 stems); and cv A30 regenerated more shoots than Sweet Laura (2.5 and 2.31 respectively). The results also showed that excised explants produced more rigorous stems compared to intact explants (Chiari and Mark 2000).

Lin et al. (1997) studied the influence of excision of leaf tissue and shoot node on regeneration rate and the number of regenerated shoots. The results indicated that excising the explant in the middle of shoot node and leaving less stem tissue with the leaf blade to be cultured, resulted in a higher rate of regeneration, but there was no significant difference between the two excision methods with respect to the number of regenerated shoots.

The Stages of *Alstroemeria* Micropropagation

Disinfection treatment

Fungal and bacterial contamination may cause devastating consequences when tissue culture is employed for commercial plant propagation (Reed et al. 1994). Application of surface disinfectants is a means for reducing the contamination of *Alstroemeria* explants. Surface disinfection is generally achieved using 70% ethanol, sodium hypochlorite, calcium hypochlorite and mercury chloride.

Antibiotics are also applicable for controlling internal contaminations which typically arise after several weeks of culture. Such contaminations may not be visible, but adversely influence growth of the explants (Reed et al. 1994). Several antibiotics have been used individually or in combinations to control internal contaminations (Reed et al. 1994). The explants however might be susceptible to antibiotics, and there is the risk of point mutations and disruptions of the electron transfer chain according to the high antibiotic concentrations used (Abedi et al. 2008).

The majority of the reports concerning *Alstroemeria* tissue culture have been concentrating on examination of different concentrations of growth regulators on stem and root regeneration from rhizome explants and, despite several reports about the intense contamination of *in vitro* cultures, few studies have aimed at determining the best methods of preparing sterile *Alstroemeria* explants (Bond and Alderson 1993). Surface sterilization has been achieved using agents such as 70% ethanol (Khaleghi et al. 2008), sodium hypochlorite (Lin and Monette 1987), calcium hypochlorite and mercury chloride (Bohluli-Zanjani et al. 2005); and addition of substances like Tween 20 to the disinfecting agent can increase its penetration (Lin and Monette 1987).

Shoot induction

Cytokinins show the highest influence on shoot induction. In a study on hybrid *Alstroemeria* micropropagation, the highest number of shoots (5–6 shoots/explant) were regenerated on MS medium containing 6 mg/l BAP, and a combination of 1.5 mg/l BAP with kinetin or 2ip highly induced shoot development; but these shoots were of inappropriate quality. The combined application of BAP with growth retardants such as paclobutrazol or flurprimidol (at 0.01–0.1 mg/l) resulted in regeneration of the highest number of shoots with the best quality (Podwyszynska 1997). Chiari et al. (1996) studied the influence of different BA concentrations (9, 18 and 36 μmol) on the number of shoots produced in *Alstroemeria in vitro* cultures and found no significant difference between the different BA concentrations. BA is known in general to have the highest influence on shoot induction compared to other cytokinins (Fig. 2b). Shahriari et al. (2012) showed that BA at a concentration of 1 mg/l induced higher shoot compared to other cytokinins such as TDZ and 2iP (Table 1).

Table 1. Effects of different cytokinins on shoot number, shoot height, bud number, root number, and shoot diameter (regenerated from rhizome-tip derived explants of *Alstroemeria*).

PGR treatments*	Shoot number	Shoot length (cm)	Bud number	Root number	Shoot diameter (mm)
BA					
0.5	3.60 bc**	5.35 abc	1.20 a	3.80 a	3.20 a
1	4.16 a	3.50 bcd	1.66 a	4.60 a	2.70 b
2	4.00 ab	3.16 cd	1.66 a	4.00 a	2.50 b
TDZ					
0.5	4.00 ab	5.73 ab	1.33 a	4.16 a	1.70 c
1	3.50 c	2.75 d	1.50 a	4.50 a	1.70 c
2	3.30 cd	2.16 d	1.33 a	4.50 a	1.69 c
2iP					
0.5	2.60 e	7.33 a	1.66 a	3.66 a	1.73 c
1	3.00 de	4.33 bcd	1.33 a	4.33 a	1.73 c
2	3.30 cd	3.10 cd	1.50 a	4.00 a	1.72 c

*Each PGR treatment contained 0.2 mg/l NAA in addition to one of the cytokinin levels listed above.

**Means with the same letters, based on the Duncan's test, showed no significant differences.

Rooting

Auxins mostly have a role in rooting. Podwyszynska (1997, 1998) could induce 100% rooting on rhizome-bearing shoots of *Alstroemeria* in a medium containing NAA and growth retardants. Pierik et al. (1997) used MS medium containing 5% sucrose and 0.5 mg/l NAA and IBA to induce rooting on the regenerated shoots, and found NAA to be more effective in root induction. In a study on *Alstroemeria* meristem culture, plantlets bearing rhizomes and aerial shoots produced roots on MS medium with 6.1 μmol /l NAA. No rooting was observed on rhizomes cultured in MS medium containing 0 to 10 g/l sucrose, either with or without NAA (Gabryszewska 1996). Han et al. (1994) cultured shoots with rhizomes developed at their base on the rooting medium containing IBA, IAA or NAA at 0.5–5 mg/l, with the best results—in terms of the number and length of the produced roots—being obtained with a medium containing 3 mg/l IBA. Lin et al. (2000) used an MS medium containing 0.5 mg/l NAA to induce rooting in the produced shoots and rhizomes, and transferred the rooted plantlets to greenhouse after 1 month of culture (Fig. 2c).

Acclimatization

Success of every *in vitro* culture system shall only be judged according to successful transfer of plantlets from *in vitro* to *in vivo* conditions (Hazarika 2003). Bohluli-Zanjani et al. (2005) studied production of *Alstroemeria* plantlets from shoot nodes and rhizome buds. The plants were transferred to greenhouse after one month of culture on rooting medium. Out of the 40 plants regenerated from shoot nodes and rhizome buds, 19 regenerated from rhizome and 15 from shoot nodes were viable. The plants regenerated from rhizomes were bigger than those from shoot nodes, and showed a higher percentage of viability (95–97%) and higher number of shoots. The plants regenerated from rhizomes produced flowers five months after being transferred to the greenhouse, whereas those from shoot buds needed only three weeks to begin flowering (Bohluli-Zanjani et al. 2005). Several authors have reported successful transfer of *Alstroemeria* plantlets regenerated from rhizome explants to the soil, and in all cases, the transferred plants produced flowers after growing.

Conclusion

When tissue culture is aimed at commercial micropropagation, contamination control and preparation of sterile explants and also optimization of *in vitro* conditions is of considerable importance. Finding the most appropriate growth regulator treatment is an important factor when optimizing *in vitro* regeneration. The literature shows that optimization of a medium containing high levels of cytokinin and low auxin levels, which will result in simultaneous regeneration of shoot and root, is highly important for *in vitro* propagation of *Alstroemeria*. The role of cultivar should not although be ignored, and different cultivars should be examined separately in order to determine the best conditions for regeneration of each one.

References

Abedi, G., H. Salehi and M. Khosh-khui. 2008. A novel nanomaterial for removal of bacterial contaminants in valerian (*Valeriana officinalis* L.) tissue culture. Acta. Physiol. Plant 30: 709–714.

Akustu, M. and H. Sato. 2002. Induction of proembryos in liquid culture increase the efficiency of plant regeneration from *Alstroemeria* calli. Plant Science 163: 475–479.

Bakken, A.K. and O.A. Bavre. 1999. Optimization of the lighting regime for *Alstroemeria* with respect to photoperiod and fluence rates. Scientia Horticulturae 80: 225–233.

Bohluli-Zanjani, S., Y. Hamidughli and A. Hatam Zade. 2005. A study of *Alstroemeria in vitro* Culture. M.Sc. thesis, Gilan University 100 pp.

Bond, S. and P.G. Alderson. 1993. The effect of explants density, temperature and light on rhizome growth *in vitro* of Alestroemeria. Journal of Horticultural Science 68: 855–859.

Buitendijk, J.H., M.S. Ramanna and E. Jaceobsen. 1992. Micropropagation ability: Towards a selection criterion in *Alestroemeria* breeding. Acta Horticulturae 325: 493–498.

Chiari, A. and M.P. Bridgen. 1996. Effect of medium and genotype on the *in vitro* growth of *Alstroemeria*. Horticulture Science 31: 629–631.

Chiari A. and P.B. Mark. 2000. Rhizome splitting: a new micropropagation technique to increase *in vitro* propagule yield in *Alstroemeria*. Plant Cell Tissue and Organ Culture 62: 39–46.

Chiari, A. and P.B. Mark. 2002. Meristem culture and virus eradication in *Alstroemeria*. Plant Cell Tissue and Organ Culture 68: 49–55.

Gabryszewska, E. 1996. The influence of temperature, daylenght and sucrose concentration on the growth and development of *Alstroemeria* Zebra *in vitro*. Acta Agrobotanica 49: 131–140.

Han, B.H., Y.J. Kim and J.K. Choi. 1994. Micropropagation of *Alstroemeria* through rhizome bud culture. Journal of the Korean Society Horticultural Science 35: 172–177.

Hazarika, B.N. 2003. Acclimatization of tissue cultured plants. Current Science 85: 1704–1712.

Khaleghi, A., A. Sahraroo, I.N. Rasoulnia and R. Ataei. 2008. *In vitro* propagation of *Alstroemeria* cv. Fuego. American Eurasian Journal of Agriculture and Environmental Science 3: 492–497.

Kristiansen, K., H. Grnstrup and K. Brandt. 1999. *In vitro* PPFD and media composition affect both in and *ex vitro* performance of *Alstroemeria* Butteryfly—Hybrid. Plant Cell, Tissue and Organ Culture 59: 145–153.

Lin, H.S., M.J. De Jeu and E. Jacobsen. 1997. Direct shoot regeneration from excised leaf explants of *in vitro* grown seedling of *Alstroemeria*. Plant Cell Reports 16: 770–774.

Lin, S.H., C. Toorn, K.J.M. Raemakers, R.G.F Visser, M.G. De Jeu and E. Jacobsen. 2000. Development of a plant regeneration system based on friable embryogenic callus in the ornamental *Alstroemeria*. Plant Cell Reports 19: 529–534.

Lin, W.C. and P.L. Monette. 1987. *In vitro* propagation of *Alstroemeria* Alsaan. Plant Cell Tissue and Organ Culture 9: 29–35.

Murashige, T. and F. Skoog. 1962. A revised medium for rapid growth and bio-assay with tobacco tissue culture. Physiologia Plant 15: 473–497.

Nobre, J. 1994. *In vitro* shoot proliferation of *Myrtus communis* from field- grown plants. Scientia Horticulturae 58: 235–258.

Pedraz Santos, M.E., M.C. Lopez Peralta, V.A. Gonzalez Hernandez, E.M. Engleman Clark and P. Sanchez Garcia. 2005. *In vitro* regeneration of *Alstroemeria* cv Yellow King by direct organogenesis. Plant Cell, Tissue and Organ Culture 84: 189–198.

Pierik, R.L.M., A. Voorst, G. Boody, C.A.M. Acker, C.L.C. Lelivelt and J.C. Wit. 1997. Vegetative propagation of Alstroemeria Hybrids *in vitro*. Acta Horticulturae 226: 143–147.

Podwyszynska, M., E. Gabryszewska and A. Przybyla. 1997. Micropropagation of *Alstroemeria* by apical meristem. Revista Brasileira de Horticultura Ornamental 2: 47–51.

Podwyszynska, M., E. Gabryszewska and A. PrzybylaA. 1998. Effect of growth retardants, cytokinins and auxins on the multiplication and rooting *in vitro* of *Alstroemeria* X hybrid Juanita. Acta Agrobotanica 51: 23–31.

Razdan, M.K. 2003. Introduction to plant tissue culture, 2nd edn. Science Publishers Inc. New Hampshire.

Reed, B.M., P.M. Buckley and T.N. Dewllde. 1994. Detection and eradication of endophytic bacteria from micropropagated mint plants. *In vitro* Cell Dev. Biol. 31: 53–57.

Shahriari, A.G., A. Bagheri, A. Sharifi and N. Moshtaghi. 2012. Efficient Regeneration of 'Caralis' *Alstroemeria* Cultivar from Rhizome Explants. Notulae Scientia Biologicae 4(2): 86–90.

Section C

Medicinal Plants

14

The Current Impact and Potential of Biotechnology to Improve the Capacity of Orange-fleshed Sweet Potato (*Ipomoea batatas*) to Prevent Vitamin A Deficiency

Betty J. Burri

ABSTRACT

Vitamin A deficiency is the leading cause of preventable blindness in the world and an important cause of premature death in young children and pregnant women. Billions of people get most of their vitamin A from plants that are rich in pro-vitamin A carotenoids such as beta-carotene. Orange-fleshed sweet potatoes (OFSP) are an excellent source of beta-carotene, and have improved vitamin A status in several small-scale interventions. The amount of OFSP that must be consumed to provide 50% of the RDA for vitamin A is reasonable, ranging from about 0.1 to 1.2 cups per day. World production is adequate to meet the needs of all of the people most at risk for vitamin A deficiency, if most sweet

USDA, ARS, Western Human Nutrition Research Center, 430 W. Health Sciences Drive, Davis, CA 95616.
Email: betty.burri@ars.usda.gov

potatoes were OFSP. OFSP is difficult to improve by conventional hybridization techniques because of their complex genetics, and this complexity has also slowed biotechnological improvements in OFSP. This research is evaluated to estimate its current and potential impact on vitamin A status. OFSP are already an excellent source of beta-carotene, but there is little research on increasing carotenoid concentrations through biotechnology. Most current research focuses on increasing the resistance of sweet potatoes to virus infestations, herbicides, and environmental stressors. Recently sweet potato production per hectare and total production have increased slightly worldwide, while the hectares upon which sweet potatoes are grown have decreased. This is especially true in China, which is the major producer of sweet potatoes worldwide, and where biotechnological efforts to improve sweet potatoes are established. However, the impact of biotechnology on this increased production is unknown because few reports of field trials, or of failures, exist. It is probable that biotechnology has had minimal impact so far because most of the research is so recent, and because much of it is targeted to prevent crop failure instead of to increase production. However, the potential impact is greater.

Keywords: beta-carotene, sweet potato, biofortified, biotechnology, human nutrition

Sweet Potato

Sweet potato (*Ipomoea batatas*) is a tuber with white or purple flowers and green or purple leaves. It can be white, cream, yellow, orange, or purple (Woolfe 1992, Bovell-Benjamin 2007). Orange, white and cream are the most commonly grown and eaten (Sandhill Preservation Center 2012). In parts of Africa, Asia, and the Pacific sweet potatoes are an important staple crop (Woolfe 1992, Bovell-Benjamin 2007). Sweet potatoes grow well in hot, humid climates, but they need a rather long growing season (90 to 150 frost-free days; Bovell-Benjamin 2007). Sweet potatoes normally flower in summer and fruit in late summer and fall.

Sweet potatoes have complex genetics. Animals and some plants have diploid (two) chromosomes. However, many plants have more, having tetraploid (four) or hexaploid (six) chromosomes. Although a few tetraploid sweet potatoes have been found in the wild, almost all sweet potatoes are hexaploid plants, with six sets of chromosomes. Not surprisingly, sweet potatoes are genetically diverse (Arizio et al. 2009, Cervantes-Flores et al. 2007). Also, because they have six sets of chromosomes, genetic inheritance is complex and traditional breeding results in many potential combinations of traits. This means that plants are difficult to breed true. Sweet potatoes also have high levels of male sterility and cross-incompatibility (Liu 2011). Their genotype reacts with their environment in ways that are difficult to predict, so that some plants give high yields in good environments, while others are better adapted to poor environments (Tekalign et al. 2007).

There are many hundreds of sweet potato varieties, many limited to countries or regions. Large collections are stored at the International Potato Center (CIP) in Lima, Peru (International Potato Center 2006) and at the United States Germplasm Repository, Plant Introduction Station, in Griffin, GA, USA. The major center of genetic diversity for sweet potatoes is in northwestern South America (northern Peru, southern Ecuador) but other important centers of diversity exist in sub-Saharan Africa, Papua New Guinea, and Indonesia. Despite the abundance of sweet potato varieties in the wild, there tend to be only a few important commercial varieties grown at any location (Scahleitner et al. 2011). For example, there are only a small number of important sweet potatoes in the United States (such as Jewel, Beauregard, and Garnet), and all of them are orange-fleshed sweet potatoes (OFSP). The varieties that dominate Africa and Asia are white-fleshed sweet potatoes. This is presumably because it has been difficult to breed in useful genetic traits so that they are inherited consistently.

Sweet Potatoes and Biotechnology

The hexaploid nature of sweet potatoes, as well as their high levels of cross-incompatability and sterility makes conventional breeding efforts based on sexual hybridization difficult and inefficient. Thus, assisting breeding through biotechnology is attractive. However, the complexity of sweet potato genetics also hampered early attempts to improve OFSP through biotechnology. An important early genetic engineering effort is a fairly well-known failure (NW Resistance Against Genetic Engineering 2004, GM Watch 2010). In that instance, Monsanto and a consortium of scientists attempted to use biotechnology to create an OFSP that had high yields and other desirable qualities and was resistant to virus. Early reports of this research were highlighted as an example of biotechnology being used to feed poor Africans, and were reported in major scientific journals such as Nature (Wambugu 1999, Qaim 2000). Notably, the failure of this experiment was only reported in a non-peer reviewed journal, New Scientist (New Scientist 2004). The details of this experiment and its results have never been published.

This failure highlights one reason why reports of OFSP improvement by biotechnology—including this one—must be viewed with caution. The scientific literature in this field consists of reports of small-scale successes. Evidence of failures is rare, and this evidence is usually not presented in a scientific journal where it can be evaluated. Furthermore, the toxicity of genetically modified plants, including modified OFSP, has seldom been studied. There is little evidence for, or against, the safety of genetically modified crops (Domingo 2007). Essentially the scientific and corporate literature available tells us what biotechnology might do to improve OFSP; not what it will do. The studies reported in this overview are of successful experiments in their preliminary stages. The actual impact of these studies on sweet potato production and nutrition—if any—is not known.

However, there have been several recent advances in sweet potato biotechnology. Significant progress has been made in developing gene maps, and standardizing useful methods (Cervantes-Flores et al. 2008, Schafleitner et al. 2010). A searchable genetic index is available at http://www.cipotato.org/genebank/. Recent advances have been made in standardizing methods, and new methods have been developed that appear to produce transgenic plants reproducibly. The most common method uses *Agrobacterium tumefaciens* mediated transformation of the embryogenic callus from meristem tissues of sweet potato to produce reproducible transgenic plants with good viability (Otani et al. 2003). Agrobacterium tumefaciens-mediated transformation has been standardized for several important cultivars, and has been used to produce transgenic plants that appear to resist diseases, stresses and herbicides. Furthermore, somatic hybridization has been used to overcome incompatibility between sweet potato and its relatives, and may have generated useful inter-specie hybrids (Liu 2012). Methods to induce potentially helpful mutations with gamma-rays (Shin et al. 2011) have also been developed. Furthermore, a visible marker for intergenic vectors has been identified (Kim et al. 2010). Embryogenic suspension cultures have been developed for many different sweet potato genotypes (Liu 2012).

Sweet Potato Nutrition

All sweet potatoes are nutritious (Burri 2011). They are low in fat and protein, and rich in carbohydrate, a variety of antioxidants (Teow and other 2007), as well as fiber, zinc, potassium, sodium, manganese, calcium, magnesium, iron, and vitamin C (Antia et al. 2006). OFSP are also very good sources of beta-carotene, which forms vitamin A (Hagenimana et al. 1999a,b, Teow et al. 2007, Wu et al. 2008, Burri 2011). Because of their high carotenoid content and good yields, OFSP have already been used in several small-scale studies to increase human vitamin A status (Haskell et al. 2004, 2005, van Jaarsveld et al. 2005, Low et al. 2007a,b). These interventions have been successful.

Vitamin A

Vitamin A (Fig. 1) is an essential nutrient, with numerous roles in genetic regulation (Blomhoff and Blomhoff 2006, McGrane 2007, Marceau et al. 2007, Wolgemuth and Chung 2007, Mark et al. 2009), the visual cycle (Blomhoff and Blomhoff 2006, Travis et al. 2007, Ritter et al. 2008), normal growth and development (Blomhoff and Blomhoff 2006, McGrane 2007, Marceau et al. 2007, Wolgemuth and Chung 2007, Mark et al. 2009), and immune function (Montrone et al. 2009, Duriancik et al. 2010).

Vitamin A deficiency is a serious health issue for between 134–219 million people in the developing world, especially the Indian subcontinent, South-East Asia and Sub-Saharan Africa (West 2002, World Health Organization (WHO) 2005, WHO 2009). It is a leading cause of preventable blindness in the world (Sommer et al. 1996, WHO 2009). It is responsible for over 600,000 deaths per year, mostly in young children or pregnant women (WHO 1995, West 2002, West et al. 2002, Black et al. 2008).

The amount of vitamin A one person needs from the diet depends on their age, sex, genetics and health status. However, data on the impact of these factors on vitamin A metabolism and requirements are scarce. Therefore, recommended daily intakes of vitamin A for healthy individuals are often based on extrapolations from a few individuals, and thus vary with the country of organization that makes the estimate. Recommended nutrient intakes for vitamin A estimated by the World Health Organization (WHO 2009), and the United States Institute of Medicine (USIOM 2000) are shown in Table 1.

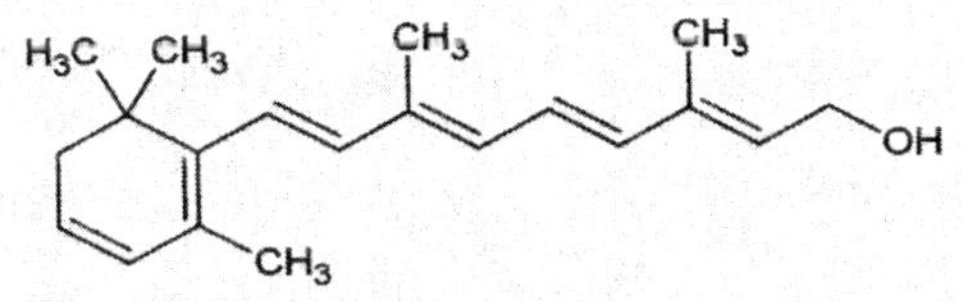

Fig. 1. Vitamin A.

Beta-carotene

Beta-carotene (Fig. 2) is the most common pro-vitamin A carotenoid in the food supply. It is found in many green or orange fruits and vegetables such as OFSP, carrots, mangos, pumpkin, spinach, and kale (Britton 1995, USDA ARS 2012).

Vitamin A is usually formed from beta-carotene by a reaction involving beta-carotene 15,15′-monooxygenase (CMO1), which cleaves beta-carotene centrally to form two molecules of retinal. There is also a

Table 1. Recommended nutrient intakes for vitamin A (WHO and USIOM).

Life Stage	Recommended Nutrient Intake (WHO 2009) (μg RE/d)	Dietary Reference Intake (USIOM 2000) (μg RAE/d)
Infants 0–6 months	375	400
Infants 7–12 months	400	500
Children 1–3 years	400	300
Children 4–6 years	450	400
Children 7–8 years	500	400
Children 9 years	500	600
Children 10–13 years	600	600
Males 14–65+ years	600	900
Females 14–18 years	600	700
Adult females 19+ years	500	700
Adult females 65+ years	600	700
Pregnant females (14–18 years)	800	750
Pregnant females (19+ years)	800	770
Lactating females (14–18 years)	850	1,200
Lactating females (19+ years)	850	1,300

RE = retinol equivalents; RAE = retinol activity equivalents.

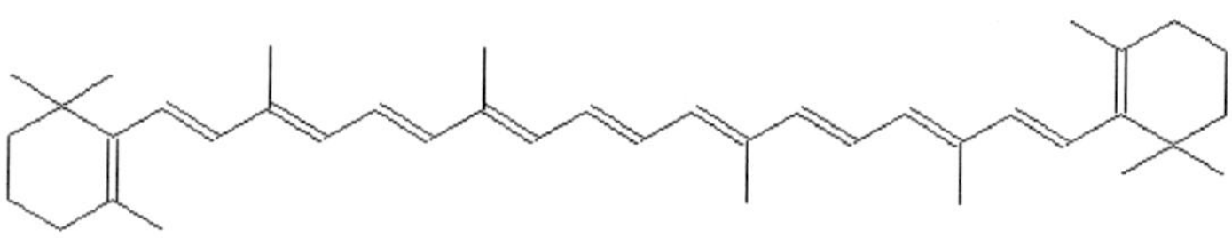

Fig. 2. Beta-carotene.

secondary mechanism, where beta-carotene 9′,10′-monooxygenase (CMO2) cleaves beta-carotene (as well as alpha-carotene and beta-cryptoxanthin) eccentrically to form one molecule of retinal and one apocarotenal (Chicili et al. 2005, Ho et al. 2007, Biesalski et al. 2007, Lietz et al. 2010). Thus, 1 μg beta-carotene should form about 0.5–1 μg of vitamin A.

Unfortunately, carotenoid conversion in the body is much less effective (Parker et al. 1999, Lin et al. 2000, Hickenbottom et al. 2002, Burri and Clifford 2004, Ho et al. 2009). The World Health Organization estimates a conversion ratio of 6 μg beta-carotene : 1 μg vitamin A (WHO 2009), while the US Institute of Medicine uses the ratio of 12 μg beta-carotene : 1 μg vitamin A (USIOM 2000).

Probably the most important reason for these poor conversion ratios is that beta-carotene is poorly absorbed from most foods (de Pee et al. 1998, Veda et al. 2006). Beta-carotene is better absorbed from orange fruits and vegetables, such as OFSP, than from leafy green vegetables (de Pee et al. 1998, O'Connell et al. 2007). Beta-carotene absorption is also much better when oil is added to the meal (Huang et al. 2000, Failla et al. 2009, Bengtsson et al. 2009).

Only one human study has estimated the vitamin A equivalency of the carotenoids found in OFSP. The vitamin A equivalency of beta-carotene fed to Bangladeshi men with moderate vitamin A stores was 13.4 μg beta-carotene to 1 μg retinol (Haskell et al. 2004), similar to the USIOM estimate.

Food-based Interventions to Prevent Vitamin A Deficiency

Vitamin A deficiency is usually prevented by distributing high-dose vitamin A supplements once or twice a year (Pedro et al. 2004, Aguayo et al. 2005, Donnen et al. 2007, Idindili et al. 2007, Bezerra et al. 2010). However, the costs of supplementing remote rural populations can be much higher than the costs of supplementing city dwellers. Therefore, extending vitamin A supplementation programs to rural areas has been problematic (Semba et al. 2010). For example, India has supported national vitamin A supplementation programmes for more than 40 years, but these programs have reached less than 25% of their population (Stein et al. 2006, supplemental material). Food fortification programs could also prevent vitamin A deficiency (Dary and Mora 2002, Kafwembe et al. 2009, Oguntibeju et al. 2009, Fiedler and Afidra 2010, Klemm et al. 2010). For example, most milk, dairy products, and some cereals are fortified with vitamin A in the United States. However, like supplementation programs, food fortification programs are most effective in cities, where residents purchase staples.

A third strategy to prevent vitamin A deficiency is to increase vitamin A or vitamin A-forming carotenoids in the diet (Ruel 2001, Faber and van Jaarsveld 2007, Tanumihardjo 2008, Tanumihardjo et al. 2008, Loechl et al. 2009, Bouis and Welch 2010). Although a few food-based interventions have used fish, which provide vitamin A directly (Kongsbak et al. 2008, Roos et al. 2002), most of these interventions rely on foods that are rich in beta-carotene (de Pee et al. 1998, Haskell et al. 2004, 2005, van Jaarsveld et al. 2005, Low et al. 2007a, b, Burri and Turner 2009, Burri 2011). These food-based intervention strategies are well-suited for rural communities. Long-term sustainability might be achieved, because fruit and vegetable seeds can be harvested and shared at a local level, instead of being provided by a national programme. An important component of food-based interventions is to modify crops, so that they contain increased amounts of carotenoids, or to improve disease resistance.

OFSP and vitamin A deficiency prevention

OFSP are an excellent source of beta-carotene, the most common vitamin A forming carotenoid. Furthermore, sweet potatoes are already a secondary staple crop in many areas with vitamin A deficiency (USDA Economic Research Service (ERS) 2009, Burri 2011). (This apparent contradiction arises because most people in areas where vitamin A deficiency is common grow and eat white sweet potatoes, which contain minimal amounts of beta-carotene). Dietary treatments using OFSP have been successful in increasing vitamin A status in several small-scale interventions (Haskell et al. 2004, 2005, van Jaarsveld et al. 2005, Low et al. 2007a, b, Burri and Turner 2009, Zeng et al. 2009). As little as one-half cup per day improves vitamin A reserves in children (van Jaarsveld et al. 2005, Burri 2011).

Beta-carotene in OFSP

The most abundant vitamin A-forming carotenoid in OFSP is beta-carotene (Kim et al. 2007, Bengtsson et al. 2008, Wu et al. 2008, USDA ARS 2012), although small amounts of alpha-carotene and beta-cryptoxanthin are also present in some varieties. The concentration of beta-carotene in OFSP depends on its variety (Hangenimana et al. 1997, Kidmose et al. 2006, Kidmose et al. 2007, Kidmose et al. 2009, Kim et al. 2007, Bengtsson et al. 2008, Wu et al. 2008, USDA ARS 2012). Beta-carotene concentrations of OFSP varieties are shown in Table 2.

Table 2. Beta-carotene content in OFSP varieties.

Reference	Beta-carotene (μg/100 g edible portion)
Lako et al. 2007	12000–14000
Kidmose et al. 2007	1240–10800
Wu et al. 2008	5320–8400
Failla et al. 2009	11120–28100
USDA 2009	5501–11509
Takahata et al. 1993	7400–18700
Fonseca et al. 1998	10120
Teow et al. 2007 (fresh weight)	
Light orange	1180–2980
Orange	4490–22900
Hangenimana et al. 1999 (beta-carotene)	
Light orange	111–2217
Orange	4085–7984

Beta-carotene concentrations also vary with OFSP growing, harvesting, and storage conditions (Bengtsson et al. 2008), farming site (K'osambo et al. 1998), season (Liu et al. 2009), root age (K'osambo et al. 1998, Hagenimana 1999), drought (van Heerden et al. 2008) and virus infestation (Kapinga et al. 2009).

Impact of Biotechnology on the Nutrition of OFSP

Biotechnology methods could improve OFSP nutrition in several ways. The most direct method would be to increase nutrient concentrations or bioavailability. One promising study induced a mutation in beta-carotene hydroxylase, a regulatory enzyme which catalyzes the hydroxylation of beta-carotene to beta-cryptoxanthin, and beta-cryptoxanthin to zeaxanthin (Kim et al. 2012). Beta-carotene concentrations were increased by down-regulating this gene. Cell cultures from a white-fleshed sweet potato transformed to down-regulate this gene increased total carotenoid, beta-carotene, beta-cryptoxanthin and zeaxanthin concentrations substantially.

In another study, differences in gene expression between Jewel, a common OFSP, and White Jewel, its white-fleshed counterpart, were compared in a large-scale gene expression analysis (McGregor et al. 2006). The only difference between cultivars that would impact carotenoid concentrations was that Isopentenyl diphosphate isomerase was down-regulated in White Jewel. Several genes associated with chloroplasts were also differentially expressed.

There have also been a few studies that attempted to enhance other nutrients or taste enhancers, such as potassium (George et al. 2002), resveritrol (Pan et al. 2012), and sucrose (Li et al. 2003). Although research studies attempting to improve beta-carotene or other nutrient concentrations in OFSP are in their infancy, these genetic manipulations might eventually result in a variety of true-breeding OFSP with higher beta-carotene concentrations and other desirable attributes.

Effects of Cooking and Storage on Carotenoid Content of Sweet Potatoes

OFSP are popular vegetables that can be baked, fried, steamed, dried, and fermented into alcohol (Woolfe 1992, Rodriguez-Amaya 1997, Emenhiser et al. 1999, Rodriguez-Amaya 2003, Sulaeman et al. 2003, Mohapatra et al. 2007, Kudoh and Matsuda 2000, Low and van Jaarsveld 2008, Ray and Sivakumar 2009). They can also be made into flour (Ahmed et al. 2010), buds (Valdez et al. 2001), curd (Mohapatra et al. 2007), or yogurt (Kudoh and Matsuda 2000). Beta-carotene is fairly stable with most cooking methods

(Table 3). Losses with boiling ranged from –32 to +26%, roasting from –27 to +6%, steaming from –20 to –32%, and frying from –5 to –28%. Dehydration causes greater losses, of up to 41% (Fonseca et al. 2008). Sweet potatoes store well and retain much of their carotenoids for at least 50 days (Watanabe et al. 1999, Chattopadhyay et al. 2006). In subsequent calculations the loss of beta-carotene from cooking and storage is assumed to be 10%.

Table 3. Effect of cooking and processing on beta-carotene in OFSP.

Reference	Varieties/Treatment	% Loss (or Gain)
Kidmose et al. 2007	Tainung Boiled Roasted	–15 –27
	SPK 004 Boiled Roasted	(+26) –25
	Salyboro Boiled Roasted	–12 (+6)
	Zapallo Boiled Roasted	(+2) (+19)
Bengtsson et al. 2008	Ejumula Boiled Steamed Deep-fried	–32 –32 –5
	SPK004/6 Boiled Steamed Deep-fried	–20 –20 –20
Vimala et al. 2011	10 clones Boiled Fried Oven dried Sun Dried	–10 to –16 –14 to –28 –4 to –11 –27 to –37

Bioaccessibility of Sweet Potatoes

OFSP are rich sources of beta-carotene, but not all of it is accessible. Beta-carotene bioaccessibility is the fraction of beta-carotene transferred from food to mixed micelles for uptake by intestinal mucosa. It depends on the food matrix, especially the fiber and fat content of the food (Veda et al. 2006, Tumuhimbise et al. 2009, Bengtsson et al. 2009, Bechoff et al. 2011). Specifically, beta-carotene bioavailability from OFSP depends on the variety and on the cooking and processing method (Bengtsson et al. 2009, Tumuhimbise et al. 2009, Bechoff et al. 2011); raw < baked < steamed/boiled < deep fried. Beta-carotene bioaccessibility increased greatly with fat. One simulated digestion showed that only 0.6–3% of OFSP beta-carotene was micellerized, increasing to 7% in highly processed baby food (Failla et al. 2009). Another study (Bengtsson et al. 2009) showed that beta-carotene bioaccessibility in the micellar phase varied from 0.5–1.1% without fat, increasing to 11–22% with 2.5% fat. Even highly processed foods prepared with OFSP, such as chapatis mandazi and porridge had higher bioavailability with fat: 10–16% without oil, 49–73% with oil (Bechoff et al. 2011). Furthermore, a study in vitamin A-depleted Mongolian gerbils fed OFSP with 3, 6, and 12% fat for three week showed that carotenoid absorption increased as the amount of fat in the diet increased (Mills et al. 2009). The higher fat (12% fat) diets improved VA status, while the lower fat diets maintained status quo. Finally, a human study feeding OFSP with fat, calculated its bioavailability as 37%, compared to 65% for beta-carotene beadlets (Huang et al. 2000).

Thus, the bioaccessibility of beta-carotene from OFSP can be very low (<1%) if fed without fat, but even a small amount of fat increases beta-carotene bioaccessibility by 2–20 fold. Even so, only about 25% (11–48%) of the beta-carotene in OFSP is bioaccessible, and thus available to be absorbed in the intestine. This is important because carotenoid metabolism studies (Burri et al. 2001, Lin et al. 2000, Hickenbottom et al. 2002, Ho et al. 2009) suggest that poor absorption of carotenoids from food is the major reason for the low conversion ratio of beta-carotene to vitamin A. In subsequent calculations, the fraction of beta-carotene that is bioaccessible is assumed to be 25%.

The Amount of OFSP Needed to Supply Vitamin A to One Person

OFSP was evaluated for its potential to prevent vitamin A deficiency on a world-wide basis. First, the beta-carotene concentration of OFSP (Table 2) was combined with the data on the effects of cooking and processing (Table 3), and carotenoid bioaccessibility for OFSP to adjust beta carotene concentrations for losses due to cooking, storage, and poor bioaccessibility. Thus, bioaccessible beta-carotene in OFSP was calculated as follows:

Bioaccessible beta-carotene from OFSP = Beta-carotene in OFSP X Fraction retained after cooking and storage X Bioaccessible Fraction

When beta carotene in OFSP = 4085–28100 by conventional breeding methods (Takahata et al. 1993, Hangenimana et al. 1999, Teow et al. 2007), Fraction retained after cooking and storage = 0.90; Bioaccessible fraction = 0.25.

Therefore, the concentrations of bioaccessible beta-carotene in traditionally-bred OFSP ranged from 919–6322 μg/100 g OFSP; or 9–63 μg bioaccessible beta-carotene/g OFSP; while the concentration in transgenic OFSP attained 18968 μg/100 g.

The concentrations of bioaccessible beta-carotene in OFSP, the weight of 1 cup of OFSP (USDA ARS 2012), and the conversion ratio of beta-carotene to vitamin A estimated by the World Health Organization (WHO) were used to calculate the amount (in grams and in cups per day) of OFSP needed to supply 50% of the vitamin A requirements of one person, at different life stages (Table 1).

Well nourished people appear to convert less vitamin A than poorly nourished people with low vitamin A status (Ribaya-Mercado et al. 2000, Howe and Tanumihardjo 2006, Tanumihadjo 2008). A vitamin A equivalency for OFSP was calculated as 13.4 μg beta-carotene: 1 μg retinol in OFSP fed to people with adequate vitamin A status (Haskell et al. 2004). Poorly nourished women and children are likely to have a smaller retinol equivalency ratio of perhaps 3 μg beta-carotene: 1 μg retinol (Tanumihardjo 2008). The conversion ratio estimated by the WHO is 6 μg beta-carotene forms 1 μg retinol, which is midway between both measurements. Therefore, we will use the WHO conversion factor in our calculations.

The OFSP in grams per day (g/d) needed to meet 50% of the requirements for one person with marginal vitamin A deficiency is calculated as:

OFSP (g/d) = ((vitamin A requirement μg RE/day)/((μg bioaccessible beta-carotene/g sweet potato) X (conversion ratio 6/1))) X 0.50

For a 4–6 year old with marginal deficiency, this ranges from:

OFSP g/d = (450 μg RE/day/(63 μg beta-carotene/g sweet potato/6)) X 0.50 = 21.4 g OFSP/d; to

OFSP g/d = (450 μg RE/day/(9.2 μg beta-carotene/g sweet potato/6)) X 0.50 = 147 g OFSP/d in conventionally bred OFSP

The ranges of OFSP/day needed to meet 50% of vitamin A requirements is shown in Fig. 3. The amounts of OFSP/d were also calculated in terms of cups per day. This allows one to determine if the portion size could reasonably be fed as part of a normal diet. Since one cup of OFSP is estimated to weigh 255 g (USDA ARS 2012), one must divide the grams of OFSP per day by 255 g/cup to calculate cups per day. Thus, the 4–6 year old child in the examples above would need 0.08–0.59 cups per day for OFSP. The daily intakes of OFSP in cups per day needed to meet 50% of the requirements for vitamin A for all individuals is also shown in Fig. 3. These results show that the daily intakes of OFSP that would be supply 50% of the vitamin A requirement are reasonable.

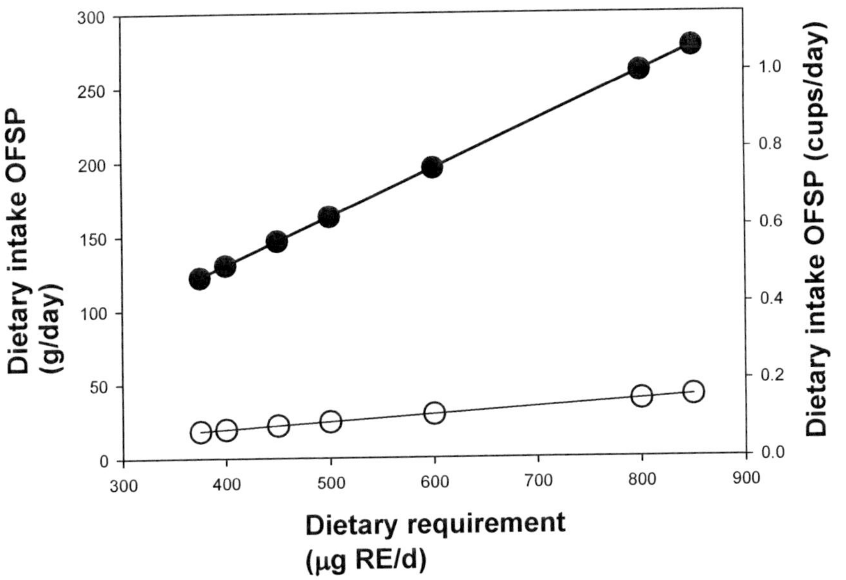

Fig. 3. High and low estimates of OFSP intakes needed to meet 50% of recommended dietary intakes at different life stages.

The Amount of OFSP that Supplies 50% of the Vitamin A Required for the Population Most at Risk for Vitamin A Deficiency

The amount of OFSP needed to supply vitamin A to all the 208 million people most at risk for vitamin A deficiency was calculated. The people most at risk for vitamin A deficiency were defined as the 190 million preschool children and 19.1 million pregnant women from countries estimated to have low vitamin A status (Table 1, WHO 2009). Since most of these women and children do not live in the United States, the vitamin A requirements determined by the World Health Organization were again used for these estimates. Thus, pregnant women had a requirement of 800 RE per day, and pregnant and lactating women had a requirement of 850 RE per day. To make these calculations, the following assumptions were used: 20% of the pregnant women were also lactating, with the higher requirement of 850 RE per day); 75% of the preschool children were aged 1–3 y (with a requirement of 400 RE per day), and 25% were aged 4–5 (with a requirement of 450 RE per day).

Therefore, 147.5 million children have an estimated requirement of 400 RE per day; and each need 19.0–131 g OFSP per day

47.5 million children had an estimated requirement of 450 RE per day; and each need 21.4–147 g OFSP per day

15.28 million pregnant women had a requirement of 800 RE per day; and each need 38.0–261 g OFSP per day

3.82 million pregnant women were also lactating and had a requirement of 850 RE per day and each need 40.4–278 g OFSP per day

Therefore, the smallest amount of OFSP needed to supply 50% of the RE for the 208.1 million people most at risk is:

(147,500,000 x 19.0 g OFSP/d x 365 d/y) + (47,500,000 x 21.4 g OFSP/d x 365 d/y) + (15,280,000 x 38.0 g OFSP/d x 365 d/y) + (3,820,000 x 40.4 g OFSP/d x 365 d/y) = 1.662×10^{12} g OFSP/y, or 1.662 million tonnes/y.

The higher, and probably more realistic amount of OFSP needed for these 208.1 million people is:

(147,500,000 x 131 g OFSP/d x 365 d/y) + (47,500,000 x 147 g OFSP/d x 365 d/y) + (15,280,000 x 261 g OFSP/d x 365 d/y) + (3,820,000 x 278 g OFSP/d x 365 d/y) = 11.44×10^{12} g OFSP/y, or 11.44 million tonnes/y.

Production, Yield, and Areas Harvested with Sweet Potato

There is little information on OFSP production, since data are given for sweet potatoes of any variety. Both white and orange sweet potatoes are popular. We found essentially no international or national data on OFSP (USDA Economic Research Service (ERS) 2012; Food and Agriculture Organization of the United Nations (FAOSTAT) 2012.). Thus, the production data available probably overestimates OFSP production by a factor of 2–3. Even so, production far exceeds the amount that would be required to supply substantial amounts of vitamin A to all individuals at risk.

The production, yield, and land area harvested for the world, Asia, Africa, China, and low income countries with food deficits, and the United States are shown in Table 4.

Current world production of sweet potatoes is 107.6 million tonnes, which is much higher than the 1.66–11.44 million tonnes that would be required to supply 50% of vitamin A for the people most at risk for vitamin A deficiency in the world. This production has increased, as has the yield per hectare, in Africa, Asia, Low income food deficit countries, and the world, even as the area harvested has decreased slightly.

Table 4. Production quantity, area harvested, and yield for sweet potatoes, and percent change from 2009 (FAOSTAT 2012, data from 2010).

	Production Tonnes (%)	Yield Tonnes/Hectare (%)	Area Harvested Hectares (%)
World	107,649,494 1.1	132,181 3.7	8,143,321 –2.6
Asia	88,555,961 0.9	200,387 3.3	4,491,253 –0.7
Africa	15,217,095 1.3	46,589 5.4	3,266,263 –4.0
Low income food deficit countries	102,031,988 1.1	133,358 4.1	7,650,986 –2.8
China	81,175,660 0.8	220,372 0.3	3,683,581 –2.6

Most sweet potatoes are produced by low-income food deficit countries, with 80% being grown in China. Other major producers include Nigeria, Uganda, Indonesia and Vietnam. China is also a center of biotechnology research (Liu 2011), and it has one of the higher yields of sweet potatoes worldwide. The production and yield of sweet potatoes in 2010 (the most recent year for which data is available), and the changes from 2009 data, are shown in Table 4. These uniformly show a small increase in production and yield, but a small decrease in area harvested.

Despite this encouraging data, many of the countries that are major producers of sweet potatoes (e.g., China, Nigeria, and Uganda) also have moderate levels of vitamin A deficiency (WHO 1995, West 2002, West et al. 2002, Black et al. 2008). The probable explanation is that most of the sweet potatoes grown and eaten in Asia and Africa are white-fleshed sweet potatoes, which contain only low concentrations of beta-carotene. This is rather surprising, since sweet potatoes are a popular food, and new varieties of OFSP appear to be accepted easily (Laurie and Magoro 2008). Consumers also enjoy new products made with sweet potato flour (Low and other 2007, Naico et al. 2010).

Surveys of farmers and scientists in developing countries have determined the most important constraints facing sweet potato farmers in developing countries. They found that there were a few key needs: control of viruses (through varietal resistance, quality planting material, and crop management); small-enterprise development for sweet potato processing; improving availability and quality of healthy sweet potato planting material; and improved cultivars with high and stable yields. Additional priorities for sub-Saharan Africa include improved control of the sweet potato weevil and cultivars with high beta-carotene content to address vitamin A deficiency (Alleman et al. 2004, Fuglie et al. 2007).

Improving Production through Biotechnology

Efforts to improve OFSP through biotechnology have focused primarily on increasing tolerance to environmental stressors (such as salt and drought), or resistance to viruses and insect pests. Most of this research appears to be in its infancy. Most reports focus on trying to identify the genes related to pest and stress resistance. For example, genotypes have been examined for resistance to North Carolina root-knot nematode (Cervantes-Flores et al. 2001). Another report suggests that a few transgenic sweet potatoes strains have been developed that have improved resistance to black rot (Muramoto et al. 2012). Resistance was increased by implanting a gene for barley alpha-hordothionin downstream of the promoter of a sweet potato gene for beta-amylase of storage roots; but the resistance level was similar to the normally grown sweet potato variety Tamayutaka.

Many studies have focused on virus resistance, especially protection from the sweet potato feathery moth virus (Nishiguchi et al. 1998, Okada et al. 2001, Okada et al. 2002, Wambugu 2003, Okada and Saito 2008), and sweet potato chlorotic stunt virus (Cuellar et al. 2011) with some success. Progress has also been made in identifying the molecular markers for sweet potato virus disease, which develops when plants are infected with both sweet potato feathery mottle virus and sweet potato chlorotic stunt virus (Miano et al. 2008), and the gene expression markers of resistant plants (McGregor et al. 2009).

Transgenic sweet potatoes have been developed that express the bar gene to glutamine synthetase inhibiters, and are 20–82-fold more resistant to some herbicides (such as glufosinate), but not others such as oxyfluorfen and paraquat (Yi et al. 2007, Shin et al. 2011). Some of the genes involved in chill-tolerance have been identified (Lin et al. 2011). Finally, transgenic varieties have been developed that are resistant to a variety of environmental stressors, through a variety of mechanisms. Approaches include introducing a gene from Spinacia oleracea encoding betaine aldehyde dehydrogenase, which is involved in the synthesis of glycine betaine (Fan et al. 2012); or a gene from Arabidopsis inducing homeodomain glabrous 11 (Ruan et al. 2012); increasing genes for superoxide dismutase and ascorbate peroxidase (Lim et al. 2007), and inducing multiple stress response gene 1 from the sweet potato itself (Seo et al. 2010).

Taken together, these results suggest that newly developed transgenic sweet potatoes may have improved resistance to important diseases and environmental stressors, which might help increase production, and also extend OFSP production to areas that have marginal climates and conditions. However, the percent increases in production and yield in China, a hotbed of sweet potato biotechnology, are lower than those for the world in general, and Africa in particular. Furthermore, there is no evidence that research on the impact of environmental stress has actually resulted in an expansion of the areas in which sweet potatoes are grown. Taken together, these results suggest that it is unlikely that biotechnology has been a major force driving the increases in production and yield seen worldwide. However, several research areas offer promise for the future.

Conclusion

OFSP are excellent sources of vitamin A. They are relatively simple to grow, durable, and are easy to prepare. Therefore, OFSP have considerable potential as a nutritious, sustainable source for vitamin A. OFSP have complex genetics, and have been difficult to improve by traditional cross-breeding methods. Biotechnology has the potential to improve OFSP quality and quantity in several ways, mostly by improving resistance to viruses and environmental stress. At present, though, this research is in its infancy and appears to have had little impact on sweet potato nutrient concentrations or yields.

References

Aguayo, V.M., S.K. Baker, X. Crespin, H. Hamani and A. Mamadoul Taïbou. 2005. Maintaining high vitamin A supplementation coverage in children: lessons from Niger. Food Nutr. Bull. 26: 26–31.

Ahmed, M., M.S. Akter and J.B. Eun. 2010. Peeling, drying temperatures, and sulphite-treatment affect physicochemical properties and nutritional quality of sweet potato flour. Food Chem. 121: 112–118.

Aina, A.J., K.O. Falade, J.O. Akingbala and P. Titus. 2009. Physicochemical properties of twenty-one Caribbean sweet potato cultivars. Internat. J. Food Sci. Tech. 44: 1696–1704.

Allemann, J., S.M. Laurie, S. Thiart and H.J. Vorster. 2004. Sustainable production of root and tuber crops (potato, sweet potato, indigenous potato, cassava) in southern Africa. South African J. Bot. 70: 60–66.

Antia, B.S., E.J. Akpan, P.A. Okon and I.U. Umoren. 2006. Nutritive and anti-nutritive evaluation of sweet potatoes (Ipomoea batatas) leaves. Pak. J. Nutr. 2: 166–168.

Arizio, C.M., N. Hompanera, E.Y. Suarez and M.M. Manifesto. 2009. Genotypic identification and diversity evaluation of a sweet potato (*Ipomoea batatas* (L). Lam) collection using microsatellites. Plant Genetic Res. Character. Utiliz. 7: 135–138.

Bechoff, A., M. Poulaert, K.I. Tomlins, A. Westby, G. Menya, S. Young and C. Dhuique-Mayer. 2011. Retention and bioaccessibility of beta-carotene in blended foods containing orange-fleshed sweet potato flour. J. Agric. Food. Chem. 59: 10373–10380.

Bengtsson, A., A. Namutebi, M. Larson Alminger and U. Svanberg. 2008. Effects of various traditional processing methods on the all-*trans*-β-carotene content of orange sweet potato. J. Food Comp. Analy. 21: 134–143.

Bengtsson, A., M. Larsson Alminger and U. Svanberg. 2009. *In vitro* bioaccessibility of beta-carotene from heat-processed orange-fleshed sweet potato. J. Agric. Food Chem. 57: 9693–9698.

Bezerra, D.S., K.F. de Araújo, G.M. Azevêdo and R. Dimenstein. 2010. A randomized trial evaluating the effect of 2 regimens of maternal vitamin a supplementation on breast milk retinol levels. J. Hum. Lact. 26: 148–156.

Biesalski, H.K., G.R. Chichili, J. Frank, J. von Lintig and D. Nohr. 2007. Conversion of beta-carotene to retinal pigment. Vitam. Horm. 75: 117–130.

Black, R.E., R.H. Allen, Z.A. Bhutta, L.E. Caulfield, M. de Onis, M. Ezzati, C. Mathers and J. Rivera. 2008. Maternal and child undernutrition: global and regional exposures and health consequences. Lancet 371: 243–260.

Blomhoff, R. and H.K. Blomhoff. 2006. Overview of retinoid metabolism and function. J. Neurobiol. 66: 606–630.

Bouis, H.E. and R.M. Welch. 2010. Biofortification—a sustainable agricultural strategy for reducing micronutrient malnutrition in the global south. Crop Sci. 50: S20–S32.

Bovell-Benjamin, A.C. 2007. Sweet potato: a review of its past, present, and future role in human nutrition. Adv. Food Nutr. Res. 52: 1–59.

Britton, G. 1995. Structure and properties of carotenoids in relation to function. F.A.S.E.B. J. 9: 1551–1558.

Burri, B.J. 2011. Evaluating sweet potato as an intervention food to prevent vitamin A deficiency. Comp. Rev. Food Sci. Food Safety 10: 118–130.

Burri, B.J. and A.J. Clifford. 2004. Carotenoid and retinoid metabolism: insights from isotope studies. Arch. Biochem. Biophys. 430: 110–119.

Burri, B.J., T.R. Neidlinger and A.J. Clifford. 2001. Serum carotenoid depletion follows first-order kinetics in healthy adult women fed naturally low carotenoid diets. J. Nutr. 131: 2096–2100.

Burri, B.J. and T. Turner. 2009. Evaluating the effectiveness of beta-carotene-rich food interventions for improving vitamin A status. *In*: L. Haugen and T. Bjornson (eds.). Beta Carotene: Dietary Sources, Cancer and Cognition. Nova Scientific Pub. Haupergne, NY, pp. 120–139.

Cervantes-Flores, J.C., G.C. Yencho and E.L. Davis. 2001. Development of sweetpotato host differential genotypes for Meloidogyne sp. Hort. Science 36: 493.

Cervantes-Flores, J.C., G.C. Yencho, A. Kriegner, K.V. Pecota, M.A. Faulk, R.O.M. Mwanga and B.R. Sosinski. 2008. Development of a genetic linkage map and identification of homologous linkage groups in sweetpotato using multiple-dose AFLP markers. Mol. Breeding 21: 511–532.

Chattopadhyay, A., I. Chakraborty, P.R. Kumar, M.K. Nanda and H. Sen. 2006. Uncontrolled storage behaviour of sweet potato (*Ipomoea batatas* L. Lam). J. Food Sci. Tech.-Mysore 43: 41–45.

Chichili, G.R., D. Nohr, M. Schäffer, J. von Lintig and H.K. Biesalski. 2005. Beta-carotene conversion into vitamin A in human retinal pigment epithelial cells. Invest. Ophthalmol. Vis. Sci. 46: 3562–3569.

Cuellar, W.J., R.K. Cruzado, S. Fuentes, M. Untiveros, M. Soto and J.F. Kreuze. 2011. Sequence characterization of a Peruvian isolate of Sweet potato chlorotic stunt virus: Further variability and a model for p22 acquisition. Virus Res. 157: 111–115.

Dary, O. and J.O. Mora. 2002. Food fortification to reduce vitamin A deficiency: international vitamin A consultative group recommendations. J. Nutr. 132: 2927S–2933S.

de Pee, S., C.E. West, D. Permaesih, S. Martuti, Muhilal and J.G. Hautvast. 1998. Orange fruit is more effective than are dark-green, leafy vegetables in increasing serum concentrations of retinol and beta-carotene in schoolchildren in Indonesia. Am. J. Clin. Nutr. 68: 1058–1067.

Domingo, J.L. 2007. Toxicity studies of genetically modified plants: A review of the published literature. Crit. Rev. Food Sci. Nutr. 47: 721–733.

Donnen, P., A. Sylla, M. Dramaix, G. Sall, N. Kuakuvi and P. Hennart. 2007. Effect of daily low dose of vitamin A compared with single high dose on morbidity and mortality of hospitalized mainly malnourished children in Senegal: a randomized controlled clinical trial. Eur. J. Clin. Nutr. 61: 1393–1399.

Duriancik, D.M., D.E. Lackey and K.A. Hoag. 2010. Vitamin A as a regulator of antigen presenting cells. J. Nutr. 140: 1395–1399.

Emenhiser, C., R.H. Watkins, N. Simunovic, N. Solomons, J. Bulux, J. Barrows and S.J. Schwartz. 1999. Packaging preservation of beta-carotene in sweet potato flakes using flexible film and an oxygen absorber. J. Food Quality 22: 63–73.

Faber, M. and P.J. van Jaarsveld. 2007. Review—The production of provitamin A-rich vegetables in home-gardens as a means of addressing vitamin A deficiency in rural African communities. J. Sci. Food Agric. 87: 366–377.

Failla, M.L., S.K. Thakkar and J.Y. Kim. 2009. *In vitro* bioaccessibility of beta-carotene in orange fleshed sweet potato (Ipomoea batatas, Lam.). J. Agric. Food Chem. 57: 10922–10927.

Fan, W., M. Zhang, H. Zhang and P. Zhang. 2012. Improved tolerance to various abiotic stresses in transgenic sweet potato (Ipomoea batatas) expressing spinach betaine aldehyde dehydrogenase. Plos One 7. AR e37344.

Fiedler, J.L. and R. Afidra. 2010. Vitamin A fortification in Uganda: comparing the feasibility, coverage, costs, and cost-effectiveness of fortifying vegetable oil and sugar. Food Nutr. Bull. 31: 193–205.

Fonseca, M.J.D., A.G. Soares, M.F. Junior, D.L. de Almeida and J.L.R. Ascheri. 2008. Effect of extrusion-cooking in total carotenoids content in cream and orange flesh sweet potato cultivars. Horticultura Brasileira 26: 112–115.

[Food and Agriculture Organization of the United Nations] FAO, FAOSTAT, Crops and Livestock Products. 2012. [Accessed 2012 Aug 16]. Available from: http://faostat.fao.org/site/535/default.aspx#ancor.

Fuglie, K.O. 2007. Priorities for sweetpotato research in developing countries: Results of a survey. Hort. Science. 42: 1200–1206.

George, M.S., G.Q. Lu and W.J. Zhou. 2002. Genotypic variation for potassium uptake and utilization efficiency in sweet potato (*Ipomoea batatas* L.). Field Crops Res. 77: 7–15.

[GM Watch]. 2010. "Millions served"—the GM sweet potato. [Accessed 2012 Jul 24]. Available from: http://www.gmwatch.org/gm-myths/11132-qmillions-servedq-the-gm-sweet-potato.

Hagenimana, V., E.E. Carey, S.T. Gichuki, M.A. Oyunga and J.K. Immungi. 1999. Carotenoid contents in fresh, dried, and processed sweet potato products. Eco. Food Nutr. 37: 455–473.

Hagenimana, V., L.M. K'Osambo and E.E. Carey. 1999. Potential of sweet potato in reducing vitamin A deficiency in Africa. *In*: [International Potato Center] Impact on a Changing World. Program Report 1997–1998. International Potato Center, Lima, Peru, pp. 287–294.

Haskell, M.J., K.M. Jamil, F. Hassan, J.M. Peerson, M.I. Hossain, G.J. Fuchs and K.H. Brown. 2004. Daily consumption of Indian spinach (Basella alba) or sweet potatoes has a positive effect on total-body vitamin A stores in Bangladeshi men. Amer. J. Clin. Nutr. 80: 705–709.

Haskell, M.J., P. Pandey, J.M. Graham, J.M. Peerson, K.A. Shreshtha and K.H. Brown. 2005. Recovery from impaired dark adaptation in nightblind pregnant Nepali women who receive small daily doses of vitamin A as amaranth leaves, carrots, goat liver, vitamin A-fortified rice, or retinyl palmitate. Amer. J. Clin. Nutr. 81: 461–471.

Hickenbottom, S.J., J.R. Follett, Y. Lin, S.R. Dueker, B.J. Burri, T.R. Neidlinger and A.J. Clifford. 2002. Variability in conversion of beta-carotene to vitamin A in men as measured by using a double-tracer study design. Am. J. Clin. Nutr. 75: 900–907.

Ho, C.C., F.F. de Moura, S.H. Kim, B.J. Burri and A.J. Clifford. 2009. A minute dose of 14C-beta-carotene is absorbed and converted to retinoids in humans. J. Nutr. 139: 1480–1486.

Ho, C.C., F.F. de Moura, S.H. Kim and A.J. Clifford. 2007. Excentral cleavage of beta-carotene *in vivo* in a healthy man. Am. J. Clin. Nutr. 85: 770–777.

Howe, J.A. and S.A. Tanumihardjo. 2006. Carotenoid-biofortified maize maintains adequate vitamin A status in Mongolian gerbils. J. Nutr. 136: 2562–2567.

Huang, C.J., Y.L. Tang, C.Y. Chen, M.L. Chen, C.H. Chu and C.T. Hseu. 2000. The bioavailability of beta-carotene in stir- or deep-fried vegetables in men determined by measuring the serum response to a single ingestion. J. Nutr. 130: 534–540.

Idindili, B., H. Masanja, H. Urassa, W. Bunini, P. van Jaarsveld, J.J. Aponte, E. Kahigwa, H. Mshinda, D. Ross and D.M. Schellenberg. 2007. Randomized controlled safety and efficacy trial of 2 vitamin A supplementation schedules in Tanzanian infants. Am. J. Clin. Nutr. 85: 1312–1319.

[International Potato Center (CIP)]. 2006. International potato center (CIP) report. McKnight Foundation CCRP East African Regional Sweetpotato Project Annual Report (April 2005–March 2006). International Potato Center, Nairobi, Kenya.

Kafwembe, E.M., J. Chipipa, E. Njunju and R. Chilengi. 2009. The vitamin A status of Zambian children in a community of vitamin A supplementation and sugar fortification strategies as measured by the modified relative dose response (MRDR) test. Int. J. Vitam. Nutr. Res. 79: 40–47.

Kapinga, R., J. Ndunguru, G. Mulokozi and S. Tumwegamire. 2006. Impact of common sweet potato viruses on total carotenoids and root yields of an orange-fleshed sweet potato in Tanzania. Scientia Horticulturae 122: 1–5.

Kidmose, U., S. Agili and S.H. Thilsted. 2009. Beta-carotene in orange fleshed sweet potato from different locations in Kenya and stored using different methods. Ann. Nutr. Metab. 55: 376.

Kidmose, U., L.P. Christensen, S.M. Agili and S.H. Thilsted. 2007. Effect of home preparation practices on the content of provitamin A carotenoids in coloured sweet potato varieties (Ipomoea batatas Lam.) from Kenya. Innov. Food Sci. Emerg. Tech. 8: 399–406.

Kidmose, U., R.Y. Yang, S.H. Thilsted, L.P Christensen and K. Brandt. 2006. Content of carotenoids in commonly consumed Asian vegetables and stability and extractability during frying. J. Food Comp. Analy. 19: 562–571.

Kim, C.Y., Y.O. Ahn, S.H. Kim, Y.H. Kim, H.S. Lee, A.S. Catanach, J.M.E. Jacobs, A.J. Conner and S.S. Kwak. 2010. The sweet potato IbMYB1 gene as a potential visible marker for sweet potato intragenic vector system. Physiologia Plantarum 139: 229–240.

Kim, S.H., Y.O. Ahn, M.J. Ahn, H.S. Lee and S.S. Kwak. 2012. Down-regulation of beta-carotene hydroxylase increases beta-carotene and total carotenoids enhancing salt stress tolerance in transgenic cultured cells of sweetpotato. Phytochem. 74: 69–78.

Klemm, R.D., K.P. West Jr., A.C. Palmer, Q. Johnson, P. Randall, P. Ranum and C. Northrop-Clewes. 2010. Vitamin A fortification of wheat flour: considerations and current recommendations. Food Nutr. Bull. 31(1 Suppl): S47–S61.

Kongsbak, K., S.H. Thilsted and M.A. Wahed. 2008. Effect of consumption of the nutrient-dense, freshwater small fish *Amblypharyngodon mola* on biochemical indicators of vitamin A status in Bangladeshi children: a randomised, controlled study of efficacy. Brit. J. Nutr. 99: 581–597.

K'osambo, L.M., E.E. Carey, A.K. Mirsa, J. Wilkes and V. Hagenimana. 1998. Influence of age, farming site, and Bowling on pro-vitamin A content of sweet potato (*Ipomonea batatas* (L.) Lam) storage roots. J. Food Comp. Analy. 11: 305–321.

Kudoh, Y. and S. Matsuda. 2000. Effect of lactic acid bacteria on color tone and anthocyanin content of sweet potato yoghurt. J. Jap. Soc. Food Sci. Tech.-Nippon Shokuhin Kagaku Kogaku Kaishi 47: 619–625.

Lako, J., V.C. Trenerry, M. Wahlqvist, N. Wattanapenpaiboon, S. Sotheeswaran and R. Premier. 2007. Phytochemical flavonols, carotenoids, and the antioxidant properties of a wide selection of Fijian fruit, vegetables and other readily available foods. Food Chem. 101: 1727–1741.

Laurie, S.M. and M.D. Magoro. 2008. Evaluation and release of new sweet potato varieties through farmer participatory selection. African J. Agric. Res. 3: 672–676.

Lietz, G., J. Lange and G. Rimbach. 2010. Molecular and dietary regulation of beta, beta-carotene 15,15'-monooxygenase 1 (BCMO1). Arch. Biochem. Biophys. 502: 8–16.

Li, X.Q. and D.P. Zhang. 2003. Gene expression activity and pathway selection for sucrose metabolism in developing storage root of sweet potato. Plant Cell Physiol. 44: 630–636.

Lim, S., Y.H. Kim, S.-H. Kim, S.-Y. Kwon, H.S. Lee, J.-S. Kim, K.-Y. Cho, K.-Y. Paek and S.S. Kwak. 2007. Enhanced tolerance of transgenic sweetpotato plants that express both

CuZnSOD and APX in chloroplasts to methyl viologen-mediated oxidative stress and chilling. Molec. Breeding 19: 227–239.

Lin, K.H., C.M. Chiang, Y.C. Lai, S.H. You and H.F. Lo. 2011. Identification of chilling-inducible genes in sweet potato by suppression subtractive hybridization. Res. J. Biotech. 6: 37–43.

Lin, Y., S.R. Dueker, B.J. Burri, T.R. Neidlinger and A.J. Clifford. 2000. Variability of the conversion of beta-carotene to vitamin A in women measured by using a double-tracer study design. Am. J. Clin. Nutr. 71: 1545–1554.

Liu, Q. 2011. Sweet potato Omics and Biotechnology in China. Plant Omics 4: 295–301.

Liu, S.C., J.T. Lin and D.J. Yang. 2009. Determination of cis- and trans- alpha- and beta-carotenoids in Taiwanese sweet potatoes (*Ipomoea batatas* (L.) Lam.) harvested at various times. Food Chem. 116: 605–610.

Loechl, C.U., C. Hotz, A. Lubowa, C. Sison, D. Gilligan and J.V. Meenakshi. 2009. Introduction of orange sweet potato in Uganda could improve vitamin A intakes of vulnerable groups. Ann. Nutr. Metab. 55: 376.

Low, J.W., M. Arimond, N. Osman, B. Cunguara, F. Zano and D.A. Tschirley. 2007a. Ensuring the supply of and creating demand for a biofortified crop with a visible trait: Lessons learned from the introduction of orange-fleshed sweet potato in drought-prone areas of Mozambique. Food Nutr. Bull. 28: S258–S270.

Low, J.W., M. Arimond, N. Osman, B. Cunguara, F. Zano and D.A. Tschirley. 2007b. A food-based approach introducing orange-fleshed sweet potatoes increased vitamin A intake and serum retinol concentrations in young children in rural Mozambique. J. Nutr. 137: 1320–1327.

Low, J.W. and P.J. van Jaarsveld. 2008. The potential contribution of bread buns fortified with beta-carotene-rich sweet potato in Central Mozambique. Food Nutr. Bull. 29: 98–107.

Marceau, G., D. Gallot, D. Lemery and V. Sapin. 2007. Metabolism of retinol during mammalian placental and embryonic development. Vitam. Horm. 75: 97–115.

Mark, M., N.B. Ghyselinck and P. Chambon. 2009. Function of retinoic acid receptors during embryonic development. Nucl. Recept. Sign. 7: e002.

McGrane, M.M. 2007. Vitamin A regulation of gene expression: molecular mechanism of a prototype gene. J. Nutr. Biochem. 18: 497–508.

McGregor, C.E. and D.R. LaBonte. 2006. Differential expression of genes between storage roots of sweet potato cultivars jewel and white jewel. J. Am. Soc. Horticult. Sci. 131: 798–805.

McGregor, C.E., D.W. Miano, D.R. LaBonte, M. Hoy, C.A. Clark and G.J.M. Rosa. 2009. Differential gene expression of resistant and susceptible sweetpotato plants after infection with the causal agents of sweet potato virus disease. J. Am. Soc. Horticult. Sci. 134: 658–666.

Miano, D.W., D.R. LaBonte and C.A. Clark. 2008. Identification of molecular markers associated with sweet potato resistance to sweet potato virus disease in Kenya. Euphytica 160: 15–24.

Mills, J.P., G.A. Tumuhimbise, K.M. Jamil, S.K. Thakkar, M.L. Failla and S.A. Tanumihardjo. 2009. Sweet potato beta-carotene bioefficacy is enhanced by dietary fat and not reduced by soluble fiber intake in Mongolian gerbils. J. Nutr. 139: 44–50.

Mohapatra, S., S.H. Panda, S.K. Sahoo, P.S. Sivakumar and R.C. Ray. 2007. Beta-carotene-rich sweet potato curd: production, nutritional and proximate composition. Internat. J. Food Sci. Tech. 42: 1305–1314.

Montrone, M., D. Martorelli, A. Rosato and R. Dolcetti. 2009. Retinoids as critical modulators of immune functions: new therapeutic perspectives for old compounds. Endocr. Metab. Immune Disord. Drug Targets 9: 113–131.

Muramoto, N., K. Tanaka, T. Shimamura, N. Mitsukawa, E. Hori, K. Koda, M. Otani, M. Hirai, K. Nakamura and T. Imaeda. 2012. Transgenic sweet potato expressing thionin from barley gives resistance to black rot disease caused by Ceratocystis fimbriata in leaves and storage roots. Plant Cell Rep. 31: 987–997.

Naico, A.T.A. and J.L. Lusk. 2010. The value of a nutritionally enhanced staple crop: results from a choice experiment conducted with orange-fleshed sweet potatoes in Mozambique. J. Afric. Econ. 19: 536–558.

[New Scientist]. 2004. Monsanto's showcase project in Africa fails. New Scientist 181: 2433.

Nishiguchi, M., M. Mori, Y. Okada, T. Murata, T. Kimura, J. Sakai, K. Hanada, C. Miyazaki and A. Saito. 1998. Virus resistant transgenic sweet potato with the CP gene: Current challenge and perspective of its use. Phytoprotection 79: 112–116.

[NW Resistance Against Genetic Engineering]. Sweet potato: Monsanto's GM technology fails in Kenya. [Accessed 25 Jun 2012]. Available from: http://nwrage.org/content/sweet-potato-monsantos-gm-technology-fails-kenya-hype-falls-apart.

O'Connell, O.F., L. Ryan and N.M. O'Brien. 2007. Xanthophyll carotenoids are more bioaccessible from fruits than dark green vegetables. Nutr. Res. 27: 258–264.

Okada, Y., M. Nishiguchi, A. Saito, T. Kimura, M. Mori, K. Hanada, J. Sakai, Y. Matsuda and T. Murata. 2002. Inheritance and stability of the virus-resistant gene in the progeny of transgenic sweet potato. Plant Breeding 121: 249–253.

Okada, Y. and A. Saito. 2008. Evaluation of resistance to complex infection of SPFMVs in transgenic sweet potato. Breeding Sci. 58: 243–250.

Okada, Y., A. Saito, M. Nishiguchi, T. Kimura, M. Mori, K. Hanada, J. Sakai, C. Miyazaki, Y. Matsuda and T. Murata. 2001. Virus resistance in transgenic sweetpotato *Ipomoea batatas* L. (Lam) expressing the coat protein gene of sweet potato feathery mottle virus. Theoret. Applied Genet. 103: 743–751.

Otani, M., Y. Wakita and T. Shimada. 2003. Production of herbicide-resistant sweet potato (*Ipomoea batatas* (L.) Lam) plants by agrobacterium tumefaciens mediated-transformation. Breeding Sci. 53: 145–148.

Pan, L.-P., S.L. Yu, C.-J. Chen, H. Li, Y.L. Wu and H.H. Li. 2012. Cloning a peanut resveratrol synthase gene and its expression in purple sweet potato. Plant Cell Rep. 31: 121–131.

Pedro, M.R., J.R. Madriaga, C.V. Barba, R.C. Habito, A.E. Gana, M. Deitchler and J.B. Mason. 2004. The national Vitamin A Supplementation Program and subclinical vitamin A deficiency among preschool children in the Philippines. Food Nutr. Bull. 25: 319–329.

Qaim, M. 2001. A prospective evaluation of biotechnology in semi-subsistence agriculture. Agric. Econ. 25: 165–175.

Ray, R.C. and P.S. Sivakumar. 2009. Traditional and novel fermented foods and beverages from tropical root and tuber crops: review. Internat. J. Food Sci. Tech. 44: 1073–1087.

Ribaya-Mercado, J.D., F.S. Solon, M.A. Solon, M.A. Cabal-Barza, C.S. Perfecto, G. Tang, J.A.A. Solon, C. Fjeld and R. Russell. 2000. Bioconversion of plant carotenoids to vitamin A in Filipino school-aged children varies inversely with vitamin A status. Am. J. Clin. Nutr. 72: 455–465.

Ritter, E., M. Elgeti and F.J. Bartl. 2008. Activity switches of rhodopsin. Photochem. Photobiol. 84: 911–920.

Rodriguez-Amaya, D.B. 1997. Carotenoids and food preparation: the retention of provitamin A carotenoids in prepared, processed, and stored food. USAID, OMNI Project. John Snow, Inc., Arlington Virginia.

Rodriguez-Amaya, D.B. 2003. Food carotenoids; analysis, composition and alterations during storage and processing foods. Forum Nutr. 56: 35–37.

Roos, N., T. Leth, J. Jakobsen and S.H. Thilsted. 2002. High vitamin A content in some small indigenous fish species in Bangladesh: perspectives for food-based strategies to reduce vitamin A deficiency. Int. J. Food Sci. Nutr. 53: 425–437.

Ruan, L., L. Chen, Y. Chen, J. He, W. Zhang, Z. Gao and Y. Zhang. 2012. Expression of Arabidopsis HOMEODOMAIN GLABROUS 11 enhances tolerance to drought stress in transgenic sweet potato plants. J. Plant Biol. 55: 151–158.

Ruel, M.T. 2001. Can food-based strategies help reduce vitamin A and iron deficiencies? A review of recent evidence. International Food Policy Research Institute. Washington, DC.

[Sandhill Preservation Center]. 2010. Sweet potatoes. [Accessed 2012 Aug 12]. Available from: http://www.sandhillpreservation.com/catalog/sweet_potatoes.html.

Semba, R.D., S. de Pee, K. Sun, N. Akhter, M.W. Bloem and V.K. Raju. 2010. Coverage of vitamin A capsule programme in Bangladesh and risk factors associated with non-receipt of vitamin A. J. Health Popul. Nutr. 28: 143–148.

Schafleitner, R., L.R. Tincopa, O. Palomino, G. Rossel, R.F. Robles, R. Alagon, C. Rivera, C. Quispe, L. Rojas, L.A. Pacheco, J. Solis, D. Cerna, J.Y. Kim, J. Hou, J. and R. Simon. 2010. A sweet potato gene index established by *de novo* assembly of pyrosequencing and Sanger sequences and mining for gene-based microsatellite markers. BMC Genomics 11: 604.

Seo, S.-G., J.S. Kim, Y.-S. Yang, B.-K. Jun, S.-V. Kang, G.-P. Lee, W. Kim, J.B. Kim, H.-U. Lee and S.H. Kim. 2010. Cloning and characterization of the new multiple stress responsible gene I (MuSI) from sweet potato. Genes Genomics 32: 544–552.

Shin, J.-M., B.-K. Kim, S.-G. Seo, S.B. Jeon, J.-S. Kim, B.-K. Jun, S.-Y. Kang, J.-S. Lee, M.-N. Chung, and S.-H. Kim. 2011. Mutation breeding of sweet potato by gamma-ray radiation. Afric. J. Agric. Res. 6: 1447–1454.

Shin, J.S., K.M. Kim, D.J. Lee, S.B. Lee, N.R. Burgos and Y.I. Kuk. 2011. Resistance levels and fitness of glufosinate-resistant transgenic sweet potato in field experiments. Field Crops Res. 121: 324–332.

Sommer, A., K.P. West, J.A. Olson and A.C. Ross. 1996. Vitamin A deficiency: health, survival and vision. New York: Oxford University Press, 438 pp.

Stein, A.J., H.P.S. Sachdev and M. Qaim. 2006. Potential impact and cost effectiveness of Golden Rice. Nature Biotech. and supplementary material. 24: 1200–1201.

Sulaeman, A., L. Keeler, D.W. Giraud, S.L. Taylor and J.A. Driskell. 2003. Changes in carotenoid, physicochemical and sensory values of deep-fried carrot chips during storage. Internat. J. Food Sci. Tech. 38: 603–613.

Takahata, Y., T. Noda and T. Nagata. 1993. HPLC determination of beta-carotene content of sweet-potato cultivars and its relationship with color values." Jap. J. Breeding 43: 421–427.

Tanumihardjo, S.A. 2008. Food-based approaches for ensuring adequate vitamin A nutrition. Comp. Rev. Food Sci. Food Safety 7: 373–381.

Tanumihardjo, S.A., H. Bouis, C. Hotz, J.V. Meenakshi and B. McClafferty. 2008. Biofortification of staple crops: An emerging strategy to combat hidden hunger. Comp. Rev. Food Sci. Food Safety 7: 329–334.

Tekalign, T. 2007. Genotype x environment interaction for root yield of elite sweet potato (*Ipomoea batatas* (L.) lam.) genotypes. South Afric. J. Plant Soil 24: 144–146.

Teow, C.C., V.D. Truong, R.F McFeeters, R.L. Thompson, K.V. Pecota and G.C. Yencho. 2007. Antioxidant activities, phenolic and beta-carotene contents of sweet potato genotypes with varying flesh colours. Food Chem. 103: 829–838.

Travis, G.H., M. Golczak, A.R. Moise and K. Palczewski. 2007. Diseases caused by defects in the visual cycle: retinoids as potential therapeutic agents. Ann. Rev. Pharmacol. Toxicol. 47: 469–512.

Tumuhimbise, G.A., A. Namutebi and J.H. Muyonga. 2009. Microstructure and *in vitro* beta carotene bioaccessibility of heat processed orange fleshed sweet potato. Plant Foods Hum. Nutr. 64: 312–318.

[United States Department of Agriculture (USDA), Agricultural Research Service (ARS)]. 2012. National Nutrient Database for Standard Reference, Release 24. [Accessed 2012 Aug 6]. Available from: http://www.nal.usda.gov/fnic/foodcomp/search/ .

[United States Institute of Medicine, Food and Nutrition Board]. 2000. Dietary reference intakes for vitamin A, vitamin K, arsenic, boron, chromium, copper, iodine, iron, manganese, molybdenum, nickel, silicon, vanadium, and zinc. National Academy Press, Washington DC.

[USDA. Economic Research Service (ERS)]. 2012. US Sweet potato statistics. [Accessed 2012 Aug 14]. Available from: http://usda.mannlib.cornell.edu/MannUsda/viewDocumentInfo.do?documentID=1492.

Valdez, C., C.Y. Lopez, S. Schwartz, J. Bulux and N.W. Solomons. 2001. Sweet potato buds: the origins of a "designer" food to combat hypovitaminosis A in Guatemala. Processing, vitamin A content and preservation characteristics. Nutr. Res. 21: 61–70.

Vallabhaneni, R., C.E. Gallagher, N. Licciardello, A.J. Cuttriss, R.F. Quinlan and E.T. Wurtzel. 2009. Metabolite sorting of a germplasm collection reveals the hydroxylase3 locus as a new target for maize provitamin A biofortification. Plant Physiol. 151: 1635–1645.

van Heerden, P.D. and R. Laurie. 2008. Effects of prolonged restriction in water supply on photosynthesis, shoot development and storage root yield in sweet potato. Physiol. Plant 134: 99–109.

van Jaarsveld, P.J., M. Faber, S.A. Tanumihardjo, P. Nestel, C.J. Lombard and A.J. Beade. 2005. Beta-carotene-rich orange-fleshed sweet potato improves the vitamin A status of primary school children assessed with the modified-relative-dose-response test. Am. J. Clin. Nutr. 81: 1080–1087.

van Jaarsveld, P.J., M. De Wet, E. Hamse, P. Nestel and D.B. Rodriguez-Amaya. 2006. Retention of beta-carotene in boiled, mashed orange-fleshed sweet potato. J. Food Comp. Anal. 19: 321–29.

Veda, S., Kamath, A., A. Kamath, K. Platel, K. Begum and K. Srinivasan. 2006. Determination of bioaccessibility of beta-carotene in vegetables by *in vitro* methods. Mol. Nutr. Food Res. 50: 1047–1052.

Vimala, B., B. Nambisan and B. Hariprakash. 2011. Retention of carotenoids in orange-fleshed sweet potato during processing. J. Food Sci. Technol. 48: 520–524.

Wambagu, F.M. 1999. Why Africa needs agricultural biotech. Nature 400: 15–16.

Wambugu, F.M. 2003. Development and transfer of genetically modified virus-resistant sweet potato for subsistence farmers in Kenya. Nutr. Rev. 61: S110–S113.

Watanabe, K., T. Toki, M. Ino, S. Hirota and B. Takahashi. 1999. Changes in beta-carotene content and beta-carotene isomer composition in sweet potato during storage. J. Jap. Soc. Hort. Sci. 68: 1044–1046.

West, K.P. Jr. 2002. Extent of vitamin A deficiency among preschool children and women of reproductive age. J. Nutr. 132: 2857S–2866S.

West, K.P. Jr., A. Rice and J.D. Sugimoto. 2002. Tables on the global burden of vitamin A deficiency and xerophthalmia among preschool aged children and low vitamin A status, vitamin A deficiency, and night blindness among pregnant women by WHO region [Accessed 2012 Aug 24]. Available from: http://www.jhsph.edu/CHN/GlobalVAD.pdf; updated August 2002 .

Wolgemuth, D.J. and S.S. Chung. 2007. Retinoid signaling during spermatogenesis as revealed by genetic and metabolic manipulations of retinoic acid receptor alpha. Soc. Reprod. Fertil. Suppl. 63: 11–23.

Woolfe, J.A. 1992. Sweet potato: an untapped food resource. Cambridge University Press. New York, New York.

World Health Organization. 1995. The global prevalence of vitamin A deficiency. Micronutrient Deficiency Information System (MDIS) Working Paper 2. Geneva, World Health Organization, (WHO/NUT/95.3). [Accessed 2012 Aug 19]. Available from: (http://www.who.int/nutrition/publications/vad_global_prevalence/en/index.html).

World Health Organization (WHO). 2009. Global prevalence of vitamin A deficiency in populations at risk 1995–2005. WHO global database on vitamin A deficiency. Geneva, WHO [Accessed 2012 Aug 19]. Available from: http://whqlibdoc.who.int/publications/2009/9789241598019_eng.pdf.

Wu, X., C.J. Sun, L.H. Yang, G. Zeng, Z.Y. Liu and Y.M. Li. 2008. Beta-carotene content in sweet potato varieties from China and the effect of preparation on beta-carotene retention in the Yanshu No. 5. Innov. Food Sci. Emerg. Tech. 9: 581–586.

Yi, G., Y.M. Shin, G. Choe, B. Shin, Y.S. Kim and K.M. Kim. 2007. Production of herbicide-resistant sweet potato plants transformed with the bar gene. Biotech. Lett. 29: 669–675.

Zeng, G., L. Lin, Z.Y. Liu, L. Yan, Z. Lan and C.J. Sun. 2009. The effect of beta-carotene rich sweet potato on serum retinol concentration in rural Chinese children. Ann. Nutr. Metab. 55: 119.

15

Tuberous Medicinal Plants of India

Biology and Biotechnology

Varsha Sharma[1] and *Kishan Gopal Ramawat*[2,*]

ABSTRACT

Tuberous plants are important for people all over the world due to their edible as well as medicinal properties. Tubers are generally underground swollen storage organs of the plants. Plant tubers have two fundamental biological roles; first, as storage organ and, second as propagule to overcome resting period and then they sprout to give rise to a new plant. Plant tubers are rich in starch and contain groups of storage protein hence used as staple food in most countries. These plants are also exploited for their medicinal properties in all the traditional systems of medicine. Removal of tubers eradicates plants forever without replenishment. Consequently, conservation of biodiversity is of paramount importance vis-a-vis maintenance of livelihoods to millions of people worldwide. The biological activities of tuberous plants range from analgesic, antifungal, antimicrobial, hypolipidemic to hepatoprotective and anticancerous. In addition to propagation, storage of these valuable genetic resources is equally important. In this chapter

[1] Department of Biochemical Engineering & Biotechnology, Indian Institute of Technology (IIT), Hauz Khas, New Delhi-110016.
[2] 221, Landmark Treasure Town, Badgaon, Udaipur 313011, India.
* Corresponding author: kg_ramawat@yahoo.com

we shall review recent works on important tuberous medicinal plants of India involving biotechnological approaches aimed at improving and conserving these plants.

Keywords: Tuber, Biology, Medicinal plants, *Chlorophytum*, *Curcuma*, Antimicrobial

Introduction

At the current growth rate the world population is likely to touch the 7.5 billion mark by the year 2020. Mostly, this increase is in the developing or underdeveloped countries, 80% of whose population still relies on a traditional system of medicine based on herbal drugs (Ramawat and Goyal 2008). The Indian traditional system of medicine includes Ayurveda, Siddha, Unani, and homeopathy (Mukherjee 2011). Herbal drugs are not only used for primary health care in rural areas in developing countries, but also in developed countries where infect modern medicines are predominantly recommended (Joshi and Rao 2011, Krishnan et al. 2011). Medicinal plants (MPs) are the main source of drugs that are used from the ancient times as herbal remedies for health care, and prevention and cure of various diseases and ailments (Kalia 2005). Globally, 77,000 plants (18.2% of total plant species) are currently put to use for medicinal purpose. Out of the 77,000 MPs used worldwide, 3000 MPs are internationally traded, only 900 species are cultivated and rest of the species are gathered from their natural habitats (Rao et al. 2012, Ramawat and Goayl 2008). Ayurveda, the oldest medicinal system in the Indian sub-continent, has alone reported approximately 2000 medicinal plant species, followed by Siddha and Unani. The *Charak Samhita*, an ancient document on herbal therapy, reports on the production of 340 herbal drugs and their indigenous uses. The market prices for medicinal plants and derived materials provide only a limited insight into the workings of the market, and not on the precise information of profits, supply and demand. Global imports and exports (2005–2010) of medicinal plants and plant parts are to the tune of US$ 2.09 and 2.08 billion/year, respectively (Ramawat and Goayl 2008, Kala et al. 2006).

Tubers, corms and bulbs are various types of modified plant structures that are enlarged to store nutrients. They are used by plants to survive the long winters or dry months and provide energy and nutrients for regrowth during the next growing season. Multiple storage organs are produced by such plants; therefore they represent a method of asexual reproduction. There are both stem and root modification to produce such organs, mostly underground root crops, tuber crops and corm crops, special terms used to refer to specific types of farm crops collectively called root and tuber crops. A bulbous plant is an ornamental or edible plant, herbaceous or perennial species which produce fleshy storage organs including true bulbs as well as corms, tubers, rhizomes, and tuberous roots (Dominy et al. 2008). Selected tuberous medicinal plants and their storage organs are shown in Fig. 1.

Plants from diverse taxa produce these structures which may provide food, medicine or flowers and saplings for the next generation. The annual consumption pattern shows that ~ 35% of the plants are used for extraction of below ground portion which include roots, root tubers, rhizomes, corms, etc. (Fig. 2). This data is for medicinal plants excluding edible plant parts. Tuberous plants have high calorific value and carbohydrate content along with medicinal properties due to the presence of bioactive molecules. Many tuberous medicinal plants (Fig. 1) are currently used in the preparation of stimulants, tonics, carminatives, expectorants and these are rich in dietary fibres, carotenoids, steroids, saponins, polyphenolics and anthocyanin (Sharma and Ramawat 2013, Edison et al. 2006). Some of the plants commonly used for the preparation of herbal formulations in India for healthcare include *Asparagus adscendens*, *Coleus forskohlii*, *Holostemma adakodien* and *Typhonium trilobatum* (Fig. 3). Thus it is evident that substantial amount of tuberous parts are collected leading to complete removal of plants and their eradication. The process of tuber formation is slow in nature and less studied.

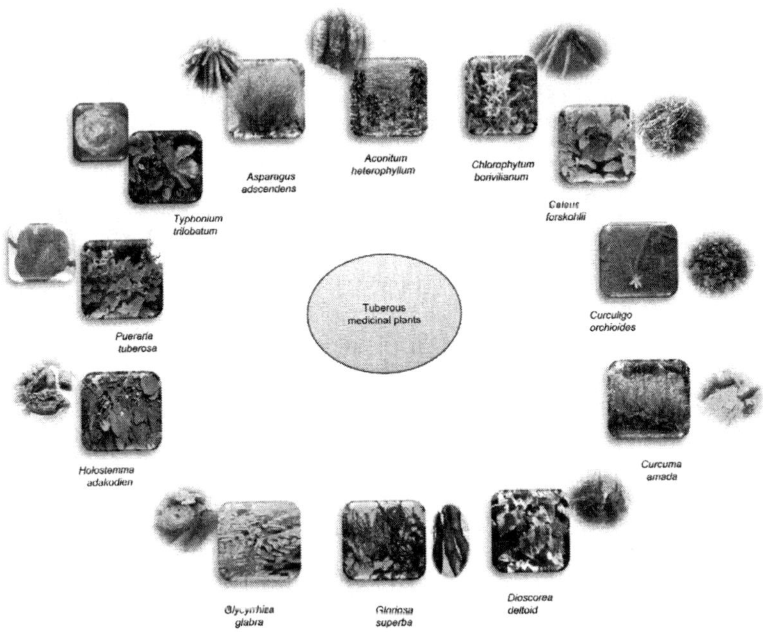

Fig. 1. Tuberous medicinal plants with their tubers.

Color image of this figure appears in the color plate section at the end of the book.

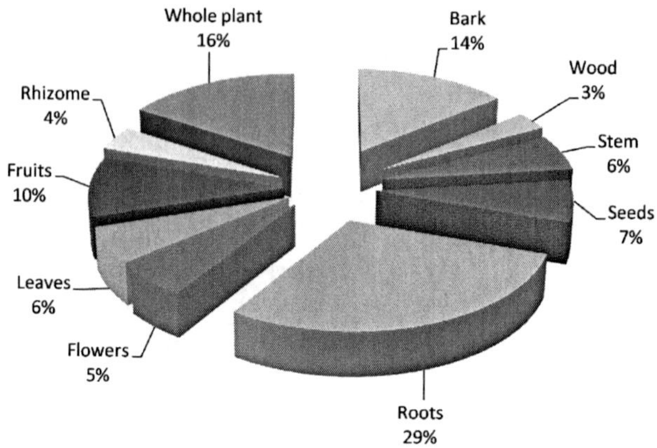

Fig. 2. Different % of plant parts used for medicinal purposes.

Color image of this figure appears in the color plate section at the end of the book.

Fig. 3. Market herbal products from tuberous plants.

In recent years underground storage organs have been studied in an attempt to understand the functional link between underground processes and aboveground mechanism (Arora et al. 2006, Neema and Ramawat 2010, Neema et al. 2008a, b). Apart from the classical role of providing mechanical support and nutrient uptake, roots and tubers also perform specialized roles as chemical factories that can synthesize, store, and secrete a diverse array of biologically active compounds such as proteins and low molecular weight secondary metabolites (Flores et al. 2003, Valls et al. 2006). Medicinal plant drug discovery continues to provide new and important leads against various pharmacological targets including cancer, malaria, cardiovascular diseases and neurological disorders due to availability of high throughput screening tools (Ramawat et al. 2009).

In this article we shall review recent developments about biotechnological approaches, promising therapeutic potential, bioactive molecules and pharmacological validation for some of the most widely used tuberous medicinal plants of India.

Biodiversity of Tuberous Medicinal Plants

Globally, medicinal plants have been unique sources of medicines and constitute the most common human use of biodiversity (Joshi and Rao 2011). India is one of the 12 mega biodiversity centres having over 45,000 plant species (Ramawat and Goyal 2008). There are five major areas of distribution of root and tuber crops in India. These are (i) South-western hilly and coastal region, (ii) Southern peninsular region, (iii) Eastern coastal region, (iv) Northeastern region and (v) North-western region. The two hot spots of global biodiversity, viz. North Eastern Himalayas and Western Ghats are particularly rich in wild relatives of tropical root and tuber crops. These hotspots also contain a large number of wild relatives of cultivated tuberous plants as well as many under-exploited tuberous plants known to tribals. Western Ghats is the centre of origin/diversification of several tuber crops like *Dioscorea*, *Alocasia*, *Colocasia*, *Amorphophallus*, etc. (Edison et al. 2006). Numerous other types of plant tubers are consumed in small amounts in different parts of the world, particularly as vegetables like *Ipomea*, *Manihot*, *Dioscorea* and *Curcuma*. Some of these plants have long history of traditional use and are established as major therapeutic agents. Estimated consumption per annum of tuberous medicinal plants ranges from ~20 tons for *Aconitum heterophyllum* to a few hundred tons for *Pueraria tuberosa* and *Asparagus adscendens* to a few thousand tons for *Glycyrrhiza glabra*. Modern scientific validation methods have confirmed the strong analgesic, antiarthritic, antifungal, antimicrobial, antiparasitic, hepatoprotective, hypolipidemic, insecticidal and anticancerous activities of several of these species (Table 1). It is evident that these plants are from diverse taxa belonging to monocot and dicot groups; herbs, shrubs and linas; having morphologically different storage organs; possess divergent bioactive molecules and show wide ranging biological and pharmacological properties.

Overexploitation and habitat destruction are two major causes of threat categories for tuberous medicinal plants. According to the International Union for Conservation of Nature and Natural Resources (IUCN), tuberous species that have become rare and need immediate attention for conservation in India are *Aconitum balfourii* and *Gloriosa superba* (Nautiyal et al. 2002, Arumugam and Gopinath 2012). *Aconitum* genus is widely distributed in the Himalaya region. About 300 species of *Aconitum* are found

Table 1. Bioactive compounds from tuberous medicinal plants and their pharmacological actions.

Plants /Family	Bioactive compounds	Pharmacological actions validated	References
Asparagus adscendens (Liliaceae)	Spirostanol glycosides (asparanin A & asparanin B) and furostanol glycosides (asparoside A & B), Saponins	Antispasmodic, appetizer, cooling and galactagogue, chemopreventive, improve the stress immune disorders	Singh et al. (2011) Sharma et al. (2012) Kanwar and Bhutani (2010)
Aconitum balfourii (Ranunculaceae)	9-Hydroxysenbusine A Balfourine, 8-O-Methylveratroylpseudaconine	Antibacterial, antifungal, Anti-inflammatory	Khetwal (2007) Shyaula (2011)
Aconitum heterophyllum (Ranunculaceae)	Alkaloids heterophyllinine-A & B, flavonoids and sterols	Anti-inflammatory activity, Hypolipidemic, Antibacterial	Nisar et al. (2009) Verma et al. (2010)
Chlorophytum borivilianum (Liliaceae)	Saponins, stigmasterol, hecogenin, steroids and flavonoids	Antiproliferative, larvicidal Aphrodisiac, Immunomodulatory effect, melanin stimulatory effects	Deore and Khadabadi (2010a,b, 2009) Bathoju and Giri (2012) Thakur et al. (2011a,b, 2009) Chaudhary et al. (2012)
Coleus forskohlii (Lamiaceae)	forskolin	Hepatotoxicity, anticoagulant action, Hypotensive effect	Virgona et al. (2012) Yokotani et al. (2012)
Curculigo orchioides (Amaryllidaceae)	Curculigoside Two triterpenoid glycosides (Curculigosaponin N, Curculigosaponin O)	Anti-osteoporosis, anticancer, Hepatoprotective, Neuroprotective effects, Cytotoxic activities	Wang et al. (2012), Banu et al. (2012), Tian et al. (2012), Zuo et al. (2012)
Curcuma amada (Zingiberaceae)	Alkaloids, Flavonoids, Terpenoids labdane diterpenoid	Cytotoxicity effect Antimycobacterial activity, Antioxidant activity	Akatheeswaran et al. (2012) Singh et al. (2010), Policegoudra et al. (2011a,b)
Dioscorea deltoida (Dioscoreaceae)	Diosgenin, steroidal saponin (orbiculatoside B)	Cancer chemoprevention, Antioxidant, hepatoprotective Activity	Raju and Mehta (2009) Amir et al. (2011)

Gloriosa superba Colchicaceae	Colchicines, gloriosine, N-formyl-deacetyl colchicines	Anticancer, Antimalarial activity, PDE inhibiting activity	Nautiyal (2011), Bagavan et al. (2011), Kuchta et al. (2012)
Glycyrrhiza glabra (Fabaceae)	Triterpenoids (glycyrrhizin, glycyrrhizinic acid), flavonoids, pterocarpans, coumarins, etc.	Hypnotic effects, Anticancer activity, Anti-inflammatory	Simons et al. (2011), Cho et al. (2012), Smolarczyk et al. (2012)
Holostemma adakodien (Asclepiadacea)	Alkaloids, flavonoids, flavanones, tannins, terpenoids	Antidiabetic activity, Aphrodisiac, galactogogue	Janapati et al. (2009)
Pueraria tuberosa (Leguminosae)	Isoflavones, pterocarpans, tuberosin and triterpenoids	Anti-osteoporosis, cardiovascular, Nephroprotective property, Anticancerous property	Tripathi et al. (2012) Hussain et al. (2012)
Typhonium trilobatum (Araceae)	Triterpenoids, glycosides	Larvicidal activity, Analgesic, anti-inflammatory and anti-diarrheal activity	Haldar et al. (2011) Ali et al. (2012)

in the world of which 24 species are just found in the great Himalayas of India (Sharma and Gaur 2012). It is very necessary to survey, document, collect and conserve the genetic wealth of tuber crops existing in India.

Biotechnological Studies

Medicinal plants are an important source of life-saving drugs for the majority of the world's population. Recent advances in molecular biology, enzymology and fermentation technology of plant cell cultures suggest that these systems may become a viable source of important secondary metabolites. Despite recent advances in biotechnology, combinatorial biochemistry, and high throughput screening, natural products from plants continue to serve as a major source of new chemical entities for pharmaceutical research (Ramawat et al. 2009, Julsing et al. 2007). Biotechnological tools are also important for multiplication and genetic improvement of the medicinal plants by adopting techniques such as *in vitro* regeneration and genetic transformations. These plants have been regenerated through somatic embryogenesis or organogenesis. *In vitro* tuber formation has been reported in a few plants like *C. borivilianim, C. orchioides, G. superba* and *D. deltoidea* (Table 2).

The important high value tuberous medicinal plants of India with current biotechnology and biological activities are described below.

Chlorophytum borivilianum

Chlorophytum borivilianum is a very popular medicinal herb commonly known as "Safed Musli" reputed for its tuberous roots (Jain et al. 2010). Generally, one plant produces only one inflorescence but sometimes 3–4 inflorescences are produced depending upon the age of the mother plant. *Chlorophytum* species are reported as diploid, triploid, tetraploid and polyploid with basic chromosome number 7 or 8. The number of flowers per inflorescence increases with the increase in size of the underground tuberous roots (Arora et al. 2004). The tuberous roots (mostly powdered) are widely used in Indian system of medicine as a non-hormonal restorative tonic, in fatigue, general debility, sterility and to enhance male potency, as cardiac and brain tonic. *Chlorophytum* species are known to contain steroidal saponins (neohecogenin, neotigogenin, stigmasterol, tokorogenin), a bibenzyl xyloside (2,4,4′-trihydroxy-2-xylopyranosyl bibenzyl), a disubstituted tetrahydrofuran (4-hydroxy-8,11-oxidoheneicosanol), nonacosone, tetracosanoic and triacontanoic acid, pentacosyl docosanoate, chlorophytoside-I, galactoglucan and sugars (arabinose, glucose and xylose) (Tandon and Shukla 1995, Kaushik 2005).

Table 2. Plant species and there *in vitro* response on different media.

Plant species	Medium	Explant	Response	References
Aconitum balfourii	MS + NAA (2.5 mg/1) + BAP (1 mg/1) MS + BA alone (0.5–4 mg/l) MS + IAA (1.8 mg/l)	Leaf segment Callus Shoots	Callus Shooting Rooting	Bist et al. (2011)
Aconitum heterophyllum	MS + BAP (0.25 mg/l) + NAA (0.5 mg/l)	Callus	Shoots multiplication	Jabeen et al. (2006)
Asparagus adscendens	MS + NAA (0.05 mg/l) + Kn (0.1 mg/l) + 0.6% agar	Nodal explant	Shoots multiplication	Mehta and Subramanian (2005)
Chlorophytum borivilianum	MS + 2,4-D (0.25 mg/l) + 2-iP (1.5 mg/l) + additives MS + BAP (200 mg/l) ¾ MS + IBA (2 mg/l)	Zygotic embryo Shoots Shoots	Tuberous roots, Somatic embryogenesis Multiple Shoots Rooting	Arora et al. (1999, 2004, 2006); Rizvi et al. (2012) Joshi and Purohit (2012)
Coleus forskohlii	MS + IBA (1mg/l) + BAP (0.5mg/l) MS + BAP (0.2 mg/l) MS + IAA (2.5 mg/)	Leaf lamina Node Shoots	Multiple shoots Multiple shoots Rooting	Praveena et al. (2012) Verma et al. (2012)
Curculigo orchioides	B5 medium + BAP (8 mg/l) MS + BAP(1.0 mg/l) + Morphactin (0.1 mg/l) MS+ BAP (1 mg/l) + morphactin (0.1 mg/l) + Growth retardants	Leaf Leaf Leaf	Bulbil formation Bulbils formation Bulbils induction	Suri et al. (1999) Neema et al. (2008) Malviya et al. (2011)
Curcuma amada	MS + BAP (3 mg/l) + IBA (0.5 mg/l) + Maltose 3% 1/2 MS + NAA (0.5 mg/l) MS + BAP (3 mg/l)	Rhizome Shoots Rhizome	Shoot multiplication Rooting Shooting	Das et al. (2010) Das et al. (2010) Mohanty et al. (2012)
Dioscorea deltoidea	MS + 2,4 –D (1ppm)	Nodal explant	Callus culture	Amir et al. (2012)

Table 2. contd....

Table 2. contd.

Plant species	Medium	Explant	Response	References
Gloriosa superba	MS + 2,4 –D (1.0 mg/l) + IAA (0.5 mg/l)	Corm bud	Callus	Arumugam and Gopinath (2012)
	1/2 MS + BAP (1.5 mg/l) + Kn (1.0 mg/l) + CW 20%	Callus	Shoots multiplication	Arumugam and Gopinath (2012)
	MS + BAP (4 mg/l) + NAA (1mg/l)	Sprouted tuber	Tubers formation	Selvarasu and Kandhasamy (2012)
Glycyrrhiza glabra	MS + NAA (0.5 mg/l) + 2,4-D (0.5 mg/l) +1000CH	Node	Callus to Suspension culture	Verma et al. (2012)
Holostemma adakodien	½ MS + 2,4 –D (0.05-0.1 mg/l)	Callus	Somatic embryogenesis	Martin (2003)
Pueraria tuberosa	MS+ TDZ (0.25 mg/l) + IBA (0.05 mg/l) MS + 2,4,5-T + Kn (0.1 mg/l) + additives	Nodal part Callus	Shoot culture, Cell culture, isoflavonoids production	Sharma et al. (2011) Goyal et al. (2008a,b, 2011), Sharma and Ramawat (2013)
Typhonium trilobatum	BAP (5.0 mg/l) + NAA (1.5 mg/l)	Corm bud	Shoots multiplication	Paul and Bari (2012)

All the medium recipes were formulated using Murashige and Skoog [1962] basal medium; 2,4-D: 2,4-diphenoxyacetic acid; 2IP: 2,isopentenyl adenine; BAP: 6-benzylamino purine; CH: casein hydrolysate; IAA: indole-3-acetic acid; IBA: indole-3-butyric acid; KN: kinetin; NAA: α naphthalene acetic acid; CH: Coconut water; AS: Adenine sulphate; PGR: Plant growth regulators; TDZ: Thidiazuron; ppm: Parts per million

Four new spirostan-type saponins named borivianosides E-H were recently isolated from ethanol extract of the roots of *C. borivilianum* (Acharya et al. 2009). The chemical structure of the secondary metabolites isolated from *Chlorophytum* species are shown in Fig. 4.

Pharmacological properties have been examined for improvement of sexual behavior in rat model as reflected by increased penile erection index and improved sperm counts and related behavior (Thakur et al. 2011a). The polysaccharides from *C. borivilianum* could be considered for cancer treatment via post-chemotherapy revival of the immune system. This study further validates and provides molecular evidence for the potential benefits of rasayan herbs of the Ayurvedic system of medicine (Thakur et al. 2011b). The root extract of *C. borivilianum* has significant powerful antioxidant activity and favorably affect atherosclerosis risk status by reducing LDL oxidation susceptibility (Visavadiya et al. 2010).

Plant regeneration via somatic embryogenesis (Arora et al. 2006, 1999) has been reported in *C. borivilianum* (Fig. 5a). *In vitro* growth and shoot proliferation were studied in controlled CO_2 environment by Joshi et al. (2009). The best response with high number of shoots was obtained at 40 g m^{-3} CO_2 in the chamber. Recently, plant regeneration and genetic fidelity has been investigated from embryogenic callus and suspension culture (Rizvi et al. 2012). Optimization of various factors like propagule size, subculture strategy, gelling agents, liquid pulse treatment and vessel type were influenced shoot multiplication in *C. borivilianum* (Joshi and Purohit 2012). Still there are gaps in the study such as identification of bioactive molecules, bioactivity based fractionation, pharmacological evaluation and large scale cultivation of uniform promising therapeutic value plants.

Curculigo orchioides

Curculigo orchioides (Kali musli) is a tiny herbaceous plant and named 'Xian-Mao' in pharmacopoeia of China. Tuberous roots are widely used as tonic for health, vigour and vitality because of the presence of flavanone glycosides and other steroidal saponins (Neema and Ramawat 2010).

Several bioactive compounds isolated from the plant include flavones, glycosides, fatty acids, steroids, saponins and triterpenoids. New curculigosides from *in vitro* tubers of *C. orchioides* (Valls et al. 2006), and orchioside D and curculigoside E from roots were isolated (Dall'Acqua et al. 2009). The dried rhizomes yielded two phenolic glycosides, curculigoside, orcinol-beta-D-glucoside, and two cycloartane saponins, curculigosaponin G, curculigosaponin I (Lee et al. 2009). Three new phenolic glycosides, curculigosides F, G and H were isolated from *Curculigo* rhizomes (Zuo et al. 2010). Recently, two new cycloartane triterpenoid glycosides, named

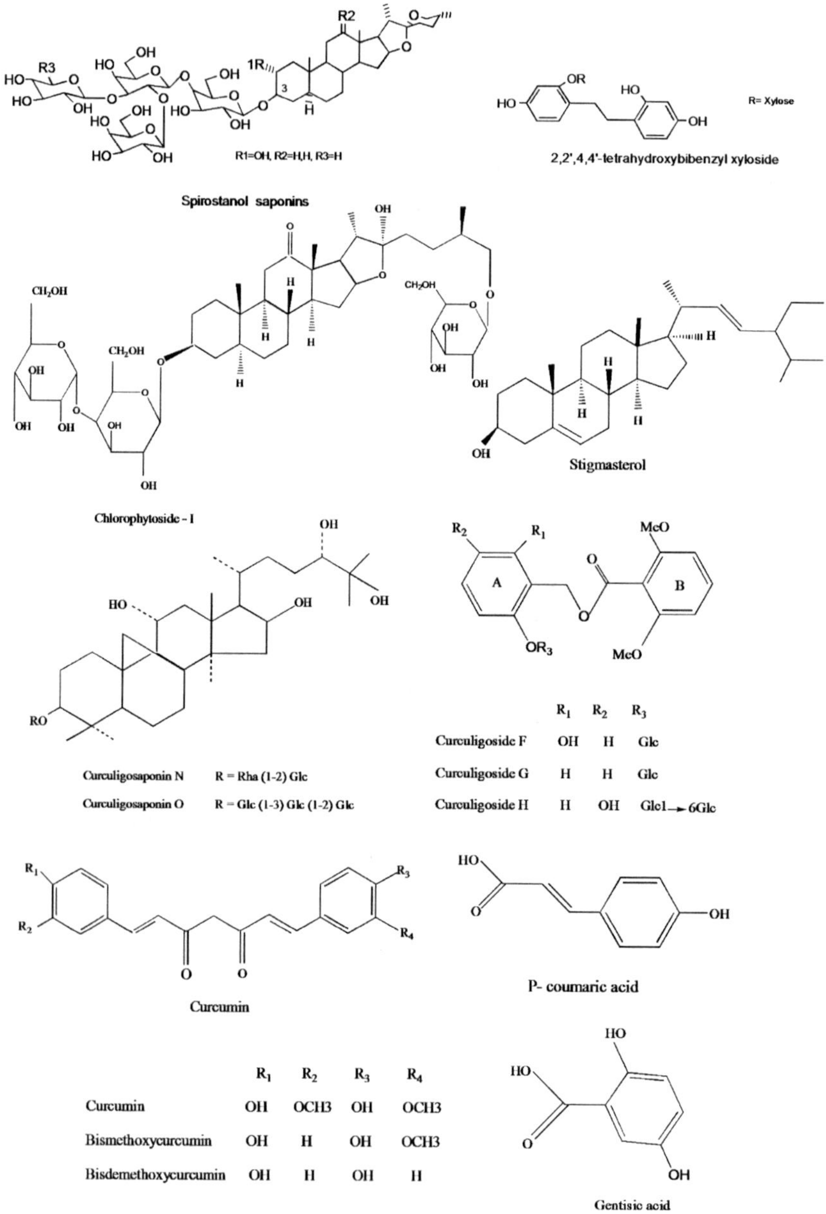

Fig. 4. Bioactive molecules of tuberous medicinal plants.

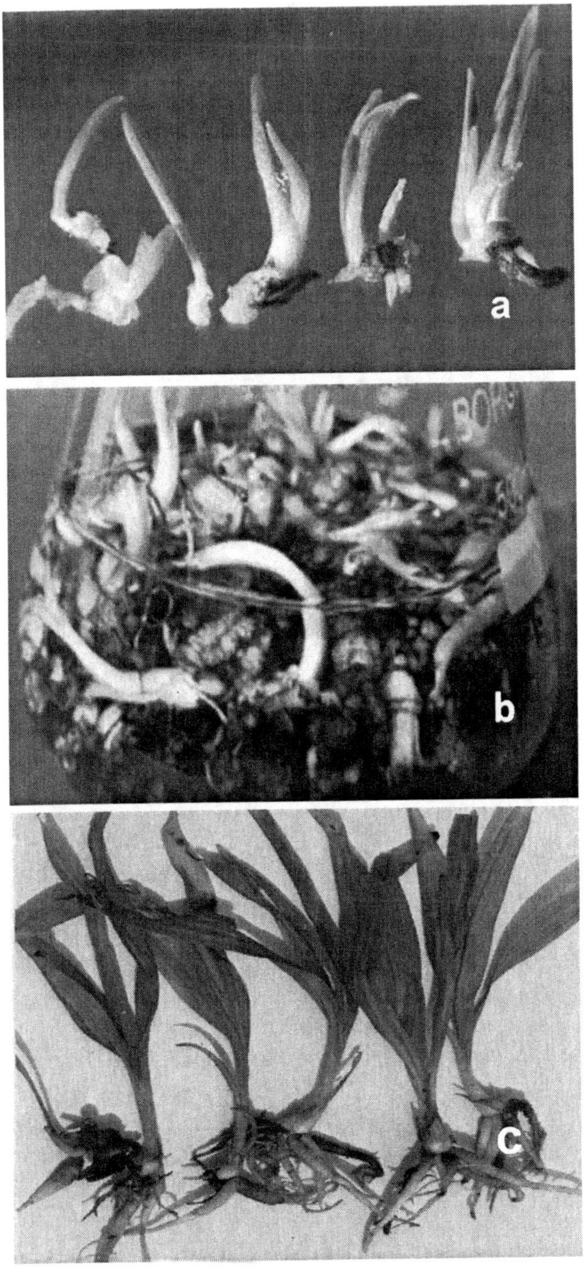

Fig. 5. Regeneration in *C. borivilianum* through somatic embryogenesis (a), and through bulbils formation (b) and germination (c) in *C. orchioides*.

Color image of this figure appears in the color plate section at the end of the book.

curculigosaponin N and curculigosaponin O, were isolated from rhizomes of *C. orchioides* (Zuo et al. 2012). The newly isolated bioactive molecules from *C. orchioides* are shown in Fig. 4.

A pharmacological study of this traditional herb includes antiosteoporosis (Jiao et al. 2009), immunostimulant and anti-inflammatory activity (Venkatesh et al. 2009). The protective effect of curculigoside was found positive against oxidative stress in calvarial osteoblasts through reducing the production of ROS (Wang et al. 2012). Neuroprotective effects of curculigoside were investigated against *N*-methyl-D-aspartate-induced neuronal cell loss and reduced the number of apoptotic and necrotic cells in a time and concentration dependent manner (Tian et al. 2012). The ameliorating effects of curculigoside were determined on learning and memory in aged rats. The study suggested that curculigoside can be possibly developed as a new drug for the treatment of Alzheimer's disease in the future (Wu et al. 2012).

Rapid multiplication of *C. orchioides* is highly desirable to meet the commercial demand for medicinal purposes (Figs. 5b, c). Influence of culture medium (Jain et al. 2007), and high frequency bulbil induction from leaf explants was recorded on MS medium with 1mg/l BAP and 0.1 mg/l morphactin (Neema et al. 2008a,b). Effect of growth retardants on bulbils formation has shown that alar and CCC markedly reduces the bulbil formation, yield and fresh weight of bulbils. Incorporation of retardants resulted in 60% germination inhibition, thereby prolonging the storage conditions (Malviya et al. 2011). This plant needs further studies to find some lead molecules responsible for biological activities and successful drug development.

Curcuma amada

Curcuma amada is one of the most important rhizomatous species having a raw mango-like flavor blended with that of ginger commonly known as "mango ginger". The rhizomes are branched, and sympodial. The rhizome has carminative, stomachic, antioxidant, antibacterial, hypotriglyceridemic, anti-tubercular and platelet-aggregation inhibitory activities (Policegoudra et al. 2011a,b).

There are more than 100 phytochemicals reported from fresh and dried extracts of *C. amada* as compiled by Jatoi et al. (2007). The volatile oils, terpenoids and free phenolic acids like caffeic, gentisic, ferulic, gallic and p-coumaric acids present in mango ginger are reviewed by Policegoudra et al. (2011a). The major bioactive compounds consist of curcuminoids, curcumin, demethoxy curcumin, bis-demethoxy curcumin (Fig. 4) and diterpene labda-8(17), 12-diene-15, 16-dial (Singh et al. 2010). Mangiferin was extracted from the *C. amada* by the help of microwave assisted extraction

(Padmapriya et al. 2012). The drug yielding qualitative traits of *Curcuma* from different agroclimatic regions were studied to find out the promising elite genotypes in terms of high curcumin, oleoresin and essential oil content (Singh et al. 2013).

Recently, Curcuminoids and their related compounds were evaluated for their larvicidal effect against the mosquito *Culex pipiens* (Sagnou et al. 2012). Preliminary evaluation of *in vitro* cytotoxicity effect of solvent extracts of *C. amada* rhizome on HeLa cell lines was studied by Akatheeswaran et al. (2012). The plant extract of *C. amada* showed highest levels of cytotoxicity among all other the species and induced significant quantitative changes in gene expression patterns (Bcl-2, Bax, Bak and p53) associated with intrinsic pathway of apoptosis (Ramachandran et al. 2012).

Efficient plant regeneration from rhizomes and assessment of genetic stability of regenerated plants was studied by Das et al. (2010). A new approach, *Luffa*-sponge based conservation strategy for regrowth of encapsulated vegetative explants was explored. In the experiment young sprouted vegetative microshoot was encapsulated in gelling mixture and resultant beads stored for six months at room temperature on *Luffa* sponge storage matrix. Synthetic seeds derived healthy plants were transferred to greenhouse analyzed for RAPD based variation in *C. amada* (Banerjee et al. 2012). This traditional medicinal plant needs more scientific validation of pharmacological properties.

Dioscorea deltoidea

Dioscorea deltiodea (yam) is a widely used tuberous medicinal plant of indigenous system of medicine. It is commercially used for the extraction of diosgenin (25R-spirost-5-en-3β-ol, the primary active ingredient, Fig. 6), a precursor of steroidal drugs. The plant tuber shows antitumor, antimicrobial, antioxidant, stomachic, and hypoglycemic activities. Spirostanol saponins, flavonoids, phenol, resin and tannins are present in the *D. deltoidea* (Subhash et al. 2012, Shen et al. 2002). Dioscorin is the major tuber storage protein of *Dioscorea*.

Diosgenin suppresses fatty acid synthase expression in HER2-overexpressing breast cancer cells (Chiang et al. 2007). Currently, Diosgenin treatment influences IL-2 levels in the brain of ovariectomized rats (Ho et al. 2012), and beneficial effect against aortic damage (Pari et al. 2012). The aqueous extract of *D. deltoidea* rhizome and its callus was demonstrated significant antioxidant and hepatoprotective potential (Amir et al. 2011).

Respiration activity of cell suspension culture of *D. deltoidea* produced biological active compounds. Maximum rate of oxygen consumption was marked in *Dioscorea* for active synthesis of secondary metabolites (Titova et al. 2011). Recently, a new simple and stability-indicating high-performance

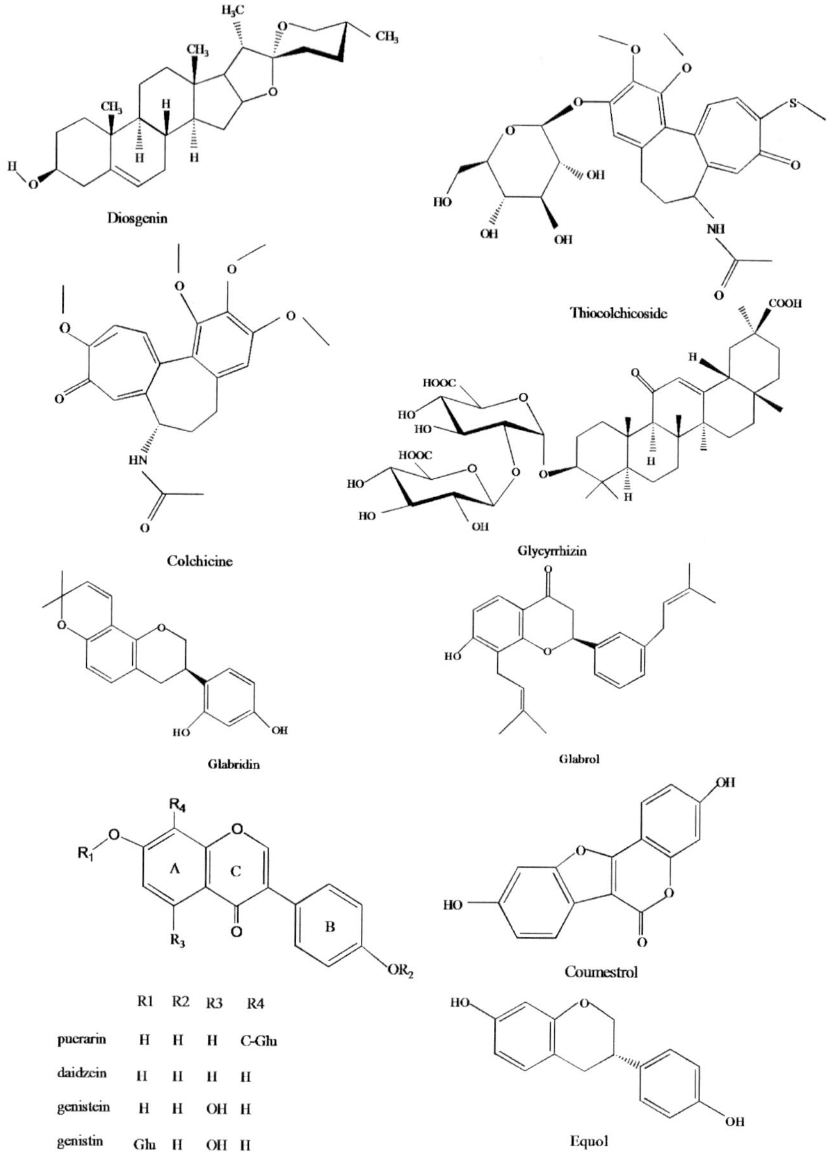

Fig. 6. Bioactive molecules of tuberous medicinal plants.

thin layer chromatographic (HPTLC) method for the analysis of diosgenin in callus and rhizome of *D. deltoidea* was developed and validated. The HPTLC method was developed on TLC aluminium plates precoated with silica gel $60F_{254}$ using solvent system petroleum ether-isopropanol (12:1, v/v) (Amir et al. 2012). This plant is an example of combined efforts of chemistry, biotechnology and pharmacology to develop the plant product at industrial level. Further research is required to develop high yielding clones of *D. deltoidea*.

Gloriosa superba

Gloriosa superba (Glory lily) is a perennial tuberous climbing herb and native to the tropics of Africa and Asia. Perenniality is by underground tubers producing new aerial plant every year. The flowers are pendulous and yellow, red, purplish, or variously bicolored in color (Amano et al. 2008). The tuberous root is used as a germicide, to cure ulcers, piles, hemorrhoids, inflammation, leprosy, dyspepsia, worm's infestation, flatulence, intermittent fevers, and debility arthritis and against snake poison (Jana and Shekhawat 2011).

Two important alkaloids colchicine and gloriosin are present in seeds and tubers (0.7% to 0.9%) with other alkaloids such as colchicoside (its semi-synthetic derivative—thiocolchicoside), superbine, lumicolchicine, 3-demethyl-N-deformyl-N deacetylcolchicine (Fig. 6). Long chain fatty acids, glumiccolchicines and 2-hydroxy-6-methoxy benzoic acid from tubers and N-formyl-deacetyl colchicines have been isolated from flower (Nautiyal 2011, Jana and Shekhawat 2011, Ghosh and Jha 2008). Recently, a few bioactive metabolites were isolated from an endophytic fungus (*Aspergillus* species) present in seeds of *G. superba*. These compounds were also evaluated for their anticancer potential against six cancer cell lines A-549, HEP-2, MCF-7, OVCAR-5, THP-1, and CVI-1 (Budhiraja et al. 2012).

Leaf extract (Choudhary et al. 2011) and crude extract of *G. superba* exhibited a significant antimicrobial and phosphodiesterase (PDE) inhibiting activities, respectively (Kuchta et al. 2012). The leaf extract of *G. superba* was also found to be active against *Plasmodium falciparum* (Bagavan et al. 2011).

The major problem in the regeneration of *G. superba* is dormancy of the tuber. The germination capacity of seeds is also very low (about 0.01%). *In vitro* tuberization of *G. superba* on basal medium (Ghosh et al. 2007) and micropropagation using corm bud explants was reported by Arumugam and Gopinath (2012). Recent work suggested large-scale multiplication of plants and extraction of colchicine from tubers are needed for industrial use and pharmaceuticals.

Glycyrrhiza glabra

Glycyrrhiza glabra (liquorice) is one of the oldest and most frequently used botanicals in the oriental herbal medicine. The roots and rhizomes of liquorice have long been used worldwide as a natural sweetener. Its root is a traditional medicine used mainly for the treatment of peptic ulcer, hepatitis C, pulmonary and skin diseases (Yin et al. 2011).

The roots are a rich source of flavonoids, in particular, prenylated flavonoids, such as the isoflavan glabridin and the isoflavene glabrene (Simons et al. 2011). In *G. glabra* Pentacyclic triterpene saponin are major constituents, which include liquiritin, liquiritigenin, isoliquiritigenin, liquiritin apioside, glycyrrhizin and glycyrrhizic acid (Yin et al. 2011). From ethanolic extract of liquorice 39 phenolics and 10 new compounds were isolated. Among the isolated compounds, 5′-formylglabridin, (2R, 3R)-3,4′,7-trihydroxy-3′-prenylflavane, echinatin, (3R)-2′,3′,7-trihydroxy-4′-methoxyisoflavan, kanzonol X, kanzonol W, shinpterocarpin, licoflavanone A, glabrol, shinflavanone, gancaonin L, and glabrone all exhibited significant PPAR-γ ligand-binding activity (Kuroda et al. 2010). Principal compounds of G. *glabra* are shown in Fig. 6.

The estrogenic (Simons et al. 2011), hepatoprotective and antioxidant activities of the root extract (Yin et al. 2011) were reported. The inhibitory property of de-glycyrrhizinated extract of *G. glabra* root and its phytoconstituents (glabridin, isoliquiritigenin and glycyrrhizin) was studied on production of pro-inflammatory mediators (Thiyagarajan et al. 2011). Glycyrrhizin inhibited intercellular adhesion in *Herpex simplex* virus infection (Huang et al. 2012). Hypnotic effects of glabrol (Cho et al. 2012) and the role of glycyrrhizin in anticancer therapy are being under investigation (Smolarczyk et al. 2012).

Liquorice was successfully established *in vitro* by Verma et al. (2012). *In vitro* grown axillary micro shoots of *G. glabra* were encapsulated in alginate beads. Re growth was characterized by the development of both shoot and root from single encapsulated micro shoot and healthy plants were established to glass house with 95% survival. ISSR and RAPD based evaluation of genetic stability of encapsulated micro shoots were reported (Mehrotra et al. 2012). The biotransformation of β-artemether by cell suspension cultures of *G. glabra* was reported by Patel et al. (2011). However, being a legume, works on glycyrrhizin metabolic mechanisms and cell cultures are required.

Pueraria tuberosa

Pueraria tuberosa is a wild liana commonly known as Vidarikand. Root tubers are large, 20–50 cm long, weighing up to 20 kg and decorticated, sliced, dried, and sold in the market. Traditionally it is used as contraceptive, cardiotonic, rejuvenating, rheumatism, laxative, and useful in disorders of liver and spleen (Dev 2006).

Characteristic secondary metabolites isolated from the tubers are isoflavones, pterocarpans, tuberosin and triterpenoids. Total flavonoids occur in the semi-ripe tubers to the extent of ~1% (on dry basis). Puerarin, genistin, genistein, daidzein, daidzin are the principal molecules of *P. tuberosa*. Isoflavonoids present in the plant and their biological activities are reviewed recently by Sharma and Ramawat (2013). In addition several minor isoflavonoids and other molecules such as isorhamnetin-3-O-glucoside, equol, hydroxypuerarin-4-O-D-glucoside, 6-O-D-xylosilpuerarin and a coumestan, are also present in the cell cultures of *P. tuberosa* (Malviya et al. 2011). Bioactive molecules of *P. tuberosa* are shown in Fig. 6.

Puerarin protects human umbilical vein endothelial cells against high glucose-induced apoptosis by upregulating heme oxygenase-1 and inhibiting calpain activation (Chen et al. 2012a). Different pharmacological properties of *P. tuberosa* include nephroprotection from tubers (Tripathi et al. 2012), prevention of rat breast cancer by genistin and selenium (Hamdy et al. 2012), inhibitory effect of genistein on human cervical cancer (Hussain et al. 2012), and inhibitory effects of daidzein on intestinal motility in normal and high contractile states (Chen et al. 2012b).

Biotechnological approaches of *P. tuberosa* include marked effect of plant growth regulators, altered nitrogen supply and sucrose concentration on their accumulation in cell cultures (Goyal and Ramawat 2007, 2008a). Increased puerarin biosynthesis in cell cultures and effect of *cuscuta* elicitor on isoflavonoids accumulation was studied in our laboratory (Sharma et al. 2011). The optimized elicitation conditions were used in vessels of varying capacity where maximum yield of 91 mg^{-1} of isoflavonoid was recorded in 2 l bioreactor which was about 19% higher than the values achieved earlier (Goyal et al. 2011, Sharma et al. 2009, Goyal and Ramawat 2008b). Development of high productive cell lines, growth of the cells in shake flasks and bioreactor as well as in organized cultures were attepted to develop technology towards production of isoflavonoids (Figs. 7a–c). These recent works suggest a great promise for exploitation of isoflavones production by cell cultures technology.

Conclusion

It is evident from the foregoing account that India is rich in the floral diversity of root and stem tuberous plants. It is estimated that biodiversity hotspots are undergoing exceptional loss of habitat resulting in the erosion or even the loss of precious endemic species. Hence there is an urgent need to check this loss and protect the plants and habitats. The recent surge in publications describing the biotechnological approaches and biological activities of tuberous plants demonstrates the potential for their application in many different areas of human medicine. An array of prototype bioactive

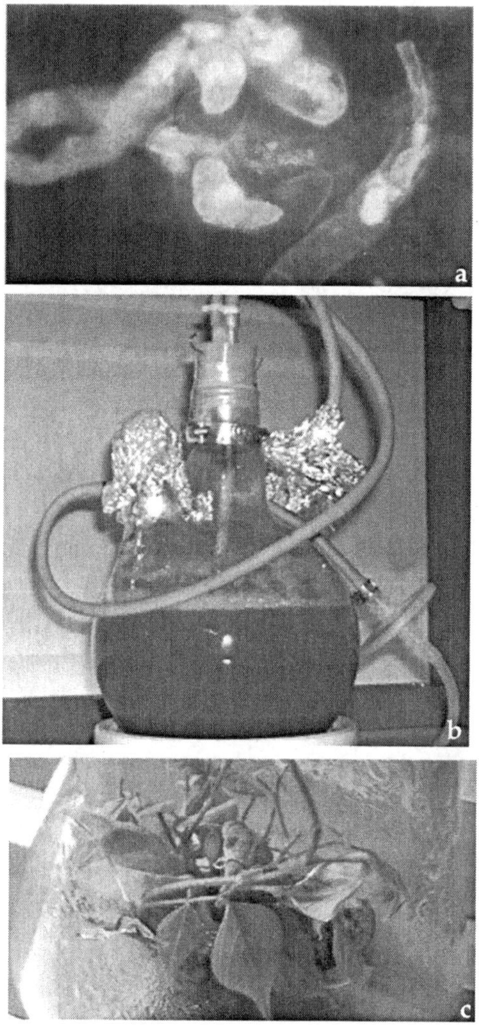

Fig. 7. Cells under fluorescence microscope (a), Cell culture in 3L bioreactor (b) and organized shoot cultures (c) of *P. tuberosa*.

Color image of this figure appears in the color plate section at the end of the book.

molecules of different classes has been obtained from these plants, some of which led to important drugs that are available in the market today. The *in vitro* propagation protocols developed for medicinal plants are satisfactory in laboratory scale but their use in commercial scale needs further field trials. The laboratory production of secondary metabolites achieved in some tuberous plants necessitates of experimentation at the bioreactor level to elevate commercial production of phytochemicals. Use of cell culture technology in fermenters will be helpful in the production of bioactive molecules where synthesis is not available.

Acknowledgements

VS thanks the Department of Biotechnology, Government of India, New Delhi for research associateship.

References

Acharya, D. et al. 2009. Cytotoxic spirostane-type saponins from the root of *Chlorophytum borivilianum*. J. Natural Products 72: 177–181.

Akatheeswaran, V.D., P.V. Rajalakshmi and S. Samiappan. 2012. Preliminary evaluation of *in vitro* Cytotoxicity of solvent extracts of *Curcuma amada* rhizome on HeLa cell lines. Asian Pacific J. Tropical Biomedicine 1–4.

Ali, K., A. Ashraf and N.N. Biswas. 2012. Analgesic, anti-inflammatory and anti-diarrheal activities of ethanolic leaf extract of *Typhonium trilobatum* L. Schott. Asian Pacific J. Tropical Biomedicine 2: 722–26.

Amano, J., S. Kuwayama, Y. Mizuta and M. Nakano. 2008. Morphological Characterization of Three Intergeneric Hybrids Among *Gloriosa superba* 'Lutea', *Littonia modesta*, and *Sandersonia aurantiaca* (Colchicaceae). Hort. science 43: 115–118.

Amir, M., M. Mujeeb, A. Sayeed, A. Aftab and M. Aqil. 2011. Antioxidant and hepatoprotective activity of rhizome and callus culture of *Dioscorea deltoidea* against d-galactosamine induced hepatotoxicity in rats. Planta Med. DOI: 10.1055/s-0031-1273670.

Amir, M., A. Ahmad, N.A. Siddique, M. Mujeeb, S. Ahmad and W.A. Siddique. 2012. Development and validation of HPTLC method for the estimation of diosgenin in *in vitro* culture and rhizome of *Dioscorea deltoidea*. Acta Chromatographica 24: 111–121.

Arora, D.K., A.K. Jain, K.G. Ramawat and J.M. Merillon. 2004. *Chlorophytum borivilianum*: an endangered aphrodisiac herb. *In:* K.G. Ramawat (ed.). Biotechnology of medicinal plants: vitalizer and therapeutic. Science, Enfield, pp. 111–128.

Arora, D.K., S.S. Suri, K.G. Ramawat and J.M. Merillon. 1999. Factors affecting somatic embryogenesis in long term callus cultures of Safed Musli (*Chlorophytum borivilianum*), an endangered wonder herb. Indian J. Exp. Biol. 37: 75–82.

Arora, D.K., S.S. Suri and K.G. Ramawat. 2006. Assessment of variability in regenerants from long term cultures of Safed Musli (*Chlorophytum borivilianum*). Indian J. Biotech. 5: 527–534.

Arumugam, A. and K. Gopinath. 2012. *In vitro* Micropropagation using Corm Bud Explants: An Endangered Medicinal Plant of *Gloriosa superba* L. Asian J. Biotechnol. 4: 120–128.

Bagavan, A., A.A. Rahuman, N.K. Kaushik and D. Sahal. 2011. *In vitro* antimalarial activity of medicinal plant extracts against *Plasmodium falciparum*. Parasitol. Res. 108: 15–22.

Banerjee, S., S. Singh, H. Pandey, P. Pandey and L. Rahman. 2012. Conservation and storage of *Curcuma amada* Roxb. synseeds on *Luffa* sponge matrix and RAPD analysis of the converted plantlets. Industrial Crops and Products 36: 383–388.

Banu, J., E. Varela and G. Fernandes. 2012. Alternative therapies for the prevention and treatment of osteoporosis. Nutrition Reviews 70: 22–40.

Bathoju, G. and A. Giri. 2012. Production of medicinally important secondary metabolites (stigmasterol and hecogenin) from root cultures of *Chlorophytum borivilianum* (Safed musli). Recent Research Science Technol. 4: 45–48.

Bist, R., K. Pathak, D. Hatwal, H. Punetha and A.K. Gaur. 2011. *In vitro* propagation of *Aconitum balfourii* Stapf: A rare medicinal herb of the alpine Himalayas. Indian J. Horticulture 68: 394–398.

Budhiraja, A., K. Nepali, S. Sapra, S. Gupta, S. Kumar and K.L. Dhar. 2012. Bioactive metabolites from an endophytic fungus of *Aspergillus* species isolated from seeds of *Gloriosa superba* Linn. Medicinal Chemistry Research DOI: 10.1007/s00044-012–0032-z.

Chaudhari, S.A., J. Peter, J.M. Galgut and S.A. Ali. 2012. Melanin inhibitory and melanin stimulatory effects of extracts of *Chlorophytum tuberosum* and *Chlorophytum borivilianum* on isolated fish scale melanophores. African J. Pharmacy Pharmacology 6: 919–923.

Choudhary, K.N. 2011. Assessment of the antimicrobial potency of leaf extracts from *Vitex nugundo* and *Gloriosa superba*. Pharmacognosy Journal 3: 80–84.

Chen, D., Y. Xiong, Z. Tang, B. Lv and Y. Lin. 2012a. Inhibitory effects of daidzein on intestinal motility in normal and high contractile states. Pharm. Biol. DOI:10.3109/13880209.2012.698285.

Chen, Y.Y., J. Chen, X.M. Zhou, X.H. Meng, J.P. Jiang and Y.L. Shen. 2012b. Puerarin protects human umbilical vein endothelial cells against high glucose-induced apoptosis by upregulating heme oxygenase-1 and inhibiting calpain activation. Fundamental Clinical Pharmacology 26: 322–331.

Chiang, C.T., T.D. Way, S.J. Tsai and J.K. Lin. 2007. Diosgenin, a naturally occurring steroid, suppresses fatty acid synthase expression in HER2-overexpressing breast cancer cells through modulating Akt, mTOR and JNK phosphorylation. FEBS Lett. 581: 5735–742.

Cho et al. 2011. Hypnotic effects and GABAergic mechanism of licorice (*Glycyrrhiza glabra*) ethanol extract and its major flavonoid constituent glabrol. Bioorganic & medicinal chemistry 20: 3493–501.

Dall'Acqua, S., B.B. Shrestha, S. Comai, G. Innocenti, M.B. Gewali and P.K. Jha. 2009. Two phenolic glycosideas from *Curculigo orchiodes* Gaertn. Fitoterapia 80: 279–282.

Das, A., V. Kesari and L. Rangan. 2010. Plant regeneration in *Curcuma* species and assessment of genetic stability of regenerated plants. Biologia Plantarum 54: 423–429.

Deore, S.L. and S.S. Khadabadi. 2009. Larvicidal activity of the saponin fractions of *Chlorophytum borivilianum* santapau and Fernandes. J. Entomol. Nematol. 1: 64–66.

Deore, S.L. and S.S. Khadabadi. 2010a. Antiproliferative activity of saponin fractions of *Chlorophytum borivilianum*. Pharmacognosy Journal 2: 29–33.

Deore, S.L. and S.S. Khadabadi. 2010b. Isolation and characterization of phytoconstituents from *Chlorophytum borivilianum*. Pharmacognosy Res. 2: 343–349.

Dev, S. 2006. A Selection of Prime Ayurvedic Plant Drugs, Ancient-Modern concordance. Anamaya publishers, New Delhi.

Dominy, N.J., E.R. Vogel, J.D. Yeakel, P. Constantino and P.W. Lucas. 2008. Mechanical Properties of Plant Underground Storage Organs and Implications for Dietary Models of Early Hominins. Evol. Biol. DOI: 10.1007/s11692-008-9026-7.

Edison, S., M. Unnikrishnan, B. Vimala, S.V. Pillai, M.N. Sheela, M.T. Sreekumari and K. Abraham. 2006. Biodiversity of Tropical Tuber Crops in India. NBA Scientific Bulletin Number -7, National Biodiversity Authority, Chennai, TamilNadu, India, p 60.

Flores, H.E., T.S. Walker, R.L. Guimaraes, H.P. Bais and J.M. Vivanco. 2003. Andean root and tuber crops: underground rainbows. Hort. Science 38: 161–167.

Ghosh, S., B. Ghosh and S. Jha. 2007. *In vitro* tuberisation of *Gloriosa superba* L. on basal medium. Scientia Horticulturae 114: 220–223.

Ghosh, S. and S. Jha. 2008. Colchicine—an Overview for Plant Biotechnologists. *In*: K.G. Ramawat and J.M. Merillon (eds.). Bioactive molecules and Medicinal Plants. Springer, Germany, pp. 215–232.

Goyal, S. and K.G. Ramawat. 2007. Effect of chemical factors on production of isoflavonoids in *Pueraria tuberosa* (Roxb. ex Willd.) DC. suspension culture. Indian J. Exp. Biol. 45: 1063–1067.

Goyal, S. and K.G. Ramawat. 2008a. Synergistic effect of morphactin on cytokinin-induced production of isoflavonoids in cell cultures of *Pueraria tuberosa* (Roxb. ex Willd.) DC. Plant Growth Regul. 55: 175–181.

Goyal, S. and K.G. Ramawat. 2008b. Improvement of isoflavonoid accumulation by ethrel in cell suspension cultures of *Pueraria tuberosa*, a woody legume. Acta Physiol. Plant 30: 849–853.

Goyal, S., V. Sharma and K.G. Ramawat. 2011. Marked effect of Cuscuta on puerarin accumulation in cell culture of *Pueraria tuberosa* grown in shake flasks and a bioreactor. Plant Biotech. Report 5: 121–126.

Haldar, K.M., P. Ghosh and G. Chandra. 2011. Evaluation of target specific larvicidal activity of the leaf extract of *Typhonium trilobatum* against Culex quinquefasciatus Say. Asian Pacific J. Tropical Biomedicine S199–S203.

Hamdy, S.M., A.K. Latif, E.A. Drees and S.M. Soliman. 2012. Prevention of rat breast cancer by genistin and selenium. Toxicol. Ind. Health 28: 746–757.

Ho, Y.J., S.Y. Tai, C.R. Pawlak, A.L. Wang, C.W. Cheng and M.H. Hsieh. 2012. Behavioral and IL-2 responses to diosgenin in ovariectomized rats. Chin. J. Physiol. 55: 91–100.

Huang, W., X. Chen, Q. Li, P. Li, G. Zhao, M. Xu and P. xie. 2012. Inhibition of Intercellular Adhesion in *Herpex Simplex* Virus Infection by Glycyrrhizin. Cell Biochem. Biophysics 62: 137–140.

Hussain, A., G. Harish, S.A. Prabhu, J. Mohsin, M.A. Khan, T.A. Rizvi and C. Sharma. 2012. Inhibitory effect of genistein on the invasive potential of human cervical cancer cells via modulation of matrix metalloproteinase-9 and tissue inhibitiors of matrix metalloproteinase-1 expression. Cancer Epidemiology http://dx.doi.org/10.1016/j.canep.2012.07.005.

Jabeen, N., A.S. Shawl, G.H. Dar, A. Jan and P. Sultan. 2006. Callus induction and organogenesis from explants of *Aconitum heterophyllum*- medicinal plant. Biotechnology 5: 287–291.

Jain, A.K., S. Dass, S. Suthar and K.G. Ramawat. 2007. Influence of culture medium constituents on bulbil formation from leaf explants of *Curculigo orchioides* in shake flask cultures. J. Herbal Medicine Toxicol. 1: 21–26.

Jain, N., S. Goyal and K.G. Ramawat. 2010. Biotechnological approaches to aphrodisiac plants of Rajasthan, India. *In*: K.G. Ramawat (ed.). Desert plants-biology and biotechnology. Springer, Berlin/Heidelberg, pp. 479–495.

Jana, S., G.S. Shekhawat. 2011. Critical review on medicinally potent plant species: Gloriosa superba Fitoterapia 82: 293–301

Janapati Y.K., R. Ahemad, K. Jayaveera and R. Reddy. 2009. Anti-diabetic activity of ethanolic extract of *Holostemma ada Kodien* Schults in alloxan induced diabetic rats. The Internet J. Endocrinol. DOI: 10.5580/24ee.

Jatoi, S.A., A. Kikuchi, S.A. Gilani and K.N Watanabe. 2007. Phytochemical, pharmacological and ethnobotanical studies in mango ginger (*Curcuma amada* Roxb. Zingiberaceae). Phytother. Res. 21: 507–516.

Jiao, L., D.P. Cao, L.P. Qin, T. Han, Q.Y. Zhang, Z. Zhu and F. Yan. 2009. Antiosteoporotic activity of phenolic compounds from *Curculigo orchiodes*. Phytomedicine 16: 874–881.

Joshi, N., A. Dave, S. Vyas and S.D. Purohit. 2009. Growth and shoot proliferation in *Chlorophytum borivilianum* Sant. et Fernand. *in vitro* under different carbon dioxide environment. Indian J. Biotechnol. 8: 323–327.

Joshi, N. and S.D. Purohit. 2012. Optimization of Factors Influencing Shoot Multiplication During Micropropagation of *Chlorophytum borivilianum* Sant. et Fernand. Proceedings of the National Academy of Sciences 82: 421–429.

Joshi, P. and N. Rao. 2011. Role of indigenous people in conservation of biodiversity of medicinal plants: an indian case study. *In*: H. Gökçekus et al. (eds.). Survival and Sustainability, Environmental Earth Sciences DOI: 10.1007/978-3-540-95991-5_10. pp 91–101.

Julsing, M.K., W.J. Quax and O. Kayser. 2007. The engineering of medicinal plants: prospects and limitations of medicinal plant biotechnology. *In:* O. Kayser and W.J. Quax (eds.).

Medicinal Plant Biotechnology: From Basic Research to Industrial Applications. Wiley -VCH Verlag GmbH & Co. KGaA, Weinheim, pp. 3–8.

Kala, C.P., P.P. Dhyani and B.S. Sajwan. 2006. Developing the medicinal plants sector in northern India: challenges and opportunities. Journal of Ethnobiology and Ethnomedicine 2: 32.

Kalia, A.N. 2005. Text Book of Industrial Pharmacognosym. Oscar Publication, New Delhi, India.

Kanwar, A.S. and K.K. Bhutani. 2010. Effects of *Chlorophytum arundinaceum*, *Asparagus adscendens* and *Asparagus racemosus* on pro-inflammatory cytokine and corticosterone levels produced by stress. Phytotherapy Research 24: 1562–1566.

Kaushik, N. 2005. Saponins of Chlorophytum species. Phytochemistry Reviews 4(2–3): 191–196.

Khetwal, S. 2007. Constituents of high altitude Himalayan herbs. A C-19 diterpenoid alkaloid from *Aconitum balfourii*. Indian J. Chem. 4: 1364.

Krishnan, P.N., S.W. Decruse and R.K. Radha. 2011. Conservation of medicinal plants of Western Ghats, India and its sustainable utilization through *in vitro* technology. *In vitro* Cell. Dev. Biol. Plant 47: 110–122.

Kuchta, K., G. Abraham and H.W. Rauwald. 2012. PDE inhibition by *Gloriosa superba* and its main alkaloid colchicines. Planta Med. 78 - PF48 DOI: 10.1055/s-0032-1320595.

Kuroda, M., Y. Mimaki, S. Honda, H. Tanaka, S. Yokota and T. Mae. 2010. Phenolics from *Glycyrrhiza glabra* roots and their PPAR-γ ligand-binding activity. Bioorganic Medicinal Chemistry 18: 962–970.

Lee, S.Y., M.R. Kim, H.S. Choi, H.I. Moon, J.H. Chung, D.G. Lee and E.R. Woo. 2009. The effect of curculigoside on the expression of matrix metalloproteinase-1 in cultured human skin fibroblasts. Arch. Pharm. Res. 32: 1433–9.

Malviya, S., N. Jain and K.G. Ramawat. 2011. Effect of alar and CCC on induction and germination of bulbils in *Curculigo orchioides* grown in shake flask cultures. Acta Physiol. Plant 33: 1543–1545.

Martin, K.P. 2003. Plant regeneration through somatic embryogenesis on *Holostemma ada-kodien*, a rare medicinal plant. Plant Cell Tissue Organ. Culture 72: 79–82.

Mehrotra, S., O. Khwaja, A.K. Kukreja and L. Rahman. 2012. ISSR and RAPD Based Evaluation of Genetic Stability of Encapsulated Micro Shoots of *Glycyrrhiza glabra* Following 6 Months of Storage. Molecular Biotechnol. DOI: 10.1007/s12033-011-9491-6.

Mehta, S.R. and R.B. Subramanian. 2005. Direct *in vitro* Propagation of *Asparagus adscendens* Roxb. Plant Tissue Cult. 15: 25–32.

Mohanty, S., R.K. Joshi, E. Subudhi, Sahoo and S.S. Nayak. 2012. Genetic Stability Assessment of Micropropagated Mango Ginger (*Curcuma amada* Roxb.) Through RAPD and ISSR Markers. Research J. Medicinal Plant 6: 529–536.

Mukherjee, P.K., S. Ponnusankar, P. Venkatesh, A. Gantait and B.C. Pal. 2011. Marker Profiling: An approach for quality evaluation of Indian medicinal plants. Drug Information Journal 45: 1–14.

Murashige, T. and F. Skoog. 1962. A revised medium for rapids growth and bioassays with tobacco tissue cultures. Physiol. Plant 15: 473–497.

Nautiyal, O.P. 2011. Isolation of 3-demethylcolchicine from *Gloriosa superba* sludge and coupling with -acetobromoglucose to yield colchicoside and thiocolchicoside. J. Natural Products 4: 87–93.

Nema, R.K. and K.G. Ramawat. 2010. Isolation and identification of a new molecule from *Curculigo orchioides* (hypoxidaceae). J. Chem. Pharm. Res. 2: 610–617.

Nema, R.K., S. Dass, M. Mathur and K.G. Ramawat. 2008a. Morphactin and cytokinin promote high frequency bulbil formation from leaf explants of *Curculigo orchioides* grown in shake flask cultures. Indian J. Biotechnol. 7: 520–525.

Nema, R.K., K.G. Ramawat and J.M. Merillon. 2008b. Rapid micropropagation of *Curculigo orchioides* in shake flask culture. Pharmacognosy Magazine 4: 315–319.

Nisar, M. et al. 2009. New diterpenoid alkaloids from *Aconitum heterophyllum* Wall: Selective butyrylcholinestrase inhibitors. J. Enzyme Inhibition Medicinal Chemistry 24: 47–51.

Padmapriya, K., A. Dutta, S. Chaudhuri and D. Dutta. 2012. Microwave assisted extraction of mangiferin from *Curcuma amada*. 3 Biotech. 2: 27–30.

Pari, L., P. Monisha and A.M. Jalaludeen. 2012. Beneficial role of diosgenin on oxidative stress in aorta of streptozotocin induced diabetic rats. Europian J. Pharmacol. 691: 143–150.

Patel, S., R. Gaur, M. Upadhyaya, A. Mathur, A.K. Mathur and R.S. Bhakuni. 2011. *Glycyrrhiza glabra* (Linn.) and *Lavandula officinalis* (L.) cell suspension cultures-based biotransformation of β-artemether. Journal of Natural Medicines 65: 646–650.

Paul, K.K. and M.A. Bari. 2012. *In vitro* multiple shoot regeneration from corm bud explant in ghet kachu, *Typhonium trilobatum* schott—a medicinal aroid. Bangladesh J. Scientific Industrial Research 47: 2.

Policegoudra, R.S., S.M. Aradhya and L. Singh. 2011a. Mango ginger (*Curcuma amada* Roxb.)—a promising spice for phytochemicals and biological activities. J. Biosci. 36: 739–748.

Policegoudra, R.S., R.H. Chandrashekhar, S.M. Aradhya and L. Singh. 2011b. Cytotoxicity, platelet aggregation inhibitory and antioxidant activity of *Curcuma amada* Roxb. extracts. Food Technol. Biotechnol. 49: 162–168.

Praveena, R., S.A. Pandian and M. Jegadeesan. 2012. *In vitro* Culture Studies on Medicinal Herb - *Coleus forskohlii* Briq. Libyan Agriculture Research Center J. International 3: 30–35.

Raju, J. and R. Mehta. 2009. Cancer chemopreventive and therapeutic effects of diosgenin, a food saponin. Nutr. Cancer 61: 27–35.

Ramachandran, C., S.M. Nair, K.W. Quirrin, E.A. Escalon and S.J. Melnick. 2012. Adjuvant therapeutic use of supercritical-ethanol extracts of *Curcuma* species with cancer drugs in rhabdomycosarcoma cell lines. Planta Med. 78-PI137 DOI: 10.1055/s-0032-1320824.

Ramawat, K.G. and S. Goyal. 2008. Indian herbal drugs scenario in global perspectives. *In*: K.G. Ramawat and J.M. Merillon (eds.). Bioactive molecules and medicinal plants. Springer, Heidelberg, pp. 323–34.

Ramawat, K.G., S. Dass and M. Mathur. 2009. The chemical diversity of bioactive molecules and therapeutic potential of medicinal plants. *In:* K.G. Ramawat (ed.). Herbal drugs: ethnomedicine to modern medicine. Springer, Heidelberg, pp. 7–321.

Rao, B.R.R., D.K. Rajput, G. Nagaraju and G. Adinarayana. 2012. Scope and potential of medicinal and aromatic plants products for small and medium enterprises. J. Pharmacognosy 3: 112–114.

Rizvi, M.Z., A.K. Kukreja and N.S. Bisht. 2012. Plant regeneration in *Chlorophytum borivilianum* Sant. et Fernand. from embryogenic callus and cell suspension culture and assessment of genetic fidelity of plants derived through somatic embryogenesis. Physiology and Molecular Biology of Plants 18: 253–263.

Sagnou, M., K.P. Mitsopoulou, G. Koliopoulos, M. Pelecanou, E.A. Couladouros and A. Michaelakis. 2012. Evaluation of naturally occurring curcuminoids and related compounds against mosquito larvae. Acta Tropica 123: 190–195.

Selvarasu, A. and R. Kandhasamy. 2012. *In vitro* tuberization of glory lily (*Gloriosa superba* L.). J. Horti. Forest 4: 81–84.

Sharma, E. and A.K. Gaur. 2012. *Aconitum balfourii* Stapf: A rare medicinal herb from Himalayan Alpine. J. Medicinal Plants Research 6: 3810–3817.

Sharma, O.P., N. Kumar, B. Singh and T.K. Bhat. 2012. An improved method for thin layer chromatographic analysis of saponins. Food Chem. 132: 671–674.

Sharma, V., S. Goyal and K.G. Ramawat. 2009. Scale up production of isoflavonoid in cell suspension cultures of *Pueraria tuberosa* grown in shake flasks and bioreactor. Eng. Lif. Sci. 9: 267–271.

Sharma, V., S. Goyal and K.G. Ramawat. 2011. Increased puerarin biosynthesis during *in vitro* shoot formation in *Pueraria tuberosa* grown in Growtek bioreactor with aeration. Physiol. Mol. Biol. Plants 17: 87–92.

Sharma, V. and K.G. Ramawat. 2013. Isoflavonoids. *In*: K.G. Ramawat and J.M. Merillon (eds.). Handbook of Natural Products, Phytochemistry, Botany, Metabolism, Vol. 2 (in press).

Shen, P., L.S. Wang, X.K. Liu, C.R. Yang, B. Cai and X.S. Yao. 2002. A New Steroidal Saponin from *Dioscorea deltoidea* Wall var. orbiculata. Chinese Chemical Letter 13: 851–854.

Shyaula, S.L. 2011. Phytochemicals, traditional uses and processing of *Aconitum* species in Nepal. Nepal J. Science Technol. 12: 171–178.

Simons, R. et al. 2012. Agonistic and antagonistic estrogens in licorice root (*Glycyrrhiza glabra*). Arch. Immunol. Ther. Exp. 60: 391–399.

Singh, M., S. Singh and R.K. Kale. 2011. Chemomodulatory potential of *Asparagus adscendens* against murine skin and forestomach papillomagenesis. European J. Cancer Prevention 20: 240–247.

Singh, S., R.K. Joshi and S. Nayak. 2013. Identification of elite genotypes of turmeric through agroclimatic zone based evaluation of important drug yielding traits. Industrial Crops and Products 43: 165–171.

Singh, S., J.K. Kumar, D. Saikia, K. Shanker, A.S. Negi and S. Banerjee. 2010. A bioactive labdane diterpenoid from *Curcuma amada* and its semi synthetic analogues as anti tubercular agents. Eur. J. Med. Chem. 45: 4379–82.

Smolarczyk, R., T. Cichon, S. Matuszczak, I. Mitrus, M. Lesiak, M. Kobusin'ska, W. Kamysz, M. Jarosz, A. Sieron and S. Szala. 2012. The Role of Glycyrrhizin, an Inhibitor of HMGB1 Protein, in Anticancer Therapy. Arch. Immunol. Ther. Exp. 60: 391–399.

Subhash, C., S. Sarla and D. Mridul. 2012. Evaluation of Garhwal Himalaya wild edible tuber *Dioscorea deltoidea*. Int. Research J. Pharm. 3: 152–156.

Suri, S.S., S. Jain and K.G. Ramawat. 1999. Plantlet regeneration and bulbil formation *in vitro* from leaf and stem explant of *Curculigo orchioides*—an endangered medicinal plant. Sci. Hortic. 1210: 1–8.

Tandon, M. and Y.N. Shukla. 1995. Phytoconstitutes of *Asparagus adscendens, Chlorophytum arundinaceum* and *Curculigo orchioides*: a review. Curr. Res. Med. Arom. Plants 17: 42–50.

Thakur, G.S. et al. 2009. *Chlorophytum borivilianum*: a white gold for biopharmaceuticals and neutraceuticals. Current Pharm. Biotechnol. 10: 650–666.

Thakur, M., D. Thompson, P. Connellan, M.A. Deseo, C. Morris and V.K. Dixit. 2011a. Improvement of penile erection, sperm count and seminal fructose levels *in vivo* and nitric oxide release *in vitro* by ayurvedic herbs. Andrologia 43: 273–277.

Thakur, M., P. Connellan, M.A. Deseo, C. Morris and V.K. Dixit. 2011b. Immunomodulatory polysaccharide from *Chlorophytum borivilianum* roots. Evidence-Based Complementary and Alternative Medicine, DOI: 10.1093/ecam/neq 012.

Thiyagarajan, P., C.V. Chandrasekaran, H.B. Deepak and A. Agarwal. 2011. Modulation of lipopolysaccharide-induced pro-inflammatory mediators by an extract of *Glycyrrhiza glabra* and its phytoconstituents. Inflammopharmacology 19: 235–241.

Tian, Z., W. Yu, H.B. Liu, N. Zhang, X.B. Li, M.G. Zhao and S.B. Liu. 2012. Neuroprotective effects of curculigoside against NMDA-induced neuronal excitoxicity *in vitro*. Food Chemical Toxicol. DOI: http://dx.doi.org/10.1016/j.fct.2012.08.006.

Titova, M.V., E.A. Berkovich, O.V. Reshetnyak, I.E. Kulichenko, A.V. Oreshnikov and A.M. Nosov. 2011. Respiration Activity of Suspension Cell Culture of *Polyscias filicifolia* Bailey, *Stephania glabra* (Roxb.) Miers, and *Dioscorea deltoidea* Wall. Applied Biochemistry and Microbiology 47: 87–92.

Tripathi, Y.B., S. Nagwani, P. Mishra, A. Jha and S.P. Rai. 2012. Protective effect of *Pueraria tuberosa* DC. embedded biscuit on cisplatin-induced nephrotoxicity in mice. J. Nat. Med. 66: 109–18.

Valls, J., T. Richard, F. Larronde, V. Leblais, B. Muller, J.C. Delaunay, J.P. Monti, K.G. Ramawat and J.M. Merillon. 2006. Two new benzylbenzoate glucosides from *Curculigo orchioides*. Fitoterapia 77: 416–419.

Venkatesh, P., P.K. Mukherjee, N.K. Nema, A. Bandyopadhyay, H. Fukui and H. Mizuguchi. 2009. Mast cell stabilization and antihistaminic potentials of *Curculigo orchiodes* rhizomes. J. Ethnopharmacol. 126: 434–436.

Verma, S., S. Ojha and M. Raish. 2010. Anti-inflammatory activity of *Aconitum heterophyllum* on cotton pellet-induced granuloma in rats. J. Medicinal Plants Research 4: 1566–1569.

Verma, P., A.K. Mathur, S.P. Jain and A. Mathur. 2012. *In vitro* Conservation of Twenty-Three Overexploited Medicinal Plants Belonging to the Indian Sub Continent. The Scientific World Journal DOI: 10.1100/2012/929650.

Virgona, N., Y. Taki, S. Yamada and K. Umegaki. 2012. Dietary Coleus forskohlii extract generates dose-related hepatotoxicity in mice. J. Appl. Toxicol. DOI: 10.1002/jat.2770.

Visavadiya, N.P., B. Soni, N. Dalwadi and D. Madamwar. 2010. *Chlorophytum borivilianum* as potential terminator of free radicals in various *in vitro* oxidation systems. Drug Chem. Toxicol. 33: 173–82.

Wang, Y. et al. 2012. Curculigoside isolated from *Curculigo orchioides* prevents hydrogen peroxide-induced dysfunction and oxidative damage in calvarial osteoblasts. Acta Biochim. Biophys. Sin. (Shanghai) 44: 431–441.

Wu, X.Y., J.Z. Li, J.Z. Guo and B.Y. Hou. 2012. Ameliorative effects of curculigoside from *Curculigo orchioides* Gaertn on learning and memory in aged rats. Molecules 17: 10108–10118.

Yin, G., L. Cao, Xu Pao, G. Jeney, M. Nakao and C. Lu. 2011. Hepatoprotective and antioxidant effects of Glycyrrhiza glabra extract against carbon tetrachloride (CCl4)-induced hepatocyte damage in common carp (*Cyprinus carpio*). Fish Physiol. Biochem. 37: 209–216.

Yokotani, K. et al. 2012. Hepatic cytochrome P450 mediates interaction between warfarin and *Coleus forskohlii* extract *in vivo* and *in vitro*. J. Pharmacy Pharmacol. DOI: 10.1111/j.2042-7158.2012.01563.x.

Zuo, A.X., Y. Shen, Z.Y. Ziang, X.M. Zhang, J. Zhou, J. Lu and J.J. Chen. 2010. Three new phenolic glycosides from *Curculigo orchioides*. G. Fitoterapia 81: 910–913.

Zuo, A.X. et al. 2012. Two new triterpenoid glycosides from *Curculigo orchioides*. J. Asian Natural Products Research 14: 407–412.

16

Medicinal Bulbous Plants
Biology, Phytochemistry and Biotechnology

Sayantani Nath, Partha Sarathi Saha and *Sumita Jha**

ABSTRACT

Bulbous plants are geophytes with an enlarged storage structure called bulb which helps them to perennate extreme conditions. True bulbs mostly constitute petaloid monocotyledons belonging to Amaryllidaceae, Alliaceae, Hyacinthaceae, Liliaceae and Iridaceae. These bulbous plants produce an array of unique biologically active compounds and have been extensively used in ethno medicine. Norbelladine alkaloids produced by Amaryllidaceae plants have anticancerous, acetylcholine esterase inhibitory and antimalarial activities while organosulphur compounds characteristic of the family Alliaceae are known for their antioxidant, anticancerous, antidiabetic and antihypertensive activities. Cardenolides and bufadienolides produced by members of Hyacinthaceae exhibit cardiotonic properties and have been used to treat heart failure since time immemorial. Such extensive medicinal properties of bulbous plants have resulted in their ruthless exploitation making their conservation an immediate necessity. Micropropagation, genetic improvements as well as elicitation of

Center of Advanced study, Department of Botany, University of Calcutta 35, Ballygunge Circular Road, Kolkata- 700019.
Emails: snsayantani1@gmail.com; parthasarathisaha11@gmail.com; sumitajha.cu@gmail.com/ sjbot@caluniv.ac.in
* Corresponding author

secondary metabolite can aid in the judicious exploitation of bulbous medicinal plants. The present review encompasses within its scope biology, phytochemistry and biotechnology of true bulbous geophytes of medicinal importance.

Keywords: Medicinal bulbous plants, Geophytes, Amaryllidaceae, Hyacinthaceae, Iridaceae, Liliaceae, Phytochemistry, Micropropagation, Biotechnology

Introduction

Bulbous plants are geophytes with a highly shortened stem along with fleshy leaves or leaf bases forming an enlarged swollen underground storage structure called bulb. Plants with other types of storage organs like corms, rhizomes and tubers are also sometimes erroneously designated as bulbous plants. Some epiphytic orchids also form above ground storage organs that superficially resemble bulbs known as pseudo bulbs.

A bulb constitutes a modified basal stem tissue with leaf bases called scales. The scales do not generally support leaves but store food for the plant to survive adverse conditions. At the centre of the bulb is the growing point from which vegetative leaves or flowering shoot emerges. Roots emerge from underside the base of the bulb. Based on morphology there are two types of bulbs—tunicate bulbs with thin, dry paper like protective covering, e.g., *Allium* sp., *Hippeastrum* sp., *Narcissus* sp., *Tulipa* sp., etc. and non-tunicate bulbs, those which are devoid of the protective tunic, e.g., *Lilium* sp., *Fritillaria* sp., etc. Each year a bulbous plant normally undergoes three phases, i.e., vegetative growth, flowering and dormancy. After vegetative growth and flowering, the plants in most cases enter into the duration of dormancy. This cycle repeats year after year. However the chronology of the three phases varies from species to species.

The present review encompasses within its scope biology, phytochemistry and biotechnology of true bulbous geophytes of medicinal importance.

Biology of Bulbous Plants

Origin of bulbs

There are several opinions regarding the origin and evolution of bulbous plants. It has been suggested that climate and day-length are prime factors for the evolution of bulbous geophytes and the onset of a hot dry climate may have played a major role in the evolution of bulbous plants, bulbs being preferred as perennating organs to overcome extreme conditions like drought, cold or heat (Savos'kin 1960). Some bulbs even have remarkable frost tolerance enabling them to overcome cold winters. The bulb habit also provides the advantage of vegetative propagation, in most species, by the formation of many lateral buds.

Taxonomy

Nearly all plants which form true bulbs are petaloid monocotyledons mostly including lilies, onions, belladonnas, and irises. The most common bulbs generally belong to three classical families—Amaryllidaceae, Iridaceae and Liliaceae. However, the recent APG III classification (Chase et al. 2009) prefers a broad view of Amaryllidaceae grouped together with Agapanthaceae and Alliaceae. Liliaceae is also split into several small homogeneous groups, viz. Alliaceae (Allioideae *sensu* APG), Hyacinthaceae (Scilloideae *sensu* APG), Asparagaceae (Asparagoideae *sensu* APG) and Xanthorrhoeaceae, etc. The only dicotyledonous genus that produces true bulbs is *Oxalis* belonging to the family Oxalidaceae.

Phytochemistry of Bulbous Plants

Although bulbous plants are studied less thoroughly than herbs and trees, they are known to synthesize an array of unique biologically active compounds and have been extensively used in ethno medicine since time immemorial. They are commonly used as anticancer, antimutagenic, antiinfective, antimalarial, analgesic, immunostimulatory, respiratory and cardiovascular drugs (Likhitwitayawuid et al. 1993, Pettit et al. 1995, Hutchings et al. 1996, Gao et al. 1999, Fennell and Staden 2001). Ethno medicinal uses of these bulbous species can provide new insight for pharmaceutical developments. However, the biochemical content and pharmacological action of a majority of these bulbous plants are still inadequately understood, future research should be invaluable in this aspect. The present section includes a family-wise description of the ethno medicinal importance, bioactive principles, their putative biosynthetic pathway and biological activity of the aforesaid bulbous plant families.

Family Amaryllidaceae

The family Amaryllidaceae constitutes a group of bulbous perennial plants whose general characteristics are very similar to those of Liliaceae. There have been widely differing views on taxonomic delimitation of the family, now placed in the monocot order Asparagales in the most recent APG classification (Chase et al. 2009). The APG III classification prefers a broad view of Amaryllidaceae which now includes 3 subfamilies—

Agapanthoideae (the old Agapanthaceae family), the Allioideae (the old Alliaceae family) and the Amaryllidoideae (the old Amaryllidaceae family), altogether includes about 75 genera and 1600 species. The former two have superior ovaries while the latter have inferior ovary. Plants are fleshy with two ranked leaves and umbel inflorescence at the apex of leafless flowering stems or scapes.

The Amaryllidaceae members are richly distributed in the tropics and have pronounced centers of diversity in South Africa and the Andean region. Some members are also found in the Mediterranean area and temperate regions of Asia (Bastida et al. 2011).

The medicinal activity of the Amaryllidaceous plants is frequently correlated to the alkaloids they produce and thus have attracted the attention of researchers over the past three decades (Cook and Loudon 1952, Wildman 1968, Martin 1987).

Norbelladine alkaloids, unique to the family are a major component responsible for the biological activity of these plants (Bastida et al. 2011). Specific mannose binding lectins are also produced by the members of the family (Van damme et al. 1991). The members of Allioideae produce allyl sulfide, compounds responsible for the characteristic smell of this group. Members of Agapanthoideae produce saponins and chalcones.

Subfamily: Amaryllidoideae

The plants belonging to the subfamily Amaryllidoideae (Chase et al. 2009) extensively used in folk medicine of different countries since ancient times are also cultivated as ornamentals and for production of volatile oils. As early as the fourth century BC, oil from *Narcissus poeticus* was employed by the Greek physician Hippocrates of Cos (the "Father of Medicine") for the treatment of cancer (Hartwell 1967). The major bioactive principles unique to Amaryllidaceae family are alkaloids along with some specific mannose binding lectins (Van damme et al. 1991).

Bioactive principles. Alkaloids, a major active principle of the extracts is investigated actively for nearly 200 years. Many such alkaloids have been isolated, screened for biological activity and chemically synthesized over the past three decades (Bastida et al. 2011).

These isoquinoline/isocarbostyril alkaloids are biogenetically related, being produced from the common precursor norbelladine (a tyrosine and L-Phenylalanine derivative) (Spenser 1970), which after varied skeletal rearrangements by intramolecular oxidative phenol-coupling form the major types (Barton et al. 1963). The Lycorine, Crinane and Galanthamine types are classically recognized although several other types (Norbelladine, Homolycorine, Haemanthamine, Narciclasine, Tazettine and Montanine) are now well known (Jin 2005, Unver 2007). However all Amaryllidaceae alkaloids have a basic structure consisting of a C1-C6 A ring (derived from L-phenylalanine) and a N-C2-C6 C ring (derived from L-Tyrosine).

Research interest in Amaryllidaceous alkaloids was hiked by the discovery of potent anticancer activity exhibited by certain Lycorine type alkaloids and selective, reversible acetylcholinesterase inhibitory activity exhibited by certain Galanthamine type alkaloids (Marco-Contelles et al. 2006). These discoveries aggravated the clinical importance of the Amaryllidaceous alkaloids and their synthetic derivatives (Rinner and Hudlicky 2005) in anti-tumour and anti-Alzheimer's therapeutics.

Study of Amaryllidaceous alkaloids was initiated with the isolation of lycorine from *Narcissus pseudonarcissus* in 1877 (Cook and Loudon 1952). The Lycorine type and the Pancratistatin type (Pettit et al. 1993) constituents and their natural and synthetic derivatives (Rinner and Hudlicky 2005) have achieved significant attention from the biomedical perspective because of their potent anticancer activity. Even placing a cut daffodil (*Narcissus* sp.) in a vase shortens the vase life of other flowers indicating that the components that ooze out from the former have powerful growth inhibitory effects.

Galanthamine type alkaloids such as galanthamine and sanguinine are incorporated in the treatment of Alzheimer's and other neuro degenerative diseases (Houghton et al. 2006). Galanthamine, obtained from *Galanthus* and *Narcissus*, exhibits acetylcholinesterase inhibitory action, and can thus boost levels of acetylcholine in the synapse of the cerebral cortex and used for symptomatic treatment of Alzheimer's disease. Galanthamine hydrobromide, marketed under the name Reminyl®, is the first Amaryllidaceae alkaloid to be approved as a prescription drug in the treatment of a human disease (Heinrich and Theo 2004).

Crinum alkaloids also exhibit a range of biological activities including cytotoxicity to malarial parasite (Likhitwitayawuid et al. 1993). Biological activity of major Amaryllaceous alkaloids is listed in Table 1.

Distinctive lectins are also found in Amaryllidoideae and Allioideae (Van Damme et al. 1991) which exhibit exclusive specificity towards mannose. Lectins have been reported from snowdrop *(Galanthus nivalis)* (Van Damme et al. 1987), daffodil *(Narcissus pseudonarcissus)* (Kaku et al. 1990) and snowflake *(Leucojum aestivum and Leucojum vernum)* (Van Damme et al. 1987). The unique carbohydrate binding specificity of Amaryllidaceae lectins enables their use for the purification and characterization of glucoconjugates (Shibuya et al. 1988) and makes them useful tools for glucoconjugate research. Because of the potent inhibition of human and animal retroviruses including HIV by the Amaryllidaceae lectins they have become important aspects in AIDS research (Balzarini et al. 1991).

Table 1. Amaryllidaceous alkaloids and their pharmacological effects.

Alkaloid Type	Alkaloid	Uses	References
Lycorine type	Lycorine (*N. pseudonarcissus*, *Amaryllis belladonna*, *Crinum amabile*, *Calostemma purpureum*)	Antimitotic, Anticancerous	Hua et al. 1997, Yui et al. 1998
		Antiviral (against poliovirus, measles virus, Herpes simplex type 1 viruses, etc.)	Ieven 1982, 1983, Renard-Nozaki 1989, Gabrielsen et al. 1992, Szlávik et al. 2004
		Antifungal (against *Candida albicans*)	Evidente 2004
		Hypotensive and Analgesic	Tanker 1996, Fennell and Staden 2001
		Antimalarial	Toriizuka et al. 2008
	Caranine (*A. belladonna*)	Anticancerous	Suzuki 1974, Pettit 1984, Weniger 1995
	Galanthine (*Narcissus* sp.)	Anticancerous	Suzuki 1974, Pettit 1984, Weniger 1995
	Pseudolycorine	Anticancerous	Suzuki 1974, Jimenez et al. 1975, 1976, Pettit 1984, Weniger 1995
		Antiviral	Ieven et al. 1982, 1983, Gabrielsen et al. 1992
Galanthamine Type	Galanthamine (*Galanthus nivalis. Calostemma purpureum*, *Chlidanthus fragrans*, *Leucojum aestivum*)	Treat neuro degenerative diseases (paralysis syndrome,schizophrenia, dementia and Alzheimer's disease)	Fulton et al. 1996, Sramek et al. 2000, Guillou et al. 2001, Marco-Contelles et al. 2006
Narciclasine Type	Narciclasine (*N. poeticus*, *A. belladonna*, *A. fmossisima*, *A. zeylanica*)	Antineoplastic, Antimitotic	Ceriotti 1967, Pettit 1995
		Antiviral	Gabrielsen et al. 1992

Table 1. contd....

Table 1. contd.

Alkaloid Type	**Alkaloid**	**Uses**	**References**
Pancratistatin type	Pancratistatin (*A.belladonna, A. fmossisima, A. zeylanica, Crinum* sp., *Hisppeastrum* sp., *Pancratium* sp., *Zephyranthes* sp.)	Anticancerous	Pettit et al. 1995, McLachlan et al. 2005
Crinane Type	Crinamine (*Crinum amabile*)	Antimalarial	Likhitwitayawuid et al. 1993, McNulty et al. 2007
	Haemanthamine (*Crinum amabile*)	Anticancerous	McNulty et al. 2007
	Buphanidrine and Distichamine (*Boophone disticha*)	Potent drug for nervous disorders	Neergaard et al. 2009

Subfamily: Allioideae

Alliaceae, initially a part of Liliaceae, was separated as a homogeneous family including *Allium* and related members according to Dahlgren et al. (1985) and ultimately reduced to subfamily Allioideae in the family Amaryllidaceae, order Asparagales according to recent APG III classification (Chase et al. 2009). The type genus *Allium* is widely distributed throughout the temperate regions of the world including Europe, Asia, North America, Africa and include several important food plants like onions (*Allium cepa*), chives (*A. schoenoprasum*), garlic (*A. sativum* and *A. scordoprasum*), and leeks (*A. porrum*). Other members like *Gilliesia* sp., *Ipheion* sp., *Leucocoryne* sp., *Nothoscordum* sp., and *Tulbaghia* sp., are cultivated as ornamentals.

Traditionally, *Allium* species, mainly garlic (*A. sativum*) and onion (*A. cepa*) has a long history of therapeutic use against cancer (Gao et al. 1999) leprosy, heart disease, hypertension and infections (Rivlin 2001, Banerjee et al. 2003). Onion was an object of worship among the ancient Egyptians. According to Indian mythology garlic plants originated from drops of nectar which was spilled in the Himalayan region of the earth from the hands of Lord Garuda, the personal carrier of Lord Vishnu while he snatched a pot of "nectar—the elixir of life" from the hands of Lord Indra, the King of heaven. All these instances depict the beneficial properties of this group.

Bioactive principles. The distinct aroma as well as many of the therapeutic properties of the genus *Allium* is due to the presence of specific phytochemicals—the organosuphur compounds which are also called allyl sulphides or cysteine sulphoxides (CS) (Jones et al. 2004).

There are two putative pathways of CS biosynthesis. In the first one, glutamyl cysteine and glutathione are the starting compounds where *S*-alk(en)ylation of the cysteine residue of glutathione, followed by the removal of the glycyl residue by transpeptidation and subsequently oxidation and loss of the glutamyl group yields the parental CS. Alternatively, direct *S*-alk(en)ylation of cysteine or thioalk(en)ylation of O-acetyl serine followed by oxidation can also yield the parental CS (Whitaker 1976, Block 1992, Jones et al. 2004).

The odorless, nonvolatile parental cysteine sulphoxide derivatives called *S*-allyl-L-cysteine sulfoxides (ACSO) such as alliin [(+)-S-(2-propenyl)-L-cysteine sulphoxide], isoalliin [(+)-S-(1-propenyl)-L-cysteine sulphoxide], methiin, propiin, etc. localized in cytosol are the precursors of the lachrymatory compounds typically found in *Allium*. Upon tissue damage the enzyme allinase, common to all *Allium* species is released from vacuoles and converts the ACSOs to respective thiosulphinates (e.g., allicin) responsible for the characteristic aroma of the members. However depending on species and conditions, thiosulphinates can further decompose to additional sulfur constituents like diallyl, methyl allyl, and alk(en)yl (poly)sulphides [diethyl mono-, di-, tri-, tetra-, penta-, and hexasulfides], the vinyldithiins and ajoene (Jones et al. 2004).

Several therapeutic activities of *A. cepa* plants have been reported such as antidiabetic (Galal et al. 1965) antioxidant (Prakash et al. 2007), antihypertensive (Sakai et al. 2003), antithrombotic (Yamada et al. 2004), hypoglycemic (Compos et al. 2003) and antihyperlipidemic (Lata et al. 1991) effects. Garlic (*A. sativum*) is listed among the best disease preventive foods on the basis of its wide range of reported therapeutic properties (Amagase 2006) viz. anticoagulant, antimicrobial, antiatherosclerotic, antihypertensive, antidote (for heavy metal poisoning), immune enhancement, chemopreventive, hypolipidaemic, hypoglycaemic and hepatoprotective effects (Rivlin 2001, Banerjee et al. 2003, Thomson and Ali 2003, Amagase 2006). An indigenous South African species, *Tulbaghia alliacea* used in folk medicine to treat infections, exhibit antiinfective activity against *Candida* sp. *in vitro*, thus can be used as a remedy for candidadides (Thamburan et al. 2006).

Neuroprotective activity of organosulfur compounds, such as the alliin and allicin, abundant in garlic and onions have been reported (Kosuge et al. 2003). Besides antioxidant activities, allyl sulfides putatively activate stress response pathways, causing upregulation of neuroprotective proteins such as mitochondrial uncoupling proteins (Oi et al. 1999). A new stress resistance system turned on by garlic organosulfur compounds is the NRF2–ARE pathway (Chen et al. 2004).

Increased consumption of *Allium* vegetables (particularly garlic and onions) have been correlated to lower incidence of cancers among human populations (Block et al. 1992). The potent anticancer activity of *Allium* chemicals in animals involves stimulation of phase 2 enzymes, a cellular stress response against toxins (Tsai et al. 2005).

Allium plants have been used to treat infections in folk medicine (Gao et al. 1999, Fleischauer and Arab 2001). Thiosulphinates, like allicin are considerated to play a significant role in the antimicrobial activities of garlic. Allicin derivatives like DATS and ajoene also have some *in vitro* antimicrobial activity. Several National Health Service Hospitals in UK use allicin against methicillin resistant *Staphylococcus aureus* which have developed resistance against penicillin drugs (Nagori et al. 2010). However, in spite of the extensive therapeutic uses of *Allium* since ancient times it is only recently that there is any significant advancement in determining their modes of action. Extensive research is required for further improvement. Biological activities exihibited by some *Allium* species are listed in Table 2.

Table 2. *Allium* species and their biological activities.

Plant	Activity	Reference
Allium sativum (Garlic)	Analgesic, Antiatherogenic, Antimycotic, Antiinflammatory	Blumenthal et al. 2000
	Anticancer	Fleischauer and Arab 2001, Thomson and Ali 2003
	Antioxidant	Banerjee et al. 2003
Allium cepa (Onion)	Antiaggregant, Antiseptic, Antitumor, Diuretic, Hypocholesterolemic	Blumenthal et al. 2000
	Antiallergic, Antiasthmatic, Antiinflammatory	Fleming et al. 1998
Allium schoenoprasum (Chives)	Anthelminthic	Gruenwald et al. 2000
	Carminative, Diuretic, Antiseptic	Steinmetz 1957
Allium ursinum (Ramson, Bear garlic)	ACE Inhibitor, Antiaggregant, Antioxidant, Antiseptic, Antispasmodic, Expectorant	Khare 2004

Subfamily Agapanthoideae

Agapanthus, commonly called "Lily of the Nile" is the only genus of the subfamily Agapanthoideae of Amaryllidaceae with all species native to South Africa (Tamura 1998). Initially it was placed in Alliaceae (Dahlgren 1985), then in Amaryllidaceae *ss* (Fay and Chase 1996) and ultimately separated as Agapanthoideae in the APG III classification (Chase et al. 2009).

The genus *Agapanthus* has been extensively used in traditional medicine. The *Agapanthus africanus* plant is used by South African ethnomedicine to treat pregnancy related problems and to augment labour. Aqueous extract of *A. africanus* have been shown to cause smooth muscle contractions in the isolated uterus and ileum preparations (Veale et al. 1992, 1999). The ethanolic extract of the rhizomes of *A. africanus* showed antifungal activity (Singh et al. 2008). In Zulu herbal medicine *Agapanthus* root infusions are used to treat chronic coughs and serious heart diseases (Watt and Breyer-Brandwijk 1962). Use of *A. praecox* bulbs as an aphrodisiac is reported in Zulu herbal medicine (Watt and Breyer-Brandwijk 1962).

Bioactive principles. Two groups of phychemicals predominantly present in these plants include saponins (agapanthagenin, yuccagenin, etc.) and chalcones (Kamara et al. 2005). Saponins exhibit antiinflammatory and antitussive properties while chalcones are known to have antibacterial, antifungal, antitumor and antiinflammatory properties. Thus, it is evident that the family Agapanthaceae can serve as an enexploited phytochemical reserve that awaits further investigation.

Family Hyacinthaceae (Subfamily Scilloideae *sensu* APG)

Hyacinthaceae formerly placed in Liliaceae was first designated as a separate family by Dahlgren (Dahlgren et al. 1985) and since then is recognized as a distinct group but was renamed as subfamily Scilloideae in the family Asparagaceae according to the most recent APG III classification (Chase et al. 2009). It comprises approximately 67 genera and 900 species worldwide with the highest diversity in South Africa and in the region from the Mediterranean to South-West Asia. It constitutes a large group bulbous plants which are chemically and morphologically diverse. Scilloideae is subdivided into four monophyletic tribes—Ornithogaleae, Urgineeae, Hyacintheae and Oziroeeae.

Hyacinthaceae plants have been employed in traditional medicine for treating various ailments like rheumatic fever, sprains, syphilis and even cancer. A number of species are known for their toxic effects following human and livestock deaths after ingestion (Pohl et al. 2000).

Drimia (syn. *Urginea*), commonly called "squill", is most extensively used for therapeutic purposes. Medicinal use of *D. maritima* (syn. *U. maritima*), European or red squill, has been reported since ancient times and was mentioned as early as 1554 B.C. in the Papyrus Ebers of the Middle Empire of Egypt as a cure for dropsy. Extracts from squill bulbs were used by ancient Romans and Egyptians as a cardiotonic (Kreig 1966). Use of squill as a diuretic, antijaundice, expectorant, anti-colic and anti-asthmatic agent is also reported in folk medicines. Squill has also been used as rodenticide since ancient times (Blumenthal et al. 1998). Methanolic extracts of red squill have reported activity against Seborrhea and dandruff and is incorporated in hair tonic preparation (Leung 1980). *D. indica* (syn. *U. indica*), the Indian squill, also exhibits similar medicinal properties and have been extensively used in Indian ethnomedicine as an expectorant, cardiac stimulant, in treating rheumatism, dropsy, edema, gout, asthma and as an anticancer agent (Leung 1980, Pascual-Villalobos 2002). Paste of *D. indica* bulbs is used in traditional medicine to cure ulcers (Tiwari and Yadav 2003), boils, tumours (Singh and Pandey 1982), pyrexia, as an antinematode (Choudary et al. 2008), as well as to relieve joint pains (Prusti and Behera 2007). Paste of the bulb of *D. indica* along with the leaf paste of *Achyranthes aspera* is used against scorpion bite (Kala 2009). *D. indica*, known as Kolakanda in Madhya Pradesh is used as an antitoxic and antihelminthic remedy (Khare 2004).

Ethnobotanical uses of other members of the family are also reported. Many species secure a prominent place among the top ten African medicinal plants in trade, of which *Bowiea volubilis*, *Eucomis autumnalis*, and *Scilla natalensis* deserve special mention. Pohl et al. (2000) in their medicinal plant survey of South Africa placed Hyacinthaceae at the second position (Aloaceae being the first) represented by 13 out of 198 taxa catalogued in their survey. Bioactive principles and therapeutic use of some members of Hyacinthaceae is listed in Table 3.

Hyacinthaceae members are consumed very rarely by human. The bulbs of *Muscari comosum* are eaten pickled in Greece and the inflorescences of *Ornithogalum pyrenaicum* are eaten as a vegetable in France (Pohl et al. 2000). Bushmen in Africa eat the bulbs of *Ledebouria apertiflora* and *L. revoluta* (Pohl et al. 2000).

Table 3. Bioactive principles and their therapeutic properties in Hyacinthaceae.

Plant source	Bioactive principle/ fraction	Medicinal Activity	Reference
Drimia maritima	Proscillaridin A, scilliroside, and scillirosidin	Anti insect	Pascual-Villalobos 2002
	2, 3-Butanediol	Cardiotonic	Kopp et al. 1996
	Proscillaridin A, Scillaren A	Immunoregulatory	Terness et al. 2001
Drimia indica	Glycoprotein, chitinase	Antifungal	Deepak et al. 2003, Shenoy et al. 2006
	Aqueous extract	Antibacterial	Chittoor et al. 2012
	Aqueous extract	Anthelmintic activity	Chittoor et al. 2012
	Salicylic acid	Relieve joint pains	Singh and Pandey 1982, Tiwari and Yadav 2003
Scilla indica	Diosgenin	Antiviral	Khan et al. 2012
Ledebouria graminifolia	3-Benzyl-4-chromanones	Cytotxic against cancer cell lines	Mutanyatta et al. 2003, Shushu et al. 2005

Bioactive principles

Phytochemical investigation of different members of the family Hyacinthaceae revealed the presence of a diverse range of chemical compounds which can be categorized into two major classes—cardiac/steroidal glycosides (cardenolides and bufadienolides) and homoisoflavones. Ornithogaleae, Urgineeae and Hyacintheae are characterized by the presence of cardenolides, bufadienolides and homoisoflavanones respectively which serve as chemotaxonomic markers for these groups (Pohl et al. 2000).

Steroidal glycosides. The cardiactive steroidal glycosides constitute an important group of natural drugs created by plants and amphibians as defence, with both toxic as well as advantageous actions. It has a history of medicinal use and plants producing cardiac glycosides have been used as poisons and heart drugs since as early as 1500 B.C. Plant extracts containing cardiac glycosides were used by Arab physicians to cure malignant diseases since the eighth century. However these compounds are extremely toxic, can even be fatal if therapeutic dose is not administered precisely. Use of cardiac glycosides is still well appreciated for treatment of congestive heart failure in spite of their extreme toxicity. Traditionally they have also been used as arrow poisons, abortifacients, emetics and diuretics. Recent findings have revealed selective antiproliferative activity of these compounds being involved in complex cell signal transduction mechanisms indicating their promising role as a chemotherapeutic drug or in novel combination therapy.

Structurally, cardiac glycosides are complex triterpene molecules consisting of two parts—the sugar moiety (glycoside) and the aglycone/genin (steroid nucleus substituted at C17 by a lactone ring) moiety. Based on the lactone group at C17 position there are two main classes of cardiac glycosides-cardenolides and bufadienolides with a butenolide (5 membered) and pentadienolide (6 membered) substitution respectively. The sugar moiety consists of one to four sugars in β conformation attached to 3β-OH group of the steroid nucleus and contributes to the water solubility of the compounds. The therapeutic action of cardiac glycosides depends on the structure of the distinct aglycones as well as on the type and number of sugar units attached at C3.

The presumed biosynthetic pathway of plant cardiac glycosides roughly constitutes three stages—Terpenoid backbone biosynthesis (Mevalonic acid and Methyl Erythritol Phosphate pathway), Phyto-steroid biosynthesis and Cardenolide/Bufadienolide biosynthesis.

Biological activity of cardiac steroids is mainly attributed by their ability to bind Na^+/K^+ ATPase pumps (Liu et al. 2000), binding capcity being largely specified by the lactone group of aglycone moiety (Keenan et al. 2005).

Bufadienolides. Bufadienolide derives its name from the toad genus *Bufo* being present in its venom glands. Hyacinthaceae and Iridaceae are the only known sources of bufadienolides among monocots. Preparations from bufadienolide containing plants and toads have been used in traditional medicine, while, on the other hand, bufadienolide containing plants cause a major problem of livestock toxicity in South Africa and Australia.

A variety of bufadienolides have been isolated from different members of Urgineeae, a brief account of which is given in Table 4.

Bufadienolides function as cardiotonic, blood pressure stimulant and also have antineoplastic activity. Bufadienolides derived from *D. maritima* and *D. indica* are cardioactive steroids with digoxin-like effects. Scillaren A and proscillaridin A are the major glycosides found in the bulbs of *D. maritima* and *D. indica* (Jha and Sen1981, Bayazit et al. 2010). *D. maritima*

Table 4. Bufadienolides isolated from Urgineeae.

Bufadienolide	Plant source	Reference
Physodine A, B, C, D	*D. physodes*	Van Heerden et al. 1988
Hellebrigenin	*D. altissima*	Shimada et al. 1979
Bovoruboside, Bovokryptoside, Bovogenin A, Bovoside A	*Bowiea volublis*	Tschesche et al. 1958
Scillirosidin	*D. sanguinea*	Krenn et al. 1993
Desacetylscillirosidin	*D. sanguinea*	Krenn et al. 1993
Scillaren A	*D. indica*	Jha and Sen 1981
Scilliphaeoside, Proscillaridin A	*D. indica*	Jha and Sen 1981
Urginin	*D. altissima*	Pohl et al. 2001
Rubellin	*D. calcarata* *D. macrocentra*	Steyn et al. 1986 Moodley et al. 2007
Riparianin	*U. riparia*	Moodley et al. 2007
Scillicyanosidin, Lydenburgenin	*U. lydenburgensis*	Crouch et al. 2006

is used in traditional medicine as an expectorant, this property being principally attributed by the bufadienolides scillaren A and proscillaridin A (Dewick 1997). Other bufadienolides present in this species like glucoscilliroside and scilliroside, have been used as a rodenticide (Dewick 1997). The toxic compound (scilliroside, a bufadienolide) is employed in poisoning rats (Pohl et al. 2000).

Cardenolides. Cardenolides are more common than bufadienolides in plants and are known to occur in Apocyanaceae and Scrophulariaceae apart from Hyacinthaceae. Within Hyacinthaceae cardenolides are characterically present in the tribe Ornithogaleae but have been recently identified in *U. fugax* (Krenn et al. 2004). Cardenolide also exhibit medicinal properties similar to bufadienolides, the most common of which are cardiotonic and cytotoxic activities (Kuroda et al. 2002). A few examples of cardenolides from the family Hyacinthaceae is given in Table 5.

Homoisoflavones. Homoisoflavonoids comprises a small class of phenolic compounds related to isoflavonoids whose occurence is so far largely restricted to Hyacinthaceae. Homoisoflavonoids were first reported from *Eucomis* in 1967 (Bohler and Tamm 1967) and eucomin and eucomol were the first one to be isolated and characterized. Since then homoisoflavonoids have been reported from many genera of Hyacinthaceae, the most predominant of which *Scilla* sp., *Eucomis* sp. and *Muscari* sp. in addition to others like *Bellevali* sp., *Ophiopogon* sp., *Veltheimia* sp., etc. The homoisoflavonoids accumulate in the waxy, scale-like layers of the Hyacinthaceae bulbs.

Structurally Homoisoflavones are very similar to isoflavones, both being oxygen heterocycles comprising the basic flavonoid ring system of a chromane (benzo-gamma-pyrone) ring bearing a second aromatic ring (benzyl or benzylidene) at position 3. However they differ from isoflavones in having 16 carbon atoms as opposed to 15 carbon atoms of the latter having an extra carbon bridge between the flavonoid ring systems.

Homoisoflavonoids can be subdivided into four basic structural types (Pohl et al. 2000), namely

a) 3-benzyl-4-chromanones (I),
b) 3-benzylidene-4-chromanones (II),
c) 3-benzyl-3-hydroxy-4-chromanones (III) and
d) scillascillins (IV).

A brief account of these homoisoflavones and their plant sources are presented in Table 6.

Flavonoids and homoisoflavones being phenolic compounds are biosynthesized via the shikimate pathway (starting from a carbohydrate) and polyketide pathway (starting from acetyl and malonyl coenzyme A). Homoisoflavones are of mixed origin, the A-ring is a polyketide pathway derivative while the B-ring is a shikimate pathway derivative. Chalcone formation is the first and the most important step of homoisoflavonoid biosynthesis, as for all other flavonoids and thus chalcone synthase is the key enzyme in the process. Conversion of chalcones to homoisoflavones has been described by Dewick (1997).

Table 5. Cardenolides from Hyacinthaceae.

Cardenolide	**Plant source**	**Reference**
Fugaxin	*Urginea fugax*	Krenn et al. 2004
Ornithogalin	*Ornithogalum magnum*	Komissarenko et al. 1971
Ornithosaponins	*Ornithogalum thyrsoides*	Kuroda et al. 2002

Table 6. Homoisoflavones from Hyacinthaceae.

Type	**Homoisoflavone**	**Plant sources**	**Reference**
I	3,9-dihydro-eucomin	*Eucomis bicolor*	Heller and Tamm 1981
	3,9-dihydropunctatin	*Eucomis comosa*	Finckh and Tamm 1970
II	eucomol	*Eucomis bicolor*	Bohler and Tamm 1967
	7-O-methyleucomol	*Eucomis bicolor*	Bohler and Tamm 1967
III	eucomin	*Eucomis bicolor*	Bohler and Tamm 1967
	punctatin	*Eucomis comosa*	Finckh and Tamm 1970
	autumnalin	*Scilla nervosa* subsp. *rigidifolia*	Silayo et al. 1999
IV	3′,5-dihydroxy-4′,7-dimethoxyspiro{2H-1-benzopyran-3(4H), 7′- bicyclo[4.2.0]octa[1,3,5]-trien)-4-one	*Drimiopsis maculata*	Koorbanally 2001

A large number of homoisoflavones have being reported from different members of Hyacinthaceae with ethnomedicinal importance. However investigations on the biological activities of homoisoflavones are still inadequate and the few reports that are available include anti-inflammatory (Tang et al. 2002, Du Toit et al. 2005), antibacterial (Du Toit et al. 2007), anticancer and angioprotective activities (Kim et al. 2008). Thus it is evident that homoisoflavones have great clinical potentials and requires further investigations on many aspects such as biological activities, mechanism of action and structure activity relationship which can open new door for synthetic drug design and development.

Family Iridaceae

The family Iridaceae named after the type genus *Iris* sp., is unique for its phytochemical diversity. It belongs to the monocot order Asparagales and is one of the important members of bulbous geophytes. This family is inimitable by its *isobilateral* and unifacial leaves, which differentiates them from typical plants having *dorsiventral* and bifacial leaves.

Almost worldwide in distribution and one of the most important families in horticulture, Iridaceae includes more than 2000 species. This family is native to the Antarctica and Australasia. According to the APG III classification (Chase et al. 2009), this family includes 66 genera, distributed among 7 subfamilies (Iridoideae, Isophysidoideae, Crocoideae, Nivenioideae, Patersonioideae, Geosiridoideae and Aristeoideae).

However, Iridaceae not only include true bulbs but also corms and rhizomes. *Eleutherine* sp. is a medicinally important genus and are true bulbs. *Gladiolus* sp. and *Crocus* sp. are corms while the genus *Iris* sp. includes rhizomes.

Bioactive principles

Members of Iridaceae are particularly rich in phenolic compounds, viz., anthocyanins, flavonols, flavones, isoflavones, biflavones, xanthones, naphthoquinones and proanthocyanidins along with saponins, fructans, non-protein amino acids, sterols, and bufadienolides.

Bulbs of the genus *Eleutherine* sp. have been widely used in traditional medicine against heart diseases, as immunostimulant, anti-inflammation, anti-tumor and, anti-bleeding agent (Babula et al. 2005, Xu et al. 2006). Chemical investigation of bulbs of *Eleutherine* sp. (*E. bulbosa* and *E. americana*) revealed the presence of naphtoquinones like elecanacine, eleutherine, eleutherol, eleutherinone, etc. (Hara et al. 1997, Alves et al. 2003, Nielsen and Wege 2006, Xu et al. 2006, Han et al. 2008) which are known to exhibit antimicrobial, antifungal, antiviral, and antiparasitic properties, as well as an anti-cancer agent andantioxidant (Babula et al. 2005, Xu et al. 2006). Medicinal uses of different members of Iridaceae are listed in Table 7.

Table 7. Medicinal uses of different members of Iridaceae.

Plant	Use	Reference
Eleutherine bulbosa	Antifungal	Alves et al. 2003, Tânia et al. 2003
	Anthelmintic	Schultes and Raffauf 1990
	Treat intestinal disorders	Lin et al. 2002
	Abortive and antifertility agent	Weniger et al. 1982
Eleutherine americana	Treat coronary disorders (angina pectoris), intestinal infection and as a diuretic and purgative	Anwar-Bruni et al. 1996, Johnson 1999
	Anti-HIV, antibacterial, and anticancererous	Hara et al. 1997, Preethi and Kenneth 2000
Gladiolus sp.	Treatment of dermal, gut, urogenital and respiratory infections	Nguedia et al. 2004
Crocus sativus (Saffron)	Stimulant, aphrodisiac and antidepressant	Rios et al. 1996, Abe and Saito 2000, Akhondzadeh et al. 2005
	Tumour-protective effects	Abdullaev 2005
	Antiviral, detoxicant, expectorant, hypotensive	Foster and Yue 1992

Family Liliaceae

The family Liliaceae constitutes a major section of bulbous geophytes having profuse medicinal properties. Delimitation of the family Liliaceae has been a subject of controversy in the history of the Monocots taxonomy (Dahlgren et al. 1985). Reassessment of various characters in recent years (Takhtajan 1997) has resulted in new concepts on delimitation of Liliaceae *sensu stricto*. According to Takhtajan (1997) Liliaceae *sensu stricto* includes very few genera, viz. *Erythronium* sp., and *Tulipa* sp. (subfamily Tulipeae); *Cardiocrinum* sp., *Lilium* sp., *Notholirion* sp., *Nomocharis* sp., *Fritillaria* sp. and *Rhinopetalum* sp. (subfamily Lilieae). However according to Tamura (1998), two subfamilies which constitute Liliaceae *sensu sticto* are- Medeoloideae and Lilioideae. According to the APG III classification (Chase et al. 2009), genera included in Liliaceae *sensu strico* are—*Amana* sp., *Calochortus* sp., *Cardiocrinum* sp., *Erythronium* sp., *Fritillaria* sp., *Gagea* sp., *Lilium* sp., *Lloydia* sp., *Medeola* sp., *Nectaroscordum* sp. and *Nomocharis* sp.

Such thorough revision of the family Liliaceae by plant taxonomists resulted in its reduction from one having 5000 species to a small group of 350 species (10 genera) while the rest were distributed to several smaller families These are predominantly bulbous plants with wide distribution from south-west Asia to China. Members from all 10 genera are cultivated as ornamentals. However, Lilies (*Lilium* sp.) and tulips (*Tulipa* sp.) are of exclusive commercial value because of their exceptional beauty and horticultural importance.

Members of Liliaceae have profuse medicinal importance and are preferred research materials of many pharmacologists because of their biologically active components, most important of which are isosteroidal alkaloids. Clinical investigation of Liliaceae was aggravated by their extensive therapeutic use in Chinese and other folk medicines in the treatment of various ailments.

Members of the genus *Lilium*, distributed mainly in the North Temperate Zone (Haw 1986) have been used as food and medicine in China and Japan since ancient times (Jiang 1977). Medicinal uses of *Lilium longiflorum*, *L. brownii* var. *colchesteri* and *L. tenuifolium* have been listed in Chinese herbal reports (Jiang 1977). *L. longiflorum* and *L. superbum* (Turk's cap lily) were also reported as irritant (Weber 1937). Ethnobotanical uses of some *Lilium* species are listed in Table 8.

The genus *Fritillaria* has been used in folk medicine of China, Japan, Turkey, Pakistan and South East Asia. In Traditional Chinese medicine (TCM), the *Fritillaria* bulbs are called "*Bei mu*" and have been used as antitussives, antiasthmatics and expectorants since ancient times and nine species, viz. *F. thunbergii*, *F. cirrhosa*, *F. unibracteata*, *F. przewalskii*, *F. delavayi*, *F. ussuriensis*, *F. walujewii*, *F. pallidiflora* and *F. hupehensis*, were included in the 2005 edition of China Pharmacopoeia.

Table 8. Ethnobotanical uses of *Lilium*.

Plant	Part used	Purpose	Reference
L. candidum (white lily, madonna lily)	Fresh and dried bulbs, flowers	Gynecological disorders, ulcer, burns, injuries and diuretic, analgesic, antiinflammatory	Gruenwald 2000
L. lancifolium (Tiger Lily)	Bulbs	Carminative, diuretic, emollient, expectorant	Steinmetz 1957
L. martagon (Martagon)	Bulbs	Carminative, diuretic, emollient, expectorant	Steinmetz 1957

Bioactive principles

Different types of steroidal alkaloids obtained from liliaceous plants are major bioactive compounds in clinical and pharmaceutical research.

Isosteroidal alkaloids constitute the main bioactive principle isolated from the genus *Fritillaria* (Li et al. 1999, 2000), types and amount being species specific, quality and quantity of which are very critical for safe clinical use (Ding et al. 1996, Li et al. 1999, 2000). Major alkaloids from *Fritillaria* and their biological activities are listed in Table 9.

The bulbs of many species of *Lilium* sp. exhibit a wide range of biological actions. Chemical investigation of the *Lilium* sp. bulbs revealed the presence of a wide range of steroidal saponins and/or phenolic glycosides, which probably contribute for their pharmacological effects (Mimaki 1993).

Tulipa sp. cultivated in many countries as ornamentals are known to cause allergic reactions, mainly in workers of horticultural farms, the irritant effect is described as a classical syndrome, the "tulip fingers" (Rook 1960). The dust in tulip sheds is also reported to cause conjunctivitis, rhinitis and asthma and reaginic responses in sensitive persons (Van der Werff 1959). The allergenic principle was isolated and identified as α-methylene-γ-butyrolactone by Brongersma-Oosterhoff (1967).

Methanolic and ethanolic extracts from *Gagea fibrosa* bulb and leaves exhibited varying degrees of total phenolic content and antioxidant activity and thus can be further exploited (Mammadov et al. 2011).

Table 9. Steroidal alkaloid from *Fritillaria* and their bioactivity.

Plant	Bioactive principle	Activity	Reference
Fritillaria ussuriensis	Steroidal alkaloids (verticinone, verticine, and peimisine)	Antihypertensive	Oh et al. 2003
Fritillaria delavayi	Chuanbeinone	Vasorelaxant	Kaneko et al. 1986
Fritillaria imperialis	Ebeinone	Anticholinergic	Rahman et al. 1994
	Forticine, delavine, persicanidine and imperialine	Anticholinergic	Rahman et al. 2002
Fritillaria puqiensis	*N*-dementhylpuqietinone	Antitumourous	Jiang et al. 1977

Biotechnological Aspects of Medicinal Bulbous Plants

From the pre-historic era human beings depend completely on natural phytomedicine to prevent diseases. A large group of these phyto healers constitutes bulbous geophytes leading to their intensive exploitation from the wild causing loss of genetic diversity and habitat destruction. Therefore, all these bulbous plants need to be conserved as their environments are being ruined. Although increased regulation and the introduction of sustainable wild harvesting methods can protect some plant species, a more practicable long-term alternative is required. *In situ* and *ex situ* conservation of such plant species can be increased by means of advanced biotechnological tools.

The growing commercial importance of the phyto chemicals, mostly secondary metabolites, necessitates their augmented production either by increasing the rate of propagation or by manipulating the synthesis of bioactive principles. Micropropagation can be employed to increase the multiplication rate as well as the *ex situ* conservation of these valuable plants. Biosynthesis of target metabolites can be altered using tissue culture and genetic transformation. These aspects can be harnessed for the synthesis of secondary metabolites using plants as bioreactors.

In vitro culture, micropropagation, elicitation of secondary metabolite and genetic improvements, have all attributed to the commercial exploitation of bulbous medicinal plants and will continue to do so in the future.

Micropropagation

Micropropagation holds remarkable potential for the rapid production of clonal plant materials. Bulbous plants belonging to the Liliaceae, Hyacinthaceae, Iridaceae and Amaryllidaceae are mostly sexually sterile and natural propagation by vegetative means is relatively slow. Tissue culture technology significantly influenced the demand of rapid multiplication of slow-growing bulbous monocots. Bulbs are commonly used as source of explants for the establishment of *in vitro* cultures. Being underground storage organs, bulbs are very prone to pathogens which are passed on to the next generation during conventional means of propagation. Stage-I contamination is a major problem in case of micropropagation of bulbous plant which becomes more evident while dealing with limited stocks. However, aerial parts like inflorescence stalks, leaves and peduncle can be a functional alternative source for overcoming this problem. In addition to this, the use of aerial parts as explants prevents the destruction of the stock plants. The study of the morphogenic potential of different types of explants is an important aspect for micropropagation of bulbous medicinal plants. Regenerated buds or embryos have been induced to form storage organs directly by the manipulation of growth regulators and environmental factors in culture.

Micropropagation systems based on clonal and non-clonal means of regeneration like axillary bud development, organogenesis and somatic embryogenesis have been reported for many bulbous medicinal plants, as listed in Table 10.

Table 10. Examples on micropropagation of different medicinally important bulbous plants.

Family	Plant	Explant/ Method	References
Amaryllid-aceae	*Narcissus* spp.	Bulb scales	Santos et al. 1998
		Callus culture	Sell'es et al. 1999, Sage et al. 2000
		Somatic embryogenesis (*N. pseudonarcissus*)	Sage et al. 2000
		Callus induction, somatic embryogenesis and organogenesis (*N. confuses*)	Sell'es et al. 1999
		Single leaf explants/Shoot clump culture	Staikidou 1994
	Amaryllis sp.	Bulb explants/*in vitro* regeneration	Smith et al. 1999
	Hippeastrum sp.	Bulb explants/*in vitro* regeneration	Smith et al. 1999, Sultana et al. 2010
	Zephyranthes sp.	Seed germination, Twin scale explants	Smith et al. 1999
	Crinum variabile	Twin scale explants/Shoot proliferation	Van Staden et al. 2001
Alliaceae	*Allium wallichii*	Shoot culture	Wawrosch et al. 2001
	A. sativum	Meristem culture	Abo El-Nil 1977, Novak 1981
		Leaf explants/Somatic embryogenesis	Fereol et al. 2002
		Primordial leaf explants/ Somatic embryogenesis	Dolezel and Novák 1985
		Bulb explants/suspension culture	Mukhopadhyay et al. 2005
	A. cepa	Bulb explants/suspension culture	Mukhopadhyay et al. 2005
	A. ampeloprasum L. (elephant garlic)	Mature zygotic embryo/ organogenesis and somatic embryogenesis	Silvertand et al. 1996
	Tulbaghia sp.	Leaf, peduncle and bulb scale explants/Shoot multiplication	Zschocke and Van Staden 2000
Liliaceae	*Lilium* sp.	Bulb scales	Stimart and Ascher 1978
		Scale leaves, stems and petals	Niimi 1984
		Thin cell layer explants from receptacle/leaf/stem in *L. longiflorum*	Nhut et al. 1998

Table 10. contd....

Table 10. contd.

Family	Plant	Explant/ Method	References
Liliaceae	*L. longiflorum*	Bulb scales/callus cultures	Priyadarshi and Sen 1992
		Anther with filaments	Arzate-Fernandez et al. 1997
		Stem nodes	Nhut 1998
	Tulipa gesneriana	Axillary buds	Riviere and Muller 1976
		Scale explants	Nishiuchi 1980
		Flower stalk explants/ Regeneration and Shoot multiplication	Podwyszynska and Marasek 2003
Hyacintha-ceae	*Albuca* sp.	Adventitious bulblet segment/ Shoot multiplication (*Albucabracteata*) Peduncle explants/Shoot multiplication (*Albucanelsonii*)	Ascough and Staden 2010
	Ledebouria graminifolia	Twin scale and scale leaf explants/ morphogenesis	Shushu et al. 2005
	Bowiea volubilis	Inflorescence segments/ Shake culture system Bulb explants/*in vitro* regeneration	Jha and Sen 1985 Cook et al. 1988
		Inflorence Axis/*in vitro* regeneration	Hannweg et al. 1996
	Eucomis autumnalis	Twin scale explants/Shoot proliferation	Ault 1995
	E. comosa	Twin scale explants/Shoot proliferation	Ault 1995
	E. zambesiaca	Twin scale explants/Shoot proliferation	Ault 1995
	U. maritima	Bulb-scales/morphogenesis	Ramadan and Backhaus 1987
	U. indica	Bulb explants/indirect morphogenesis Bulb and inflorescence segments/indirect morphogenesis	Jha et al. 1984

Enhancement of active principles

Production of bioactive principles can be enhanced in *in vitro* cultures using various biotechnological strategies, viz. somaclonal variation, screening of high yielding cell line, media modification, precursor feeding, elicitation, large scale cultivation in bioreactor system, hairy root culture, plant cell immobilization, biotransformation, etc. A few examples are listed in Table 11.

Bulbous geophytes are the source of a wide range of biologically active compounds and have been extensively used in ethnomedicine. Blooming trade in medicinal bulbous plants is having a negative impact on the natural populations of these valuable resources and is threatening the survival many of these species. Several biotechnological aspects can be incorporated to supplement the high demand and reduce the burden on natural population. Further research in this field and judicious exploitation is necessary to conserve these precious plants.

Table 11. Biotechnological approaches for production of biologically active principles in medicinally important bulbous plants.

Plant	Active principles	Methods	References
Narcissus	Galanthamine	Liquid-shake culture of shoot-clumps using different elicitors like methyl jasmonate/ arachidonic acid/ chitosan/ salicylic acid as elicitors	Colque et al. 2004
Leucojum aestivum	Galanthamine and related alkaloids	Shoot culture using a temporary immersion technology	Ivanov et al. 2011
	Galanthamine	Shoot culture in a modified bubble column bioreactor	Georgiev et al. 2012
	Galanthamine	Shoots grown in different bioreactor systems	Schumann et al. 2012

Acknowledgement

SN is grateful to the Department of Science and Technology (GOI) for the award of DST INSPIRE Fellowship and PSS is grateful to University Grant Commission for the award of RFSMS Fellowship.

References

Abdullaev, F.I. 2005. Cancer chemopreventive and tumoricidal properties of saffron (*Crocus sativus* L.). Exp. Biol. Med. 227: 20–25.

Abe, K. and H. Saito. 2000. Effects of saffron extract and its constituent crocin on learning behaviour and long-term potentiation. Phytother. Res. 14: 149–152.

Abo El-Nil, M.M. 1977. Organogenesis and embryogenesis in callus cultures of garlic (*Allium sativum* L.). Plant Sci. Lett. 9: 259–264.

Akhondzadeh, S., N. Tahmacebi-Pour, A.A. Noorbala, H. Amini, H. Fallah-Pour, A.H. Jamshidi and M. Khani. 2005. *Crocus sativus* L. in the treatment of Mild to Moderate depression: a double-blind, randomized and placebo-controlled trial. Phytother. Res. 19: 148–151.

Alves, T.M.A., K. Helmut and L.Z. Carlos. 2003. Eleutherinone, a novel fungitoxic naphtoquinone from *Eleutherine bulbosa* (Iridaceae). Mem. Inst. Oswaldo Cruz. 98: 709–712.

Amagase, H. 2006. Clarifying the real bioactive constituents of garlic. J. Nutr. 136: 7165–7255.

Anwar-Bruni, D.M., S.E. Hogan, D.A. Schwartz, C.M. Wilcox, R.T. Bryan and J.L. Lennox. 1996. Atovaquone is effective treatment for the symptoms of gastrointestinal microsporidiosis in HIV-1-infected patients. AIDS 10: 619–623.

Arzate-Fernandez, A.M., T. Nakazaki, Y. Okumoto and T. Tanizaka. 1997. Efficient callus induction and plant regeneration from filaments with anther in lily (*Lilium longiflorum* Thunb.). Plant Cell Rep. 16: 836–840.

Ascough, G.D. and J. Van Staden. 2010. Micropropagation of *Albuca bracteata* and *A. nelsonii* —Indigenous ornamentals with medicinal value. S. Afr. J. Bot. 76: 579–584.

Ault, J.R. 1995. *In vitro* Propagation of *Eucomis autumnalis*, *E. comosa*, and *E. zambesiaca* by Twin-scaling. Hort. Science 30: 1441–1442.

Babula, P., R. Mikelová, D. Potěšil, V. Adam, R. Kizek, L. Havel and Z. Sladky. 2005. Simultaneous determination of 1,4- naphtoquinone, lawsone, juglone and plumbagin by liquid chromatography with UV detection. Biomed. Paper 149: 25–28.

Balzarini, J., D. Schols, J. Neyts, E. Van Damme, W. Peumans and E. De Clercq. 1991. α-(1–3)- and α-(1–6)-D-mannose-specific plant lectins are markedly inhibitory to Human Immunodeficiency Virus and cytomegalovirus infections *in vitro*. Antimicrob. Agents Chemother. 35: 410–416.

Banerjee, S.K., P.K. Mukherjee and S.K. Maulik. 2003. Garlic as an antioxidant: the good, the bad and the ugly. Phytother. Res. 17: 97–106.

Barton, D.H.R, G.W. Kirby, J.B. Taylor and G.M. Thomas. 1963. Phenol oxidation and biosynthesis. Part VI. The biogenesis of Amaryllidaceae alkaloids. J. Chem. Soc. 4545–4558.

Bastida, J., S. Berkov, L. Torras, N.B. de Andrade, V. Martínez, C. Codina and F. Viladomat. 2011. Chemical and biological aspects of Amaryllidaceae alkaloids. *In*: D.M. Torrero (ed.). Recent Advances in Pharmaceutical Sciences. Transworld Research Network. Kerala, India pp. 65–100.

Bayazit, V. and V. Konara. 2010. Analgesic effects of scilliroside, proscillaridin—A and taxifolin from squill bulb (*Urginea maritima*) on pains. Dig. J. Nanomater. Bios. 5: 457–465.

Block, E. 1992. The organosulfur chemistry of the genus *Allium*-implications for the organic chemistry of sulfur. Angew. Chem. Int. Ed. Engl. 31: 1135 1178.

Blumenthal, M. 1998. The complete German Commission E. Monographs: Therapeutic guide to herbal medicines. American Botanical Council, Austin, Texas.

Blumenthal, M., A. Goldberg and J. Brinckmann. 2000. Herbal Medicine—Expanded Commission E Monographs. Integrative Medicine Communications, Newton, USA.

Böhler, P. and C. Tamm. 1967. The homo-isoflavones, a new class of natural product. Isolation and structure of eucomin and eucomol. Tetrahedron Lett. 36: 3479–3483.

Brongersma-Oosterhoff, U.W. 1967. Structure determination of the allergenic agent isolated from tulip bulbs. Recl. Trav. Chim. Pays Bas 86: 705–708.

Ceriotti, G. 1967. Narciclasine: An antimitotic substance from *Narcissus* bulbs. Nature 213: 595–596.

Chao, R.B. and L. Hu. 1993. The quantitative determination of peimissine in *Fritillaria* hulbs by HPLC. Acta Pharm. Sin. 28: 705–708.

Chase, M.W., J.L. Reveal and M.F. Fay. 2009. A subfamilial classification for the expanded asparagalean families Amaryllidaceae, Asparagaceae and Xanthorrhoeaceae. Bot. J. Linn. Soc. 161: 132–136.

Chen, C., D. Pung, V. Leong, V. Hebbar, G. Shen, S. Nair, W Li and A.N.T. Kong. 2004. Induction of detoxifying enzymes by garlic organosulfur compounds through transcription factor Nrf2: effect of chemical structure and stress signals. Free Radic. Biol. Med. 37: 1578–1590.

Chittoor, M.S., A.J.R. Binny, S.K. Yadlapalli, A. Cheruku, C. Dandu and Y. Nimmanapalli. 2012. Anthelmintic and antimicrobial studies of *Drimia indica* (Roxb.) Jessop. bulb aqueous extracts. J. Pharm. Res. 5: 3677–3686.

Choudary, K., M. Singh and U. Pillai. 2008. Ethnobotanical survey of Rajasthan—an update. Am. -Eurasian J. Bot. 1: 38–45.

Colque, R., F. Viladomat, J. Bastida and C. Codina. 2004. Improved production of galanthamine and related alkaloides by methyl jasmonate in *Narcissus confuses* shoot-clumps. Planta Med. 70: 1180–1188.

Compos, K.E., Y.S. Diniz, A.C. Cataneo, L.A. Faine, M.J.Q.F Alves and E.L. Nobeli. 2003. Hypoglycaemic and antioxidant effect of onion, *Allium cepa*: diatary onion addition, antioxidant activity and hypoglycaemic effect on diabetic rats. Int. J. Food Sci. Nutr. 54: 241–246.

Cook, E.L., A. Cunningham and J. Van Staden. 1988. The tissue culture of an exploited medicinal plant *Bowiea volubilis*. S. Afr. J. Bot. 54: 509–510.

Cook, J.W. and J.D. Loudon. 1952. Alkaloids of the Amaryllidaceae. *In:* R.H.F. Manske and H.L. Holmes (eds.). The Alkaloids: Chemistry and Physiology. Academic Press Inc., New York, pp. 331–352.

Crouch, N.R., V. Bangani and D.A. Mulholland. 1999. Homoisoflavanones from three South African Scilla species. Phytochemistry 51: 943–946.

Crouch, N.R., K. du Toit, D.A. Mulholland and S.E. Drewes. 2006. Bufadienolides from bulbs of *Urginea lydenburgensis* (Hyacinthaceae: Urgineoideae). Phytochemistry 67: 2140–2145.

Dahlgren, R.M.T., H.T. Clifford and P.F. Yeo. 1985. The families of the monocotyledons: structure, evolution and taxonomy. Springer-Verlag, Berlin.

Deepak, A.V., G. Thippeswamy, M.N. Shivakameshwari and B.P. Salimath. 2003. Isolation and characterization of a 29 kDa glycoprotein with antifungal activity from bulbs of *Urginea indica*. Biochem. Biophys. Res. Commun. 311: 735–742.

Dewick, P.M. 1997. Medicinal Natural Products: A Biosynthetic Approach. John Wiley and Sons, Chichester, UK.

Ding, K., G. Lin, Y.P. Ho, T.Y. Chen and P. Li. 1996. Prederivatization and high-performance liquid chromatographic analysis of alkaloids of bulbs of *Fritillaria*. Journal of Pharmaceutical Sciences. J. Pharm. Sci. 85: 1174–1179.

Doležel, J. and F.J. Novák. 1985. Karyological and cytophotometric study of callus induction in *Allium sativum* L. J. Plant Physiol. 118: 421–429.

Du Toit, K., E. Elgoraschi, S. Malan, S. Drewes, J. Van Staden, N. Crouch and D. Mulholland. 2005. Anti-inflammatory activity and QSAR studies of compounds isolated from Hyacinthaceae species and *Tachiadernus longiflorus* Griseb (Gentianaceae). Bio. org. Med. Chem. 13: 2561–2568.

Du Toit, K., E.E. Elgorashi, S.F. Malan, D.A. Mulholland, S.E. Drewes and J. Van Staden. 2007. Antibacterial activity and QSAR of homoisoflavanones isolated from six Hyacinthaceae species. S. Afr. J. Bot. 73: 236–241.

Evidente, A., A. Andolfi, A.H. Abou-Donia, S.M. Touema, H.M. Hammoda, E. Shawky and A. Motta. 2004. (–)-Amarbellisine, a lycorine-type alkaloid from *Amaryllis belladonna* L. growing in Egypt. Phytochemistry 65: 2113–2118.

Fay, M.F. and M.W. Chase. 1996. Resurrection of Themidaceae for the *Brodiaea* alliance, and recircumscription of Alliaceae, Amaryllidaceae and Agapanthoideae. Taxon 45: 441–451.

Fennell, C.W. and J. van Staden. 2001. *Crinum* species in traditional and modern medicine. J. Ethnopharmacol. 78: 15–26.

Fereol, L., V. Chovelon, S. Causse, N. Michaux-Ferriere and R. Kahane. 2002. Evidence of a somatic embryogenesis process for plant regeneration in garlic (*Allium sativum* L.). Plant Cell Rep. 21: 197–203.

Finckh, R.E. and C. Tamm. 1970. The homo-isoflavones III. Isolation and structure of punctatin, 3,9-dihydro-punctatin,4′-O-methyl-3,9-dihydro-punctatin, 4′-demethyl-eucomin and 4′-demethyl-5-O-methyl-3,9-dihydro-eucomin. Experientia 26: 472–473.

Fleischauer, A.T. and L. Arab. 2001. Garlic and cancer: a critical review of the epidemiologic literature. J. Nutr. 131: 1032S–1040S.

Fleming, T. 1998. PDR for Herbal Medicine. Medical Economics Co., Montvale, NJ.

Foster, S. and C. Yue. 1992. Herbal Emissaries: Bringing Chinese Herbs to the West: A Guide to Gardening, Herbal Wisdom, and Well-being. Healing Arts Press, Rochester, Vermont.

Fulton, B. and P. Benfield. 1996. Galanthamine. Drugs Aging 9: 60–65.

Gabrielsen, B., T.P. Monath, J.W. Huggins, D.F. Kefauver, G.R. Pettit, G. Groszek, M. Hollingshead, J.J Kirsi, W.M. Shannon, E.M. Schubert, J. Dare, B. Ugarkar, M.A. Ussery and M.J. Phelan. 1992. Antiviral (RNA) activity of selected Amaryllidaceae isoquinoline constituents and synthesis of related substances. J. Nat. Prod. 55: 1569–1581.

Galal, E.E. and M.A. Gawad. 1965. Antidiabetic activity of Egyptian onion "*Allium cepa*" extract. J. Egypt Med. Assoc. 48: 14–45.

Gao, C.M., T. Takezaki, J.H. Ding, M.S. Li and K. Tajima. 1999. Protective effect of *Allium* vegetables against both esophageal and stomach cancer: a simultaneous case-referent study of a high-epidemic area in Jiangsu province. China. Jpn. J. Cancer Res. 90: 614–621.

Georgiev, V., I. Ivanov, S. Berkov, M. Ilieva, M. Georgiev, T. Gocheva and A. Pavlov. 2012. Galanthamine production by *Leucojum aestivum* L. shoot culture in a modified bubble column bioreactor with internal sections. Eng. Life Sci. 12: 534–543.

Gruenwald, J., T. Brendler and C. Jaenicke. 2000. PDR for herbal medicines. Medical Economics Co. Montvale, NJ.

Guillou, C., J.L. Beunard, E. Gras and C. Thal. 2001. An efficient total synthesis of (±)-galanthamine. Angew. Chem. Int. Ed. Engl. 113: 4881–4882.

Han, A.R., H.Y. Min, J.W. Nam, N.Y. Lee, A. Wiryawan, W. Suprapto, S.K. Lee, K.R. Lee and E.K. Seo. 2008. Identification of a new naphthalene and its derivatives from the bulb of *Eleutherine americana* with inhibitory activity on lipopolysaccharide- induced nitric oxide production. Chem. Pharm. Bull. 56: 1314–1316.

Hannweg, K., M.P. Watt and P. Berjak. 1996. A simple method for the micropropagation of *Bowiea volubilis* from inflorescence explants. Bot. Bull. Acad. Sin. 37: 213–218.

Hara, H., N. Maruyama, S. Yamashita, Y. Hayashi, K.H. Lee, K.F. Bastow, Chairul, R. Marumoto and Y. Imakura. 1997. Elecanacin, a novel new naphthoquinone from the bulb of *Eleutherine americana*. Chem. Pharm. Bull. 45: 1714–1716.

Hartwell, J.L. 1967. Plants Used Against Cancer: A Survey. Quarterman Publications, Lawrence, Massachusetts, USA.

Haw, S.G. and S.Y. Liang. 1986. The Lilies of China: The genera *Lilium*, *Cardiocrinum*, *Nomocharis* and *Notholirion*. Timber Press, Portland, USA.

Heinrich, M. and H.L. Teoh. 2004. Galanthamine from snowdrop—the development of a modern drug against Alzheimer's disease from local Caucasian knowledge. J. Ethnopharmacol. 92: 147–162.

Heller, W. and C. Tamm. 1981. Homoisoflavanones and biogenetically related compounds. Fortschr. Chem. Org. Naturst. 40: 105–152.

Houghton, P.J., Y. Ren and M.J. Howes. 2006. Acetylcholinesterase inhibitors from plants and fungi. Nat. Prod. Rep. 23: 181–199.

Hua, D.H., S. Saha and D.J. Takemoto. 1997. Anticancer activities of 2,5,8,9-substituted 6-oxo-1,2,3,4,5,6-hexahydrophenanthridines on multi-drug-resistant phenotype cells. Anticancer Res. 17: 2435–2441.

Hutchings, A. and A.H. Scott. 1996. Zulu Medicinal Plants: An Inventory. University of Natal Press, Pietenmaritzburg, South Africa.

Ieven, M., A.J. Vlietinck, D.A. van den Berghe, J. Totte, R. Dommisse, E. Esmans and F. Alderweireldt. 1982. Plant antiviral agents. III. Isolation of alkaloids from *Clivia miniata* Regel (Amaryllidaceae). J. Nat. Prod. 45: 564–573.

Ieven M., D.A. van den Berghe and A.J. Vlietinck. 1983. Plant antiviral agents. Planta Med. 49: 109–114.

Ivanov, I., V. Georgiev, M. Georgiev, M. Ilieva and A. Pavlov. 2011. Galanthamine and related alkaloids production by *Leucojum aestivum* L. shoot culture using a temporary immersion technology. Appl. Biochem. Biotech. 163: 268–277.

Jha, S., G.C. Mitra and S. Sen. 1984. *In vitro* regeneration from bulb explants of Indian squill, *Urginea indica* Kunth. Plant Cell Tissue Organ Cult. 3: 91–100.

Jha, S., N.P. Sahu and S.B. Mahato. 1991. Callus induction, organogenesis and somatic embryogenesis in three chromosomal races of *Urginea indica* and production of bufadienolides. Plant Cell Tissue Organ Cult. 25: 85–90.

Jha, S. and S. Sen. 1981. Bufadienolides in different chromosomal races of Indian squill. Phytochemistry 20: 524–526.

Jha, S. and S. Sen. 1983. Quantitation of principal Bufadienolides in different cytotypes of *Urginea indica*. Planta Med. 47: 43–45.

Jha, S. and S. Sen. 1985. Regeneration and rapid multiplication of *Bowiea volubilis* Harv. in tissue culture. Plant Cell Rep. 4: 12–14.

Jiang, S. 1977. Dictionary of Traditional Chinese Crude Drugs. Shanghai Scientific Technologic, Shanghai.

Jiang, Y., H. Li, P. Li, Z. Cai and W. Ye. 2005. Steroidal alkaloids from the bulbs of *Fritillaria puqiensis*. J. Nat. Prod. 68: 264–267.

Jimenez, A., A. Santos, G. Alonso and D. Vazquez. 1976. Inhibitors of protein synthesis in eukarytic cells. Comparative effects of some Amaryllidaceae alkaloids. Biochim. Biophys. Acta. 425: 342–348.

Jimenez, A., L. Sanchez and D.Vazquez. 1975. Yeast ribosomal sensitivity and resistance to the Amaryllidaceae alkaloids. FEBS Lett. 60: 66–70.

Jin, Z. 2005. Amaryllidaceae and *Sceletium* alkaloids. Nat. Prod. Rep. 22: 111–126.

Johnson, T. 1999. CRC ethnobotany desk reference. CRC Press, Boca Raton, Florida.

Jones, M.G., J. Hughes, A. Tregova, J. Milne, A.B. Tomsett and H.A. Collin. 2004. Biosynthesis of the flavour precursors of onion and garlic. J. Exp. Bot. 55: 1903–1918.

Kaku, H., E.J.M. Van Damme, W.J. Peumans and I.J. Goldstein. 1990. Carbohydrate-binding specificity of the Daffodil (*Narcissus pseudonarcissus*) and Amaryllis (*Hippeastrum hybr.*) bulb lectins. Arch. Biochem. Biophys. 279: 298–304.

Kala, C.P. 2009. Aboriginal uses and management of ethnobotanical species in deciduous forests of Chhattisgarh state in India. J. Ethnobiol. Ethnomed. 5: 1–9.

Kamara, B.I., D.T.L. Manong and E.V. Brandt. 2005. Isolation and synthesis of a dimeric dihydrochalcone from *Agapanthus africanus*. Phytochemistry 66: 1126–1132.

Kaneko, K., T. Katsuhara and H. Mitsuhashi. 1986. Chuanbeinone, a novel D/E cis-(22R, 26S)-5- α-cevanine alkaloid from chinese herbal drug, chuan-bei-mu. Tetrahedron Lett. 27: 2387–2390.

Keenan, S.M., R.K. DeLisle, W.J. Welsh, S. Paula and W.J. Ball Jr. 2005. Elucidation of the Na^+, K+-ATPase digitalis binding site. J. Mol. Graph. Model. 23: 465–475.

Khan, M.M.A.A., S.N.H. Zaidi, S.A. Musanna and N. Singh. 2012. Identification of a New Semliki Forest Virus Inhibitory Diosgenin from *Scilla Indica* Baker Plant. Asian. J. Exp. Biol. Sci. 3: 476–480.

Khare, C.P. 2004. Indian Herbal Remedies: Rational Western Therapy. Ayurvedic and other Traditional usage, Botany. Springer Verlag, Berlin Heidelberg, New York.

Kim, J.H., J.H. Kim, Y.S. Yu, H.O. Jun, H.J. Kwon, K.H. Park and K.W. Kim. 2008. Inhibition of choroidal neovascularization by homoisoflavanone, a new angiogenesis inhibitor. Mol. Vis. 14: 556–561.

Komissarenko, N.F. 1971. Ornithogalin—a cardenolide glycoside from *Ornithogalum magnum*. Chem. Nat. Compd. 7: 33–35.

Koorbanally, C., N.R. Crouch and D.A. Mulholland. 2001. Scillascillin-type homoisoflavanones from *Drimiopsis maculata* (Hyacinthaceae). Biochem. Syst. Ecol. 29: 539–541.

Kopp, B., L. Krenn, M. Draxler, A. Hoyer, R. Terkola, P. Vallaster and W. Robien. 1996. Bufadienolides from *Urginea maritima* from Egypt. Phytochemistry 42: 513–522.

Kosuge, Y., Y. Koen, K. Ishige, K. Minami, H. Urasawa, H. Saito and Y. Ito. 2003. S-allyl-L-cysteine selectively protects cultured rat hippocampal neurons from amyloid b-protein- and tunicamycininduced neuronal death. Neuroscience 122: 885–895.

Kreig, M.B. 1966. Green Medicine. The search for plant that heal. Chicago, New York.

Krenn, L., B. Kopp, M. Bamberger, E. Brustmann and W. Kubelka. 1993. Bufadienolides and a steroidal sapogenin from *Urginea sanguinea* (Hyacinthaceae). Nat. Prod. Lett. 3: 139–143.

Krenn, L., H. Antje, K. Anna and S. Franz. 2004. Chemotaxonomic relevance of cardenolides in *Urginea fugax*. Phytochemistry 65: 2881–2884.

Kuroda, M., Y. Mimaki, A. Yokosuka, F. Hasegawa and Y. Sashida. 2002. Cholestane glycosides from the bulbs of *Ornithogalum thyrsoides* and their cytotoxic activity against HL-60 leukemia cells. J. Nat. Prod. 65: 1417–1423.

Lata, S., K.K. Saxena, V. Bhasin, R.S. Saxena, A. Kumar and V.K. Shrivastva. 1991. Beneficial effect of *Allium sativum*, *Allium cepa* and *Comiphora mukul* on experimental hyperlipidemia and arthrosclerosis—a comparive evaluation. J. Postgrad. Med. 37: 132–135.

Leung, A.Y. 1980. Encyclopedia of common natural ingredients used in Food, Drugs and Cosmetics. Wiley, New York.

Li, S.L., P. Li, G. Lin, S.W. Chan and Y.P. Ho. 2000. Simultaneous determination of seven major isosteroidal alkaloids in bulb of *Fritillaria* by gas chromatography J. Chromatogr. A. 873: 221–228.

Li, S.L., S.W. Chan, P. Li, G. Lin, G.H. Zhou, Y.J. Ren and F.C.K. Chiu. 1999. Pre-column derivatization and gas chromatographic determination of alkaloids in bulbs of *Fritillaria*. J. Chromatogr. A. 859: 183–192.

Likhitwitayawuid, K., C.K. Angerhofer, H. Chai, J.M. Pezzuto, G.A. Cordell and N. Ruangrungsi. 1993. Cytotoxic and antimalarial alkaloids from the bulbs of *Crinum amabile*. J. Nat. Prod. 56: 1331–1338.

Lin, J., T. Puckree and T.P. Mvelase. 2002. Anti-diarrhoeal evaluation of some medicinal plants used by Zulu traditional healers. J. Ethnopahrmacol. 79: 53–56.

Liu, J., J. Tian, M. Haas, J.I. Shapiro, A. Askari and Z. Xie. 2000. Ouabain interaction with cardiac Na+/K+-ATPase initiates signal cascades independent of changes in intracellular Na+ and Ca2+ concentrations. J. Biol. Chem. 275: 27838–27844.

Mammadov, R., P. Ili, V.H. Ertem and M.A. Afacan. 2011. Antioxidant activity and total phenolic content of *Gagea fibrosa* and *Romulea ramiflora*. Iran. J. Chem. Chem. Eng. 30: 57–62.

Marco-Contelles, J., R. Leon, C. de Los Rios, A. Guglietta, J. Terencio, M.G. Lopez, A.G. Garcia and M. Villarroya. 2006. Novel multipotent tacrine-dihydropyridine hybrids with improved acetylcholinesterase inhibitory and neuroprotective activities as potential drugs for the treatment of Alzheimer's disease. J. Med. Chem. 49: 7607–7610.

Martin, S.F. 1987. The Amaryllidaceae alklaoids. *In*: A. Brossi (ed.). The Alkaloids. Academic Press, New York, pp. 251–376.

McLachlan, A., N. Kekre, J. McNulty and S. Pandey. 2005. Pancratistatin: a natural anti-cancer compound that targets mitochondria specifically in cancer cells to induce apoptosis. Apoptosis 10: 619–630.

McNulty, J., J.J. Nair, C. Codina, J. Bastida, S. Pandey, J. Gerasimoff and C. Griffin. 2007. Selective apoptosis-inducing activity of crinum-type Amaryllidaceae alkaloids, Phytochemistry 68: 1068–1074.

Mimaki, Y., Y. Sashida, O. Nakamura, T. Nikaido and T. Ohmoto. 1993. Steroidal saponins from the bulbs of *Lilium regale* and *L. henryi*. Phytochemistry 33: 675–682.

Moodley, N., N.R. Crouch and D.A. Mulholland. 2007. Bufadienolides from *Drimia macrocentra* and *Urginea riparia* (Hyacinthaceae: Urgineoideae). Phytochemistry 68: 2415–2419.

Mukhopadhyay, M.J., P. Sengupta, S. Mukhopadhyay and S. Sen. 2005. *In vitro* stable regeneration of onion and garlic from suspension culture and chromosomal instability in solid callus culture. Scientia Horticult. 104: 1–9.

Mutanyatta, J., B.G. Matapa, D.D. Shushu and B.M. Abegaz. 2003. Homoisoflavonoids and xanthones from the tubers of wild and *in vitro* regenerated *Ledebouria graminifolia* and cytotoxic activities of some of the homoisoflavonoids. Phytochemistry 62: 797–804.

Nagori, B.P., R. Solanki and N. Sharma. 2010. Natural Healing Agent: Garlic, an Approach to Healthy Life. IJRAP 1: 358–366.

Neergaard, J.S., J. Andersen, M.E. Pedersen, G.I. Stafford, J. Van Staden and A.K. Jager. 2009. Alkaloids from *Boophone disticha* with affinity to the serotonin transporter. S. Afr. J. Bot. 75: 371–374.

Nguedia, J.C.A., F.X. Etoa, V.P. Benga, D. Lontsi, Y. Kuete and R.S. Moyou. 2004. Anti-candidal property and acute toxicity of *Gladiolus gregasius* Baker (Iridaceae). Pharm. Méd. Trad. Afr. 13: 149–159.

Nhut, D.T. 1998. Micropropagation of lily (*Lilium longiflorum*) via *in vitro* stem node and pseudo-bulblet culture. Plant Cell Rep. 17: 913–916.

Nielsen, L.B. and D. Wege. 2006. The enantioselective synthesis of elecanacin through an intramolecular naphtoquinone-vinyl ether photochemical cycloaddition. Org. Biomol. Chem. 4: 868–876.

Niimi, Y. 1984. Bulblet productivity of explants from scales, leaves, stems and petals of *Lilium rubellum* Baker. Sci. Hort. 22: 391–394.

Nishiuchi, Y. 1980. Studies on vegetative propagation in tulips. IV. Regeneration of bulblets from bulb scale segments cultured *in vitro*. J. Jpn. Soc. Hortic. Sci. 49: 235–240.

Novak, F.J. and Ev.A. Petru. 1981. Tissue culture propagation of *Lilium* hybrids. Scientia Hort. 14: 191–199.

Oh, H., D.G. Kang, S. Lee, Y. Lee and H.S. Lee. 2003. Angiotensin converting enzyme (ACE) inhibitory alkaloids from *Fritillaria ussuriensis*. Planta Med. 69: 564–565.

Oi, Y., T. Kawada, C. Shishido, K. Wada, Y. Kominato, S. Nishimura, T. Ariga and K. Iwai. 1999. Allyl-containing sulfides in garlic increase uncoupling protein content in brown adipose tissue, and noradrenaline and adrenaline secretion in rats. J. Nutr. 129: 336–342.

Pascual-Villalobos, M.J. 2002. Anti insect activity of bufadienolides from *Urginea maritime*. *In:* J. Janick and A. Whipkey (eds.). Trends in New Crops and New Uses. ASHS Press, Alexandria, VA, pp. 564–566.

Pettit, G.R., G.R. Pettit 3rd, R.A. Backhaus and F.E. Boettner. 1995. Antineoplastic agents, 294. Variations in the formation of pancratistatin and related isocarbostyrils in *Hymenocallis littoralis*. J. Nat. Prod. 58: 37–43.

Pettit, G.R., G.R. Pettit 3rd, R.A. Backhaus, M.R. Boyd and A.W. Meerow. 1993. Antineoplastic agents, 256. Cell growth inhibitory isocarbostyrils from *Hymenocallis*. J. Nat. Prod. 56: 1682–1687.

Pettit, G.R., V. Gaddamidi, A. Goswami and G.M. Cragg. 1984. Antineoplastic agents, 99. *Amaryllis belladonna*. J. Nat. Prod. 47: 796–801.

Podwyszynska, M. and M. Agnieszka. 2003. Effects of thidiazuron and paclobutrazol on regeneration potential of Tulip flower stalk explants *in vitro* and subsequent shoot multiplication. Acta Soc. Bot. Pol. 72: 181–190.

Pohl, T., C. Koorbanally, N.R. Crouch and D.A. Mulholland. 2001. Bufadienolides from *Drimia robusta* and *Urginea altissima* (Hyacinthaceae). Phytochemistry 58: 557–561.

Pohl, T.S., N.R. Crouch and D.A. Mulholland. 2000. Southern African Hyacinthaceae chemistry, bioactivity and ethnobotany. Curr. Org. Chem. 4: 1287–1324.

Prakash, D., B.N. Singh and G. Upadhyay. 2007. Antioxidant and free radical scavenging activities of phenols from onion (*Allium cepa*). Food Chem. 102: 1389–1393.

Preethi, K. and F.B. Kenneth. 2000. Novel mechanisms of DNA topoisomerase II inhibition by pyranonaphthoquinone derivatives-eleutherin, alpha lapachone, and beta lapachone. Biochem. Pharmacol. 60: 1367–1379.

Priyadarshi, S. and S. Sen. 1992. A revised scheme for mass propagation of Easter lily. Plant Cell Tissue Organ Cult. 30: 193–197.

Prusti, A.B. and K.K. Behera. 2007. Ethnobotanical exploitation of Malkangiri District of Orissa, India. Ethnobot. Leaflets 11: 122–140.

Rahman, A.U., A.A. Farooq, M.I. Choudhary, A.H. Gilani, F. Shaheen, R.A. Ali, F.F. Noor-e-ain and B. Sener. 1994. A new anticholinergic steroidal alkaloid from *Fritillaria imperialis* of Turkish origin. Planta Med. 60: 377–379.

Rahman, A.U., M.N. Akhtar, M.I. Choudhary, Y. Tsuda, B. Sener, A. Khalid and M. Parvez. 2002. New steroidal alkaloids from *Fritillaria imperialis* and their cholinesterase inhibiting activities. Chem. Pharm. Bull. 50: 1013–1016.

Ramadan, E.G. and R.A. Backhaus. 1987. *In vitro* propagation of red squill, *Urginea maritima* Baker. Plant Cell Tissue Organ Cult. 10: 65–71.

Renard-Nozaki, J., T. Kim, Y. Imakura, M. Kihara and S. Kobayashi. 1989. Effect of alkaloids isolated from Amaryllidaceae on Herpes Simplex Virus. Res. Virol. 140: 115–128.

Rinner, U. and T. Hudlicky. 2005. Synthesis of Amaryllidaceae constituents—an update. Synlett 3: 365–387.

Rios, J.L., M.C. Recio, R.M. Giner and S. Manez. 1996. An update review of Saffron and its active constituents. Phytother. Res. 10: 189–193.

Riviere, S. and J.F. Muller. 1976. Vegetative propagation of *Tulipa gesneriana* cv. Paul Richter, by *in vitro* culture of scale axillary buds. C.R. Acad. Sci. Hebd. Seances Acad. Sci. D. 282: 533–536.

Rivlin, R.S. 2001. Historical perspective on the use of garlic. J. Nutr. 131: 951S–954S.

Rook, A. 1960. Plant dermatitis. Br. Med. J. 2: 1771–1774.

Sage, D.O., J. Lynn and N. Hammatt. 2000. Somatic embryogenesis in *Narcissus pseudonarcissus* cvs. Golden Harvest and St. Keverne. Plant Sci. 150: 209–216.

Sakai, Y., T. Murakami and Y. Yamamoto. 2003. Antihypertensive effect of onion on NO synthase inhibitor-induced hypertensive rats. Biosc. Biotecnol. Biochem. 67: 1305–1311.

Santos, J., I. Santos and R. Salema. 1998. *In vitro* production of bulbs of *Narcissus bulbocodium* flowering in the first season of growth. Sci. Hort. 76: 205–217.

Savos'kin, I.P. 1960. Specific biological characteristics of bulbous geophytes as related to their past and present ecology. Bot. Zhur. 45: 1073–1078.

Schultes, R.E. and R.F. Raffauf. 1990. The Healing Forest. Medicinal and Toxic Plants of the Northwest Amazonia, Dioscorides Press, Portland, USA.

Schumann, A., S. Berkov, D. Claus, A. Gerth, J. Bastida and C. Codina. 2012. Production of galanthamine by *Leucojum aestivum* shoots grown in different bioreactor systems. Appl. Biochem. Biotech. 167: 1907–1920.

Sell'es, M., F. Viladomat, J. Bastida and C. Codina. 1999. Callus induction, somatic embryogenesis and organogenesis in *Narcissus confusus*: correlation between the state of differentiation and the content of galanthamine and related alkaloids. Plant Cell Rep. 18: 646–651.

Shenoy S.R., M.N.S. Kameshwari, S. Swaminathan and M.N. Gupta. 2006. Major antifungal activity from the bulbs of Indian squill *Urginea indica* is a chitinase. Biotechnol. Prog. 22: 631–637.

Shibuya, N., I.J. Goldstein, E.J.M. Van Damme and W.J. Peumans. 1988. Binding properties of a mannose-specific lectin from the Snowdrop (*Galanthus nivalis*) bulb. J. Biol. Chem. 263: 728–734.

Shimada, K., E. Umezawa, T. Nambara and S.M. Kapchan. 1979. Isolation and characterization of cardiotonic steroids from the bulbs of *Urginea altissima* Baker. Chem. Pharm. Bull. 27: 3111–3114.

Shushu, D.D., J. Mutanyatta and B.M. Abegaz. 2005. *In vitro* propagation and phytochemical investigation of *Ledebouria graminifolia*, an important medicinal bulb in Southern Africa. J. Herbs, Spices Med. Plants 11: 97–107.

Silayo, A., B.T. Ngadjuib and B.M. Abegaz. 1999. Homoisoflavanones and stilbenes from the bulbs of *Scilla nervosa* subsp. *rigidifolia*. Phytochemistry 52: 947–955.

Silvertand, B., A. van Rooyen, P. Lavrijsen, A.M. van Harten and E. Jacobsen. 1996. Plant regeneration via organogenesis and somatic embryogenesis in callus cultures derived from mature zygotic embryos of leek (*Allium ampeloprasum* L.). Euphytica 91: 261–270.

Singh, D.N., N. Verma, S. Raghuwanshi, P.K. Shukla and D.K. Kulshreshtha. 2008. Antifungal activity of *Agapanthus africanus* extractives. Fitoterapia 79: 298–300.

Singh, V. and R.P. Pandey. 1982. Plants used in religion and magico-religious beliefs in Rajasthan. J. Eeon. Tax. Bot. 3: 273–298.

Smith R.H., J. Burrow and K. Kurten. 1999. Challenges associated with micropropagation of *Zephyranthes* and *Hippestarum* sp. (Amaryllidaceae). *In-vitro* Cellular and Develop. Biol. Plant. 35: 281–282.

Spenser, I.D. 1970. Biosynthesis of alkaloids. *In*: S.W. Pelletier (ed.). Chemistry of the Alkaloids. Van Nostrand Reinhold Company. New York, pp. 69–695.

Sramek, J.J., E.J. Frackiewicz and N.R. Cutler. 2000. Review of the acetylcholinesterase inhibitor galanthamine. Expert Opin. Inv. Drugs 9: 2393–2402.

Staikidou, C.S. and B.M.R. Harvey. 1994. Stimulation by auxin and sucrose of bulbil formation *in vitro* by single leaf cultures of *Narcissus*. New Phytol. 127: 315–320.

Steinmetz, E.F. 1957. Codex Vegetabilis. All Botanical Drugs. Keizersgracht 347, Amsterdam.

Steyn, P.S., F.R. Van Heerden and R. Vleggaar. 1986. Application of highfield NMR spectroscopy to the structural elucidation of natural products. The structure of rubellin, a novel bufadienolide glycoside from *Urginea rubella*. S. Afr. J. Chem. 39: 143–146.

Stimart, D.P. and P.D. Ascher. 1978. Tissue culture of bulb scale sections for asexual propagation of *Lilium longiflorum* Thunb. J. Am. Soc. Hortic. Sci. 103: 182–184.

Sultana, J., N. Sultana, M.N.A. Siddique, A.K.M.A. Islam, M.M. Hossain and T. Hossain. 2010. *In vitro* bulb production in Hippeastrum (*Hippeastrum hybridum*). J. Cent. Eur. Agric. 11: 469–474.

Suzuki, N., S. Tani, S. Furusawa and E. Furusawa. 1974. Therapeutic activity of *Narcissus* alkaloids on Rauscher leukemia: antiviral effect *in vitro* and rational drug combination *in vivo*. Proc. Soc. Exp. Biol. Med. 145: 771–777.

Szlávik, L., A. Gyuris, J. Minárovits, P. Forgo, J. Molnár and J. Hohmann. 2004. Alkaloids from *Leucojum vernum* and antiretroviral activity of Amaryllidaceae alkaloids. Planta Med. 70: 871–873.

Takhtajan, 1997. Diversity and Classification of Flowering Plants. Columbia University Press, New York.

Tamura, M.N. 1998. Liliaceae. *In*: The Families and Genera of Vascular plants, Springer, Berlin, pp. 343–353.

Tang, Y., B. Yu, J. Hu, T. Wu and H. Hui. 2002. Three new homoisoflavanone glycosides from the bulbs of *Ornithogalum caudatum*. J. Nat. Prod. 65: 218–220.

Tânia, M.A.A., K. Helmut and L.Z. Carlos. 2003. Eleutherinone, a novel fungitoxic naphthoquinone from *Eleutherine bulbosa* (Iridaceae). Mem. Inst. Oswaldo Cruz. Rio de Janeiro. 98: 709–712.

Tanker, M., G. Citoglu, B. Gumusel and B. Sener. 1996. Alkaloids of *Sternbergia clusiana* and their analgesic effects. Int. J. Pharmacogn. 34: 194–197.

Terness, P., D. Navolan, C. Dufter, B. Kopp and G. Opelz. 2001. The T-cell suppressive effect of bufadienolides: structural requirements for their immunoregulatory activity. Int. Immunopharmacol. 1: 119–134.

Thamburan, S., J. Klaasen, W.T. Mabusela, J.F. Cannon, W. Folk and Q. Johnson. 2006. *Tulbaghia alliacea* phytotherapy: a potential anti-infective remedy for candidiasis. Phytother. Res. 20: 844–850.

Thomson, M. and M. Ali. 2003. Garlic (*Allium sativum*): a review of its potential use as an anti-cancer agent. Curr. Cancer Drug Targets 3: 67–81.

Tiwari, D.K. and A. Yadav. 2003. Ethno botanical investigation of some medicinal plants availed by Gond tribe of Naoradehi wild life sanctuary, Madhya Pradesh. Anthropologist 5: 201–202.

Toriizuka, Y., E. Kinoshita, N. Kogure, M. Kitajima, A. Ishiyama, K. Otoguro, H. Yamada, S. Omura and H. Takayama. 2008. New lycorine-type alkaloid from *Lycoris traubii* and evaluation of antitrypanosomal and antimalarial activities of lycorine derivatives. Bio. org. Med. Chem. 16: 10182–10189.

Tsai, C.W., J.J. Yang, H.W. Chen, L.Y. Sheen and C.K. Lii. 2005. Garlic organosulfur compounds upregulate the expression of the Π-class of glutathione S-transferase in rat primary hepatocytes. J. Nutr. 135: 2560–2565.

Tschesche, R. and U. Dolberg. 1958. Plant cardiac poisons. XXXV. The isolation of two new bufadienolide glycosides from *Bowiea volubilis* Harvey. Chem. Ber. 91: 2512–2523.

Unver, N. 2007. New skeletons and new concepts in Amaryllidaceae alkaloids. Phytochem. Rev. 6: 125–135.

Van Damme, E.J.M., A.K. Allen and W.J. Peumans. 1987. Isolation and characterization of a lectin with exclusive specificity toward mannose from snowdrop (*Galanthus nivalis*) bulbs. FEBS Lett. 215: 140–144.

Van Damme, E.J.M., I.J. Goldstein and W.J. Peumans. 1991. A comparative study of mannose-binding lectins from the Amaryllidaceae and Alliaceae. Phytochemistry 30: 509–514.

Van der Werff, P.J. 1959. Occupational diseases among workers in the bulb industries (preliminary report). Acta Allergol. 14: 338–355.

Van Heerden, F.R., R. Vleggaar and L.A.P. Anderson. 1988. Bufadienolide glycosides from *Urginea physodes*. First report of natural 14-deoxybufadienolides. S. Afr. J. Chem. 41: 145–151.

Van Staden, J., C.W. Fennell and N.R. Crouch. 2001. Micropropagation of the river lily, *Crinum variabile* (Amaryllidaceae). S. Afr. J. Bot. 67: 74–77.

Veale, D.J.H., I. Havlik, D.W. Oliver and T.G. Dekker. 1999. Pharmacological effects of *Agapanthus africanus* on the isolated rat uterus. J. Ethnopharmacol. 66: 257–262.

Veale, D.J.H., K.I. Furman and D.W. Oliver. 1992. South African traditional herbal medicines used during pregnancy and childbirth. J. Ethnopharmacol. 36: 185–191.

Watt, J.M. and M.G. Breyer-Brandwijk. 1962. The Medicinal and Poisonous Plants of Southern and Eastern Africa. E. & S. Livingstone Ltd., Edinburgh.

Wawrosch, C., P.R. Malla and B. Kopp. 2001. Micropropagation of *Allium wallichii* kunth, a threatened medicinal plant of Nepal. *In vitro* Cell. Dev. Biol.-Plant 37: 555–557.

Weber, L.F. 1937. External causes of dermatitis—a list of irritants. Arch. Dermatol. Syphilol. 35: 29–179.

Weniger, B., L. Italiano, J.P. Beck, J. Bastida, S. Bergoñon, C. Codina, A. Lobstein and R. Anton. 1995. Cytotoxic activity of Amaryllidaceae alkaloids. Planta Med. 61: 77–79.

Weniger, B., M. Haag-Berrurier and R. Anton. 1982. Plants of Haiti used as antifertility agents. J. Ethnopharmacol. 6: 67–84.

Whitaker, R.J. 1976. Development of flavor, odor and pungency in onion and garlic. Adv. Food. Res. 22: 73–133.

Wildman, W.C. 1968. The Amaryllidaceae alkaloids. *In*: R.H.F. Manske (ed.). The Alkaloids, Vol. XI. Academic Press, New York, pp. 307–405.

Xu, J., F. Qiu, W. Duan, G. Qu, N. Wang and X. Yao. 2006. New bioactive constituents fom *Eleutherine americana*. Front. Chem. China 3: 320–323.

Yamada, K., A. Namaura, N. Sawashita, Y. Naguchi and J. Yamamoto. 2004. An onion variety has natural antithrombotic effect as assessed by thrombosis/thromolysis models in rodents. Thromb. Res. 114: 213–220.

Yui, S., M. Mikami, M. Kitahara and M. Yamazaki. 1998. The inhibitory effect of lycorine on tumor cell apoptosis induced by polymorphonuclear leukocyte-derived calprotectin. Immunopharmacology 40: 151–162.

Zschocke, S. and J. Van Staden. 2000. *In vitro* propagation of *Tulbaghia simmleri*. S. Afr. J. Bot. 66: 86–89.

17

Conservation of Tuberous Medicinal Plants of Western Ghats through Biotechnological Approach

P.N. Krishnan, Dhanya B. Pillai* and *K. Satheeshkumar*

ABSTRACT

Plants are a vital part of global biodiversity, a veritable chemical goldmine and a major life supporting system for all life forms. Conservation and sustainable utilization of plant genetic resources is essential to meet the future requirement of natural resources. Medicinal plants play a vital role in the treatment of a number of diseases, and form the mainstay of 75–80% of the world population, mainly in the developing countries. Tropical countries of the world supply the bulk of medicinal raw materials in demand, resulting in depletion of the population and genetic diversity. India, one among the 12 major megabiodiversity regions of the world, is a major centre of origin and diversity of medicinal plants. In India, Western Ghats is one of the major repositories of medicinal plants, harbouring about 4,500 species of higher plants of which 10% are threatened and 30% are endemic. It is also recorded that 30% of the medicinal raw material collected

Biotechnology and Bioinformatics Division, Jawaharlal Nehru Tropical Botanic Garden and Research Institute, Pacha, Palode, Thiruvananthapuram, Kerala, India-695 562.

* Corresponding author: peringattulli@yahoo.co.uk

comprises plant roots as compared to other parts. Tropically most of the endangered plants of Western Ghats belong to the tuber yielding medicinal plant species, due to their destructive extraction from the wild. Advances in biotechnology have generated new opportunities for genetic resource conservation and utilization. These include (i) *ex situ* conservation of RET species through micropropagation, reintroduction and restoration (ii) bioproduction of plant specific compounds in culture systems to avoid collection from nature (iii) bioprospecting to identify better classes of compounds and (iv) cryopreservation of shoot tips, seeds, embryo, etc. This paper addresses some of the recent developments in these areas on conservation of tuber yielding medicinal plants of Western Ghats and their sustainable utilization.

Keywords: Western Ghats, Medicinal plants, Conservation, Micropropagation, Genetic transformation, Secondary metabolites, Elicitation, Encapsulation, Cryopreservation

Introduction

Since the pre-historic times plants are universally recognized as a vital part of the planet's biological diversity and an essential natural resource having immeasurable ecological, biological, therapeutic, social, cultural, economic, spiritual and aesthetic significance. But in pursuit of benefits, we have somewhere, through the course of time, overlooked the need to judiciously use and safeguard our fathomless treasure. Extinction and decline in plant diversity have been attributed to a range of factors including population growth, high rates of habitat modification, deforestation, over-exploitation, spread of invasive alien species, pollution and climate change (Anonymous 2009). The distribution of bioresources on planet Earth is not uniform. The tropics that harbour the vast majority of the world's biota are thought to behold the key to the future. In 1988, ecologist Norman Myers introduced the term 'biodiversity hotspots' to distinguish a global set of high-priority terrestrial ecoregions for conservation (Myers 1988). 'Biodiversity hotspots' are areas where exceptional concentrations of endemic species are undergoing exceptional loss of habitat and where wise conservation investments could help minimize future extinctions (Myers et al. 2000, Myers 1988). As many as 44% of all species of vascular plants and 35% of all species in four vertebrate groups are confined to the world's 27 hotspots comprising only 1.4% of the land surface of the earth (Myers et al. 2000).

Conservation of plants particularly RET (rare, endangered and threatened) species is an outstanding global priority and calls for urgent new initiatives and actions to save them from extinction. It is particularly so since as many as 33,798 out of an estimated 2,50,000 higher plant species are currently in the endangered list and as many as two-thirds of the world's plant species are reported to be in danger of extinction during the course of the 21st century (Jackson 2002). Among the non-wood forest products (NWFPs), medicinal plants are a special group of plants forming key focus segment of the developmental paradigm in developing countries and are of great relevance with nearly 80% of the world's population depending on them for primary healthcare needs.

The increasing realization of the health hazards and toxicity associated with the indiscriminate use of synthetic drugs or antibiotics has renewed the interest in the use of plants and plant based drugs. Subsequent inclination towards herbal medicine has resulted in the expansion of plant-based pharmaceutical industries world over. The use of herbal medicines in the developing countries is also growing and 75% of UK's population takes herbal medicine regularly (Vines 2004). Approximately two-third of the 50,000 medicinal plant species in use are collected from the wild (Edwards 2004). Such species are often collected illegally or unsustainably and as a result, may lose genetic diversity and face the threat of extinction (Reed et al. 2011). In Europe only 10% of medicinal plant species used commercially is cultivated. In India more than 95% of the plant species used in the production of medicine in industries are being harvested from the wild (Uniyal et al. 2000).

The concept of conservation of medicinal plants and its genetic resources comprise of collection, conservation, management, identification of potentially valuable material by characterization and evaluation for subsequent use. Advances in biotechnology, especially *in vitro* culture techniques and molecular biology provide some important tools for improved conservation and management of plant genetic resources (Rao 2004, Krishnan et al. 2011). Tropical countries which are the treasure house of valuable medicinal species are persistently under pressure and are progressively losing these resources due to several man-made as well as climatic factors (Sharma and Pandey 2013).

Western Ghats and Medicinal Plant diversity

The region along the western passive continental margin of India, running almost parallel to the Arabian sea coast with Vindhya and Satpura mountain ranges on the northern side and the semi-arid Deccan plateau on the eastern side, is known as the Sahyadri (Sanskrit for "benevolent mountain") or the Western Ghat (Ghat—broad flight of steps leading to a river). Lying between 8°20′–40′ N and 73°77′ E, the Western Ghats traverses through the states of Maharashtra, Goa, Karnataka, Tamil Nadu, and Kerala over a 1,600 km distance at an average width of 200 km. This unique land mass of India with high levels of topographic and climatic heterogeneity as well as edaphic conditions designs a variety of micro-climatic/ecological niches that nurture one of the richest endemic flora of peninsular India. With its extremely environmentally heterogeneous biogeographic area, the Western Ghats as a whole is an abode of rich diverse flora, including a multitude of medicinal plants in use for curative purposes since the Vedic era (Krishnan et al. 2011, Gunawardene et al. 2007, Kale et al. 2010).

The natural forests of Western Ghats region are the source of nearly 500 medicinal plants, some of which have been used for traditional and folk medicinal practices (Joy et al. 2001). Many endemic as well as threatened plants grow naturally in this area with a large number of these being used to cure illnesses by traditional healers (Deshmukh 2010). Of the numerous species available in the Western Ghats, about 50 species hold immense value in the folk and herbal health forms for the treatment of different forms of ailments (Suja 2005). Whilst several healing plants are known to the scientific world, several medicinal uses remain unknown to science due to lack of proper documentation. All such valuable medicinal plants are being exploited and their survival may become threatened due to improper management, lack of awareness, harvesting, grazing, shifting cultivation and uprooting of plant species for the purpose of food, fodder, medicines or forest clearance (land utilization) by tribal and local people (Deshmukh 2010).

Of the more than 2,50,000 species of higher plants existing on this planet, India harbours about 17,000 species in which 8,000 are medicinal species. India regularly exports 40,000 to 50,000 tonnes of dried herbs and plant parts to western World (USA, Europe, Japan) in addition to the local and household consumption of 27,000–28,000 tonnes/year. Considering that fresh herbs conatin 75% moisture, nearly a million tonnes of fresh biomass of herbs are harvested every year. In this 45% biomass comes from less than 40 cultivated species and 37% of the biomass obtained from 77% of the species harvested from forests (Ved and Goraya 2008), a substantial portion from the Western and Eastern Ghats. Additionally, rural communities in a good number of these tropical developing economies are intensely dependent on plant-based remedies produced locally from raw materials mostly harvested from wild sources (Sharma and Pandey 2013). According to Pushpangadan et al. (1994) most of the known types of floristic components comprising nearly 7,000 plant species with 30% endemics are represented in the Western Ghats. The bulk of these plants with their intraspecific variations represent a chemical and medicinal goldmine as is evidenced from the strong tradition of natural drug use.

India harbours two of the richest global biodiversity hotspots the Western Ghats and North-Eastern region comprising a part of the Indo-Burmese hotspot. These hotspots harbour a large number of tuberous root-bearing medicinal plants like *Rauwolfia* spp., *Dioscorea* spp.,

Coscinium fenestratum, Coleus forskohlii, Mucuna pruriens, Zingiber spp., *Curcuma* spp., etc. which are commercially valuable both in modern and traditional systems. In addition it is estimated that wild relatives of about 160 species of cultivated plants are found in the forests of Western Ghats in which many are medicinal and aromatic in nature. A list of tuberous medicinal plants found in the Western Ghats with scientific name, conservation status and its uses are given in Table 1. Several traditional medicinal preparations like Dashamoolarishta, Asvas and Kashayas in Ayurvedic system uses roots/tubers of medicinal plants. In addition, in modern medicine, several low volume high value compounds are extracted from tubers. For example, reserpine, ajmaline, ajmalicine, etc. from the species belonging to *Rauwolfia* and vincristine, vinblastine from *Catharanthus roseus,* etc. Indian ginseng is *Withania somnifera* where tubers are the medicinally important part. Ramakrishnappa (2002) observed that 30% of the medicinal raw material is of roots of plants when compared to flowers, fruits, leaves, bark, wood, etc. Extensive and destructive collection of roots of these medicinal plants causes severe problems on existence of their genetic diversity resulting to endangerment. So in this juncture, there is a need for conservation of these medicinal plants and sustainable utilization through appropriate technologies.

According to Shah and Seth (2010), about 90% of medicinal plants used by the industries are collected from the wild sources and more than 70% plant collection involved destructive harvesting because of parts used like root, bark, stem, wood and whole plant. When a portion of plant material is harvested from an individual, the nature and quantity of nutrients and/or photosynthetic capacity removed, and the potential for survival and effective propagation, will depend on the kind of material harvested. This makes it evident that the type of plant part harvested affects the potential for species to tolerate harvest and sustain in the face of exploitation (Ticktin 2004). Although gums, oleoresins, barks as well as roots and bulbs, represent a large proportion of commercial wild-harvested NWFPs (Cunningham 1993, Vantomme et al. 2002), the ecological consequences of harvesting these plant parts remain poorly studied. Ability of a plant to tolerate destructive harvesting varies with its life history, climatic conditions, reproductive strategy, habitat specificity, growth rates, resources available to stimulate plant growth and rate of recuperation after harvest (Ticktin 2004). When the life form and plant parts collected are viewed together, species susceptible to over-collection can be distinguished well. It was found out by Schippmann et al. (2002) that the species most susceptible to over-harvest are habitat specific, slow growing and those destructively harvested for their bark, roots or the whole plant. These species suffer most from harvesting and many of them have been seriously depleted (Schippmann et al. 2002). Destructive harvest of roots may not only cause severe problems to the sustenance of

Table 1. Medicinal root-bearing plants of Western Ghats with their uses.

Sl. No.	Plant	Family	Status	Habit	Medicinal uses of the roots/rhizome/tubers
1.	*Abrus precatorius* L.	Papilionaceae/ fabaceae	Rare	Shrub	Jaundice, gonorrhea, night blindness, dysentery, diarrhoea
2.	*Acorus calamus* L.	Araceae	Rare	Herb	Dyspepsia, flatulence, cough, fever, piles and asthma
3.	*Alangium salvifolium* (L.f.) Wang.	Alangiaceae	Common	Shrub	Piles, epilepsy, fever, skin diseases, diarrhea, rheumatism, dropsy
4.	*Alpinia galanga* (L.) Sw.	Zingiberaceae	Common	Herb	Improve appetite and voice, respiratory diseases, rheumatic pains and as aphrodisiac
5.	*Aristolochia indica* L.	Aristolochiaceae	Common	Climbing shrub	Diarrhea, fever, dropsy, poisoning, abortifacient
6.	*Artanema longofolia* (L.) Vatke	Scrophulariaceae	Common	Herb	Rheumatism, kidney stones, syphilis, ophthalmia
7.	*Asparagus racemosus* Willd.	Liliaceae	Rare	Spinous under-shrub	Female reproductive disorders, decreased lactation, arthritis, diarrhoea, dysentery, edema, rheumatism, fevers, aphrodisiac, cooling tonic, sunstroke, cough , peptic ulcers and as rejuvenator
8.	*Boerhavia diffusa* L.	Nyctaginaceae	Rare	Herb	Sudation, fever, skin disease, head ache, dropsy, heart disease, disease of head and neck, musculoskeletal disorder and gout
9.	*Caesalpinia crista* L.	Caesalpinaceae	Common	Shrub	Piles
10.	*Canthium parviflorum* Lam.	Rubiaceae	Common	Shrub	Intestinal disorders
11.	*Cardiospermum halicacabum* L.	Sapindaceae	Rare	Climber	Rheumatism, nervous disorders, piles, arthritis, diarrhoea

Table 1. contd....

Table 1. contd.

Sl. No.	Plant	Family	Status	Habit	Medicinal uses of the roots/rhizome/tubers
12.	*Ceropegia bulbosa* Roxb.	Asclepiadaceae	Rare	Twining herb	Kidney stones, analgesic, diarrhoea, dysentery, increase fertility and vitality
13.	*Ceropegia juncea* Roxb.	Asclepiadaceae	Rare	Herb	Stomach and gastric disorders
14.	*Ceropegia intermedia* Wight	Asclepiadaceae	Rare	Herb	Diarrhoea and dysentery
15.	*Ceropegia sahyadrica* Ans. & Kulk.	Asclepiadaceae	Rare (Endemic to Western Ghats)	Herb	Used as food
16.	*Chlorophytum tuberosum* (Roxb.) Baker	Liliaceae/ Asparagaceae	Rare	Herb	Fever, leucorrhoea and as aphrodisiac
17.	*Clerodendrum serratum* (L.) Moon	Verbenaceae	Rare	Shrub	Malaria, cough, asthma, rheumatism, fistula, cholera, paralysis, fever, indigestion
18.	*Coccinia grandis* (L.) Voigt.	Cucurbitaceae	Common	Herb (climber)	Diabetes mellitus, bronchitis, skin diseases, tongue-sores and ear-ache
19.	*Coleus forskohlii* (Willd.) Briq	Lamiaceae	Common	Herb	High blood pressure, spasmolysis, obesity and constipation
20.	*Coscinium fenestratum* Colebr.	Menispermaceae/ amaranthaceaea	Common	Shrub	Dressing wounds, ulcers, cutaneous leishmaniasis, colic and stomachache
21.	*Costus speciosus* (Koenig) Smith	Costaceae	Common	Herb	Leprosy, skin disease, cough, respiratory diseases, anaemia, blood diseases, fever and inflammation
22.	*Curculigo orchioides* Gaertn.	Liliaceae	Common	Herb	Muscle and tissue augumentor, tumours, post operative wounds in cosmetic surgery
23.	*Curcuma amada* Roxb.	Zingiberaceae	Common	Herb	Appetizer, antipyretic, aphrodisiac, diuretic, emollient, expectorant, laxative, skin diseases, bronchitis, asthma and inflammation

24.	*Curcuma domestica* Val.	Zingiberaceae	Common	Herb	Chest pain, skin disease, piles, insect and rabid animal bite, gout, tumours and wound
25.	*Curcuma longa*	Zingiberaceae	Common	Herb	Flatulence, inflammation and as food additive
26.	*Curcuma pseudomontana* Grah.	Zingiberaceae	Near threatened	Herb	Leprosy, dysentery, cardiac diseases, general debility, jaundice and for blood purification
27.	*Cyathula prostrata* (L.) Blume	Amaranthaceae	Common	Herb	Expectorant, tuberculosis, dysentery and skin diseases
28.	*Cyclea peltata* (Lam.) Hook.f. & Thoms.	Menispermaceae	Rare	Herb	Dysentery, jaundice, fever, leprosy, asthma, and stomachache
29.	*Decalepis arayalpathra* (Joseph and Chandras) Venter.	Asclepiadaceae	Critically endangered	Shrub	Peptic ulcer and as rejuvenating tonic
30.	*Decalepis hamiltonii* Wight & Arn.	Asclepiadaceae	Common	Climber	Loss of appetite, skin diseases, diarrhoea, epilepsy and central nervous system disorders
31.	*Desmodium triflorum* (L.) DC.	Papilionaceae	Common	Herb	Piles, ophthalmia and bilious complaints
32.	*Elephantopus scaber* L.	Asteraceae	Common	Herb	Vomiting, toothache, snake-bite, filariasis, rheumatism, urinary troubles, fever, diarrhea, dysentery and liver troubles
33.	*Embelia ribes* Burm. f.	Myrsinaceae	Abundant	Shrub	Toothache, pneumonia, diarrhea and cough
34.	*Flacourtia indica* (Burm.f.) Merr.	Flacourtiaceae	Rare	Shrub/ tree	Body pain, bronchial allergy, blood pressure, skin diseases, dysentery and snake bite
35.	*Gloriosa superba* L.	Liliaceae/ Colchicaceae	Rare Least concern globally	Herb	Arthritis, intestinal worms, gout, bruises, infertility and parasitic skin diseases
36.	*Glycosmis pentaphylla* (Retz.) A. DC.	Rutaceae	Common	Shrub	Toothache, boils and fever

Table 1. contd....

Table 1. contd.

Sl. No.	Plant	Family	Status	Habit	Medicinal uses of the roots/rhizome/tubers
37.	*Glycrrhiza glabra* L.	Leguminosae/ Fabaceae -Papilionoideae	Not threatened	Herb	Gastric problem, ulcers, indigestion, coughs, and sore throat
38.	*Gymnema sylvestre* (Retz.) R. Br. ex. Schult.	Asclepiadaceae	Common	Shrub	Expectorant, emetic, diabetes, eye disease, cataract and piles
39.	*Hedychium coronarium* Koenig	Zingiberaceae	Near threatened	Herb	Eye diseases
40.	*Helicteres isora* L.	Sterculiaceae	Common	Shrub	Amoebiasis, diabetes, flatulence, snake bite, dog bite, dysentery, asthma, cough and wounds
41.	*Hemidesmus indicus* (L.) R. Br.	Periplocaceae	Common	Undershrub	Wound, bronchial asthma, fever, intrinsic haemorrhage, skin diseases, poisoning, rejuvenation therapy in children and during pregnancy
42.	*Holostemma annulare* (Roxb) Schum.	Asclepiadaceae	Rare	Climber	Eye diseases, stimulant, aphrodisiac, expectorant and as tonic
43.	*Homonoia riparia* Lour.	Euphorbiaceae	Common	Shrub	Laxative, emetic, ulcers, chest pain, toothache, and bladder stone
44.	*Ipomoea mauritiana* Jacq	Convolvulaceae	Common	Climbing shrub	Diseases of spleen , liver, debility, aphrodisiac and as tonic
45.	*Kaempferia galanga* L.	Zingiberaceae	Common	Herb	Cough, pectoral affections and nasal blocks
46.	*Kaempferia rotunda* L.	Zingiberaceae	Common	Herb	Cataract, night blindness and in cosmetics
47.	*Leea indica* (Burm. f.) Merr.	Leeacee	Common	Shrub	Cooling, expectorant, diarrhea, dysentery, and boils

48.	*Mirabilis jalapa* L.	Nyctaginaceae	Common	Herb	Diabetes, constipation, genitourinary system disorders and injuries
49.	*Mucuna pruriens* (L.) DC.	Papilionaceae	Common	Herb	Dropsy, falling hair, elephantiasis, prickly heat, wounds, cholera and kidney trouble
50.	*Ocimum gratissimum* L.	Lamiacea	Common	Herb	As sedative
51.	*Piper longum* L.	Piperaceae	Scarce	Undershrub	Asthma, fever, jaundice, throat infection, epilepsy, cough, stomach pain, paralysis, and snake bite
52.	*Plumbago indica* L.	Plumbaginaceae	Rare	Shrub	Leucoderma, leprosy, diarrhoea, piles, dyspepsia, intestinal worms and skin diseases
53.	*Plumbago zeylanica* L.	Plumbaginaceae	Rare	Shrub	Fever, diarrhoes, malaria, dyspepsia, leprosy, skin diseases and piles
54.	*Pueraria tuberosa* (Roxb. ex Willd.) DC	Papilionaceae	Near threatened	Shrub	Constipation, rejuvenation and as aphrodisiac
55.	*Rauvolfia serpentina* (L.) Benth. ex Kurz	Apiaceae	Rare	Undershrub	Fever, ophthalmia, epilepsy-applied to eye lid, poisonous bites and spider bite
56.	*Rubia cordifolia* L.	Rubiaceae	Common	Herb	Fever, premature grey hair, dropsy, herpes, wounds, skin disease, gout, breast milk impurity, wounds, fistula and fractures
57.	*Salacia chinensis* L.	Hippocrateaceae	Common	Shrub	Diabetes, gonorrhoea, rheumatism, asthma and skin diseases
58.	*Salacia oblonga* Wall.	Hippocrateaceae	Endangered-regionally	Climbing shrub	Diabetes and obesity
59.	*Sida acuta* N. Burm.	Malvaceae	Rare	Herb	Blood and pile disorders, general debility, nervous and urinary disorders
60.	*Sida cordifolia* L.	Malvaceae	Common	Undershrub	Skin diseas, fever, headache, herpes, wounds, gout, breast milk disorder, postnatal care and tumour

Table 1. contd....

Table 1. contd.

Sl. No.	Plant	Family	Status	Habit	Medicinal uses of the roots/rhizome/tubers
61.	*Tinospora cordifolia* (Willd.) Hook. F. & Thoms.	Menispermaceae	Common	Climbing shrub	Anti-stress activity
62.	*Tinospora sinensis* (Lour.) Merr.	Menispermaceae	Rare or vulnerable	Shrub	Debility and malaria
63.	*Tylophora indica* (Burm. f.) Merr.	Asclepiadaceae	Rare	Herb	Asthma, bronchitis, whooping cough, dysentery, rheumatic gouty pains and hydrophobia
64.	*Typhonium flagelliforme* (Lodd.) Blume	Araceae	Rare	Herb	Piles, snake venom and as stimulants
66.	*Urena lobata* L.	Malvaceae	Rare or vulnerable	Shrub	Colic, hyperacidity, dysentery, dog bite, fever, syphilis and urinary troubles
67.	*Withania somnifera* (L.) Dunal	Solanaceae	Rare or vulnerable	Shrub	Constipation, debility, rheumatism, , nervous exhaustion, memory loss and spermatorrhoea
68.	*Zingiber officinale* Roscoe	Zingiberaceae	Common	Herb	Dropsy, otitis musculoskeletal disorder, ophthalmia dental disease, ascites
69.	*Zingiber zerumbet* (L.) Smith	Zingiberaceae	Common	Herb	Appetizer, ulcers, stomachaches, diarrhea, asthma, rheumatism and inflammation

Source:

Sahyadri: Western Ghats Biodiversity Information System Flora Database http://www.ces.iisc.ernet.in/biodiversity/sahyadri_database/database1/allspecies7.php

http://envis.frlht.org/botanical_search.php?

Sood SK, Sudershna "Root drug plants of ethnic india" 2008 satish serial publishing house, Delhi. ISBN: 8189304577

Anonymous, 1997. The Wealth of India. Raw materials. CSIR, New Delhi, India.

Anonymous, 2001. The Wealth of India: A Dictionary of Indian Raw Materials and Industrial Products. Vol. 2, National Institute of Science Communication, New Delhi, ISBN: 81-7236-213-7.

the plants thereby posing a threat to biodiversity, but also disturb vital ecological processes. For instance, plants such as *Asparagus racemosus* act as good soil binders in the Indian forests in which they are found, and intense harvest of their roots has led to large-scale soil erosion (Ramakrishnappa 2002). Similarly, for medicinal root bearing plants with modifications such as tuberous roots/ rhizomes, the propagation is often through vegetative propagules (like tubers, rhizomes, etc.), which is a reason for the destructive harvest being more detrimental to the survival of such species. Additionally, the genetic diversity among different individuals of such species will be fundamentally less due to propagation by vegetative means, adding to the toll of destruction caused to biodiversity. Since lesser options remain for enhancing genetic diversity of these species by merely preventing further deterioration of the existing population and allowing them to propagate naturally, strategies need to be devised for improving the genetic base of such species. This is possible by employing several biotechnological tools such as plant tissue culture, genetic transformation, etc.

Conservation Strategies

There are two approaches for the conservation of plant biodiversity, viz. *in situ* and *ex situ* conservation. The former takes into account the preservation of natural resources in the wild/where they naturally occur, while the latter consists of conservation outside the natural habitat and is generally used to safeguard populations in danger of destruction or deterioration.

In situ conservation practices involving biosphere reserves, sanctuaries, and national parks can preserve the breeding system, but this alone is insufficient for the sustained use of genetic diversity. Exploration, collection, characterization, evaluation, domestication/cultivation, and *ex situ* conservation in field gene banks in botanic garden and conservatories, seed banks, etc. eventually support their sustained use, supplying quality planting material and certified raw drugs. Modern biotechnological tools including *in vitro* culture, cryopreservation techniques, DNA fingerprinting and bioreactor mediated bioproduction of phytochemicals are mostly relevant in this context. Micropropagation and ecorestoration in fact support the *in situ* conservation activities, facilitating population enhancement in species where natural propagation is hindered because of destructive harvesting or reproductive barriers (Krishnan et al. 2011, Rao 2004). In certain cases as in vegetatively propagated species, conservation alone would not be sufficient and broadening the genetic base is necessary to safeguard biodiversity as well as to sustain the species in the face of growing threat to diversity.

Gene bank

During the past few decades, new conservation strategies have been devised by converging *in situ* and *ex situ* conservation approaches for better fulfilment of the objectives of conservation (Chandel 1996). As an addition to Field Gene Banks and Seed Banks, cell and tissue culture techniques with other biotechnological aids have done much to ensure the creation and safe transfer of healthy germplasm around the world through gene banks (Innes 1992).

In India, the National Bureau of Plant Genetic Resources (NBPGR) has tried to incorporate both these strategies leading to the establishment of India's first National Gene bank in 1985, for preserving germplasm diversity of agri-horticultural crops. A National Facility for Plant Tissue Culture Repository (NFPTCR) has also been established at NBPGR with several new approaches such as preservation of cell, suspension cultures, meristems and shoot tips, embryos, embryonic axes being implemented for allowing conservation and sustainable utilization of germplasm resources (Chandel 1996). As a result of the programme initiated by DBT, Government of India in 1992, presently there are four National Gene Bank facilities specifically devoted to the conservation of medicinal and aromatic species of the country. National Gene Bank at JN TBGRI is one among the four, working with the objective of conserving medicinal and aromatic plants of Southern Peninsular India through Field Gene Bank, Seed Bank and biotechnological intervention, including *in vitro* and cryopreservation techniques (Krishnan et al. 2011).

Conservation through *in vitro* propagation

Micropropagation

Cultivation is recognized as one of the most effective ways of conservation and reducing the pressure on natural population as it assures production of plants or plant parts with both qualitatively and quantitatively uniform chemical constituents required for the preparation of standard formulations. However, the ever-increasing demand of plant raw drugs calls for mass production of medicinal plants and their cultivation often faces shortcomings of withering land holdings, lack of proper agrotechnology, dependence on climate, attack from pests/diseases, etc. But, majority of the medicinal plants from Western Ghats extensively used in traditional medicine are seldom cultivated (Chaturvedi et al. 2007, Ved and Goraya 2007, Krishnan et al. 2011). Also, cultivation of some herbs has proved difficult because of low germination rates or specific ecological requirements (Vines 2004). There may be lack of knowledge about specific requirements for pollination, seed germination and growth. In such cases *in vitro* methods have lately gathered increasing importance as an alternative option for conservation of plants, being the most efficient way to clonally propagate a species without depleting wild resources. Depending on the need of the species, *in vitro* methods offer a variety of tools such as *in vitro* seed germination, *in vitro* propagation or micropropagation, culture of cells and tissues, genetic transformation, bioreactor based production of plants, short-term storage/slow growth cultures and cryopreservation (Reed et al. 2011). The development of effective *in vitro culture* protocols are of great importance for conservation of rare/threatened plant species and especially medicinal plants, by virtue of producing uniform planting material to offset the pressure on natural populations (Kapaia et al. 2010). In addition, *in vitro* culture techniques are being used to study the growth habits of plants, as well as the ecological factors that influence their growth and development (i.e., *in vitro* ecology) which may again serve as an asset to conservation efforts (Reed et al. 2011). There are several reports on wild grown plants which are difficult to propagate conventionally, being multiplied through micropropagation, as *Adathoda beddomei* (Sudha and Seeni 1994), *Decalepis arayalpatra* (Sudha et al. 2005), *Embelia ribes* (Preetha et al. 2012), *Holostemma annulare* (Sudha et al. 1998, 2000), *Kaempferia galangal* (Preetha et al. 2008), *Trichopus zyelanicus* (Krishnan et al. 1995), etc. Many of the above species are tuber producing medicinal plants.

Micropropagation is also helpful when populations have decreased due to over exploitation by destructive harvesting, as in the case of several medicinal plants. Additionally, in such cases it helps to produce clonally uniform elite plants to meet the growing demand of the industries (Kapaia et al. 2010). The *in vitro* technologies and micropropagation also aid in producing genetically diverse and high quality propagules for reintroduction and translocation projects aimed at the restoration of damaged ecosystems, as and when required (Reed et al. 2011). A survey conducted recorded that approximately 40 tuber yielding medicinal plants from Western Ghats have been successfully propagated through tissue culture during the lasts two decades (Table 2). Different explants like node, stem-tip, leaf, rhizome, embryo, coteledonary node, etc. were used with different pathways for obtaining plantlets. Though technologies for many plants have been standardized, only a few are being commercially utilized, while many species were used for restoration in their natural habitat to enrich the population.

Apart from production of clonally uniform plants, micropropagation can also cater the needs for inducing genetic variability in plant species. This becomes necessary as in the case of medicinal root bearing and vegetatively propagated plants, where the threat to loss of the species is greater due to reduced intraspecific variability in addition to destructive harvesting, both

Table 2. Clonal propagation strategies developed for various medicinal root bearing plants of Western Ghats.

Sl no.	Species	*In vitro* strategy	Explant	Reference
1.	*Abrus precatorius*	Micropropagation (Indirect regeneration of shoots form callus)	Node	Biswas et al. 2007
2.	*Acorus calamus*	Shoot organogenesis	Rhizome buds	Anu et al. 2001
		Micropropagation	Rhizome tip	Ahmed et al. 2007
		Micropropagation	Seeds	Lee and Han 2011
		Micropropagation	Rhizome	Bhagat 2011
3.	*Alpinia galanga*	Micropropagation	Rhizome bud	Borthakur et al. 1998
		Direct regeneration of shoots	Rhizome	Rao et al. 2011
		Indirect regeneration of shoots form callus	Rhizome	
4.	*Aristolochia indica*	Micropropagation (Direct regeneration)	Node and shoot tip	Manjula et al. 1997
		Micropropagation Indirect regeneration	Leaf and internode	
		Micropropagation (Direct regeneration)	Node and shoot tip	Soniya and Sujitha 2006
		Micropropagation Indirect regeneration	Leaf and internode	
5.	*Aristolochia tagala*	Micropropagation (Indirect Shoot organogenesis)	Leaf	Remya et al. 2013
6.	*Asparagus racemosus*	Micropropagation	Node	Bopana and Saxena 2008
7.	*Boerhavia diffusa*	Indirect shoot organogenesis	Stem	Gupta et al. 2004
		Micropropagation	Node	Biswas et al. 2009

8.	*Cardiospermum halicacabum*	Micropropagation (Indirect shoot organogenesis)	Cotyledon, hypocotyl, cotyledonary node, leaf, internode and node	Babber et al. 2001
		Micropropagation	Node	Kalita et al. 2007
		Micropropagation	Axillary bud	Jahan and Anis 2009
9.	*Ceropegia bulbosa*	Micropropagation	Node	Patil 1998
		Micropropagation	Node	Goyal and Bhadauria 2006
		Micropropagation (Indirect organogenesis)	Epicotyl, hypocotyl, cotyledonary node and root	Phulwaria et al. 2013
10.	*Ceropegia sahyadrica*	Organogenesis	Node	Nikam and Savant 2007
11.	*Ceropegia intermedia*	Micropropagation	Node	Karuppusamy et al. 2009
12.	*Clerodendrum serratum*	Micropropagation	Node	Sharma et al. 2009
13.	*Coleus forskohlii*	Micropropagation	Stem tip	Bhattacharyya and Bhattacharyya 2001
		Micropropagation (Indirect organogenesis)	Leaf	Reddy et al. 2001
14.	*Costus speciosus*	Micropropagation	Embryo	Pal and Roy 1991
		Micropropagation	Rhizome	Roy and Pal 1991
15.	*Curculigo orchioids*	Micropropagation	Apical meristem	Francis et al. 2007
		Somatic embryogenesis and Micropropagation	Rhizome	Nagesh et al. 2010
16.	*Curcuma amada*	Micropropagation (Direct and indirect organogensis)	Rhizome and leaf sheath	Prakash et al. 2004
		Micropropagation	Rhizome	Das et al. 2010
17.	*Curcuma longa*	Micropropagation	Rhizome	Das et al. 2010
18.	*Cyclea peltata*	Micropropagation	Node	Abraham et al. 2010

Table 2. contd....

Table 2. contd.

Sl no.	Species	*In vitro* strategy	Explant	Reference
19.	*Decalepis arayalpathra*	Micropropagation	Node and Cotyledonary node	Sudha et al. 2005
20.	*Embelia ribes*	Micropropagation	Inflorescence segment	Shankaramurthy and Krishna 2006
		Micropropagation	Axillary shoot	Dhavala and Rathore 2010
		Micropropagation	Node/leaf	Preetha et al. 2012
21.	*Gloriosa superba*	Micropropagation	Apical and axillary bud	Hassan and Roy 2005
22.	*Glycrrhiza glabra*	Micropropagation	Apical and axillary bud	Thengane et al. 1998
23.	*Gymnema sylvestre*	Somatic embryogensis and organogenesis	Hypocotyl, cotyledon and seedling leaf	Kumar et al. 2002
		Micropropagation	Apical bud	Sharma and Bansal 2010
24.	*Hemidesmus indicus*	Micropropagation	Nodes and root segment	Sreekumar et al. 2000
		Micropropagation (Indirect organogenesis)	Axillary shoot	Siddique and Bari 2010
25.	*Holostemma annulare*	Micropropagation	*In vitro* root segment	Sudha et al. 2000
		Micropropagation (Direct and indirect organogenesis)	Axillary bud, node, internode and leaf	Martin 2002
		Micropropagation	Shoot	Geetha et al. 2009
26.	*Ipomoea mauritiana*	Micropropagation (Indirect organogenesis)	Node	Priya et al. 2007

27.	*Kaempferia galanga*	Micropropagation	Rhizome bud	Vincent et al. 1992
		Micropropagation	Rhizome bud	Chirangini et al. 2005
		Micropropagation (Somatic embryogenesis)	*In vitro* leaf base segment	Preetha et al. 2008
		Micropropagation	Rhizome bud	Chitra et al. 2005
		Micropropagation	Rhizome bud	Mohanty et al. 2011
28.	*Kaemvferia rotunda*	Micropropagation	Rhizome bud	Chirangini et al. 2005
29.	*Mucuna pruriens*	Micropropagation (Indirect organogenesis)	*In vitro* seedlings	Chattopadhyay et al. 1995
		Micropropagation	Coteledonary node of *in vitro* seedlings	Faisal et al. 2006
		Micropropagation (Indirect organigensis)	Node and internode	Lahiri et al. 2012
30.	*Ocimum gratissimum*	Micropropagation	Node	Gopi et al. 2006
		Micropropagation	Node	Saha et al. 2012
31.	*Plumbago rosea*	Micropropagation (Somatic embryogenesis form callus)	Leaf	Das and Rout 2002
		Micropropagation	Node	Binoy et al. 2007, Binoy and Satheshkumar 2010
32.	*Plumbago zeylanica*	Micropropagation	Node	Selvakumar et al. 2001
		Micropropagation	Node	Verma et al. 2002
33.	*Pueraria tuberosa*	Micropropagation	Shoot sprouts from tubers	Rathore and Shekhawat 2009
34.	*Rauwolfia serpentina*	Micropropagation	Shoot tip and node	Jain et al. 2003
35.	*Rubia cordifoia*	Micropropagation	Axillary bud	Ghatge et al. 2011

Table 2. contd....

Table 2. contd.

Sl no.	Species	*In vitro* strategy	Explant	Reference
36.	*Tinospora cordifolia*	Micropropagation	Mature nodes	Raghu et al. 2006
		Micropropagation	Node and internode	Khanapurkar et al. 2012
37.	*Tylophora indica*	Micropropagation (Indirect organogenesis)	Leaf	Thomas and Philip 2005
		Micropropagation	Node	Faisal et al. 2007
38.	*Typhonium flagelliforme*	Micropropagation	Rhizome bud	Sai 2000
39.	*Withania somnifera*	Micropropagation (Indirect organogenesis)	Stem	Manickam et al. 2000
		Micropropagation (Indirect organogenesis)	Hypocotyl, root and cotyledonary leaf segments	Rani et al. 2003
		Micropropagation (Indirect organogenesis)	*In vitro* and *in vivo* leaf	Rout et al. 2011
40.	*Zingiber officinale*	Micropropagation	Rhizome buds	Sharma and Singh 1997
41.	*Zingiber zerumbet*	Micropropagation	Rhizome buds	Stanly and Keng 2007
		Micropropagation	Rhizome buds	Mohanty et al. 2012

of which increase the risk of sustainability of a small/declining population. Presently, for selection of populations requiring priority for conservation, intraspecific genetic variation is widely accepted as an important factor. Loss of genetic diversity within populations of a species can lead to inbreeding depression which may cause reduced fitness and finally put their existence at risk. Moreover, in the case of species already in red-listed status, the populations will generally be small and here the requirement of sufficient genetic variation will be more important to help sustain the species. Also, only if a certain amount of genetic variability exists in a population, can natural selection mould species to adapt and evolve based on the changing environment. Hence, induction of genetic variation and establishment of variants is as necessary as the production of clones for efficient conservation of medicinal root bearing plants and especially vegetatively propagated ones.

Reintroduction of in vitro propagated plants

The reintroduction of plants is becoming an increasingly utilized strategy in plant conservation. Reintroduction refers to the deliberate establishment of individuals of an extinct/endangered species into an area and/or habitat where it has become extirpated with an aim to establish a viable self-sustaining population for conservation purposes. When conventional propagation is insufficient or impaired, micropropagation and reintroduction are recommended. Reintroduction and translocation projects aimed at the restoration of damaged ecosystems need genetically diverse, high quality propagules in large numbers that can be provided by employing different tissue culture methods. Additionally, on introducing individuals representing genetically diverse populations of a species, care should be taken to maintain genetic stability of the cloned materials and these considerations are more significant in the case of plants which have highly threatened status (Reed et al. 2011, Wochok 1981, Fay 1992, Fay 1994, Maunder 1992, Krishnan et al. 2011).

Experimental reintroductions of eight medicinal plants, including the endemic species of Western Ghats such as *Decalepis arapyalpathra*, were attempted at the Biotechnology and Bioinformatics Division of JNTBGRI during 2000–2007. About 100–350 plants were reintroduced into their native habitats, recording 78–95% establishment after 1–2 yr. In addition, a few micropropagated plants planted in the natural forest segments of JNTBGRI campus also showed luxuriant growth without any special care as reintroduced plants. The other medicinal plants included in this are *Acorus calamus, Blepharestemma membranifolia, Calophyllum apetalum, Celastrus paniculatus, Heracleum candolleanum, Mahonia leschenaultii, Morinda umbellate*, etc. in addition to terrestrial orchid species *E. cullenia*, a medicinally important tuber yielding plant has been successfully multiplied and reintroduced into native habitat as well as JNTBGRI campus. Thus the utility of micropropagation system was proven effective for the eco-restoration of threatened medicinal plants of Western Ghats. However, the reintroduction carried out at JNTBGRI is on the experimental scale. Thus, to realize ecological restoration, extended planting of more plants in more than one locality may be affected. The exact number of plants that needs to be reintroduced varies with species and heterogeneity of source population (Krishnan et al. 2011).

Conservation through secondary metabolites production *in vitro*

Other in vitro tissue culture systems and genetic transformation

To cope with the alarming situation of eroding natural ecosystem and dwindling resources of medicinal plants, the biotechnological tools are important for multiplication and genetic enhancement of medicinal plants by adopting techniques such as *in vitro* regeneration and genetic transformation (Abdin and Kamaludeen 2006). Advances in tissue culture combined with improvement in genetic engineering techniques, especially transformation technology, have opened new avenues for high volume production of pharmaceuticals, neutraceuticals and other beneficial substances (Khan et al. 2009). These could also be utilized for the production of secondary metabolites using bioreactors. Establishment of *in vitro* strategies such as callus, cell suspension, root cultures, etc. for production of medicinal plant derived secondary metabolites helps to decrease the harvest of the natural population for the medicinal bioactives and thereby facilitates their conservation (Reed et al. 2011).

Despite the progress made in chemical synthesis of a wide range of compounds similar to those produced by the plants, extraction of secondary metabolites from plants is still of considerable commercial importance. Economic synthesis of a large number of these metabolites is difficult or virtually impossible and in several cases consumer acceptance is greater for the natural product in comparison to an artificially produced one.

Lately, plant cell and tissue culture technology has gradually replaced whole plant cultivation, as a source of useful secondary metabolites. Cell suspension and callus cultures often produce a desired compound initially, but after a few passages production decreases which may be due to either inadequate nutrition or genetic variations. Plant cells in suspension cultures often undergo spontaneous genetic variation in terms of accumulation of secondary metabolites, which leads to heterogeneous population of cells in a suspension culture. These observations have led researchers to use organized plant tissue cultures or even cultures of organs such as roots for the production of secondary metabolites. Additionally, for plants accumulating medicinal bioactive in the roots, root cultures are a preferred option as differentiation has shown to be important for metabolite production when compared to cell suspension/callus cultures. But, root cultures normally need an exogenous phytohormone supply and grow very slowly, resulting in the poor or negligible synthesis of secondary metabolites (Chattopadhyay et al. 2002, Sasson 1991, Hu and Du 2006). The different culture systems developed for various medicinal root bearing plants is shown in Table 3.

Genetic transformation using *Agrobacterium tumefaciens* and *A. rhizogenes* is another alternative to enhance production of secondary metabolites. Considering the medicinal root containing plants, *A. tumefaciens* transformed systems (like callus/crown gall) are less preferred in comparison to hairy root cultures obtained through *A. rhizogens* transformation. The hairy root culture system is a potential approach for the production of secondary metabolites, especially pharmaceuticals, because it has several beneficial traits, such as rapid growth rate, easy culture and genetic manipulation, high productivity under hormone-free culture conditions, and most importantly, an increased ability to synthesize useful metabolites that cannot be produced by unorganized cells even higher than plant roots. Thus, many medicinally rooted plants have been transformed successfully by *A. rhizogenes* and the hairy roots induced show a relatively high productivity of secondary metabolites (Chattopadhyay et al. 2002, Sasson 1991, Hu and Du 2006) and the different transformed cultures developed for the medicinal root bearing plants of Western Ghats are shown in Table 4.

Presently, different methods are known for increasing secondary metabolites in the various plant culture systems including hairy root cultures, namely use of biotic or abiotic elicitors that could stimulate the metabolic pathways as in the intact plant or addition of a precursor of the desired compound in the culture medium. Elicitation and precursor treatment are being attempted extensively with the objective of achieving an immensely enhanced level of production that can make the scaling up strategy economically feasible (Chattopadhyay et al. 2002, Sasson 1991, Hu and Du 2006). Application of elicitor/precursor treatment in enhancing secondary metabolite production in different *in vitro* culture systems developed for the medicinal root bearing plants are given in Table 5.

Bioreactor and scaling-up

With respect to large-scale cultivation, plant cell suspension culture is desirable for industrial production of plant-derived biochemicals, as the production technology is comparable to that used for microbial systems. Due to the same reasons, scaling up of cell suspension cultures has achieved

Table 3. Other *in vitro* cultures developed for medicinal root bearing plants.

Sl. No.	Species	Strategy	Explant	Compound	Reference
1.	*Asparagus racemosus*	Callus culture	Root and node	Shatavarin	Pise et al. 2011
		Cell culture	-	Shatavarin	Pise et al. 2012
2.	*Boerhavia diffusa*	Callus culture	Leaf and stem	-	Jain et al. 2003
		Root culture	-	-	Jenifer et al. 2012
3.	*Ceropegia juncea*	Callus culture	-	Cerpegin	Nikam and Savant 2009
4.	*Coleus forskohlii*	Root culture	-	Forskolin	Krombholz et al. 1992
5.	*Coscinium fenestratum*	Callus culture	Leaf and petiole	Berberine	Khan et al. 2009
6.	*Costus speciosus*	Callus culture	Rhizome	Diosgenin	Jain et al. 1984
		Shoot culture	Embryo	Diosgenin	Pal and Roy 1991
7.	*Embelia ribes*	Embryogenic callus	Leaf	Embelin	Raghu et al. 2011
8.	*Glycrrhiza glabra*	Cell suspension	*In vitro* seedling derived callus	-	Arias-Castro et al. 1993
		Callus culture	Leaf	Glycrrhizin	Kakutani et al. 1999
9.	*Gymnema sylvestre*	Callus	Leaf, stem and petioles	Gymnemic acid	Ahmed et al. 2012
10.	*Hemidesmus indicus*	Root culture	Node	2-hydroxy 4-methoxy benzaldehyde	Sreekumar et al. 1998
11.	*Mucuna pruriens*	Cell suspension	-	3,4-dihydroxyphenylalanine (L-DOPA)	Raghavendra et al. 2012
12.	*Plumbago rosea*	Callus and Cell suspension	Leaf	Plumbagin	Komaraiah et al. 2001
		Callus and cell suspension	-----	Plumbagin	Satheeshkumar and Seeni 2002
		Cell suspension and somatic embryogensis	Leaf	Plumbain	Komaraiah et al. 2004

13.	*Pueraria tuberosa*	Callus	Leaf and petiole	(Isoflavanoids) puerarin, genistein, genistin, daidzein	Vaishnav et al. 2006
		Cell suspension	-	Puerarin, genistein, genistin, daidzein	Goyal and Ramawat 2008
		Shoot culture	-	Puerarin	Sharma et al. 2011
14.	*Rubia cordifolia*	Cell suspension	-	Antraquinones	Suzuki et al. 1985
		Shoot culture	Seedlings	Antraquinones	Radha et al. 2010
15.	*Tinospora cordifolia*	Callus and cell suspension		Berberine and jatrorrhizine	Chintalwar et al. 2003
		Cell suspension	-	Berberine	RamaRao et al. 2006
16.	*Withania somnifera*	Root culture	-	Withanolide-A	Murthy and Praveen 2012
		Callus and shoot culture	-	-	Sabir et al. 2012

Table 4. Genetically transformed *in vitro* systems for production of secondary metabolites.

Sl. No.	Species	Method	Objective	Compound of interest	Reference
1.	*Abrus precatorius*	Hairy root culture	Transformation	Glycyrrhizin	Karsawa and Dixit 2009
		Hairy root culture	Enhanced production	Glycyrrhizin	Dixit and Vaidya 2010
		A. tumefaciens Transformed cell suspension	Enhanced production	Glycrrhizin	Karsawa et al. 2011
2.	*Alpinia galanga*	Hairy root culture	Enhanced production	Acetoxychavicol acetate (ACA)	Rao et al. 2012
3.	*Boerhavia diffusa*	Hairy root culture	Transformation	-	Jenifer et al. 2012
4.	Coleus forskohlii	*A. tumefaciens* Transformed callus and shooty teratomas	Compound production	Forskolin	Mukherjee et al. 1996
		Hairy root culture	Enhanced production	Forskolin	Sasaki et al. 1998
		A. tumefaciens Transformed cell suspension	Enhanced production	Forskolin	Mukherjee et al. 2000
		Hairy root culture	Enhanced production	Rosmarinic acid	Li et al. 2005
5.	*Decalepis arayalpatra*	Hairy root culture	Transformation and compound production	2-hydroxy-4-methoxy benzaldehyde	Sudha et al. 2012
6.	*Glycrrhiza glabra*	Hairy root culture	Compound production	Phenolic substances	Toivonen and Rosenqvist 1995
		Hairy root culture	Biomass enhancement	-	Mehrotra et al. 2008
7.	*Holostemma annulare*	Hairy root culture	Transformation	-	Karmarkar et al. 2001
8.	*Mucuna pruriens*	*A. tumefaciens* transformed shoots	Transformation	-	Raghavendra et al. 2012

9.	*Plumbago rosea*	Hairy root culture	Transformation and compound production	Plumbagin	Gangopadhyay et al. 2008
		A. rhizogenes transformed shoots	Transformtion	-	Satheeshkumar et al. 2009
10.	*Plumbago zeylanica*	Hairy root culture	Enhanced production	Plumbagin	Verma et al. 2002
11.	*Rauvolfia serpentina*	Hairy root and *A.rhizogens* transformed shoots	Transformation	Total alkaloids	Benjamin et al. 1993
		Hairy roots and *A. rhizogens* transformed cell suspension	Compound production	Vinorine, perakine, vomilenine and ajmaline	Falkenhagen et al. 1993
12.	*Rubia cordifolia*	*A. rhizogens* transformed callus culture	Influence of rol genes on Compound production	Anthraquinones	Shkryl et al. 2008
		A. rhizogens transformed cell suspension	Enhanced production	Anthraquinones	Bulgakov et al. 2010
13.	*Tylophora indica*	Hairy root culture	Transformation and Compound production	Tylophorine	Chaudhuri et al. 2005
		A. rhizogenes transformed shoots	Compound production	Tylophorine	Chaudhuri et al. 2006
14.	*Withania somnifera*	Hairy root culture	Transformation and compound production	Withanaloids	Kumar et al. 2005
		Hairy root culture	Compound production	Withanolide A	Murthy et al. 2008

Table 5. Elicitation and precursor treatment for enhanced metabolite production.

Sl. No.	Species	Culture	Elicitor/precursor	Metabolite	Details of Enhancement	Reference
1.	*Abrus precatorius*	Transformed cell suspension	Various Fungal elicitor	Glycrrhizin	3.8–4.9 fold increase	Karsawa et al. 2011
2.	*Gloriosa superba*	Root culture	*p*-coumaric acid, tyramine, trans-cinnamic acid	Colchicine	7.92 fold	Ghosh et al. 2002
		Root culture	Methyl jasmonate and aluminium chloride	Colchicine	50-fold and 63-fold	Ghosh et al. 2006
		Callus	Phenylalanine and tyrosine	Colchicine	67.5 fold	Sivakumar et al. 2004
3.	*Glycrrhiza glabra*	Plantlets	Methyl jasmonate and salicylic acid	Glycyrrhizin	3.8–4.1 fold	Shabani et al. 2009
4.	*Gymnema sylvestre*	Cell suspension	Fungal elicitor	Gymnemic acid	9 fold	Devi and Srinivasan 2011
5.	*Mucuna pruriens*	Cell suspension	Methyl jasmonate, chitin, pectin, L-tyrosine	3,4-dihydroxyphenylalanine (L-DOPA)	Several fold	Raghavendra et al. 2012
6.	*Plumbago rosea*	Immobilized cell culture	Chitosan	Plumbagin	2.5–21 fold	Komaraiah et al. 2003
		Hairy root culture	Yeast carbohydrate fraction, chitosan, methyl jasmonate	Plumbagin	1.2–2 fold	Gangopadhay et al. 2011
		Hairy root culture	Methyl jasmonate	Plumbagin	4.6 fold	Martin et al. 2011
7.	*Rauvolfia serpentina*	Embryogenic suspension cultures	Methyl jasmonate	Reserpine	1.4–14.9 fold	Harisaranraj et al. 2009
8.	*Withania somnifera*	Hairy root culture	Chitosan, nitric oxide, acetyl salicylic acid, jasmonic acid, triadimefon	Withaferin A and withanolide A	Several fold	Doma et al. 2012
		Adventitious root culture	Salicylic acid	Withanolides	20–48 fold	Sivanandhan et al. 2012

commercial level application, though in limited cases, with the production of some important compounds. The first success was Mitsui Petrochemical Industry, Japan producing shikonin (a dye) on commercial scale using cell cultures of *Lithospermum erythrorhizon*. The company also succeeded in using cell culture technology to produce berberine. Recently, a milestone has been achieved by successful commercial production of 20 kg paclitaxel (under the trade name of Genexol) per year by culturing plant cells in large bioreactors of 32kL capacity. However, genetic variation in terms of accumulation of secondary metabolites has posed a commercial hurdle to the large-scale production of secondary metabolites (Chattopadhyay et al. 2002, Sasson 1991).

Though several decades have passed since the hairy root culture system was first established, the system has not been utilized significantly in practical industry production. However, recent developments indicate that the hairy root culture technology for the production of pharmaceuticals has moved from small-scale laboratory to large-scale industrial production: the German company ROOTec has specialized into broad-scale cultures of hairy roots, producing valuable metabolites such as camptothecin and podophyllotoxin (http://www.rootec.com). To achieve a further breakthrough in the application of hairy root culture systems, future research should focus on the establishment of effective and economical systems of up-scaling that can reduce the consumption together with obtaining the biggest benefits (Hu and Du 2006, Guillon et al. 2006).

Cell cultures of *Pueraria tuberosa* were grown in vessels of different sizes and scaled-up in 2L stirred tank bioreactor to obtain stable growth and total isoflavonoid yield of 76.6 mg l^{-1} (Sharma et al. 2009). *P. tuberosa* shoot cultures when grown in growtek bioreactor with aeration (20% w/v) produced 2.3 fold higher puerarin content than the shoots grown without aeration (Sharma et al. 2011). When embryogenic suspension cultures of *Rauwolfia serpentina* were grown in bioreactor and exposed to different concentrations (50–500 µM) of methyl jasmonate (MJ), reserpine content increased significantly. However, the fresh weight, dry weight and growth ratio of embryos was strongly inhibited by increasing MJ concentrations. The results of Harisaranraj et al. (2009) suggested MJ elicitation to be beneficial for reserpine accumulation in the embryogenic cell suspension cultures (Harisaranraj et al. 2009). Adventitious root cultures and *A. rhizogenes* transformed root cultures of *Coleus forskohlii* were grown in Gamborg's B5 medium containing 1 mg/l IBA and hormone free media respectively and scaled up in 20L glass jar bioreactors (Krombholz et al. 1992). A fast growing hairy root clone of *Glycrrhiza glabra* grown in a 5L capacity mechanically agitated bioreactor provided with a nylon mesh septum produced 20 times increased biomass over initial inoculum after 30 days of sterile run (Mehrotra et al. 2008).

Micropropagation by conventional techniques being labour intensive, automation of using bioreactors has been proposed by several workers (Preil 1991, Takayama and Akita 1994, Leathers et al. 1995) to reduce the cost of micropropagation. Bioreactors provide more precise control of plant growth, gas exchange, illumination, medium, agitation, temperature, pH, etc. and are designed in such a way to large scale cell, tissue and organ culture in liquid media. The bioreactor system was applied to embryogenic and organogenic cultures of several plant species (Levin et al. 1988, Preil 1988).

Conservation through *in vitro* storage and preservation

Slow growth cultures for short to medium term storage

Though plants successfully initiated into culture can be maintained indefinitely at 25 ± 2°C with regular subculturing, compiling and maintaining these comprehensive *in vitro* collections is time consuming and remains relatively costly compared to other forms of macro-propagation (cutting, division, seed) (Engelmann 1991, Withers and Engelmann 1997, Rao 2004, Engelmann 2011, Bunn et al. 2011). However, for better preservation the clonal lines can be (a) cool stored (5–10°C) for months to (rarely) year with reduced subculture frequency or (b) cryogenically stored at −196°C in liquid nitrogen (LN) (or −180°C in LN vapour phase). The *in vitro* conservation methods employed vary depending on the duration of storage required. Short- and medium-term storage is performed by reducing the growth and increasing the interval between subcultures. Slow growth procedures allow clonal plant material to be held for 1–15 years under tissue culture conditions with periodic sub-culturing, depending on species (Engelmann 1991, Withers and Engelmann 1997, Rao 2004, Engelmann 2011, Bunn et al. 2011).

There are several methods by which slow growth can be achieved, by lowering the temperature with low light intensity or even darkness. Regeneration and successful propagation of genetically stable seedlings from cultures are prerequisite in any *in vitro* conservation effort. Organized cultures such as shoots, embryos, etc. are used for slow growth storage since undifferentiated tissues are more vulnerable to somaclonal variation. Generally, for tropical species which are cold sensitive, 15° and 20°C are employed for slow growth (Engelmann 1991, Withers and Engelmann 1997, Rao 2004, Engelmann 2011, Bunn et al. 2011) at short-term storage, while 0–5°C are used for the cold-tolerant species. It is also possible to limit growth by modifying the culture medium, mainly by reducing the sugar and/or mineral elements concentration, change in the nature and/or concentration of growth regulators, addition of osmotically active compounds and reduction of oxygen level available to cultures by covering explants with a layer of liquid medium or mineral oil. *In vitro* conservation strategies developed for the medicinal root bearing plants of Western Ghats is given in Table 6.

Mannitol and sorbitol are the most common osmoregulants used in cultures for slow growth preservation. For maintaining *Coleus forskholii* culture under slow growth conditions, 22 ± 2°C with 16 h photoperiod and 3M mannitol in the media was found best (Dube et al. 2011). However, germplasm storage of *Curcuma amada* and *Curcuma domestica* was attempted at relatively high ambient temperatures (24–29°C) by inhibiting growth using mannitol (25 g l^{-1}) and extended subculture period. When stress conditions are coupled with an overlay of liquid paraffin, storage can be prolonged upto 1 year as observed in the case of *Zingiber officinale and Zingiber* sp. with high survival rates upon revival (70–100%) (Dekkers et al. 1991).

Tyagi et al. (2006) tested a total of 33 genotypes of cultivated and wild species of *Zingiber* for conservation through *in vitro* rhizome formation on a control medium under 16-h light condition. Following this protocol around 160 genotypes of cultivated and wild species of *Zingiber*, collected from different geographical regions of India, are presently being conserved at *in vitro* Genebank of National Bureau of Plant Genetic Resources, New Delhi (Tyagi et al. 2006). *In vitro* conservation of *Tylophora indica* shoots was carried out in full strength MS Medium (FMS) under Standard Culture Conditions for six months without subculture with full potential to regenerate, producing viable shoots and nodes. The root production remained unaffected due to conservation, showing high rooting activity in mannitol and low temperature treatments (Rajasekharan et al. 2009). Though slow growth procedures have been developed for a wide range of species, their use in conservation of genetic resources has been done for only a few species which are agriculturally important like *Musa* sp., potato, sweet potato, cassava, yam, allium, etc. (Rao 2004).

However, the *in vitro* storage of large quantities of material induces various problems: laboratory management of plant material which needs to be regularly subcultured, risks of genetic variation, which increase with *in vitro* storage duration, and can lead to the loss of trueness to type (Engelmann 1991, Withers and Engelmann 1997, Rao 2004, Engelmann 2011, Bunn et al. 2011, Ramakrishnappa 2002, Ticktin 2004, Schippmann et al. 2002, Edwards 2004, Uniyal et al. 2000).

Encapsulation/Synthetic seeds

Though, the *in vitro* plant establishment and maintenance is practically a feasible approach, it is done manually and at the risk of microbial contamination. Further, the stress due to regular sub-culturing and

Table 6. *In vitro* conservation strategies for medicinal root containing plants of Western Ghats.

Sl no.	Species	Stored material	Optimum period of storage (months)	Culture conditions	Reference
1.	*Curcuma* spp.	Multiple shoot clumps	9–12.5	MS + 24.6μM 2iP	Tyagi et al. 2004
2.	*Decalepis arayalpatra*	Shoot tip cuttings	>7	1/2 MS+4.44 μmBA+2% mannitol;25°C	TBGRI 1997a
3.	*Hemidesmus indicus*	Shoot cultures	18–22	1/2 MS + 20 g/L sucrose	George et al. 2010
4.	*Holstemma annulare*	*In vitro* nodes/Apical buds	12	1/2 MS+2% mannitol; 25°C; Culture tubes with pp caps	TBGRI 1997a
5.	*Rauvolfia serpentina*	*In vitro* nodes with 2 axillary buds	15	MS+4.44 μM BA+0.54 μM NAA; 15°C; Culture bottles with pp caps	Neelam and Chandel 1992
6.	*Zingiber* sp.	*In vitro* rhizome	16–20	MS; 16 h light	Tyagi et al. 2006

(a) Unpublished data from TBGRI. National gene bank for medicinal and aromatic plants- consolidated report for phase I, pp 41–50; 1997.

prolonged *in vitro* maintenance may sometimes cause cryptic genetic defects, ultimately leading to the loss of original culture and biosynthetic capacity. To overcome such constraints, synthetic seeds or alginate encapsulation of plant material offers an approach for germplasm storage and conservation under *in vitro* conditions. It can conserve the cultures in terms of both viability and productivity as it shares the benefits of clonal multiplication with those of seed propagation and storage. The efficiency of using alginate encapsulated propagules for storage lies in their small size and relative ease of handling and transportation. The increasing attention on synthetic seeds for germplasm conservation and propagation necessitates the assessment of genetic stability of stored plant material. Therefore, it is imperative to establish the genetic uniformity of alginate encapsulated and conserved propagules following their conversion (Ara et al. 2000, Reddy et al. 2012, Rao et al. 2004, Singh et al. 2006, Mehrotra et al. 2012).

Alginate encapsulation for germplasm storage and conservation is reported for various industrially important plant species such as *Rauvolfia serpentina* (Ray and Bhattacharya 2008), *Withania somnifera* (Singh et al. 2006), different species of *Ocimum* (Mandal et al. 2000) and *Tylophora indica* (Faisal and Anis 2007), etc. as shown in Table 7. Alginate encapsulated micro shoots of *Glycrrhiza glabra* stored in a moist-environment for 6 months at 25 ± 2°C attained 98% re-growth within 30 days of incubation on MS medium supplemented with 0.1 mg/l IAA and healthy plants were established in a glass house with 95% survival (Mehrotra et al. 2012). Artificial seeds of *in vitro* nodes of *Aristolochia tagala* and *in vitro* shoot tips of *Withania somnifera* were also developed by encapsulation in 3% (m/v) sodium alginate with 1% (m/v) and 75 mM Calcium chloride respectively. Shoots were successfully

Table 7. Synthetic seeds technology developed for preservation of medicinal plants.

Sl. No.	Species	Explants	Reference
1.	*Aristolochia tagala*	*In vitro* nodes	Remya et al. 2012
2.	*Curcuma longa*	Adventitious buds	Sajina et al. 1997
3.	*Glycrrhiza glabra*	Microshoots	Mehrotra et al. 2012
4.	*Hemidesmus indicus*	Somatic embryos	Cheruvathur et al. 2013
5.	*Ocimum gratissimum*	Axillary buds	Mandal et al. 2000
6.	*Plumbago rosea*	Hairy root	Gangpopadhayay et al. 2011
7.	*Rauvolfia serpentine*	Shoot tips	Ray and Bhattacharya 2008
		Microshoots	Faisal et al. 2012
8.	*Tylophora indica*	Shoots	Faisal and Anis 2007
9.	*Withania somnifera*	Shoot tips	Singh et al. 2006
		Somatic embryos	Thobunluepop et al. 2009
10.	*Zingiber officinale*	Shoot buds	Sharma et al. 1994
		Micro shoots	Sundararaj et al. 2010

regenerated from the encapsulated nodes cultured on MS medium supplemented with 3μM BAP and 0.5μM kinetin for *A. tagala* (Remya et al. 2012) maximum percentage response (87%) for conversion of encapsulated shoot tips into plantlets was achieved on MS medium supplemented with 0.5 mg/l IBA after 5 weeks of culture for *W. somnifera* (Singh et al. 2006). Even transformed tissues have been encapsulated in alginate beads for future use towards regeneration or production of secondary metabolites. Selected elite hairy root clones of *Plumbago rosea* encapsulated with sodium alginate matrix and conserved up to 6 months exhibited high regeneration potential without disturbing the innate plumbagin content. After 6 months, regeneration potential of hairy roots decreased sharply and a sudden decrease in fresh root biomass was recorded in 8 months old alginate encapsulated roots and a slight decrease in plumbagin content was noted in roots regenerated from the 8 month stored seeds (Gangopadhyay et al. 2011).

Significant commercial application of hydrated calcium alginate-based or dessicated polyoxyethylene glycol-based artificial seeds can come about in germplasm conservation through cryopreservation, provided the technology is perfected without drawbacks (Ara et al. 2000).

Cryopreservation

For long-term storage, cryopreservation, i.e., storage at very low temperatures, usually that of liquid nitrogen (–196°C), is the only current method. At this temperature, all cellular division, metabolic and biochemical events are arrested. The plant material can be stored without alteration or modification, for a theoretically unlimited period of time. Moreover, the cultures are stored in a small volume, protected from contamination, requiring a very limited maintenance (George et al. 2010). *In vitro* cultures as well as seeds of field grown plants can be successfully cryopreserved without deleterious effects on regeneration and with an opportunity for conservation of endangered medicinal plants. The first report on successful cryopreservation of plants was published in 1960 (Sakai 1960).

Cryopreservation was successfully established with seeds of *Abrus precatorius, Coleus forskohlii, Hemidesmus indicus, Ocimum gratissimum* and *Tylophora indica* and germination was tested after storage in liquid nitrogen for one week. These seeds contained 7–15% moisture at seed shedding and after desiccation to 4–7% moisture they exhibited 80–95% germination which was not affected significantly by one week storage in liquid nitrogen. Seeds isolated from ripe berries of *Rauvolfia micrantha* and *Embelia ribes* contained 40% moisture with germination of 88% and 78% respectively. The seeds of *R. micrantha* desiccated to 5–7% moisture level recorded 73% germination after cryopreservation. However, seeds of *E. ribes* desiccated down to 4%

moisture germinated well (77%) under ambient conditions but failed to germinate after storage in liquid nitrogen. The study reveals a possible link between seed moisture content and liquid nitrogen tolerance in some, but not all medicinal plants of India (Decruse et al. 1999).

Unpublished results from our laboratory indicate that in *Coscinium fenestratum* cryopreservation is not promising, as cryopreserved embryos recorded only 34% germination against 56% in desiccation control (60 min desiccation) (Krishnan et al. 2011). The work describing cryopreservation of *Rauvolfia serpentina*, analyzes the effects of type and size of explants, sucrose preculture (duration and concentration) and vitrification treatment. It was observed that sucrose preculture had a positive role in survival and subsequent regrowth of the cryopreserved explants. Seven days on 0.5 M sucrose solution significantly improved the viability of nodal segments. PVS2 incubation for 45 minutes combined with a 7-day preculture gave the optimum result of 66 percent. Plantlets derived after cryopreservation resumed growth and regenerated normally (Ray and Bhattacharya 2008).

An efficient cryopreservation technique for *in vitro* grown shoots of ginger (*Zingiber officinale*) was developed based on encapsulation dehydration, encapsulation vitrification and vitrification procedures. Pregrowth and serial preculture were needed to obtain the best re-growth for all techniques. The vitrification procedure resulted in higher re-growth (80%) when compared to encapsulation vitrification (66%) and encapsulation dehydration (41%). In all the cryopreservation procedures tested, shoots grew from cryopreserved shoot tips without intermediary callus formation (Yamuna et al. 2007). The various cryopreservation strategies developed for medicinal root bearing plants are explained in Table 8.

Molecular markers

Molecular markers are very important with respect to plant conservation as the analysis of genetic stability is unavoidable before utilization of the preserved samples. Genetic variability assessment is also necessary to differentiate between the accessions collected for conservation. Presently, DNA techniques, especially those based on the Polymerase Chain Reaction (PCR) such as microsatellites or Simple Sequence Repeats (SSR), Random Amplified Polymorphic DNA (RAPD), Inter Simple Sequence Repeats (ISSR) and Amplified Fragment Length Polymorphism (AFLP) are preferred for the genetic analyses partly because these molecular markers provide a larger number of potentially polymorphic loci than allozymes. Tissue and environment independence in expression of DNA based markers have made them more reliable over morphological and isozyme markers. Among various DNA-based markers, though RFLP can be used for screening genetic stability of tissue cultured plants, the method involves use of

Table 8. Cryopreservation strategies for medicinal root containing plants of Western Ghats.

Sl no.	Species	Category	Stored material	Method	% Germination/regeneration after LN storage	Reference
1.	*Coscinium fenestratum*	Intermediate seeds	Zygotic embryos	Dessication	34(56)	TBGRI 2008b
2.	*Dioscorea bulbifera*	Clonal	Shoot tips	Vitrification	58	Papiya et al. 2009
		Clonal	Embryogenic callus	Pregrowth+ Encapsulation-dehydration	67.8	Mandal et al. 2009
3.	*Dioscorea wallichi*	Clonal	Shoot tips	Vitrification	37(71)	Mandal et al. 1996
4.	*Holostemma annulare*	Clonal/Orthodox seeds	Shoot tips	Pregrowth+ Encapsulation-dehydration	54.2(60)	Decruse et al. 1999, 2002, 2004
5.	*Kaempferia galanga*	Clonal	Shoot tips	Pregrowth+ Vitrification	50(64)	TBGRI 2008b
		Clonal	Somatic embryos	Pregrowth+ desiccation	42.8(57.1)	TBGRI 2008b
6.	*Rauvolfia serpentine*	Intermediate seeds	Nodal segments	Pregrowth+ Vitrification	66	Avik and Sabita 2008
7.	*Zingiber officinale*	Clonal	*In vitro* Shoot tips	Pregrowth+ Vitrification	80	Yamuna et al. 2007

2008b Unpublished data from TBGRI. National gene bank for medicinal and aromatic plants—project completion report, pp. 41–78.

expensive enzymes, radioactive labelling and extensive care, therefore, appears unsuitable. RAPD and ISSR markers, on the other hand, require only small amount of DNA sample without involving radioactivity tests and are simpler as well as faster (Lakshmanan et al. 2007).

Tissue cultured *Alpinia galanga* plantlets were subjected to assessment of genetic stability considering the fact that associated *in vitro* stress might result in breakdown of control mechanism causing instability of the genome. Two DNA based molecular markers were used to assess the genetic fidelity of *in vitro* regenerated *Alpinia galanga* L. through axillary buds from unsprouted rhizomes (Parida et al. 2011).

A protocol for *in vitro* propagation of *Asparagus racemosus* by axillary branching method was developed followed by inter simple sequence repeat marker assay to establish the clonal fidelity of the regenerants (Saxena and Bopana 2009). The genetic variability in *Boerhavia diffusa* between accessions of different geographical origin within the Indian territory was assessed through RAPD markers, the plants being extensively used in herbal medicines as well as in the Ayurvedic system due to its clinically important compounds (Shukla et al. 2003). The genetic fidelity of *in vitro* raised field grown plants (from nodal explants) and their mother plant was ascertained by random amplified polymorphic DNA markers, and micropropagation method established for *Clerodendrum serratum* (Sharma et al. 2009). Karyological changes in the callus tissue of *Costus speciosus* at different ages have been analysed both by counting chromosome number and quantitating DNA content through cytophotometry (Mukhopadhyay and Sharma 1990). RAPD analysis of regenerated plantlets of *Curcuma amada* revealed 103 scorable bands from 10 primers, including nine polymorphic bands, which were absent in control. However, most of the regenerated plantlets were similar to the mother plants (Prakash et al. 2004). The genetic purity of micropropagated *C. amada* and *Curcuma longa* plantlets was analyzed using RAPD and protein profiles (Das et al. 2010).

Molecular profiling of micropropagted plantlets of *Kaempferia galanga* based on RAPD and ISSR markers revealed uniform banding pattern similar to those of mother plants (Mohanty et al. 2011). Cytological analysis of *Mucuna pruriens* regenerants revealed chromosome number stability (2n=22) with complete absence of aneuploidy. RAPD and ISSR analyses of total genomic DNA supported stability at the molecular level among all the clones when compared with their mother plants (Lahiri et al. 2012). Mehrotra et al. (2012) conducted ISSR and RAPD analysis to test the genetic stability of stored alginate encapsulated micro shoots of *Glycrrhiza glabra*. Healthy plants obtained through after 6 months of storage were established in glass house with 95% survival and the genetic fidelity of these plants was ascertained through 10 RAPD and 13 ISSR primers (Mehrotra et al. 2012). The genetic stability of cryopreserved ginger shoot buds was confirmed using ISSR and RAPD profiling (Yamuna et al. 2007). Genetic fidelity was assessed by chromosome analysis and DNA fingerprinting using RAPD in *in vitro* and *in vivo* plants of *Ocimum gratissimum*. Nine arbitrary decamers displayed same banding profile showed no genomic alterations, indicating homogeneity among the tissue culture regenerates and genetic uniformity with that of donor plants (Saha et al. 2012). The *in vitro* derived *Withania somnifera* plantlets were successfully acclimatized and established in soil. RAPD and ISSR analysis revealed a homogeneous amplification profile for all micropropagated plants validating the genetic fidelity of the *in vitro* regenerated plants (Nayak et al. 2012).

Despite these, the application of molecular markers for the resolution of problems of genetic resources conservation is at an early state and requires extensive collaboration among conservation workers and molecular biologists (Karp et al. 1997). The other areas where molecular biology studies can be used in medicinal plant biotechnology include combinatorial biosynthesis of phytomolecules, direct gene transfer, transgenic medicinal plants, engineering the pathway for increased production of active principles, genomic profiling and DNA microarray techniques, etc., which are the highly sophisticated techniques.

Conclusion

The *in vitro* plant, cell and organ culture technologies have the potential for exploitation for conservation and sustainable utilization of the tuber yielding medicinal germplasm of Western Ghats. Though micropropagation protocols were standardized for a large number of species, only few of them are cultivated in large scale like *Coleus forskholi*. On the other hand, JNTBGRI has demonstrated multiplication and reintroduction of more than 10 RET species. Such rapid progress made has helped in conservation of problem species. Slow growth culture and cryopreservation techniques require improvement before they can be used routinely for the preservation. The basic problem existing in this case is the moisture content which needs to be brought down to a level where the ice formation during cryopreservation is nullified.

The production of secondary metabolites and their enhancement using bioreactors has got tremendous potential for large scale synthesis of active principles which otherwise depend on the wild plants collected. Remarkable progress made in some species helped to conserve these wild resources.

Rapid progress in DNA technology helped in molecular genetic studies and development of novel molecular technologies to assess and screen medicinal germplasm for functional diversity and identify variations at early stages.

Hence the present-day need is to use the technologies for conservation and sustainable utilization of medicinal plant germplasm of Western Ghats as a whole.

References

Abdin, M.Z. and A. Kamaluddin. 2006. Improving quality of medicinal herbs through physic-chemical and molecular approaches. *In*: M.Z. Abdin and Y.P. Abrol (eds.). Traditional Systems of Medicine. Narosa Publishing House Pvt. Ltd. Delhi, pp. 30–39.

Abraham, J., M.K. Cheruvathur, B. Mani and T.D. Thomas. 2010. A rapid *in vitro* multiplication system for commercial propagation of pharmaceutically important *Cyclea peltata* (Lam) Hook & Thoms. based on enhanced axillary branching. Ind. Crops Prod. 31(1): 92–98.

Ahmed, B.A., A.S. Rao, M.V. Rao and R.M. Taha. 2012. Production of gymnemic acid depends on medium, explants, PGRs, color lights, temperature, photoperiod, and sucrose sources in batch culture of *Gymnema sylvestre*. The Scientific World J. Article ID 897867, 11 pages DOI: 10.1100/2012/897867 Research Article.

Ahmed, M.B., S. Ahmed, M. Salahin, R. Sultana, M. Khatun, M.A. Razvy, M.M. Hannan, R. Islam and M.M. Hossain. 2007. Standardization of a suitable protocol for *in vitro* clonal propagation of *Acorus calamus* L. an important medicinal plant in Bangladesh. Amer Euras J. Sci. Res. 2(2): 136–140.

Annapurna, D. and T.S. Rathore. 2010. Micropropagation of *Embelia ribes* Burm f. through proliferation of adult plant axillary shoots. *In vitro* Cell Dev-Pl. 46(2): 180–191.

Anonymous. 2009. The Convention on Biological Diversity Plant Conservation Report: A Review of Progress in Implementing the Global Strategy of Plant Conservation (GSPC), 48 pages, Published by the Secretariat of the Convention on Biological Diversity, Montreal, Canada. ISBN: 92-9225-095-7.

Anu, A., K. Babu, N.C.Z. John and K.V. Peter. 2001. *In vitro* clonal multiplication of *Acorus calamus* L. J. Plant Biochem. Biotechnol. 10: 53–55.

Ara, H., U. Jaiswal and V.S. Jaiswal. 2000. Synthetic seed: Prospects and limitations. Curr. Sci. 78: 1438–1444.

Arias-Castro, C., A.H. Scragg, A. Stafford and M. Rodriguez-Mendiola. 1993. Growth characteristics of *Clycyrrhiza glabra* cell suspension cultures" Plant Cell Tissue Organ Cult. 34(1): 77–82.

Avik, R. and B. Sabita. 2008. Cryopreservation of *in vitro* grown nodal segments of *Rauvolfia serpentina* by PVS2 vitrification. Cryo. Lett. 29: 321–328.

Babber, S., K. Mittal, R. Ahlawat and T.M. Varghese. 2001. Micropropagation of *Cardiospermum Halicacabum*. Biol. Plantarum 44(4): 603–606.

Banerjee, S., S. Singh and H. Pandey. 2012. Conservation and storage of *Curcuma amada* Roxb. synseeds on *Luffa* sponge matrix and RAPD analysis of the converted plantlets. Ind. Crops Prod. 36(1): 383–388.

Benjamin, B.D., G. Roja and M.R. Heble. 1993. *Agrobacterium rhizogens* mediated transformation of *Rauvolfia serpentina*: regeneration and alkaloid synthesis. Plant Cell Tissue Organ Cult. 35(3): 253–257.

Bhagat, N. 2011. Conservation of endangered medicinal plant (*Acorus calamus*) through plant tissue culture. J. Pharmacog. 2(1): 21–24.

Bhattacharyya, R. and S. Bhattacharya. 2001. *In vitro* multiplication of *Coleus forskohlii* Briq.: an approach towards shortening the protocol. *In vitro* Cell Dev-Pl. 37(5): 572–575.

Binoy, J., K. Satheeshkumar and S. Seeni. 2007. A protocol for high frequency regeneration through nodal explant cultures and *ex vitro* rooting of *Plumbago rosea* L. Pakistan J. Biol. Sci. 10: 349–355.

Binoy, J. and K. Satheeshkumar. 2010. Tuberous roots an ideal system for high frequency regeneration in *Plumbago rosea* L. Plant Tissue Cult & Biotech. 20(20): 203–209.

Biswas, A., M.A. Bari, M. Roy and S.K. Bhadra. 2009. Clonal propagation through nodal explant culture of *Boerhaavia diffusa* L.—a rare medicinal plant. Plant Tissue Cult. Biotech. 19(1): 53–59.

Biswas, A., M. Roy, M.A. Bari and S.K. Bhadra. 2007. *In vitro* Propagation of *Abrus precatorius* L.—a rare medicinal plant of Chittagong Hill Tracts. Plant Tissue Cult. Biotech. 17(1): 59–64.

Bopana, N. and S. Saxena. 2008. *In vitro* propagation of a high value medicinal plant: *Asparagus racemosus* Willd. *In vitro* Cell Dev-Pl. 44(6): 525–532.

Borthakur, M., J. Hazarika and R.S. Singh. 1998. A protocol for micropropagation of *Alpinia galangal*. Plant Cell Tissue Organ Cult. 55(3): 231–233.

Bulgakov, V.P., Y.N. Shkryl and G.N. Veremeichik. 2010. Engineering high yields of secondary metabolites in *Rubia* cell cultures through rransformation with *Rol* genes. *In*: A.G. Fett-Neto (ed.). Plant Secondary Metabolism Engineering. Series title: Methods in Molecular Biology, Humana Press Chapter 16, pp. 229–242.

Bunn, E., S.R. Turner and K.W. Dixon. 2011. Biotechnology for saving rare and threatened flora in a biodiversity hotspot. *In vitro* Cell Dev-Pl. 47: 188–200.

Chandel, K.P.S. 1996. Biodiversity and strategies for plant germplasm conservation in India. Trop. Ecol. 37(1): 21–29.

Chattopadhyay, S., S. Farkya, A.K. Srivastava and V.S. Bisaria. 2002. Bioprocess considerations for production of secondary metabolites by plant cell suspension cultures. Biotechnol. Bioprocess Eng. 7: 138–149.

Chattopadhyay, S., S.K. Datta and S.B. Mahato. 1995. Rapid micropropagation for *Mucuna pruriens* f. *pruriens* L. Plant Cell Rep. 15(3–4): 271–273.

Chaturvedi, H.C., M. Jain and N.R. Kidwai. 2007. Cloning of medicinal plants through tisuue culture—a review. Indian J. Exp. Biol. 45: 937–948.

Chaudhuri, K.N., B. Ghosh, D. Tepfer and S. Jha. 2006. Spontaneous plant regeneration in transformed roots and calli from *Tylophora indica* : changes in morphological phenotype and tylophorine accumulation associated with transformation by *Agrobacterium rhizogenes*. Plant Cell Rep. 25(10): 1059–1066.

Chaudhuri, K.N., B. Ghosh, D. Tepfer and S. Jha. 2005. Genetic transformation of *Tylophora indica* with *Agrobacterium rhizogenes* A4: growth and tylophorine productivity in different transformed root clones. Plant Cell Rep. 24(1): 25–35.

Cheruvathur, M.K., N. Najeeb and T.D. Thomas. 2013. *In vitro* propagation and conservation of Indian sarsaparilla, *Hemidesmus indicus* L. R. Br. through somatic embryogenesis and synthetic seed production. Acta Physiol. Plantarum. 35(3): 771–779.

Chintalwar, G.J., S. Gupta, G. Roja and V.A. Bapat. 2003. Protoberberine alkaloids from callus and cell suspension cultures of *Tinospora cordifolia*. Pharma. Biol. 41(2): 81–86.

Chirangini, P., S.K. Sinha and G.J. Sharma. 2005. *In vitro* propagation and microrhizome induction in *Kaempferia galanga* Linn. and *K. rotunda* Linn. Indian J. Biotech. 4(3): 404–408.

Chithra, M., K.P. Martin, C. Sunandakumari and P.V. Madhusoodanan. 2005. Protocol for rapid propagation, and to overcome delayed rhizome formation in field established *in vitro* derived plantlets of *Kaempferia galangal* L. Sci. Hortic. 104(1): 113–120.

Cunningham, A.B. 1993. African Medicinal Plants. Setting Priorities at the Interface between Conservation and Primary Healthcare. People and Plant Working Paper 1. UNESCO, Paris, France.

Das, A., V. Kesari and L. Rangan. 2010. Plant regeneration in *Curcuma* species and assessment of genetic stability of regenerated plants. Biol. Plantarum. 54(3): 423–429.

Das, G. and G.R. Rout. 2002. Plant regeneration through somatic embryogenesis in leaf derived callus of *Plumbago rosea*. Biol. Plantarum. 45(2): 299–302.

Decruse, S.W. and S. Seeni. 2002. Ammonium nitrate in the culture medium influences regeneration potential of cryopreserved shoot tips of *Holostemma annulare*. Cryo Lett. 23(1): 55–60.

Decruse, S.W., S. Seeni and P. Pushpangadan. 1999. Effect of cryopreservation on seed germination of selected rare medicinal plants of peninsular India. Seed Sci. Tech. 27: 501–505.

Decruse, S.W., S. Seeni and P. Pushpangadan. 1999. Cryopreservation of alginate coated shoot tips of *in vitro* grown *Holostemma annulare* (Roxb.) K. Schum, an endangered medicinal plant: influence of preculture and DMSO treatment on survival and regeneration. Cryo. Lett. 20: 243–250.

Decruse, S.W., S. Seeni and G.M. Nair. 2004. Preparative procedures and culture media effect on the success of cryostorage of *Holostemma annulare* shoot tips. Plant Cell Tissue Organ Cult. 76: 179–182.

Decruse, S.W. and S. Seeni. 2002. Ammonium nitrate in the culture medium influences regeneration potential of cryopreserved shoot tips of *Holostemma annulare*. Cryo. Lett. 23: 55–60.

Decruse, S.W., S. Seeni and P. Pushpangadan. 1999. Effects of cryopreservation on seed germination of selected rare medicinal plants of India. Seed Science and Technology. 27(2): 501–505.

Dekkers, A.J., A.N. Rao and C.J. Goh. 1991. *In vitro* storage of multiple shoot cultures of gingers at ambient temperatures of 24–29°C. Sci. Hortic. 47(1–2): 157–167.

Deshmukh, B.S. 2010. *Ex situ* conservation studies on ethno- medicinal rare, endemic Plant species from western ghats of Maharashtra. Int. J. Pharma. and Biosci. v1(2): 1–6.

Devi, C.S. and V.M. Srinivasan. 2011. *In vitro* studies on stimulation of gymnemic acid production using fungal elicitor in suspension and bioreactor based cell cultures of *Gymnema sylvestre* R. Br. Recent Res. Sci. Technol. 3(4): 101–104.

Dharmar, K. and A.J. De Britto. 2011. RAPD analysis of genetic variability in wild populations of *Withania somnifera* (L.) Dunal. Int. J. Biol. Technol. 2(1): 21–25.

Dixit, A.K. and S. Vaidya. 2010. *Agrobacterium rhizogenes* induced hairy root development and its effect on production of glycyrrhizin in *Abrus precatorius* (L.). Int. J. Current Res. 6: 033–038.

Doma, M., G. Abhayankar, V.D. Reddy and P.B. KaviKishor. 2012. Carbohydrate and elicitor enhanced withanolide (withaferin A and withanolide A) accumulation in hairy root cultures of *Withania somnifera* (L.). Indian J. Exp. Biol. 50(7): 484–490.

Dube, P., M. Gangopadhyay, S. Dewanjee and M.N. Ali. 2011. Establishment of a rapid multiplication protocol of *Coleus forskohlii* Briq. and *in vitro* conservation by reduced growth. Indian J. Biotech. 10(2): 228–231.

Edwards, R. 2004. No remedy in sight for herbal ransack. New Sci. 181: 10–11.

Engelmann, F. 1991. *In vitro* conservation of tropical plant germplasm—a review. Euphytica 57: 227–243.

Engelmann, F. 2011. Use of biotechnologies for the conservation of plant Biodiversity. *In vitro* Cell Dev-Pl. 47: 5–16.

Faisal, M. and M. Anis. 2007. Regeneration of plants from alginate encapsulated shoots of *Tylophora indica* (Burm. f.) Merrill, an endangered medicinal plant. J. Hortic. Sci. Biotech. 82: 351–354.

Faisal, M., I. Siddique and M. Anis. 2006. An efficient plant regeneration system for *Mucuna pruriens* L. (Dc.) using cotyledonary node explants. *In vitro* Cell Dev-Pl. 42: 59–64.

Faisal, M., N. Ahmad and M. Anis. 2007. An efficient micropropagation system for *Tylophora indica*: an endangered, medicinally important plant. Plant Biotechnol. Rep. 1(3): 155–161.

Faisal, M., A.A. Alatar, N. Ahmad, M. Anis and A.K. Hegazy. 2012. Assessment of genetic fidelity in *Rauvolfia serpentina* plantlets grown from synthetic (encapsulated) seeds following *in vitro* storage at 4°C. Molecules. 17(5): 5050–5061.

Falkenhagen, H., I.N. Kuzovkina, I.E. Alterman, L.A. Nikolaeva and J. Stockigt. 1993. Alkaloid formation in hairy roots and cell suspensions of *Rauwolfia serpentina* Benth. Nat. Product Lett. 3(2): 107–112.

Fay, M.F. 1992. Conservation of rare and endangered plants using *in vitro* methods. *In vitro* Cell Dev-Pl. 28: 1–4.

Fay, M.F. 1994. In what situations is *in vitro* culture appropriate to plant conservations? Biodivers. Conserv. 3: 176–183.

Francis, S.V., S.K. Senapati and G.R. Rout. 2007. Rapid clonal propagation of *Curculigo orchioides* Gaertn., an endangered medicinal plant. *In vitro* Cell Dev-Pl. 43(2): 140–143.

Gangaprasad, A., S.W. Decruse, S. Seeni and G.M. Nair. 2005. Micropropagation and ecorestoration of *Decalepis arayalpathra* (Joseph & Chandra.) Venter—an endemic and endangered ethnomedicinal plant of Western Ghats. Indian J. Biotech. 04(2): 265–270.

Gangopadhyay, M., D. Sircar, A. Mitra and S. Bhattacharya. 2008. Hairy root culture of *Plumbago indica* as a potential source for plumbagin. Biol. Plantarum. 52(3): 533–537.

Gangopadhyay, M., S. Dewanjee and S. Bhattacharya. 2011. Enhanced plumbagin production in elicited *Plumbago indica* hairy root cultures. J. Biosci. Bioeng. 111(6): 706–710.

Gangopadhyay, M., S. Dewanjee, D. Chakraborty and S. Bhattacharya. 2011. Role of exogenous phytohormones on growth and plumbagin accumulation in *Plumbago indica* hairy roots and conservation of elite root clones via synthetic seeds. Ind. Crops Prod. 33: 445–450.

Geetha, S.P., A.V. Raghu, G. Martin, S. George and I. Balachandran. 2009. *In vitro* propagation of two tuberous medicinal plants: *Holostemma ada-kodien* and *Ipomoea mauritiana*. Methods Mol. Biol. 547: 81–92.

George, S., S.P. Geetha, A. Anu and B. Indira. 2010. *In vitro* conservation studies in *Hemidesmus indicus*, *Decalepis hamiltonii* and *Utleria salicifolia*. *In:* Proceedings of the 22nd Kerala Science Congress, Kerala State Council for Science Technology and Environment, Thiruvanthapuram, Kerala, pp. 258–259.

Ghatge, S., S. Kudale and G. Dixit. 2011. An improved plant regeneration system for high frequency multiplication of *Rubia cordifolia* L.: a rare medicinal plant. Asian J. Biotechnol. 3(4): 397–405.

Ghosh, B., S. Mukherjee, T.B. Jha and S. Jha. 2002. Enhanced colchicine production in root cultures of *Gloriosa superba* by direct and indirect precursors of the biosynthetic pathway. Biotechnol. Lett. 24(3): 231–234.

Ghosh, S., B. Ghosh and S. Jha. 2006. Aluminium chloride enhances colchicine production in root cultures of *Gloriosa superba*. Biotechnol. Lett. 28(7): 497–503.

Gopi, C., Y.N. Sekhar and P. Ponmurugan. 2006. *In vitro* multiplication of *Ocimum gratissimum* L. through direct regeneration. Afr. J. Biotechnol. 5(9): 723–726.

Goyal, D. and S. Bhadauria. 2006. *In vitro* propagation of *Ceropegia bulbosa* using nodal segments. Indian J. Biotech. 05(4): 565–567.

Goyal, S. and K.G. Ramawat. 2008. Synergistic effect of morphactin on cytokinin-induced production of isoflavonoids in cell cultures of *Pueraria tuberosa* (Roxb. ex. Willd.) DC. Plant Growth Regul. 55(3): 175–181.

Guillon, S., J. Tre´mouillaux-Guiller, P.K. Pati, M. Rideau and P. Gantet. 2006. Hairy root research: recent scenario and exciting prospects. Curr. Opin. Plant Biol. 9: 341–346.

Gunawardene, N.R., A.E.D. Daniels, I.A.U.N. Gunatilleke, C.V.S. Gunatilleke, P.V. Karunakaran, K.G. Nayak, S. Prasad, P. Puyravaud, B.R. Ramesh, K.A. Subramanian and G. Vasanthy. 2007. A brief overview of the Western Ghats—Sri Lanka biodiversity hotspot. Curr. Sci. 93(11): 1567–1572.

Gupta, R.K., A. Srivastava and H.N. Verma. 2004. Callus culture and organogenesis in *Boerhaavia diffusa*: a potent antiviral protein containing plant. Physiol. Mol. Biol. Plants 10(2): 263–268.

Harisaranraj, R., K. Suresh and S.S. Babu. 2009. Production of reserpine in somatic embryos of *Rauwolfia serpentina* cultured in bioreactors by the induction of elicitor (methyl jasmonate). Global J. Biotechnol. Biochem. 4(2): 143–147.

Hassan, A.K.M.S and S.K. Roy. 2005. Micropropagation of *Gloriosa superba* L. through high frequency shoot proliferation. Plant Tissue Cult. 15(1): 67–74.

Hu, Z.B. and M. Du. 2006. Hairy root and its application in plant genetic engineering. J. Integr. Plant Biol. 48(2): 121–127.

Innes, N.L. 1992. Gene banks and their contribution to the breeding of disease resistant cultivars. Euphytica. *In*: R. Johnson and G.J. Jellis (eds.). Breeding for Disease Resistance. Kluwer Academic Publishers, Netherlands, 63: 23–31.

Jackson, P.S.W. 2002. Botanic Garden Conservation News. 3(8): 25–32.

Jahan, A.A. and M. Anis. 2009. *In vitro* rapid multiplication and propagation of *Cardiospermum halicacabum* L. through axillary bud culture. Acta Physiol. Plantarum. 31(1): 133–138.

Jain, M., A.K. Rathore and P. Khanna. 1984. Influence of kinetin and auxins on the growth and production of diosgenin by *Costus speciosus* (Koen.) Sm. callus derived from rhizome. Agric. Biol. Chem. 48: 529.

Jain, V., D. Singh and S. Saraf. 2003. *In vitro* culture studies for callus and root generation of boerhaavia diffusa linn. Anc. Sci. Life. 22(4): 134–139.

Jayanthi, M. and P.K. Mandal. 2001. Plant regeneration through somatic embryogenesis and RAPD analysis of regenerated plants in *Tylophora indica* (Burm. F. Merrill.). *In vitro* Cell Dev-Pl. 37: 576–580.

Jenifer, U., F.K. Cecilia and R. Ravindhran. 2012. *In vitro* adventitious root and hairy root cultures in *Boerhaavia diffusa* L. Int. J. Current Res. 4(1): 65–67.

Joy, P.P., J. Thomas, S. Mathew and B.P. Skaria. 2001. Scope of medicinal, aromatic and dye yielding crops in Kerala. In: National seminar cum workshop on Scope of medicinal, aromatic and dye yielding crops in dry land agriculture (11–13 Dec) Central Research Institute for Dryland Agriculture, Hyderabad, India.

Kakutani, K., K. Ozaki, H. Watanabe and K. Tomoda. 1999. Somatic embryogenesis and plantlet regeneration on several species of *Licorice*. Plant Biotechnol. 16(3): 239–241.

Kale, V.S. 2010. The Western Ghat: the great escarpment of India. pp. 257–264. *In*: Piotr Migon (ed.) Geomorphological Landscapes of the World, DOI: 10.1007/978-90-481-3055-9_26, ISBN: 978-90-481-3055-9.

Kalita, M.C., P. Deka, A. Boro and S. Phukan. 2007. Tissue culture studies and *in vitro* clonal propagation of three (*Cardiospermum halicacabum* L., *Spilanthes acmella* L. and *Boerhaavia diffusa* L.) herbal ethnomedicinal plants of North-East India and establishment of the plantlets in a natural environment. *In*: J.N. Govil, V.K. Singh and R. Bhardwaj (eds.). Phytomedicines. Title Recent Progress in Medicinal Plants (Vol. 16), Studium Press (India) Pvt. Ltd. pp. 587–600.

Kapaia, V.Y., P. Kapoor and I.U. Rao. 2010. *In vitro* propagation for conservation of rare and threatened plants of India—a review. Int. J. Biol. Technol. 1(2): 1–14.

Karmarkar, S.H., R. Keshavachandran, P.A. Nazeem and D. Girija. 2001. Hairy root induction in adapathiyan (*Holostemma ada-kodien* K. Schum.). J. Trop. Agri. 39: 102–107.

Karp, A., S. Kresovich, K.V. Bhat, W.G. Ayad and T. Hodgkin. 1997. Molecular tools in plant genetic resources conservation: a guide to the technologies. *In:* IPGRI Technical Bulletin No. 2. International Plant Genetic Resources Institute, Rome, Italy.

Karuppusamy, S., C. Kiranmai, V. Aruna and T. Pullaiah. 2009. *In vitro* conservation of *Ceropegia intermedia*—an endemic plant of south India. Afr. J. Biotechnol. 8(17): 4052–4057.

Karwasara, V.S. and V.K. Dixit. 2009. *Agrobacterium rhizogenes* mediated genetic transformation of *Abrus precatorius* L. Pharmacogn. Mag. 5(20): 336–342.

Karwasara, V.S., P. Tomar and V.K. Dixit. 2011. Influence of fungal elicitation on glycyrrhizin production in transformed cell cultures of *Abrus precatorius* Linn. Pharmacogn. Mag. 7(28): 307–313.

Khan, M.Y., S. Aliabbas, V. Kumar and S. Rajkumar. 2009. Recent advances in medicinal plant biotechnology. Indian J. Biotech. 8: 9–22.

Khanapurkar, R.S., N.S. Paul, D.M. Desai, M.R. Raut and A.K. Gangawane. 2012. *In vitro* propagation of *Tinospora cordifolia* (Wild.) Miers Ex Hook.F. Thoms. J. Bot. Res. 3(1): 17–20.

Komaraiah, P., G. Jogeswar, S.V. Ramakrishna and P.B. KaviKishore. 2004. Acetylsalicylic acid and ammonium-induced somatic embryogenesis and enhanced plumbagin production in suspension cultures of *Plumbago rosea* L. *In vitro* Cell Dev-Pl. 40(2): 230–234.

Komaraiah, P., P.B. KaviKishor and S.V. Ramakrishna. 2001. Production of plumbagin from cell cultures of *Plumbago rosea* L. Biotechnol. Lett. 23: 1269–1272.

Krishnan, P.N., S.W. Decruse and R.K. Radha. 2011. Conservation of medicinal plants of Western Ghats, India and its sustainable utilization through *in vitro* technology. *In vitro* Cell Dev-Pl. 47: 110–122.

Krombholz, R., R. Mersinger, W. Kreis and E. Reinhard. 1992. Production of forskolin by axenic *Coleus forskohlii* roots cultivated in shake flasks and 20-1 glass jar bioreactors. Planta Med. 58(4): 328–333.

Kumar, H.G.A., H.N. Murthy and K.Y. Paek. 2002. Somatic embryogenesis and plant regeneration in *Gymnema sylvestre* Plant Cell Tissue Organ Cult. 71(1): 85–88.

Kumar, V., K.N. Murthy, S. Bhamid, C.G. Sudha and G.A. Ravishankar. 2005. Genetically modified hairy roots of *Withania somnifera* Dunal: a potent source of rejuvenating principles. Rejuvenation Res. 8(1): 37–45.

Lahiri, K., M.J. Mukhopadhyay, Y. Desjardins and S. Mukhopadhyay. 2012. Rapid and stable *in vitro* regeneration of plants through callus morphogenesis in two varieties of *Mucuna pruriens* L.—an anti Parkinson's drug yielding plant. The Nucleus 55(1): 37–43.

Lakshmanan, V., S.R. Venkataramareddy and B. Neelwarne. 2007. Molecular analysis of genetic stability in long-term micropropagated shoots of banana using RAPD and ISSR markers. Electron. J. Biotechn. 10(1), DOI: 10.2225/vol10-issue1-fulltext-12.

Leathers, R.R., M.A.L. Smith and J. Aitken-Christie. 1995. Automation of the bioreactor process for mass propagation and secondary metabolism. *In:* J. Aitken-Christie, T. Kozai and M.A.L. Smith (eds.). Automation and Environment Control in Plant Tissue Culture. Kluwer Academic Publ., Dordrecht, the Netherlands, pp. 87–214.

Lee, J.H. and T.H. Han. 2011. Micropropagation of the plantlets derived from seeds in the genus *Acorus* (*A. calamus* and *A. gramineus*). Hort. Environ. Biotechnol. 52(1): 89–94.

Levin, R., V. Gaba, B. Tal, S. Hirsch and D. De-Nola. 1998. Automated plant tissue culture for mass propagation. Biotechnol. 6:1035–1040.

Li, W., K. Koike, Y. Asada, T. Yoshikawa and T. Nikaido. 2005. Rosmarinic acid production by *Coleus forskohlii* hairy root cultures. Plant Cell Tissue Organ Cult. 80(2): 151–155.

Mandal, B.B., D.S. Sharma and P.S. Srivastava. 2009. Embryogenic Cultures of *Dioscorea bulbifera* L. by encapsulation-dehydration. Cryo. Lett. 30: 440–448.

Mandal, B.B., K.P.S. Chandel and S. Dwivedi. 1996. Cryopreservation of yams (*Dioscorea* spp.) shoot apices by encapsulation-dehydration. Cryo. Lett. 17: 165–174.

Mandal, J., S. Pattnaik and P.K. Chand. 2000. Alginate encapsulation of axillary buds of *Ocimum americanum* L. (holy basil), *O. basilicum* L. (sweet basil), *O. gratissimum* L. (shrubby basil), and *O. sanctum* L. (sacred basil). *In vitro* Cell Dev-Pl. 36: 287–292.

Manickam, V.S., R.E. Mathavan and R. Antonisamy. 2000. Regeneration of Indian ginseng plantlets from stem callus. Plant Cell Tissue Organ Cult. 62(3): 181–185.

Manjula, S., A. Thomas, B. Daniel and G.M. Nair. 1997. *In vitro* plant regeneration of *Aristolochia indica* through axillary shoot multiplication and organogenesis. Plant Cell Tissue Organ Cult. 51(2): 145–148.

Martin, K. 2002. Rapid propagation of *Holostemma ada-kodien* Schult., a rare medicinal plant, through axillary bud multiplication and indirect organogenesis. Plant Cell Rep. 21(2): 112–117.

Martin, K.P., A. Sabovljevic and J. Madassery. 2011. High-frequency transgenic plant regeneration and plumbagin production through methyl jasmonate elicitation from hairy roots of *Plumbago indica* L. J. Crop Sci. Biotechnol. 14(3): 205–212.

Maunder, M. 1992. Plant reintroduction: an overview. Biodivers. Conserv. 1: 51–62.

Mehrotra, S., A.K. Kukreja, S.P.S. Khanuja and B.N. Mishra. 2008. Genetic transformation studies and scale up of hairy root culture of *Glycyrrhiza glabra* in bioreactor. Electron. J. Biotechn. 11(2): DOI: 10.2225/vol. 11. No. 2, fulltext-6.

Mehrotra, S., O. Khwaja, A.K. Kukreja and L. Rahman. 2012. ISSR and RAPD based evaluation of genetic stability of encapsulated micro shoots of *Glycyrrhiza glabra* following 6 Months of storage. Mol. Biotechnol. 52: 262–268.

Mohanty, S., R. Parida, A. Kuanar, S. Sahoo and S. Nayak. 2012. Evaluation of genetic fidelity of *in vitro* propagated shampoo ginger (*Zingiber zerumbet* (L.) Smith) using DNA based markers. J. Med. Plants Res. 6(16): 3143–3147.

Mohanty, S., R. Parida, S. Singh, R.K. Joshi, E. Subudhi and S. Nayak. 2011. Biochemical and molecular profiling of micropropagated and conventionally grown *Kaempferia galangal.* Plant Cell Tissue Organ Cult. 106(1): 39–46.

Mukherjee, S., B. Ghosh and S. Jha. 1996. Forskolin synthesis in *in vitro* cultures of *Coleus forskohlii* Briq transformed with *Agrobacterium tumefaciens*. Plant Cell Rep. 15(9): 691–694.

Mukherjee, S., B. Ghosh and S. Jha. 2000. Establishment of forskolin yielding transformed cell suspension cultures of *Coleus forskohlii* as controlled by different factors. J. Biotechnol. 76(1): 73–81.

Mukhopadhyay, S. and A.K. Sharma. 1990. Chromosome number and DNA content in callus culture of *Costus speciosus* (Koen.) Sm. Genetica. 80(2): 109–114.

Murthy, H.N. and N. Praveen. 2012. Influence of macro elements and nitrogen source on adventitious root growth and withanolide-A production in *Withania somnifera* (L.) Dunal. Nat. Prod. Res. 26(5): 466–473.

Murthy, H.N., C. Dijkstra, P. Anthony, D.A. White, M.R. Davey, J.B. Power, E.J. Hahn and K.Y. Paek. 2008. Establishment of *Withania somnifera* hairy root cultures for the production of withanolide A. J. Integr. Plant Biol. 50(8): 975–981.

Myers, N. 1998. Threatened biotas: hotspots in tropical forests. Environmentalist 8: 178–208.

Myers, N., A.R. Mittermeier, C.G. Mittermeier, G.A.B. da-Fonseca and J. Kent. 2000. Biodiversity hotspots for conservation priorities. Nature. 403: 853–858.

Nagesh, K.S., C. Shanthamma and T. Pullaiah. 2010. Somatic embryogenesis and plant regeneration from callus cultures of *Curculigo orchioides* Gaertn. Indian J. Biotechnol. 9: 408–413.

Nayak, S.A., S. Kumar, K. Satapathy, A. Moharana, B. Behera, D.P. Barik, L. Acharya, P.K. Mohapatra, P.K. Jena and S.K. Naik. 2013. *In vitro* plant regeneration from cotyledonary nodes of *Withania somnifera* (L.) Dunal and assessment of clonal fidelity using RAPD and ISSR markers. Acta Physiol. Plantarum. 35(1): 195–203.

Neelam, S. and K.P.S. Chandel. 1992. Low-temperature storage of *Rauvolfia serpentina* Benth. Ex Kunrz.—an endangered, endemic medicinal plant. Plant Cell Rep. 11: 200–203.

Nikam T.D. and S.R. Savant. 2009. Multiple shoot regeneration and alkaloid cerpegin accumulation in callus culture of *Ceropegia juncea* Roxb. Physiol. Mol. Biol. Plants. 15: 71–77.

Nikam, T.D. and R.S. Savant. 2007. Callus culture and micropropagation of *Ceropegia shyadrica* Ans and Kulk.: an edible starchy tuberous rare *Asclepidaceae.* Indian J. Plant Physi. 12(2): 108–114.

Nikam, T.D. and R.S. Savant. 2009. Multiple shoot regeneration and alkaloid cerpegin accumulation in callus culture of *Ceropegia juncea* Roxb. Physiol Mol. Biol. Plants 15(1): 71–77.

Pal, A. and A. Roy. 1991. Embryo culture of *Costus speciosus* (Koen.) Sm. to regenerate variable diosgenin yielding clones. Plant Cell Rep. 10(11): 565–568.

Papiya, M., B.B. Mandal, K.V. Bhat and A.K. Biswas. 2009. Cryopreservation of Asian *Dioscorea bulbifera* L. and *D. alata* L. by vitrification: importance of plant growth regulators. Cryo. Lett. 30: 110–111.

Parida, R., S. Mohanty and S. Nayak. 2011. Evaluation of genetic fidelity of *in vitro* propagated greater galangal (*Alpinia galanga* l.) using DNA based markers. Int. J. Plant Animal Environ. Sci. 1(3): 123–133.

Patil, V.M. 1998. Micropropagation studies in *Ceropegia spp. In vitro* Cell Dev-Pl. 34(3): 240–243.

Phulwaria, M., N.S. Shekhawat, J.S. Rathore and R.P. Singh. 2013. An efficient *in vitro* regeneration and *ex vitro* rooting of *Ceropegia bulbosa* Roxb.—a threatened and pharmaceutical important plant of Indian Thar Desert. Ind. Crops Prod. 42: 25–29.

Pise, M., J. Rudra, S. Bundale, D. Begde, N. Nashikkar and A. Upadhyay. 2012. *Asparagus racemosus* cell cultures: a source for enhanced production of shatavarins and sarsapogenin. *In vitro* Cell Dev-Pl. 48(1): 85–91.

Pise, M., J. Rudra, S. Bundale, D. Begde, N. Nashikkar and A. Upadhyay. 2011. Shatavarin production from *in vitro* cultures of *Asparagus racemosus* Wild. J. Med. Plants Res. 5(4): 507–513.

Prakash, S., R. Elangomathavan, S. Seshadri, K. Kathiravan and S. Ignacimuthu. 2004. Efficient regeneration of *Curcuma amada* Roxb. plantlets from rhizome and leaf sheath explants. Plant Cell Tissue Organ Cult. 78(2): 159–165.

Preetha, T.S., A.S. Hemanthakumar and P.N. Krishnan. 2012. Effect of plant growth regulators on high frequency *in vitro* multiplication of a vulnerable woody medicinal climber *Embelia ribes* Burm. f. J. Med. Plants Res. 6(23): 4011–4018.

Preetha, T.S., A.S. Hemanthkumar, S.W. Decruse and P.N. Krishnan. 2008. Effect of auxins on somatic embryogeneis from leaf derived callus of *Kaempferia galangal* L.—an endangered medicinal plant. Phytomorphology. 58(3–4): 117–124.

Preil, J.M., P. Florek, U. Wix and A. Beck. 1988. Towards mass propagation by use of bioreactors. Acta Hort. 226: 99–105.

Preil, W. 1991. Application of bioreactors in plant propagation. *In*: P.C. Debergh and R.H. Zimmerman (eds.). Micropropagation: Technology and application. Kluwer Academic Publ., Dordrecht, the Netherlands, pp. 425–455.

Priya, L., S. Anita, K. Rajmohan and S. Alex. 2007. *In vitro* shoot regeneration in *Ipomoea mauritiana* (Linn.). *In*: P. Kesavachandran, P.A. Nazeem, D. Girija, P.S. John and K.V. Peter (eds.). Recent trends in Horticultural Biotechnology. New India Publishing Agency, New Delhi, India.

Pushpangadan, P., S. Seeni and T. Jacob. 1994. Conservation of plant genetic resources in India- establishment of gene bank for medicinal and aromatic plants at Tropical Botanic Garden and Reserach Institute. G-15 GEMAB Newsletter 5&6: 10–12.

Radha, R.K., S.R. Shereena, K. Divya, P.N. Krishnan and S. Seeni. 2010. *In vitro* propagation from seedling explants and anthaquinone production from *in vitro* shoots of *Rubia cordifolia* Linn., a medicinal plant. 97th Indian Science Congress, University of Kerala, Kariavattom, India. 3–7 Jan 2010.

Raghavendra, S., V. Kumar and C.K. Ramesh. 2012. A preliminary attempt for efficient genetic transformation and regeneration of legume *Mucuna pruriens* L. mediated by *Agrobacterium tumefaciens*. Turk. J. Biol. 36: 285–292.

Raghavendra, S., V. Kumar, C.K. Ramesh and M.H. Khan. 2012. Enhanced production of L-DOPA in cell cultures of *Mucuna pruriens* L. and *Mucuna prurita* H. Nat. Prod. Res. 26(9): 792–801.

Raghu, A.V., K. Unnikrishnan, S.P. Geetha, G. Martin and I. Balachandran. 2011. Plant regeneration and production of embelin from organogenic and embryogenic callus cultures of *Embelia ribes* Burm. f.—a vulnerable medicinal plant. *In vitro* Cell Dev-Pl. 47(4): 506–515.

Raghu, A.V., S.P. Geetha, G. Martin, I. Balachandran and P.N. Ravindran. 2006. *In vitro* clonal propagation through mature nodes of *Tinospora cordifolia* (Willd.) Hook. F. & Thoms.: an important ayurvedic medicinal plant. *In vitro* Cell Dev-Pl. 42(6): 584–588.

Rajasekharan, P.E., S.R. Ambika and S. Ganeshan. 2009. *In vitro* conservation of *Tylophora indica*: a threatened medicinal plant. The IUP J. Genetics Evol. II(3): 26–35.

Ramakrishnappa, K. 2002. Impact of Cultivation and Gathering of Medicinal Plants on Biodiversity: Case Studies from India. Biodiversity and the Ecosystem Approach in Agriculture, Forestry and Fisheries. Satellite event on the occasion of the Ninth Regular Session of the Commission on Genetic Resources for Food and Agriculture. Rome, 12-13 October 2002. Inter-Departmental Working Group on Biological Diversity for Food and Agriculture, Rome, Italy, Url http://www.fao.org/DOCREP/005/AA021E/AA021E00.HTM.

Rani, G., G.S. Virk and A. Nagpal. 2003. Callus induction and plantlet regeneration in *Withania somnifera* (L.) dunal. *In vitro* Cell Dev-Pl. 39(5): 468–474.

Rao, K., B. Chodisetti, L.N. Mangamoori and A. Giri. 2012. *Agrobacterium*-mediated transformation in *Alpinia galanga* (Linn.) Willd. for enhanced acetoxychavicol acetate production. Appl. Biochem. Biotechnol. 168(2): 339–347.

Rao, K., B. Chodisetti, S. Gandi, L.N. Mangamoori and A. Giri. 2011. Direct and indirect organogenesis of *Alpinia galanga* and the phytochemical analysis. Appl. Biochem. Biotechnol. 165(5–6): 1366–1378.

Rao, N.K. 2004. Plant genetic resources: advancing conservation and use through biotechnology. Afr. J. Biotechnol. 3: 136–145.

Rathore, M.S. and N.S. Shekhawat. 2009. Micropropagation of *Pueraria tuberosa* (Roxb. Ex Willd.) and determination of puerarin content in different tissues. Plant Cell Tissue Organ Cult. 99(3): 327–334.

Ray, A. and S. Bhattacharya. 2008. Cryopreservation of *in vitro* grown nodal segments of *Rauvolfia serpentina* by PVS2 vitrification. Cryo. Lett. 29(4): 321–328.

Reddanna and P.B. Kavi Kishor. 2003. Enhanced production of plumbagin in immobilized cells of *Plumbago rosea* by elicitation and *in situ* adsorption. J. Biotechnol. 101(2): 181–187.

Reddy, M.C., K.S. RamaMurthy and T. Pullaiah. 2012. Synthetic seeds: a review in agriculture and forestry. Afr. J. Biotechnol. 11(78): 14254–14275.

Reddy, P.S., R. Rodrigues and R. Rajasekharan. Shoot organogenesis and mass propagation of *Coleus forskohlii* from leaf derived callus. Plant Cell Tissue Organ Cult. 66(3): 183–188.

Reed, B.M., V. Sarasan, M. Kane, E. Bunn and V.C. Pence. 2011. Biodiversity conservation and conservation biotechnology tools. *In vitro* Cell Dev-Pl. 47: 1–4.

Remya, M., V.N. Bai and V.N. Mutharaian. 2013. *In vitro* regeneration of *Aristolochia tagala* and production of artificial seeds. Biol. Plantarum. 57(2): 210–218.

Rout, G.R. 2006. Evaluation of genetic relationship in *Typhonium* species through random amplified polymorphic DNA markers. Biol. Plantarum. 50(1): 127–130.

Rout, G.R., S.L. Sahoo and R. Das. 2011. An attempt to conserve *Withania somnifera* (L.) Dunal —a highly essential medicinal plant, through *in vitro* callus culture. Pak. J. Bot. 43(4): 1837–1842.

Roy, A. and A. Pal. 1991. Propagation of *Costus speciosus* (Koen.) Sm. through *in vitro* rhizome production. Plant Cell Rep. 10(10): 525–52.

Sabir, F., R.S. Sangwan, R. Kumar and N.S. Sangwan. 2012. Salt stress-induced responses in growth and metabolism in callus cultures and differentiating *in vitro* shoots of Indian Ginseng (*Withania somnifera* Dunal.). J. Plant Growth Regul. 31(4): 537–548.

Saha, S., A. Kader, C. Sengupta and P. Ghosh. 2012. *In vitro* propagation of *Ocimum gratissimum* L. (Lamiaceae) and its evaluation of genetic fidelity using RAPD Marker. American J. Plant Sci. 3(1): 64–74.

Sai, S.T., C.L. Keng, N. Pargini and K.H. Teo. *In vitro* propagation of *Typhonium flagelliforme* (Lodd) blume. *In vitro* Cell Dev-Pl. 36(5): 402–406.

Sakai, A. 1960. Survival of the twigs of woody plants at -196 °C. Nature. 185: 392–394.

Sasson, A. Production of useful biochemicals by higher-plant cell cultures: biotechnological and economic aspects. Options Méditerranéennes-Série Séminaires. 14: 59–74.

SatheeshKumar, K. and S. Seeni. 2002. Production of plumbagin (5-hydroxy- 2methoxy 1, 4-naphtoquinone) in callus and cell suspension cultures of *Plumbago indica* Linn. Indian J. Biotechnol. 1: 305–308.

Saxena, S. and N. Bopana. 2009. *In vitro* clonal propagation of *Asparagus racemosus*, a high value medicinal plant. Methods Mol. Biol. 547: 179–189.

Schippmann, U., D.J. Leaman and A.B. Cunningham. 2002. Impact of Cultivation and Gathering of Medicinal Plants on Biodiversity. Global Trends and Issues Published in FAO, 2002. Biodiversity and the Ecosystem Approach in Agriculture, Forestry and Fisheries. Satellite event on the occasion of the Ninth Regular Session of the Commission on Genetic Resources for Food and Agriculture. Rome, 12–13 October 2002. Inter-Departmental Working Group on Biological Diversity for Food and Agriculture. Rome.

Selvakumar, V., P.R. Anbudurai and T. Balakumar. 2001. *In vitro* propagation of the medicinal plant *Plumbago zeylanica* L. through nodal explants. *In vitro* Cell Dev-Pl. 37: 280–284.

Shabani, L., A.A. Ehsanpour, G. Asghari and J. Emami. 2009. Glycyrrhizin production by *in vitro* cultured *Glycyrrhiza glabra* elicited by methyl jasmonate and salicylic acid. Russian Journal of Plant Physiology. 56(5): 621–626.

Shah, B. and A. Seth. 2010. Textbook of Pharmacognosy and Phytochemistry. *In:* S.K. Chauhan (ed.). Elsevier Health Sciences. Elsevier India, New Delhi.

Shankaramurthy, K. and V. Krishna. 2006. Micropropagation of *Embelia ribes* Burm.f. using inflorescence segments. Indian J. Biotechnol. 5: 551–554.

Sharma, B. and Y.K. Bansal. 2010. *In vitro* propagation of *Gymnema sylvestre* Retz. R.Br through apical bud culture. J. Med. Plants Res. 4(14): 1473–1476.

Sharma, M., S.K. Rai, D.K. Purshottam, M. Jain, D. Chakrabarty, A. Awasthi, K.N. Nair and A.K. Sharma. 2009. *In vitro* clonal propagation of *Clerodendrum serratum* (Linn.) Moon (barangi): a rare and threatened medicinal plant. Acta Physiol. Plantarum. 31(2): 379–383.

Sharma, N. and R. Pandey. 2013. Conservation of medicinal plants in the Tropics. *In*: M.N. Normah, H.F. Chin and B.M. Reed (eds.). Conservation of Tropical Plant Species. ISBN: 978-1-4614-3776-5, pp. 437–487.

Sharma, T.R. and B.M. Singh. 1997. High-frequency *in vitro* multiplication of disease-free *Zingiber officinale* Rosc. Plant Cell Rep. 17(1): 68–72.

Sharma, T.R., B.M. Singh and R.S. Chauhan. 1994. Production of disease-free encapsulated buds of *Zingiber officinale* Rosc. Plant Cell Rep. 13(5): 300–302.

Sharma, V., S. Goyal and K.G. Ramawat. 2009. Scale up production of isoflavonoids in cell suspension cultures of *Pueraria tuberosa* grown in shake flasks and bioreactor. Eng. Life Sci. 9(3): 267–271.

Sharma, V., S. Goyal and K.G. Ramawat. 2011. Increased puerarin biosynthesis during *in vitro* shoot formation in *Pueraria tuberosa* grown in growtek bioreactor with aeration. Physiol. Mol. Biol. Plants. 17(1): 87–92.

Shimomura, K. 1998. High forskolin production in hairy roots of *Coleus forskohlii*. Plant Cell Rep. 17(6–7): 457–459.

Shkryl, Y.N., G.N. Veremeichik, V.P. Bulgakov, G.K. Tchernoded, N.P. Mischenko, S.A. Fedoreyev and Y.N. Zhuravlev. 2008. Individual and combined effects of the *rol A*, *B*, and *C* genes on anthraquinone production in *Rubia cordifolia* transformed calli. Biotechnol. Bioeng. 100(1): 118–125.

Shukla, N., N.S. Sangwan, H.O. Misra and R.S. Sangwan. 2003. Genetic diversity analysis in *Boerhavia diffusa* L. of different geographic locations in India using RAPD markers. Genetic Res. Crop Evol. 50(6): 587–601.

Siddique, N.A. and M.A. Bari. 2010. Plant regeneration from axillary shoot segments derived callus in *Hemidesmus indicus* (L.) R. Br. (Anantmul) an endangered medicinal plant in Bangladesh. J. Plant Sci. 5(1): 61–67.

Singh, A.K., M. Sharma, R. Varshney, S.S. Agarwal and K.C. Bansal. 2006. Regeneration of plants from alginate- encapsulated shoot tips of *Withania somnifera* (L.) Dunal, a medicinally important plant species. J. Plant Physi. 163: 220–223.

Sivakumar, G., K.V. Krishnamurthy, J. Hao and K.Y. Paek. 2004. Colchicine production in *Gloriosa superba* calluses by feeding precursors. 40(5): 499–502.

Sivanandhan, G., M. Arun, S. Mayavan, M. Rajesh, M. Jeyaraj, G.K. Dev, M. Manickavasagam, N. Selvaraj and A. Ganapathi. 2012. Optimization of elicitation conditions with methyl jasmonate and salicylic acid to improve the productivity of withanolides in the adventitious root culture of *Withania somnifera* (L.) Dunal. Appl. Biochem. Biotechnol. 168(3): 681–696.

Soniya, E.V. and M. Sujitha. 2006. An efficient *in vitro* propagation of *Aristolochia indica*. Biol. Plantarum. 50(2): 272–274.

Sreekumar, S., S. Seeni and P. Pushpangadan. 1998. Production of 2-hydroxy 4-methoxy benzaldehyde using root cultures of *Hemidesmus indicus*. Biotechnol. Lett. 20(7): 631–635.

Sreekumar, S., S. Seeni and P. Pushpangadan. 2000. Micropropagation of *Hemidesmus indicus* for cultivation and production of 2-hydroxy 4-methoxy benzaldehyde. Plant Cell Tissue Organ Cult. 62(3): 211–218.

Stanly, C. and C.L. Keng. 2007. Micropropagation of *Curcuma zedoaria* Roscoe and *Zingiber zerumbet* Smith. Biotechnol. 6: 555–560.

Sudha, C.G., P.N. Krishnan, P. Pushpangadan and S. Seeni. 2005. *In vitro* propagation of *Decalepis arayalpathra*, a critically endangered ethnomedicinal plant. *In vitro* Cell Dev-Pl. 41(5): 648–654.

Sudha, C.G., P.N. Krishnan, S. Seeni and P. Pushpangadan. 2000. Regeneration of plants from *in vitro* root segments of *Holostemma annulare* (Roxb.), K. Schum., a rare medicinal plant. Curr. Sci. 78(4): 503–506.

Sudha, C.G., T.V. Sherina, V.P. Anu Anand, J.V. Reji, P. Padmesh and E.V. Soniya. 2012. *Agrobacterium rhizogenes* mediated transformation of the medicinal plant *Decalepis arayalpathra* and production of 2-hydroxy-4-methoxy benzaldehyde. Plant Cell Tissue Organ Cult. 112(2): 217–226.

Suja, A. 2005. Medicinal plants of Western Ghats. *In:* E. Sahyadri E. News: Issue XII http://www.ces.iisc.ernet.in/biodiversity/sahyadri_enews/newsletter/issue12/Index.htm

Sundararaj, S.G., A. Agrawal and R.K. Tyagi. 2010. Encapsulation for *in vitro* short-term storage and exchange of ginger (*Zingiber officinale* Rosc.) germplasm. Sci. Hortic. 125(4): 761–766.

Suzuki, H., T. Matsumoto and Y. Mikami. 1985. Effects of physical factors and surface active agents on the formation of anthraquinones by *Rubia cordifolia* cells in suspension culture. Agric. Biol. Chem. 49: 519–520.

Takayama, S. and M. Akita. 1994. The types of bioreactors used for shoots and embryos. Plant Cell Tissue Organ Cult. 39: 147–156.

Thengane, S.R., D.K. Kulkarni and K.V. Krishnamurthy. 1998. Micropropagation of licorice (*Glycyrrhiza glabra* L.) through shoot tip and nodal cultures. *In vitro* Cell Dev-Pl. 34(4): 331–334.

Thomas, T.D. and B. Philip. 2005. Thidiazuron-induced high-frequency shoot organogenesis from leaf-derived callus of a medicinal climber, *Tylophora Indica* (Burm. F.) Merrill. *In vitro* Cell Dev-Pl. 41(2): 124–128.

Ticktin, T. 2004. The ecological implications of harvesting non-timber forest Products. J. Appl. Ecol. 41: 11–21.

Toivonen, L. and H. Rosenqvist. 1995. Establishment and growth characteristics of *Glycyrrhiza glabra* hairy root cultures. Plant Cell Tissue Organ Cult. 41(3): 249–258.

Tyagi, R.K., A. Agrawal and A. Yusuf. 2006. Conservation of *Zingiber* germplasm through *in vitro* rhizome formation. Sci. Hortic. 108(2): 210–219.

Tyagi, R.K., A. Yusuf, P. Dua and A. Agrawal. 2004. *In vitro* plant regeneration and genotype conservation of eight wild species of *Curcuma*. Biol. Plantarum. 48: 129–132.

Uniyal, R.C., M.R. Uniyal and P. Jain. 2000. Cultivation of Medicinal Plants in India. A Reference Book. TRAFFIC India and WWF India, New Delhi, India.

Vaishnav, K., S. Goyal and K.G. Ramawat. 2006. Isoflavonoids production in callus culture of *Pueraria tuberosa*, the Indian kudzu. Indian J. Exper. Biol. 44(12): 1012–1017.

Vantomme, P., A. Markkula, and R.N. Leslie. 2002. Non-Wood Forest Products in 15 Countries of Tropical Asia. An Overview. Food and Agriculture Organization, Bangkok, Thailand.

Ved, D.K. and G.S. Goraya. 2007. Demand and Supply of Medicinal Plants in India. NMPB, New Delhi & FRLHT, Bangalore, India.

Verma, P.C., D. Singh, L. Rahman, M.M. Gupta and S. Banerjee. 2002. *In vitro* studies in *Plumbago zeylanica*: rapid micropropagation and establishment of higher plumbagin yielding hairy root cultures. J. Plant Physi. 159: 547 552.

Vincent, K.A., K.M. Mathew and M. Hariharan. 1992. Micropropagation of *Kaempferia galanga* L.—a medicinal plant. Plant Cell Tissue Organ Cult. 28(2): 229–230.

Vines, G. 2004. Herbal harvests with a future: towards sustainable sources for medicinal plants, Plant life Integration; www.plantlife.org.uk.

Withers, L.A. and F. Engelmann. 1997. *In vitro* conservatipon of plant genetic resources. *In*: A. Altman (ed.). Biotechnology in Agriculture. Marcel Dekker Inc. New York, pp. 57–88.
Wochok, Z.S. 1981. The role of tissue culture in preserving threatened and endangered plant species. Biol. Conserv. 20: 83–89.
Yamuna, G., V. Sumathi, S.P. Geetha, K. Praveen, N. Swapna and K.N. Babu. 2007. Cryopreservation of *in vitro* grown shoots of ginger (*Zingiber officinale* Rosc.). Cryo. Lett. 28(4): 241–252.

Index

Color Plate Section

Chapter 2

A

B

Fig. 1. Microtubers grown in the Liquid Lab™ (L-L) system. A) Seedlings with developing microtubers in the L-L™ vessel after 14 weeks. B) Harvested microtubers on a Petri dish. Scale bar = 1 cm. Photo credit Terttu Kämäräinen-Karppinen.

Chapter 3

Fig. 1. Adult plants of *Dioscorea cayenensis* (left and above), *D. trifida* (right and above), *D. alata* (left and below) and *D. bulbifera* (right and below) showing a few aerial tubers.

Chapter 6

Fig. 1. Direct adventitious roots from a leaf lamina disc (top) and direct adventitious shoots from an intact leaf (bottom).

Fig. 5. Green and compact callus with small-sized cells (top) and white and friable callus with large-sized cells.

Chapter 7

Fig. 1. Non-embryogenic callus: (a) Non-embryogenic callus of *C. sativus* (b) Non-embryogenic callus of *C. cancellatus*.

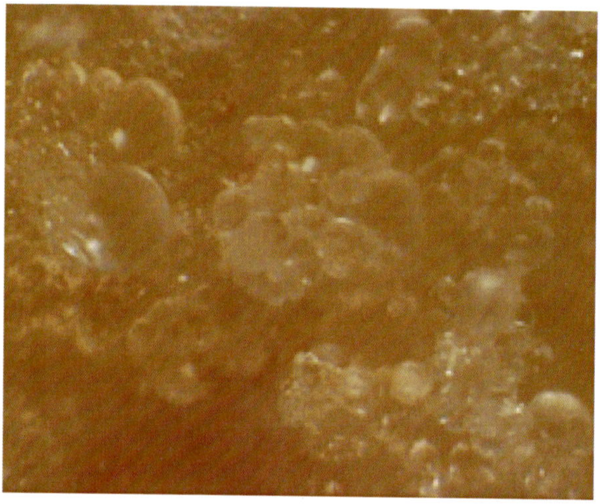

Fig. 2. Embryogenic callus of *C. sativus*.

Fig. 3. Direct micro-corms formation in *C. cancellatus*.

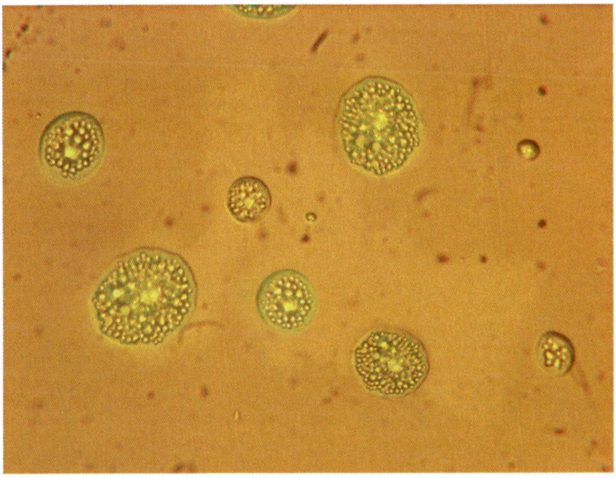

Fig.4. Isolation of protoplasts derived from embryogenic callus of *C. sativus*.

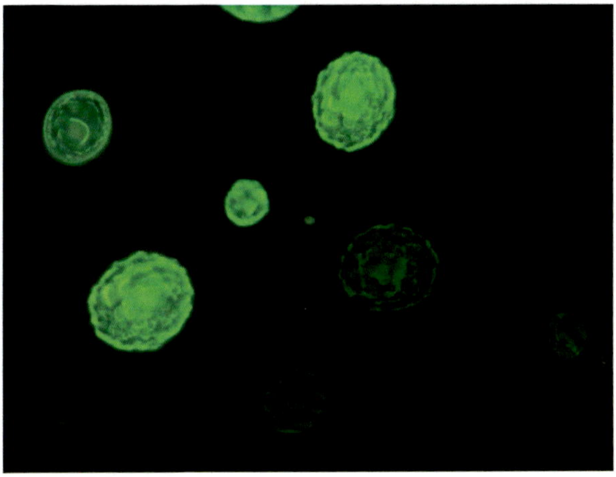

Fig. 5. *C. sativus* protoplasts after staining with FDA (Fluorescein diacetate).

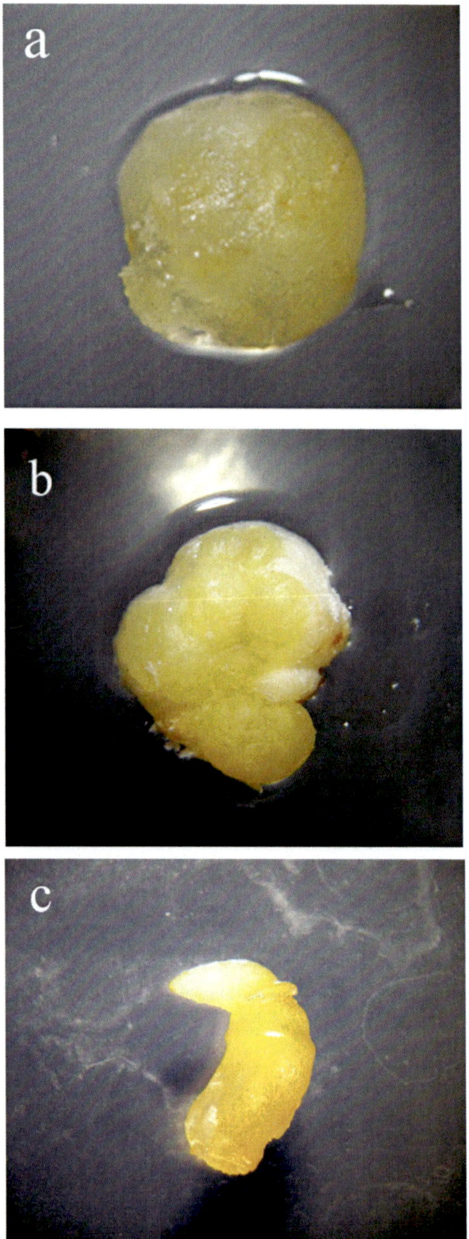

Fig. 6. Genesis stages of somatic embryogenesis in *C. sativus*: (a) globular stage (b) scutellar stage (c) coleoptilar stage.

Chapter 9

Fig. 2. Production of transgenic orange calli and regeneration of transgenic orange plants and expression of carotenoid gene cassettes in *Lilium* × *formolongi* (left). Transgenic plant from meristematic nodular calli 15 months after transfer to regeneration medium (right).

Chapter 10

Fig. 1. Caladium plants in the landscape or containers. A: multiple cultivars grown in the landscape; B and C: 'Tapestry' and 'Florida Blizzard' caladium grown in containers; D: 'Royal Flush' caladium in the landscape; E and F: 'Fairytale Princess' and 'White Butterfly' caladium grown in containers.

Chapter 11

Fig. 1. Dahlia flowers used in our experiment. a: "Michael J," b: "MJOr," c: "MJY," d: "Super Girl," e: "Yukino," g: "Cupid," f: "Evelyn Rumbold," h: "Jyunn-ai," i: "Saffron," j: "Hakuba," k: "Hakuyo," l: "Yuino," m: "Matsuribayashi," n: "Kazusa-shiranami," o: "Ori2," p: "Fidalgo Blacky," q: "Ms. Noir," r: "Kokucho," s: "Black Cat," and t: "Kokucho" purple mutant.

Chapter 12

Fig. 2. *Sparaxis* corms (Ehrich 2007).

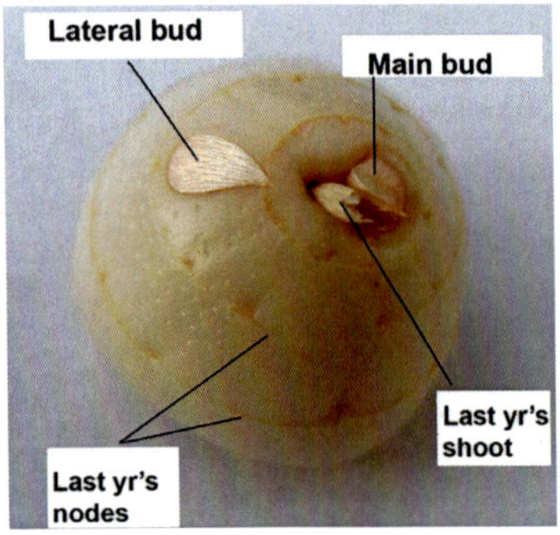

Fig. 3. *Sparaxis* corm in detail (Ehrich 2007).

Fig. 4. *Sparaxis* mother and daughter corm illustrated (Ehrich 2007).

Fig. 5. Pupation of *Tritonia deusta* corm (Ehrich 2007).

Chapter 13

Fig. 1. Alsteromeria Flowers; Beautiful and durable cut flowers.

Fig. 2. Explants preparation and in vitro growth (A) Preparation of rhizome-tip explants. (B) Explants growth and shoot emergence in the 5th week. (C) Rooting in the 7th week.

Chapter 15

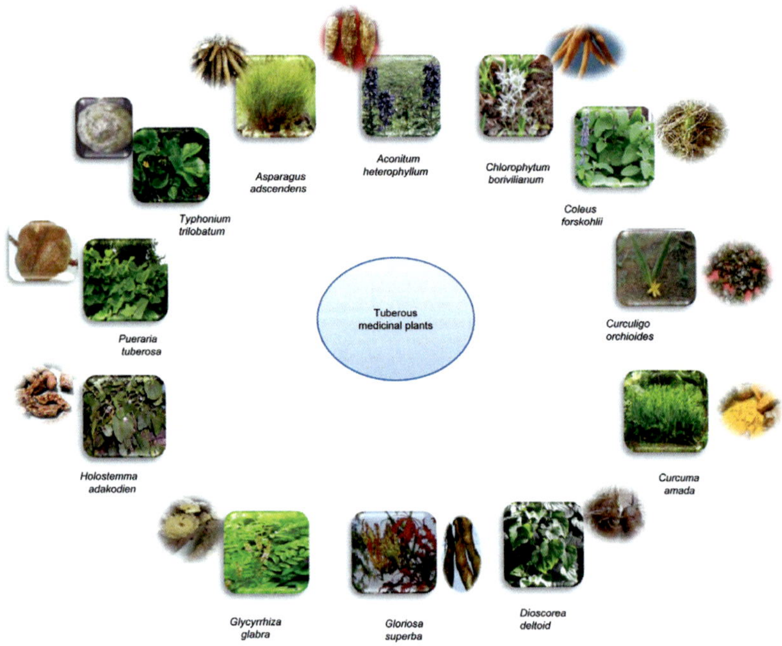

Fig. 1. Tuberous medicinal plants with their tubers.

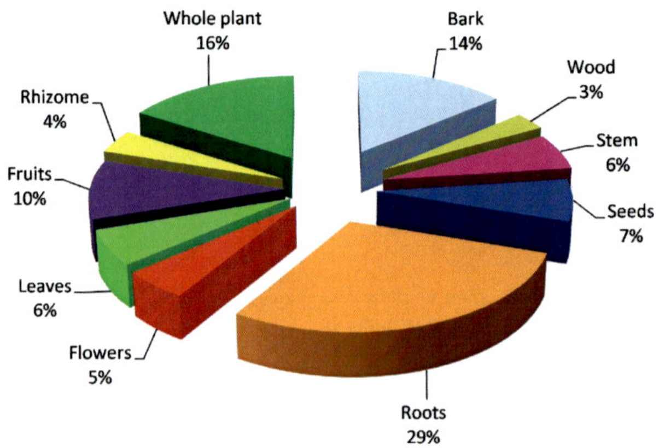

Fig. 2. Different % of plant parts used for medicinal purposes.

Fig. 5. Regeneration in *C. borivilianum* through somatic embryogenesis (a), and through bulbils formation (b) and germination (c) in *C. orchioides*.

Fig. 7. Cells under fluorescence microscope (a), Cell culture in 3L bioreactor (b) and organized shoot cultures (c) of *P. tuberosa*.